LECTURES ON THE PHYSICS OF STRONGLY CORRELATED SYSTEMS XI

Proceedings in the Series of
Lectures on the Physics of Highly Correlated Electron Systems

Year	Title	Publisher	ISBN
2007	Eleventh Training Course	AIP Conf. Proceedings Vol. 918	978-0-7354-0428-1
2005	Tenth Training Course	AIP Conf. Proceedings Vol. 846	0-7354-0340-6
2004	Ninth Training Course	AIP Conf. Proceedings Vol. 789	0-7354-0279-5
2003	Eighth Training Course	AIP Conf. Proceedings Vol. 715	0-7354-0194-2
2002	Seventh Training Course	AIP Conf. Proceedings Vol. 678	0-7354-0147-0
2001	Sixth Training Course	AIP Conf. Proceedings Vol. 629	0-7354-0083-0
2000	Fifth Training Course	AIP Conf. Proceedings Vol. 580	0-7354-0019-9
1999	Fourth Training Course	AIP Conf. Proceedings Vol. 527	1-56396-950-5
1998	Third Training Course	*not published*	
1997	Second Training Course	AIP Conf. Proceedings Vol. 438	1-56396-789-8
1996	First Training Course	*not published*	

To learn more about these titles, or the AIP Conference Proceedings Series, please visit the webpage **http://proceedings.aip.org**

LECTURES ON THE PHYSICS OF STRONGLY CORRELATED SYSTEMS XI

Eleventh Training Course in the Physics of Strongly Correlated Systems

Salerno, Italy 2 - 13 October 2006

EDITORS
Adolfo Avella
Ferdinando Mancini
Università degli Studi di Salerno
Salerno, Italy

All papers have been peer reviewed.

SPONSORING ORGANIZATIONS
Università degli Studi di Salerno, Italy
International Institute for Advanced Scientific Studies
Campania Region, Italy
SUPERMAT, Regional Laboratory CNR-INFM
COST P16

AMERICAN INSTITUTE OF PHYSICS

Melville, New York, 2007
AIP CONFERENCE PROCEEDINGS ■ VOLUME 918

Editors:

Adolfo Avella
Ferdinando Mancini

Dipartimento di Fisica "E. R. Caianiello"
Università degli Studi di Salerno
Via S. Allende
I-84081 Baronissi (SA)
Italy

E-mail: avella@physics.unisa.it
mancini@physics.unisa.it

L.C. Catalog Card No. 2007928606
ISBN 978-0-7354-0428-1
ISSN 0094-243X

Printed in the United States of America

CONTENTS

PART II. PARTICIPANT CONTRIBUTIONS

Preface

The present volume contains the notes of the lectures delivered at the "Eleventh Training Course in the Physics of Strongly Correlated Systems" held in Vietri sul Mare (Salerno, Italy) in October 2006. It also contains contributions by some of the participants, who delivered seminars in the afternoon sessions.

Following the tradition of previous years, the meeting was devoted to the training of young scientists in one of the most intriguing fields of condensed matter physics: strongly correlated systems. The intent was to bring together for two weeks four senior and about thirty junior researchers in a close location with an informal atmosphere, paying special attention to foster the active participation of the young researchers.

The course consisted of four lectures every morning, held by Professors Michele Fabrizio, Didier Poilblanc, Richard T. Scalettar, Dirk van der Marel, and of afternoon activities (seminars delivered by the junior researchers, solving of specific problems, round table on hot topics...) aimed principally at promoting discussions between the attendees and the lecturers. The outcome of this type of course was a significant interchange of ideas among the participants thanks to both the enlightening morning lectures and the long afternoon sessions devoted to discussions.

It is hoped that both the meeting, which brought together leaders in the field as well as bright and eager beginners, and the present volume, which may be useful as an up-to-date book for researchers interested in this field, shall be considered as firm points by the scientific community.

We wish to acknowledge the support of those institutions that made the course possible. The main sponsor of the event has been the University of Salerno. Finally, we wish to thank Professor Maria Marinaro, President of the International Institute of Advanced Scientific Studies, who hosted the event in the wonderful and warm venue of Vietri sul Mare.

Adolfo Avella
Ferdinando Mancini
Salerno, May 8th, 2007

PART I

LECTURES

What can be learned about the Mott transition from the physics of Anderson impurities

Michele Fabrizio

International School for Advanced Studies (SISSA), and CNR-INFM Democritos National Simulation Centre, Via Beirut 2-4, I-34014 Trieste, Italy and The Abdus Salam International Centre for Theoretical Physics (ICTP), P.O. Box 586, I-34014 Trieste, Italy

Abstract. In these lecture notes we review the physical properties of Anderson impurity models in connection with the physics of the Mott transition as revealed by Dynamical Mean Field Theory. In particular, we discuss the novel features that emerge in clusters of Anderson impuritities with respect to single-impurity models and how they may translate into the physics of lattice models across a Mott transition within the cluster extensions of Dynamical Mean Field Theory.

Keywords: Mott transition, Kondo effect
PACS: 71.10.Fd, 71.30.+h, 72.10.Fk

1. INTRODUCTION

More than fifty years after its proposal, [1, 2] the Mott transition remains an issue of current and broad interest, continually revived by the discovery of correlated materials which display anomalous phenomena in the vicinity of a Mott insulating phase, high T_c superconductivity in cuprates being just the most spectacular example.

The Mott transition in known to escape any simple single-particle-like description, including the Hartree-Fock and the DFT-LDA theories. Indeed these theories are only able to deal with band-insulators characterized by an energy gap separating totally filled from unfilled bands. Therefore they can describe a Mott insulator only if a symmetry breaking order parameter is assumed, which turns effectively a Mott- into a conventional band-insulator. The typical example is the single-band Hubbard model at half-filling. This model should always have an insulating phase at sufficiently large repulsion. Yet, in order to make this phase accessible for instance to Hartree-Fock theory, one is obliged to assume an antiferromagnetic order parameter that doubles the unit cell so to fulfil the necessary requirement for a band-insulator to have an even number of electrons per unit cell. In this way, the Mott phenomenon, which solely involves freezing of charge fluctuations, is completely masked by magnetism. Since also conventional perturbation theory is inadequate, being the Mott transition inherently non-perturbative, *ad hoc* schemes have been developed during the years, starting from Hubbard himself. [3, 4, 5] Among them, Dynamical Mean Field Theory (DMFT) [6] and numerical quantum-simulations [7, 8, 9] have been the most successful in uncovering the basic features of the Mott transition disentangled from un-necessary complications as magnetism.

However, once we accept the idea that the Mott phenomenon is well distinguishable

CP918, *Lectures on the Physics of Strongly Correlated Systems XI: Eleventh Training Course*, edited by A. Avella and F. Mancini

from the on-set of symmetry breaking that usually arises in the insulating phase, then even the latter becomes far less trivial, as we are going to discuss.

1.1. The Mott transition

As we know the interaction-driven Mott metal-to-insulator transition is due to the competition between the tendency of the electrons to delocalize throughout the crystal so to maximize the band-energy gain and the Coulomb repulsion among the same electrons which, on the contrary, tends to suppress valence fluctuations by localizing the carriers. If the band-energy gain, which can be identified with the "bare" bandwidth W, is small enough with respect to the short-range Coulomb repulsion, commonly parametrized by an on-site Hubbard U, and the electron-density per site is integer, charges get localized and the system is a Mott insulator. On the contrary, if the bandwidth is larger than a critical value or if the electron-density is incommensurate, the pure system without disorder has to remain a metal, although a strongly-correlated one. Let us discuss on very general grounds the behavior of the entropy in both cases.

1.1.1. The Mott insulator

Let us first consider how the entropy should behave well inside a Mott insulating phase, $U \gg W$. At high temperature $T \sim U$, the entropy related to valence fluctuations is suppressed and the system enters a regime with approximately constant entropy that could be defined as the *ideal* Mott insulator. Here only charge degrees of freedom are locked while all other degrees of freedom, like the spins and eventually the orbitals, are still completely free. However, at some lower temperature, other energy scales come into play, whose role is to lock these additional degrees of freedom so to rid the Mott insulator of their entropy. These energy scales include for instance the on-site Coulomb exchange, responsible of the Hund's rules, the inter-site direct- or super-exchange, the crystal field, the coupling to the lattice and so on and so forth. We will denote these energy scales collectively as J, which has to be identified as the temperature at which the entropy of the residual degrees of freedom of the ideal Mott insulator starts to be suppressed. Consequently, at low temperature, a *realistic* insulating phase is established, which is usually accompanied by a symmetry breaking phase transition at $T = T_c \leq J$, for instance a magnetic ordering. Below T_c, the entropy decreases to zero as $T \to 0$ either exponentially or with a power law with an exponent generically greater than one.

Recently a lot of research activity has focused on the possibility that different symmetry broken phases may compete in the insulator, leading to exotic low temperature phenomena. [10] Here we will discard completely this event and concentrate on a different competition which emerge in the metallic phase adjacent the Mott insulator.

1.1.2. The strongly-correlated metal

Let us move now on the metallic side of the Mott transition, where the band-energy gain W is slightly larger than its critical value $W_c \simeq U$ or the density is not commensurate, and discuss its entropy. As before we expect that the charge-fluctuation entropy is, this time only partially, reduced at some high temperature of order $W \simeq U$. The rest of the charge-fluctuation entropy as well as the entropy of the other degrees of freedom are insted suppressed by the formation of the degenerate quasi-particle gas. This occurs below a temperature T_F^*, that can be identified as the effective quasi-particle Fermi temperature. Since quasiparticles carry the same quantum numbers of the electrons, the entropy quenching involves all degrees of freedom at the same time, including the charge. Below T_F^* the entropy should therefore vanish linearly, $S(T) \simeq \gamma_* T$, with a specific heat coefficient γ_* generically larger than its non-interacting value $\gamma \sim 1/W$.

Let us suppose that the Mott metal-to-insulator transition (MIT) is continuous and try to argue how that can happen. Obviously, since quasi-particles disappear at the transition, we should expect that $T_F^* \to 0$ at the MIT. Therefore at some value of $W \geq W_c$, the quasi-particle Fermi temperature T_F^* should become smaller than J. By continuity with the insulating side, we should expect that part of the spin and eventually orbital entropy is already suppressed at temperatures of order J before the on-set of degeneracy. This would imply some kind of pseudo-gap opening above T_F^*, which is a bit at odds with the conventional Landau-Fermi-liquid theory. One way out, apart from a first-order MIT, is that something new happens when $T_F^* \simeq J$. Indeed the presence of J provides the metallic phase with an alternative mechanism to freeze spin and, eventually, orbital degrees of freedom independently of the charge ones, which becomes competitive with the on-set of a degenerate quasiparticle gas when $T_F^* \simeq J$. This competition more likely leads to an instability of the Landau-Fermi-liquid towards a low-temperature symmetry-broken phase, like in the insulating side, but it may also signal a real break-down of Fermi-liquid theory.

Notice that, unlike the competition between different symmetry broken Mott-insulating phases, which requires fine tuning of the Hamiltonian parameters that may only accidentally occur in real materials, this new type of competition should be accessible whenever it is possible to move gradually from a Mott insulator into a metallic phase, for instance by doping or pressure. We also know several examples where this competition is argued to be the origin of very appealing phenomena. For instance in heavy fermion materials the Kondo effect, favoring the formation of a coherent band of heavy quasiparticles, competes with the RKKY interaction. Here this competition might be the key to understand the anomalies that appear at the transition between the heavy fermion paramagnet and the magnetically ordered phase. [11, 12, 13] Another popular example are cuprates, where the role of J is presumably played by the super-exchange and is believed by many people to be the origin of superconductivity.

Since this competition occurs nearby the Mott transition, it is very difficult to investigate with conventional many-body techniques. Fortunately a lot of insights can still be gained by resorting to Dynamical Mean Field Theory. [6]

In what follows we are going therefore to start by briefly reviewing Dynamical Mean Field Theory (DMFT) in Section 2. Since these are lectures notes, we do not assume that all readers are expert in the whole field, so that we will present in some detail how DMFT maps lattice models onto impurity models self-consistently coupled to a conduction bath, although in the simple case of a Bethe lattice. The emergening Anderson-impurity physics in the DMFT description of the Mott transition pushes us to examine first the impurity model *per se*, without self-consistency. Therefore in Section 3 we introduce and discuss the Anderson impurity in the original context of magnetic alloys, topics that is further developed in the Appendices. In Section 4 we describe two quite powerful tools to study impurity models, the Wilson Numerical Renormalization Group and the Conformal Field Theory. Finally, in Section 5 we move to discuss our recent works that are more directly related to the aforementioned competition between T_F^* and J that arises in the metallic phase next to the Mott transition. We argue that this competition can be accessed within DMFT only if one extends the single-site formulation to a cluster one. Here the lattice model maps nomore into a single impurity but into a cluster of impurities coupled one to another and subject to a self-consistency requirement. In the impurity-cluster language, that competition translates into the competition between the Kondo effect and the tendency of the cluster to lock by its own into a low-degenerate state because of the coupling among the impurities. We will first discuss the consequences of this competition in impurity cluster without DMFT slef-consistency, in particular in the simple case of an impurity dimer. Next, we will argue what might happen when self-consistency is carried out. At the end, we will present an full DMFT analysis of a two-band Hubbard model that strongly supports our speculations.

2. BRIEF INTRODUCTION TO DYNAMICAL MEAN FIELD THEORY

In these lectures we discuss some general aspects of Dynamical Mean Field Theory which are needed for what will follow. This theory was formulated by Georges and Kotliar in 1992 [14, 6] starting from previous observations by Metzner and Vollhardt [15, 16] about the properties of models of interacting electrons in infinite-coordination lattices.

2.1. Properties of tight-binding models in infinite coordination lattices

Since we do not pretend to give here a complete overview, for which one may refer to Ref. [6], let us discuss for simplicity the case of electrons hopping between nearest neighbor sites on a Bethe Lattice, which is the limit of infinite connectivity $z \longrightarrow \infty$ of a Cayley tree, shown for instance with $z = 3$ in Fig. 1. The results, apart from some details, are generic to any infinite-coordination lattices. Let us start by calculating the density of states (DOS) of a Cayley tree with nearest neighbor hopping $-t/\sqrt{z}$ in the limit $z \longrightarrow \infty$ (the normalization of the hopping is needed to obtain a meaningful result, as shown below). The on-site Green's function of a given site, denoted as site 0, satisfies the Dyson equation with complex frequency ζ

$$G_{00}^{-1}(\zeta) = \zeta - \frac{t^2}{z} \sum_i G_{ii}^{(0)}(\zeta),$$

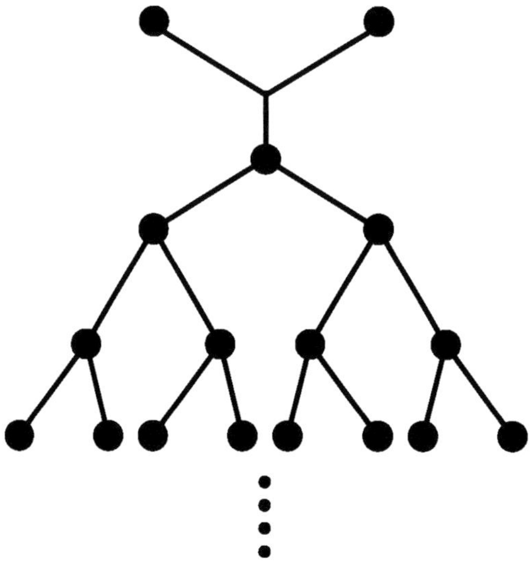

FIGURE 1. The Cayley tree for connectivity $z = 3$

where the sum is over the z nearest neighbors of 0, and $G_{ii}^{(0)}(\zeta)$ denotes the local Green's function at site i with site 0 removed. In the limit $z \to \infty$, $G_{ii}^{(0)}$ should become equal to the true local Green's function $G = G_{00}$, leading to the self-consistency equation

$$G^{-1}(\zeta) = \zeta - t^2 G(\zeta).$$

Indeed the Dyson equation has a well defined limit for $z \to \infty$ thanks to the proper normalization of the hopping. The solution is

$$G(\zeta) = \frac{1}{2t^2}\left(\zeta - \sqrt{\zeta^2 - 4t^2}\right), \tag{1}$$

with the convention that

$$\lim_{|\zeta| \to \infty} \sqrt{\zeta^2 - 4t^2} = \zeta.$$

The DOS is obtained through

$$D(\varepsilon) = -\frac{1}{\pi} \Im m\, G(\zeta \to \varepsilon + i0^+) = \frac{\sqrt{4t^2 - \varepsilon^2}}{2\pi t^2}. \tag{2}$$

The advantage of using the Bethe lattice is that its DOS is non-zero only in a finite energy-range, hence it is a good representation of the DOS of a generic tight-binding model in finite-dimensional lattices (apart from accidental model-dependent features as Van Hove singularities). For instance the DOS of a d-dimensional hypercubic lattice with nearest neighbor hopping $-t/\sqrt{2d}$ (the coordination number $z = 2d$) in the limit of infinite dimensions $d \to \infty$ is a gaussian

$$D(\varepsilon) = \frac{1}{\sqrt{2\pi t^2}} e^{-\varepsilon^2/2t^2},$$

which is non-zero for any value of the energy.

Two sites, i and j, are connected in the Cayley tree only by a single path, whose length is denoted as $R(i,j)$. In other words, an electron at i has to hop $R(i,j)$ times to reach j, each hopping bringing with itself a factor $1/\sqrt{z}$. As a result the Green's function G_{ij} scales with z as

$$G_{ij} \sim \left(\sqrt{\frac{1}{z}}\right)^{R(i,j)}. \tag{3}$$

Since this scaling behavior with the coordination number is just consequence of the normalization of the hopping amplitude, it is generic to any d-dimensional lattices when $d \to \infty$. Moreover we do not expect that this behavior does change if a short-range interaction is included.

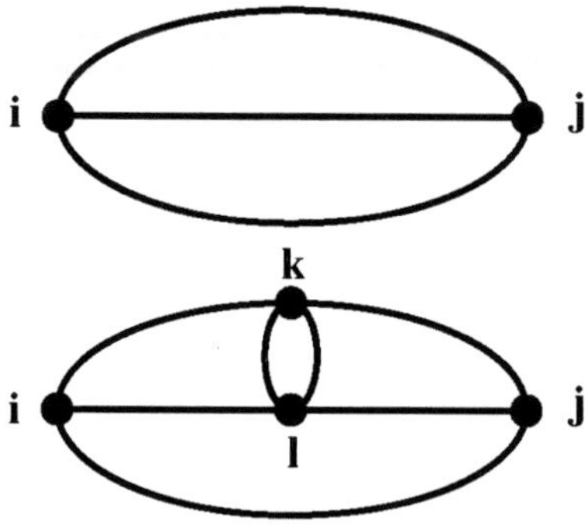

FIGURE 2. Example of two self-energy diagrams which collapse to a single site

Let us analyse the consequences of (3) within the skeleton perturbation expansion of the single-particle self-energy $\Sigma_{ij}(i\varepsilon_n)$. In Fig. 2 we draw the second- and the fourth-order skeleton diagrams of the self-energy Σ_{ij}. Each dot corresponds to the on-site interaction vertex and is labelled by the site-index, while the lines are the fully-interacting Green's functions. In the second order diagram, site i and j are connected by three lines, which means through (3) a factor $z^{-3R(i,j)/2}$. If we were to do a Fourier transform, we should take fixed for instance i and sum over all j's. Since the number of sites j which have the same distance $R(i,j) = R$ from i scales as z^R, we get to the conclusion that only the term with $R = 0$, namely $i = j$, survives. Let us consider now the fourth-order diagram. Again we can realize that each pair of sites can be joined at least by three independent paths. As a result the only non-zero contribution for $z \to \infty$ arises again when $i = j = k = l$. The same conclusion can be shown to hold at any order in perturbation theory. The consequence is that the self-energy in the skeleton expansion is only functional of the local Green's function, hence it is itself local, namely

$$\Sigma_{ij}(i\varepsilon_n) = \delta_{ij}\,\Sigma(i\varepsilon_n), \tag{4}$$

so that its Fourier transform is momentum independent,

$$\Sigma(\mathbf{k}, i\varepsilon_n) = \Sigma(i\varepsilon_n). \tag{5}$$

As a by-product, let us consider the irreducible vertex $\Gamma^0_{ij;kl}$ in some scattering channel. We know that, considering the self-energy as functional of the fully-interacting Green's function, the following relation holds:

$$\Gamma^0_{il;kj} = \frac{\delta\,\Sigma_{ij}[G]}{\delta G_{kl}}.$$

Since we just demonstrated that the self-energy is local and functional only of the local Green's function, it derives that all irreducible vertices are local too, i.e.

$$\Gamma^0_{il;kj} = \Gamma^0\,\delta_{ij}\,\delta_{ik}\,\delta_{il}. \tag{6}$$

We conclude by some definitions. Since the self-energy is only frequency dependent, the local Green's function is the Hilbert transform

$$G(i\varepsilon_n) = \widetilde{D}(\zeta) \equiv \int d\varepsilon\, D(\varepsilon)\,\frac{1}{\zeta - \varepsilon}, \tag{7}$$

where $\zeta = i\varepsilon_n + \mu - \Sigma(i\varepsilon_n)$, μ being the chemical potential. In other words ζ is the inverse Hilbert transform of G, namely

$$\zeta = R\left(\widetilde{D}(\zeta)\right) = R(G). \tag{8}$$

In the case of the Bethe lattice, one can easily show that

$$\widetilde{D}(\zeta) = \frac{1}{2t^2}\left(\zeta - \sqrt{\zeta^2 - 4t^2}\right), \tag{9}$$

with the same convention

$$\lim_{|\zeta|\to\infty} \sqrt{\zeta^2 - 4t^2} = \zeta,$$

as before. Inverting this formula leads to

$$\zeta = i\varepsilon_n + \mu - \Sigma(i\varepsilon_n) = R(G) = t^2 G + \frac{1}{G}. \tag{10}$$

2.2. Reduction to a single site problem

Let us now consider the interacting Hamiltonian on the Cayley tree

$$\mathscr{H}\left[c^\dagger, c\right] = -\frac{t}{\sqrt{z}} \sum_{<ij>,a} \left(c^\dagger_{ia} c_{ja} + H.c.\right) + \sum_i \mathscr{H}_{int}\left[c^\dagger_i, c_i\right],$$

where $c^\dagger_{ia}$ and c_{ia} are creation and annihilation operators at site i, a is a multilabel for spin and eventually orbitals, and $\mathscr{H}_{int}$ is an on-site interaction term. Within the path-integral formalism the imaginary-time action is

$$\mathscr{S} = \int_0^\beta d\tau \sum_{i,a} \bar{c}_{ia}(\tau)\,(\partial_\tau - \mu)\,c_{ia}(\tau) + \mathscr{H}\left[\bar{c}(\tau), c(\tau)\right],$$

where $\bar{c}(\tau)$ and $c(\tau)$ are Grassmann variables at the imaginary time τ. Let us focus on a particular site, for instance site 0, and write

$$\mathscr{S} = \mathscr{S}_0 + \mathscr{S}_{0-bath} + \mathscr{S}_{bath},$$

where

$$\mathscr{S}_0 = \int_0^\beta d\tau \sum_a \bar{c}_{0a}(\tau)\,(\partial_\tau - \mu)\,c_{0a}(\tau) + \mathscr{H}_{int}\left[\bar{c}_0(\tau), c_0(\tau)\right],$$

is the action of the isolated site 0,

$$\mathscr{S}_{0-bath} = -\frac{t}{\sqrt{z}} \sum_{<i0>,a} \int_0^\beta d\tau\, \bar{c}_{ia}(\tau) c_{0a}(\tau) + \bar{c}_{0a}(\tau) c_{ia}(\tau),$$

is the coupling bewteen site 0 and the other sites, described by the action $\mathscr{S}_{bath}$. The effective action for site 0 can be obtained by integrating over all the other sites within a perturbative expansion in $\mathscr{S}_{0-bath}$. By means of the linked-cluster theorem one finds that

$$\mathscr{S}_0^{eff} = \mathscr{S}_0 + \Delta\mathscr{S}_0,$$

where (we drop for simplicity the multilabel a)

$$\begin{aligned}\Delta\mathscr{S}_0 \;=\; & \sum_{n=1}^\infty \Delta\mathscr{S}_0^{(n)} = \sum_{n=1}^\infty \left(\frac{t}{\sqrt{z}}\right)^{2n} \sum_{i_1,\dots,i_n,i'_1,\dots,i'_n}^{\text{n.n.'s of } 0} \int_0^\beta \prod_{j=1}^n d\tau_j\, d\tau'_j \\ & \bar{c}_0(\tau_1)\,\bar{c}_0(\tau_2)\,\dots\,\bar{c}_0(\tau_n)\,c_0(\tau'_n)\,\dots\,c_0(\tau'_2)\,c_0(\tau'_1) \\ & G^{(0)}_{i_1,\dots,i_n,i'_n,\dots,i'_1}(\tau_1,\tau_2,\dots,\tau_n,\tau'_n,\tau'_{n-1},\dots,\tau'_1), \end{aligned} \tag{11}$$

being $G^{(0)}_{i_1,\dots,i_n,i'_n,\dots,i'_1}(\tau_1,\tau_2,\dots,\tau_n,\tau'_n,\tau'_{n-1},\dots,\tau'_1)$ the n-particle connected Green's function of the fully interacting model with site 0 removed. Notice that, in the Cayley tree, once site 0 has been removed there is no path connecting two different nearest neighbor sites of 0. Therefore the only non-zero fully-connected n-particle Green's functions appearing in (11) are those with $i_1 = i_2 = \dots = i_n = i'_n = \dots = i'_1$. As a result, the summation reduces only to the sum over the z nearest neighbors. Since the n-th order term is proportional to z^{-n}, the only one which survives for $z \to \infty$ has $n = 1$, namely

$$\Delta\mathscr{S}_0 = \Delta\mathscr{S}_0^{(1)} = \frac{t^2}{z} \sum_{i\,\text{n.n.'s of } 0} \int_0^\beta d\tau\, d\tau'\, \bar{c}_0(\tau)\, G^{(0)}_{ii}(\tau,\tau')\, c_0(\tau').$$

In the limit $z \to \infty$ the local Green's function with site 0 removed should not differ from the local Green's function G with site 0 present, which is obviously site-independent. Therefore

$$\Delta\mathscr{S}_0 \to t^2 \int_0^\beta d\tau\, d\tau'\, \bar{c}_0(\tau)\, G(\tau,\tau')\, c_0(\tau'),$$

hence it follows that

$$\mathscr{S}_0^{eff} = \int_0^\beta d\tau \sum_a \bar{c}_{0a}(\tau)\,(\partial_\tau - \mu)\, c_{0a}(\tau) + \mathscr{H}_{int}\left[\bar{c}_0(\tau), c_0(\tau)\right]$$
$$+t^2 \int_0^\beta d\tau\, d\tau'\, \bar{c}_0(\tau)\, G(\tau,\tau') c_0(\tau'). \tag{12}$$

This action describes a single site with a non-interacting single-particle Green's function

$$\mathscr{G}_0^{-1}(i\varepsilon_n) = i\varepsilon_n + \mu - t^2 G(i\varepsilon_n), \tag{13}$$

in the presence of an instantaneous interaction $\mathscr{H}_{int}$. $\mathscr{G}_0$ plays actually the same role of the Weiss field in classical Mean-Field Theory. Let us suppose we were able to solve this single-site problem and find the fully-interacting single-particle Green's function

$$\mathscr{G}^{-1}(i\varepsilon_n) = \mathscr{G}_0^{-1}(i\varepsilon_n) - \Sigma(i\varepsilon_n), \tag{14}$$

where Σ is the self-energy of the single-site problem. Since site 0 is equivalent to any other site, it immediately follows that

$$\mathscr{G}(i\varepsilon_n) \equiv G(i\varepsilon_n), \tag{15}$$

namely that

$$G^{-1}(i\varepsilon_n) \equiv i\varepsilon_n + \mu - t^2 G(i\varepsilon_n) - \Sigma(i\varepsilon_n). \tag{16}$$

Comparing (16) with (10) we discover that the self-energy of the single-site problem coincides with the momentum independent self-energy of the lattice model. The same equivalence holds for any infinite-coordination-lattice model, where the single-site non-interacting Green's function, also called the Weiss field, takes the general form

$$\mathscr{G}_0^{-1}(i\varepsilon_n) = i\varepsilon_n + \mu + G(i\varepsilon_n)^{-1} - R(G(i\varepsilon_n)), \tag{17}$$

being $R(G)$ the inverse Hilbert transform with the appropriate DOS. Eq. (17) reduces to (13) in the Bethe lattice.

2.3. Mapping onto an impurity model

We have shown that for infinite coordination lattices one can determine the single-particle self-energy, which is purely local, by solving an auxiliary single-site problem with action (12). The self-energy of this problem is identical to the lattice self-energy. In order to gain some physical insights, let us try to reformulate the Lagrangian problem (12) into an Hamiltonian one.

Let us consider the following Anderson impurity model:

$$\mathcal{H}_{AIM} = \mathcal{H}_{bath} + \mathcal{H}_{imp} + \mathcal{H}_{imp-bath}, \tag{18}$$

where

$$\mathcal{H}_{bath} = \sum_{\varepsilon a} \varepsilon\, c^\dagger_{\varepsilon a} c_{\varepsilon a}, \tag{19}$$

describes a non-iteracting bath,

$$\mathcal{H}_{imp} = -\mu \sum_a d^\dagger_a d_a + \mathcal{H}_{int}\left[d^\dagger, d\right], \tag{20}$$

is the Hamiltonian of the isolated impurity, and

$$\mathcal{H}_{imp-bath} = \sum_{\varepsilon a} V_\varepsilon\, d^\dagger_a c_{\varepsilon a} + H.c., \tag{21}$$

is the hybridization of the impurity with the bath. Within the path-integral formalism, upon integrating out the conduction electrons, we would arrive to an action for the impurity of the same form as (12) with the non-interacting Green's function given by

$$\mathcal{G}_0^{-1}(i\varepsilon_n) = i\varepsilon_n + \mu - \Delta(i\varepsilon_n), \tag{22}$$

where $\Delta(i\varepsilon)$ is the so-called hybridization function,

$$\Delta(i\varepsilon_n) = \sum_\varepsilon \frac{|V_\varepsilon|^2}{i\varepsilon_n - \varepsilon}.$$

Therefore, if we impose that the hybridization function is related to the fully-interacting impurity Green's function by the self-consistency requirement

$$\Delta(i\varepsilon_n) = \sum_\varepsilon \frac{|V_\varepsilon|^2}{i\varepsilon_n - \varepsilon} = t^2 \mathcal{G}(i\varepsilon_n), \tag{23}$$

then the solution of the impurity model coincides with the solution of the lattice model.

A big advantage of the mapping onto an Anderson impurity model is that the latter is very well known and can provide a useful guidance to interpret the behavior of the lattice model. Therefore it is convenient to begin our analysis from the Anderson impurity model *per se*, before discussing the behavior of a strongly correlated lattice models as it emerges out of DMFT. However, before we start presenting the physics of the Anderson impurity model, it is convenient to briefly overview the extension of DMFT from the single-site to a cluster.

2.4. Cluster extensions of DMFT

We have shown that, in the limit of infinite-coordination lattices, a lattice model of electrons interacting with an on-site potential can be mapped onto a single Anderson impurity model subject to a self-consistency requirement. Although in finite coordination lattices the mapping is nomore exact, one can still use DMFT as a mere approximation. Indeed, if one is convinced that what really matters close to a Mott transition is the frequency dependence of the self-energy and not its momentum dependence, then one can assume $\Sigma(\mathbf{k}, i\varepsilon_n) \simeq \Sigma(i\varepsilon_n)$ and search for an Anderson impurity model built so to have the same self-energy as the local-one in the lattice model. If one builds up this procedure in a consistent way, the resulting self-consistency requirement is just Eq. (17). However, we already mentioned that an interesting piece of physics may emerge in the metallic phase when the quasi-particle Fermi temperature, T_F^*, becomes of the same order as the processes, J, that rid the *ideal* Mott insulator of its residual entropy. Since in the most relevant cases these processes are inter-site ones, like the super-exchange in the Hubbard model, this physics is lost in the single-site DMFT approximation.

There have been recently several attempts to extend DMFT approximation to include partially spatial correlations. [17, 18, 19, 20, 21] The idea underneath is the same for all these extensions. Essentially, instead of focusing on a single site, within cluster-DMFT one considers a cluster of sites and solve an auxiliary model of a cluster of Anderson impurities embedded into a bath, with an hybridization function to be determined self-consistently. However, since there is no obvious limiting case in which a cluster extension of DMFT becomes exact, as it was the infinite-coordination limit in the single-site formulation, the self-consistency requirement is not uniquely defined and differs from one version of cluster-DMFT to another.

It is not our purpose to discuss in detail all these extensions. We only want to point out that, as the single impurity *per se* can be used as a guideline to interpret the single-site DMFT results, analogously the knowledge of the general properties of clusters of Anderson impurities may turn useful in connection with cluster-DMFT. In particular we will show that the above mentioned competition T_F^* versus J, that we aim to explore, has a very simple and inspiring analogy in impurity-clusters.

3. THE KONDO EFFECT AND THE PHYSICS OF THE ANDERSON IMPURITY MODEL

The Anderson impurity model was originally introduced by Anderson in the 60's [22] to explain the behavior of magnetic impurities (Fe, Mn, Cr) diluted into non magnetic metals. For a detailed review on the subject one can refer to the book by Hewson. [23]

We know that the thermodynamics and transport properties of good metals with a large Fermi temperature, $T_F \sim 10^4$ K, are dominated at low temperatures, $T \ll T_F$, by the Pauli principle. For instance, the magnetic susceptibility is roughly constant, $\chi \sim 1/T_F$, and one should in principle heat the sample to very high temperatures $T \gg T_F$ to release the spin entropy and recover a Curie-Weiss behavior $\chi \sim 1/T$. Moreover, the resistivity is an increasing function of temperature, since the channels that may dissipate current, the coupling to the lattice as well as to multi particle-hole excitations, become available only upon heating.

At odds with this expectation, if a very diluted (even few part per million) concentration of magnetic impurities is introduced, the above behavior changes drastically. We just mention three distinct features.

(1) The magnetic susceptibility shows a Curie-Weiss behavior well below T_F, proportional to the impurity concentration n_i. Around a very low temperature, called Kondo-temperature, T_K, the Curie-Weiss behavior turns into a logarithmic one and finally the susceptibility saturates at low temperature to a value $\chi(0) \sim n_i/T_K$, with $\chi(T) - \chi(0) \sim -T^2$.
(2) The resistivity $R(T)$ displays a minimum around T_K, followed at $T < T_K$ by a logarithmic increase. At very low temperatures, $R(T)$ approaches a constant value $R(0) \propto n_i$ with $R(T) - R(0) \sim -T^2$. The value of the residual resistivity $R(0)$ suggests very strong scattering potential, near the so-called *unitary limit*.
(3) The entropy released above the Kondo temperature indicates that the magnetic impurities behave as local moments above T_K.

Explaining this behavior amounts actually to understand (1) how it is possible to sustain a local moment in a metal; and (2) why and how, once the local moment is formed, it disappears at low temperatures.

The earlier attempts, mainly due to Friedel [24], to tackle these questions using the conventional scattering theory of metals partially failed. As we will discuss in Appendix A, the only possible way to explain within scattering theory a Pauli-like behavior $\chi \sim 1/T_K$ which turns into a Curie-Weiss one at $T_K \ll T \ll T_F$ is through the existence of a narrow resonance, of width T_K, pinned nearby the chemical potential, still on a scale T_K. This explanation is however not satisfying mainly for two reasons. First of all it does not justify why this resonance should always be pinned at the chemical potential for many different host metals and magnetic impurities. Moreover it contradicts the measured entropy-release above T_K. In fact, for temperatures bigger then its width, the full entropy of a resonant state should be recovered, that would include charge degrees of freedom as well.

3.1. The Anderson Impurity Model

Something that is obviously missing in the scattering-theory analysis are local electron-electron correlations around the magnetic ions. A narrow resonance induced by an impurity would be quite localized around the same impurity, being essentially an atomic level broadened by the hybridization with the conduction electrons of the host metal. Therefore, like an atomic level, such a resonance is likely to accomodate only a fixed number of electrons, say N, paying a finite amount of energy for adding or removing electrons. As usual, it is convenient to define a so-called Hubbard repulsion U through

$$U = E(N+1) + E(N-1) - 2E(N), \tag{24}$$

where $E(M)$ is the total energy when the resonance is forced to have M electrons, N being the equilibrium value. In principle one might include this additional ingredient by adding to the Hamiltonian a term

$$\frac{U}{2}(\hat{n} - N)^2, \tag{25}$$

with $\hat{n}$ the occupation number operator of the resonance level. If it were possible to define such an operator, then (25) would actually solve one of the puzzles. Indeed, if $U \gg T_K$ and for temperatures $T_K \ll T \ll U$, valence fluctuations with respect to the equilibrium value N would be suppressed, and the only degrees of freedom contributing to the entropy would be those related to the degeneracy of the N-electron configurations at the resonance. In other words the resonance would actually behave like a local moment. Unfortunately the resonance occupation number operator is an ill defined object. To overcome this difficulty, Anderson [22] had the idea to represent the resonance as an additional electronic-level inside the conduction band, even if this would be, rigorously speaking, an overcomplete basis. This leads to the so-called Anderson Impurity Model (AIM) defined in the simplest case of a single-orbital impurity as

$$\begin{aligned}\mathscr{H}_{AIM} &= \sum_{\mathbf{k}\sigma} \varepsilon_{\mathbf{k}}\, c^{\dagger}_{\mathbf{k}\sigma} c_{\mathbf{k}\sigma} + \sum_{\mathbf{k}\sigma} V_{\mathbf{k}}\, c^{\dagger}_{\mathbf{k}\sigma} d_{\sigma} + H.c. \\ &\quad + \frac{U}{2}\left(\hat{n}_d - 1\right)^2 + \varepsilon_d \hat{n}_d . \end{aligned} \tag{26}$$

This model describes a band of conduction electrons with energy dispersion $\varepsilon_{\mathbf{k}}$, hybridized with a single level, d_σ, with energy ε_d, both $\varepsilon_{\mathbf{k}}$ and ε_d being measured with respect to the chemical potential, and $\hat{n}_d = \sum_\sigma d^{\dagger}_{\sigma} d_\sigma$. This level suffers from an interaction U which tends to lock no more no less than a single electron.

Let us start from the non-interacting model, $U = 0$, to show that it indeed describes a resonance. Before switching the hybridization $V_{\mathbf{k}}$, the conduction electron Green's function in complex frequency z is

$$G^{(0)}_{\mathbf{k}}(z) = \frac{1}{z - \varepsilon_{\mathbf{k}}} = \dashrightarrow \tag{27}$$

and is represented as a dashed tiny line, while the impurity one is

$$\mathscr{G}^{(0)}(z) = \frac{1}{z - \varepsilon_d} = \longrightarrow \tag{28}$$

and is represented as a solid tiny line. The corresponding Green's functions in the presence of $V_{\mathbf{k}}$,

$$\begin{aligned} G_{\mathbf{kp}}(z) &= \boldsymbol{\dashrightarrow} \\ \mathscr{G}(z) &= \boldsymbol{\longrightarrow}, \end{aligned}$$

are represented as bold lines. Notice that the conduction electron Green's function is nomore diagonal in momentum. In addition one has to introduce the mixed Green's functions

$$\begin{aligned} G_{\mathbf{k}d}(\tau) &= -\langle T\left(c_{\mathbf{k}\sigma}(\tau)\, d^{\dagger}_{\sigma}\right)\rangle = \dashrightarrow\!\!\longrightarrow \\ G_{d\mathbf{k}}(\tau) &= -\langle T\left(d_{\sigma}(\tau)\, c^{\dagger}_{\mathbf{k}\sigma}(\tau)\right)\rangle = \longrightarrow\!\!\dashrightarrow \end{aligned}$$

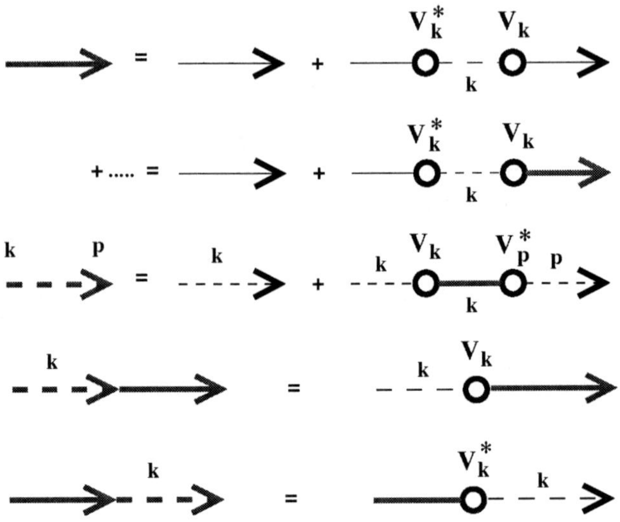

FIGURE 3. Dyson equations for the Green's functions. Bold (red) lines are the Green's functions in the presence of hybridization, represented as a circle, while tiny (black) lines are unperturbed Green's functions.

Through the Dyson equation shown diagramatically in Fig. 3, we find that

$$\mathscr{G}(z) = \frac{1}{z - \varepsilon_d - \Delta(z)}, \tag{29}$$

where the so-called hybridization function is defined through

$$\Delta(z) = \sum_{\mathbf{k}} \frac{|V_{\mathbf{k}}|^2}{z - \varepsilon_{\mathbf{k}}}. \tag{30}$$

The conduction electron Green's function is instead

$$G_{\mathbf{kp}}(z) = \delta_{\mathbf{kp}}\, G^{(0)}_{\mathbf{k}}(z) + G^{(0)}_{\mathbf{k}}(z)\, T_{\mathbf{kp}}(z)\, G^{(0)}_{\mathbf{p}}(z), \tag{31}$$

where the T-matrix, see Appendix A, is

$$T_{\mathbf{kp}}(z) = V_{\mathbf{k}}\, \frac{1}{z - \varepsilon_d - \Delta(z)}\, V^*_{\mathbf{p}}. \tag{32}$$

In addition

$$G_{\mathbf{k}d}(z) = \frac{V_{\mathbf{k}}}{z - \varepsilon_{\mathbf{k}}} \frac{1}{z - \varepsilon_d - \Delta(z)}, \tag{33}$$

$$G_{d\mathbf{k}}(z) = \frac{V_{\mathbf{k}}^*}{z-\varepsilon_{\mathbf{k}}} \frac{1}{z-\varepsilon_d-\Delta(z)}. \tag{34}$$

We notice that the hybridization function has a branch cut on the real axis since

$$\Im m\, \Delta(z \to \varepsilon \pm i0^+) = \mp\pi \sum_{\mathbf{k}} |V_{\mathbf{k}}|^2 \,\delta\left(\varepsilon - \varepsilon_{\mathbf{k}}\right) \equiv \mp\Gamma(\varepsilon). \tag{35}$$

In terms of $\Gamma(\varepsilon) > 0$ we can write

$$\Delta(z) = \int \frac{d\varepsilon}{\pi} \frac{\Gamma(\varepsilon)}{z-\varepsilon}.$$

Since we are interested in temperatures/frequencies much smaller than the conduction electron Fermi energy, we will assume for simplicity that $\Gamma(\varepsilon) = \Gamma$ for $|\varepsilon| < W$ and $\Gamma(\varepsilon) = 0$ otherwise, with W of order T_F. Then, if $|z| \ll W$ one readily finds that

$$\Delta(z) = -i\Gamma \,\mathrm{sgn}\, \Im m\, z. \tag{36}$$

In this approximation the impurity DOS is

$$\rho_d^{(0)}(\varepsilon) = -\frac{1}{\pi} \Im m\, \mathscr{G}(\varepsilon + i0+) = \frac{1}{\pi} \frac{\Gamma}{(\varepsilon-\varepsilon_d)^2 + \Gamma^2}, \tag{37}$$

which is just a resonant state trapping a number of electrons of given spin that is

$$N_d = \int_{-\infty}^{+\infty} d\varepsilon\, f(\varepsilon)\, \rho_d^{(0)}(\varepsilon). \tag{38}$$

According to the scattering theory analysis of Appendix A, this resonance corresponds to a scattering phase shift $\pi/2$ at the chemical potential.

3.1.1. Perturbation theory

Let us consider the case $\varepsilon_d = 0$ and switch on the Hubbard repulsion

$$\frac{U}{2}\left(\hat{n}_d - 1\right)^2 = \frac{U}{2} + U\, \hat{n}_{d\uparrow}\, \hat{n}_{d\downarrow} - \frac{U}{2} \hat{n}_d,$$

to be treated in perturbation theory. To be consistent the perturbation expansion is carried out both in the interaction term as well as in the last term on the right-hand side.

First of all notice that, since the conduction electrons are non-iteracting, their Green's function is still of the form

$$G_{\mathbf{kp}}(z) = \delta_{\mathbf{kp}}\, G_{\mathbf{k}}^{(0)}(z) + G_{\mathbf{k}}^{(0)}(z)\, T_{\mathbf{kp}}(z)\, G_{\mathbf{p}}^{(0)}(z), \tag{39}$$

where the interacting T-matrix is

$$T_{\mathbf{kp}}(z) = V_{\mathbf{k}}\, \mathscr{G}(z)\, V_{\mathbf{p}}^*. \tag{40}$$

with $\mathscr{G}(z)$ the fully-interacting impurity Green's function. It is convenient, for what we are going to show in the following, to introduce the combination of conduction electrons that is actually coupled to the impurity, hence the only one which scatters off the impurity. This is simply

$$c_{0\sigma} = \frac{1}{V} \sum_{\mathbf{k}} V_{\mathbf{k}}^* c_{\mathbf{k}\sigma}, \tag{41}$$

where

$$V^2 = \sum_{\mathbf{k}} |V_{\mathbf{k}}|^2,$$

so that the hybridization becomes

$$V \sum_{\sigma} \left(c_{0\sigma}^{\dagger} d_{\sigma} + H.c. \right) \tag{42}$$

The Green's function of $c_{0\sigma}$ in the absence of the impurity is given by

$$G_0^{(0)}(z) = \frac{1}{V^2} \sum_{\mathbf{k}} \frac{|V_{\mathbf{k}}|^2}{z - \varepsilon_{\mathbf{k}}} = \frac{1}{V^2} \Delta(z), \tag{43}$$

and implies a DOS,

$$\rho_0(\varepsilon) = \frac{1}{\pi V^2} \Gamma(\varepsilon) \to \frac{1}{\pi V^2} \Gamma \, \theta(W - |\varepsilon|). \tag{44}$$

the last equality being valid within the approximation leading to (36). On the contrary, in the presence of the impurity, the above Green's function transforms, through Eqs. (40) and (39), into

$$\begin{aligned} G_0(z) &= \frac{1}{V^2} \sum_{\mathbf{kp}} V_{\mathbf{k}}^* V_{\mathbf{p}} \, G_{\mathbf{kp}}(z) \\ &= G_0^{(0)}(z) + G_0^{(0)}(z) \, T(z) \, G_0^{(0)}(z), \end{aligned} \tag{45}$$

with a T-matrix

$$T(z) = V^2 \mathscr{G}(z). \tag{46}$$

It then follows that the phase-shift in the c_0-scattering-channel, see Appendix A, can be defined as

$$\delta(\varepsilon) = \Im m \ln \mathscr{G}\left(z = \varepsilon + i0^+\right). \tag{47}$$

Moreover, the S-matrix at the chemical potential is readily found to be

$$S(0) = 1 - 2 \frac{\rho_d(0)}{\rho_d^{(0)}(0)}, \tag{48}$$

where $\rho_d^{(0)}(0)$ is the non-interacting impurity DOS at the chemical potential, see Eq. (37), and

$$\rho_d(\varepsilon) = -\frac{1}{\pi}\Im m\mathscr{G}(z = \varepsilon + i0^+),$$

the interacting DOS. Notice that, in the absence of interaction, $S(0) = -1$ which once more implies a $\pi/2$-phase shift.

Since everything depends on the impurity Green's function, let us calculate up to second order its self-energy. The first and second order corrections are drawn in Fig. 4. One can easily show that the first order correction vanishes, namely the resonance stays pinned at the chemical potential. The calculation of the second order correction

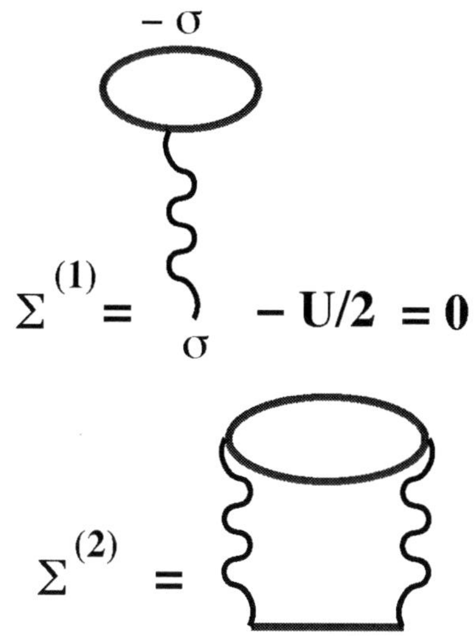

FIGURE 4. First and second order self-energy corrections.

is more involved. One can show that, for real frequencies $z \to \varepsilon + i0^+$, $\Re e\Sigma^{(2)}(\varepsilon) \sim -\varepsilon\left(U\rho_d^{(0)}(0)\right)^2$, namely it vanishes at $\varepsilon = 0$. This result remains true at any order in perturbation theory, and it is a consequence of the particle-hole symmetry that we have implicitly assumed using (36) and fixing $\varepsilon_d = 0$. The imaginary part

$$\Im m\Sigma^{(2)}(\varepsilon) \sim -U^2\left(\rho_d^{(0)}(0)\right)^3\varepsilon^2.$$

This result is equivalent to the well known behavior of the quasiparticle decay rate and derives from simple phase-space arguments. Indeed one can easily show that, at any order in perturbation theory, the imaginary part of the self-energy decays at least as ε^2. As a result, provided perturbation theory remains valid for any value of U, it should follow that

$$\rho_d(0) = -\frac{1}{\pi}\lim_{\varepsilon\to 0}\Im m\frac{1}{\varepsilon + i\Gamma - \Sigma(\varepsilon)} = \frac{1}{\pi\Gamma} = \rho_d^{(0)}. \quad (49)$$

Were this result true, it would imply that $S(0) = -1$, namely that the phase shift remains $\pi/2$ even in the presence of the Hubbard repulsion. In other words, the resonance that exists right at the chemical potential at $U = 0$, should survive for any U. This perturbative argument is compatible with the low-temperature behavior of diluted magnetic impurities if one assumes that the role of U is to shrink the resonance to a small value of order T_K. Indeed, if we assume that, at small ε, the $\Im m\Sigma$ is negligible and

$$\Re e\,\Sigma(\varepsilon) \simeq -\varepsilon\left(1 - \frac{1}{Z}\right), \tag{50}$$

with $Z < 1$ [$1 - Z \simeq \left(U\rho_d^{(0)}(0)\right)^2$ at second order in U], then

$$\mathscr{G}(\varepsilon + i0^+) \simeq \frac{Z}{\varepsilon + iZ\Gamma}, \tag{51}$$

which implies that the effective resonance-width is $Z\Gamma$, which can become very small if $Z \ll 1$.

In addition, this simple argument can also justify why the resonance remains pinned close to the chemical potential. Indeed, if we introduce back a small ε_d, the $\Re e\Sigma(\varepsilon = 0)$ becomes finite. If we define a corrected resonance energy as

$$\varepsilon_d^* = \varepsilon_d + \Re e\Sigma(\varepsilon = 0), \tag{52}$$

then

$$\mathscr{G}(\varepsilon + i0^+) \simeq \frac{Z}{\varepsilon - Z\varepsilon_d^* + iZ\Gamma}.$$

In other words, the resonance is centered at $Z\varepsilon_d^*$, which, provided ε_d^* is of the same order as ε_d, does imply a resonance pinned near the chemical potential.

These perturbative arguments are very suggestive, yet they rely on the assumption that perturbation theory does not break at some critical value of U.

3.1.2. Mean-field analysis of the interaction

In order to understand whether the small U regime is adiabatically connected to the large U one, it is convenient to adopt a different approach more appropriate for strong coupling.

The first question which we address is how strong U should be to move appreciably the model away from the $U = 0$ resonant-level behavior. Let us consider again the case $\varepsilon_d = 0$ and switch on the interaction

$$\frac{U}{2}(\hat{n}_d - 1)^2 = U\hat{n}_{d\uparrow}\hat{n}_{d\downarrow} - \frac{U}{2}\hat{n}_d.$$

The second term gives an $\varepsilon_d = -U/2$. Let us treat the first term within Hartree-Fock approximation, searching for a magnetic solution. The self-energy for the spin σ impurity

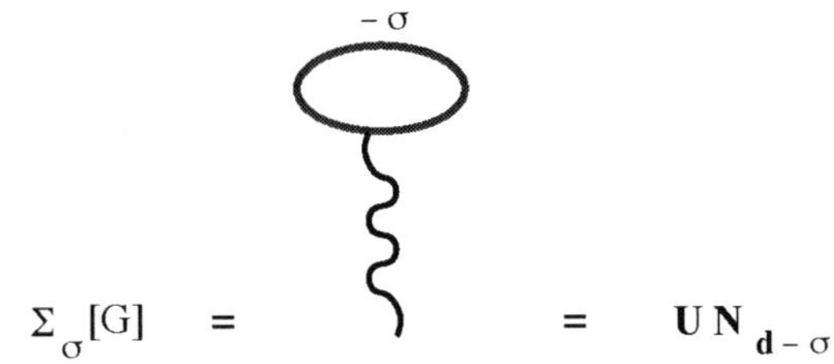

FIGURE 5. Self-energy within the Hartree-Fock approximation.

electrons is given by, see Fig. 5,

$$\Sigma_\sigma = U N_{d-\sigma}, \tag{53}$$

where

$$N_{d-\sigma} = \langle \hat{n}_{d-\sigma} \rangle ,$$

is the average value, which has to be evaluated self-consistently. Therefore the impurity Green's function is

$$\mathscr{G}_\sigma(z) = \frac{1}{z - \varepsilon_d - \Delta(z) - \Sigma_\sigma(z)} = \frac{1}{z + \frac{U}{2} - \Delta(z) - U N_{d-\sigma}}, \tag{54}$$

as if $\varepsilon_{d\sigma} = -U/2 + U N_{d-\sigma}$. By writing $N_{d\sigma} = 1/2 + m\sigma$ we find through (160) the self-consistency condition

$$m = \frac{1}{\pi} \tan^{-1} \frac{Um}{\Gamma}. \tag{55}$$

There are two possibilities. If $U \leq \pi\Gamma$ the only solution is $m = 0$. It describes a paramagnetic impurity and corresponds to the previous analysis at $U = 0$. If $U > \pi\Gamma$, there are two stable solutions at finite $m = \pm m_0$, which describe a partially polarized impurity. Since these two solutions are degenerate, the entropy $S = \ln 2$ just reflects the spin degrees of freedom. In other words, the magnetic state $m \neq 0$ describes indeed a local moment. If we assume a density matrix which is the average of the two degenerate ground states, then the impurity Green's function reads:

$$\mathscr{G}(z) = \frac{1}{2}\left[\frac{1}{z + Um + i\Gamma \mathrm{sign}\, \Im m\, z} + \frac{1}{z - Um + i\Gamma \mathrm{sign}\, \Im m\, z}\right], \tag{56}$$

and the DOS is composed by two lorentians, one below and one above the chemical potential, which are called the Mott-Hubbard bands. In particular, if $U \gg \Gamma$, $m \to 1/2$ and the two lorentians are separated by an energy roughly of order U, see Fig. 6.

The mean-field solution for large U of the Anderson impurity model shows that, above some value of the repulsion U, the mean-field solution looks indeed very different from the small-U resonant-level behavior. Does this imply an actual breakdown of perturbation theory?

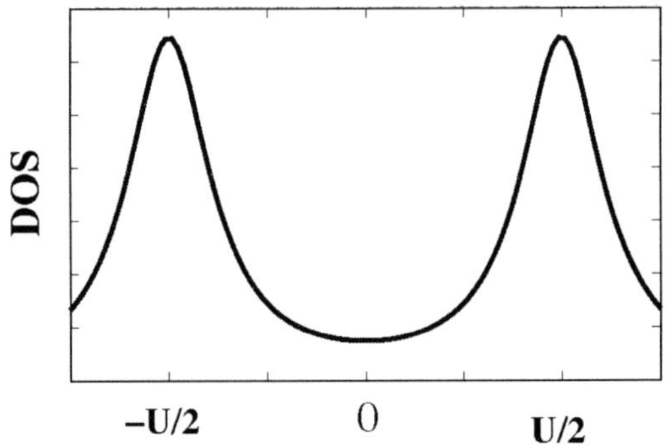

FIGURE 6. DOS of the AIM within the Hartree-Fock approximation.

3.2. From the Anderson to the Kondo model

The Hartree-Fock solution of the Anderson impurity model does explain in the large U/Γ-limit the existence of local free moments in a metal well below its Fermi temperature. Nevertheless the degeneracy of the mean-field solution is an artifact of mean-field theory, since the two solutions in the above example can be coupled through the conduction electrons.

Let us use the fermionic field (41) so that the impurity Hamiltonian plus the hybridization become

$$V\sum_{\sigma}\left(c^{\dagger}_{0\sigma}\,d_{\sigma}+H.c.\right)+\frac{U}{2}\left(\hat{n}_d-1\right)^2. \tag{57}$$

In the large U limit the impurity traps a single electron, either with spin up or down, and we can treat the hybridization as a small perturbation lifting the degeneracy between the two spin directions. By second order perturbation theory through intermediate states in which the impurity is empty or doubly occupied, with energy difference equal to $U/2$ for large U, one gets an effective operator lifting the impurity degeneracy given by, apart from constant terms,

$$\begin{aligned}&-\frac{2V^2}{U}\sum_{\sigma\sigma'}c^{\dagger}_{0\sigma}d_{\sigma}\,d^{\dagger}_{\sigma'}c_{0\sigma'}+d^{\dagger}_{\sigma'}c_{0\sigma'}\,c^{\dagger}_{0\sigma}d_{\sigma}\\&=J_K\,\mathbf{S}_0\cdot\mathbf{S}_d\end{aligned}$$

where

$$J_K=\frac{8V^2}{U},$$

and

$$\begin{aligned}\mathbf{S}_0 &= \frac{1}{2}\sum_{\alpha\beta}c^{\dagger}_{0\alpha}\,\sigma_{\alpha\beta}\,c_{0\beta}\\ &= \frac{1}{2V^2}\sum_{\mathbf{k}\mathbf{p}\alpha\beta}V_{\mathbf{k}}V^{*}_{\mathbf{p}}\,c^{\dagger}_{\mathbf{k}\alpha}\,\sigma_{\alpha\beta}\,c_{\mathbf{p}\beta},\end{aligned} \tag{58}$$

$$\mathbf{S}_d = \frac{1}{2}\sum_{\alpha\beta} d^\dagger_\alpha \,\sigma_{\alpha\beta}\, d_\beta. \tag{59}$$

In other words, in the large U limit, the AIM reduces to the so-called Kondo model describing conduction electrons antiferromagnetically coupled to a local moment, a spin-1/2 in our single-orbital AIM example. The Kondo Hamiltonian is therefore

$$\mathcal{H}_K = \sum_{\mathbf{k}\sigma} \varepsilon_{\mathbf{k}}\, c^\dagger_{\mathbf{k}\sigma} c_{\mathbf{k}\sigma} + J_K\, \mathbf{S}_0 \cdot \mathbf{S}_d. \tag{60}$$

We notice that this model has built-in a local moment, hence by construction correctly describes the local moment regime for $T_K \ll T \ll T_F$. This local moment provides a scattering potential for the conduction electrons, with the major difference with respect to a conventional scalar potential that the impurity has internal degrees of freedom. We will show now the consequences of this difference.

3.2.1. The emergence of logarithmic singularities and the Kondo temperature

Let us analyse the role of the Kondo exchange

$$J_K\, \mathbf{S}_0 \cdot \mathbf{S}_d, \tag{61}$$

in perturbation theory. Since (61) conserves independently the number of conduction and d-electrons, we use a trick due to Abrikosov [25] and treat the impurity spin in terms of d-electrons with an unperturbed Green's function

$$\mathcal{G}^{(0)}(z) = \frac{1}{z},$$

namely as a resonant state right at the chemical potential with a vanishing broadening, which does contain in the "unperturbed state" just one electron, number which is not going to be changed by (61). Since a single electron has a well defined spin, it acts indeed like a local spin-1/2 moment. The Green's function for the conduction electron c_0 is given in (43). For convenience let us rewrite (61) in a spin-asymmetric form as

$$\sum_{i=x,y,z} J_i\, S^i_0\, S^i_d, \tag{62}$$

and calculate the first order corrections to the exchange as given by the diagrams in Fig. 7. For simplicity we assume that all external lines are at zero frequencies. Each diagram is multiplied by (-1) because of first order perturbation theory. The explicit expression of the diagram (a) is

$$(\mathrm{a}) = -\sum_{i,j=x,y,z} J_i J_j \sum_{\gamma c} S^i_{\alpha\gamma} S^i_{ac} S^j_{\gamma\beta} S^j_{cb}\, T \sum_n G^{(0)}_0(i\varepsilon_n)\,\mathcal{G}^{(0)}(-i\varepsilon_n)$$

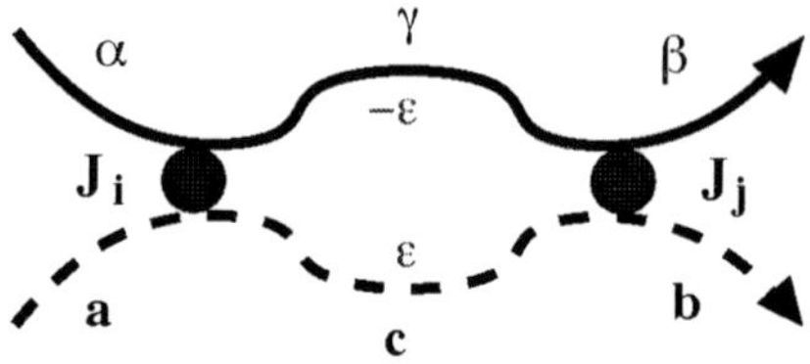

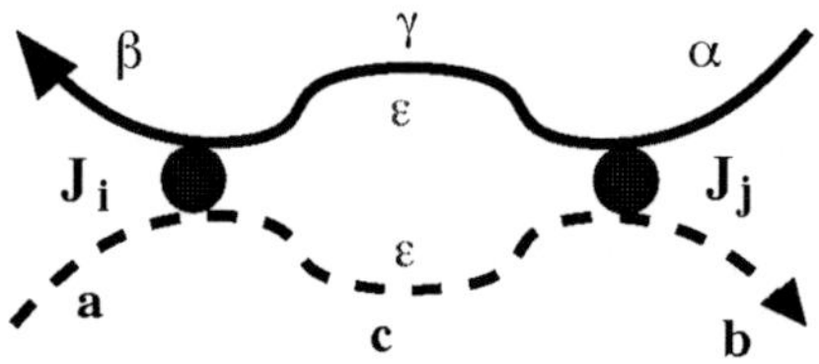

FIGURE 7. First order corrections to the Kondo exchange. Solid and dashed lines indicate impurity and conduction electron Green's functions. Also indicated are the spin labels. All external lines are assumed to be at zero frequency $\varepsilon = 0$. The internal frequency, which is going to be summed over, is also indicated.

$$
\begin{aligned}
&= -\sum_{i,j=x,y,z} \frac{J_i J_j}{V^2} \sum_{\gamma c} S^i_{\alpha\gamma} S^i_{ac} S^j_{\gamma\beta} S^j_{cb}\, T \sum_n \frac{\Delta(i\varepsilon_n)}{-i\varepsilon_n} \\
&= -I(T) \sum_{i,j=x,y,z} \frac{J_i J_j}{V^2} \sum_{\gamma c} S^i_{\alpha\gamma} S^j_{\gamma\beta} S^i_{ac} S^j_{cb},
\end{aligned}
$$

where the function of temperature $I(T)$ is defined through

$$
I(T) = -T \sum_n \frac{\Delta(i\varepsilon_n)}{i\varepsilon_n}.
$$

The diagram (b) is analogously

$$
\begin{aligned}
(\mathrm{b}) &= -\sum_{i,j=x,y,z} J_i J_j \sum_{\gamma c} S^i_{\alpha\gamma} S^i_{cb} S^j_{\gamma\beta} S^j_{ac}\, T \sum_n G^{(0)}_0(i\varepsilon_n) \mathscr{G}^{(0)}(i\varepsilon_n) \\
&= I(T) \sum_{i,j=x,y,z} \frac{J_i J_j}{V^2} \sum_{\gamma c} S^i_{\alpha\gamma} S^j_{\gamma\beta} S^j_{ac} S^i_{cb},
\end{aligned}
$$

hence the sum is

$$
\begin{aligned}
(\mathrm{a}) + (\mathrm{b}) &= I(T) \sum_{i,j=x,y,z} \frac{J_i J_j}{V^2} \sum_{\gamma} S^i_{\alpha\gamma} S^j_{\gamma\beta} \sum_c \left(S^j_{ac} S^i_{cb} - S^i_{ac} S^j_{cb} \right) \\
&= I(T) \sum_{i,j,k=x,y,z} \frac{J_i J_j}{V^2} \sum_{\gamma} S^i_{\alpha\gamma} S^j_{\gamma\beta}\, i\, \varepsilon_{jik} S^k_{ab},
\end{aligned}
$$

where we used the commutation relations between the spin operators

$$
\left[S^i, S^j \right] = i\, \varepsilon_{ijk} S^k,
$$

where ε_{ijk} is the Levi-Civita antisymmetric tensor. Since we sum over i and j, we can also write

$$\begin{aligned}
(a) + (b) &= \frac{1}{2} I(T) \sum_{i,j,k=x,y,z} \frac{J_i J_j}{V^2} \sum_\gamma S^i_{\alpha\gamma} S^j_{\gamma\beta}\, i\,\varepsilon_{jik} S^k_{ab} + (i \leftrightarrow j) \\
&= \frac{1}{2} I(T) \sum_{i,j,k=x,y,z} \frac{J_i J_j}{V^2}\, i\,\varepsilon_{jik} S^k_{ab} \sum_\gamma \left(S^i_{\alpha\gamma} S^j_{\gamma\beta} - S^j_{\alpha\gamma} S^i_{\gamma\beta} \right) \\
&= \sum_{k=x,y,z} S^k_{ab} S^k_{\alpha\beta} \left[\frac{1}{2} I(T) \sum_{i,j} \frac{J_i J_j}{V^2} |\varepsilon_{ijk}|^2 \right],
\end{aligned}$$

which provides the first order correction to the exchange constants according to

$$J_k + \delta J_k = J_k + \frac{1}{2} I(T) \sum_{i,j} \frac{J_i J_j}{V^2} |\varepsilon_{ijk}|^2. \tag{63}$$

In order to evaluate $I(T)$ we use the same integration contour as in Fig. 22. We recall that the hybridization function has a branch cut approximately given by

$$\Delta(\varepsilon + i0^+) - \Delta(\varepsilon - i0^+) = -2i\Gamma\,\theta(W - |\varepsilon|),$$

where $\theta(W - |\varepsilon|)$ is the Heavyside function and W a cut-off of the order of the conduction bandwidth. We then find that

$$I(T) = -\int_{-W}^{W} \frac{d\varepsilon}{\pi} f(\varepsilon) \frac{\Gamma}{\varepsilon} \simeq \frac{\Gamma}{\pi} \ln \frac{W}{T},$$

so that *e.g.* in the isotropic case $J_x = J_y = J_z = J$ we finally get

$$J + \delta J = J + \frac{\Gamma}{\pi V^2} J^2 \ln \frac{W}{T}.$$

This result, originally derived by Kondo in 1964, [26] implies that perturbation theory in J generates logarithmic singularities which become visible roughly around a temperature, that has to be identified with the Kondo temperature, when the correction becomes of the same order as J, namely

$$J = \frac{\Gamma}{\pi V^2} J^2 \ln \frac{W}{T_K},$$

leading to

$$T_K = W \exp\left(-\frac{\pi V^2}{\Gamma J} \right) = W \exp\left(-\frac{\pi U}{8\Gamma} \right) \ll W \sim T_F, \tag{64}$$

since $U \gg \Gamma$. In agreement with experiments, we do find that the Kondo temperature is much smaller than the host-metal Fermi temperature. Since perturbation theory becomes meaningless below T_K, the next obvious question is how to proceed further.

3.2.2. *Anderson's scaling theory*

What we have calculated so far is the first order correction to the Kondo exchange for conduction and impurity electrons at zero frequency, which we found to be

$$J_z[W] + \delta J_z[W] = J_z[W] + \frac{\Gamma}{\pi V^2} J_\perp[W]^2 \ln\frac{W}{T}, \tag{65}$$

$$J_\perp[W] + \delta J_\perp[W] = J_\perp[W] + \frac{\Gamma}{\pi V^2} J_\perp[W] J_z[W] \ln\frac{W}{T}. \tag{66}$$

In (65) and (66) we have assumed an easy-axis asymmetry $J_z \neq J_x = J_y \equiv J_\perp$, and we have explicitly indicated the dependence upon the bandwidth cut-off W. Clearly the result does not change for finite external frequencies provided they are much smaller than the temperature, otherwise they would cut-off the log-singularity instead of T. Now suppose we have another model with a smaller conduction electron bandwidth cut-off $W(s) = W/s$ with $s > 1$, different J's but equal Γ/V^2. In this case, for electrons close to the chemical potential, specifically much closer than the temperature, we would get the first order correction

$$J_z[W(s)] + \delta J_z[W(s)] = J_z[W(s)] + \frac{\Gamma}{\pi V^2} J_\perp[W(s)]^2 \ln\frac{W}{sT},$$

$$J_\perp[W(s)] + \delta J_\perp[W(s)] = J_\perp[W(s)] + \frac{\Gamma}{\pi V^2} J_\perp[W(s)]\, J_z[W(s)] \ln\frac{W}{sT}.$$

On the other hand, it makes not really a big difference for the low energy behavior whether these electrons which lie so close to the chemical potential derive from a band with width W or $W(s)$ provided they suffer the same scattering off the impurity. Therefore we can ask the following question:

what should $J[W(s)]$ be in order for the effective exchange up to first order to be the same as that one with bandwidth W ?

The answer is quite simple, since we can re-write e.g. (65) as

$$\begin{aligned} J_z[W] + \delta J_z[W] &= J_z[W] + \frac{\Gamma}{\pi V^2} J_\perp[W]^2 \ln(s) + \frac{\Gamma}{\pi V^2} J_\perp[W]^2 \ln\frac{W}{sT} \\ &\simeq \left(J_z[W] + \frac{\Gamma}{\pi V^2} J_\perp[W]^2 \ln(s)\right) \\ &\quad + \frac{\Gamma}{\pi V^2}\left(J_z[W] + \frac{\Gamma}{\pi V^2} J_\perp[W]^2 \ln(s)\right)^2 \ln\frac{W}{sT} + O\left(J^3\right), \end{aligned}$$

the equality being valid up to the order at which we have stopped the expansion. Therefore the two models with W and W/s have the same spin exchange up to first order in perturbation theory provided the bare exchange constants satisfy

$$J_z[W(s)] = J_z[W] + \frac{\Gamma}{\pi V^2} J_\perp[W]^2 \ln(s),$$

$$J_\perp[W(s)] = J_\perp[W] + \frac{\Gamma}{\pi V^2} J_\perp[W] J_z[W] \ln(s),$$

which can be cast in a differential form

$$\frac{dj_z(s)}{d\ln s} = j_\perp(s)^2, \tag{67}$$

$$\frac{dj_\perp(s)}{d\ln s} = j_\perp(s)\, j_z(s), \tag{68}$$

having defined

$$J_i\,[W(s)]\,\frac{\Gamma}{\pi V^2} \equiv j_i(s).$$

These equations describe how the bare exchange constants have to be modified in order for the model with a reduced bandwidth W/s to have the low-energy scattering amplitudes equal to those of the original model. The idea behind is that, if we are able to follow the evolution of $j_i(s)$ until $W/s \sim T$, at this point we could rely on perturbation theory since $\ln(W/sT) \simeq 0$. This is more or less how Anderson formulated his *poor man's scaling theory* for the Kondo problem in 1970, [27] which is in essence the first implementation of a Renormalization Group transformation.

Let us study the scaling equations. We readily find that, being $\ln s = x$,

$$\frac{d^2 j_z}{dx^2} = 2j_\perp \frac{dj_\perp}{dx} = 2\, j_z \frac{dj_z}{dx} = \frac{dj_z^2}{dx}.$$

By integrating from $x = \ln s = \ln 1 = 0$ to same $x > 0$ we get

$$\frac{dj_z(x)}{dx} - \frac{dj_z(0)}{dx} = \frac{dj_z(x)}{dx} - j_\perp(0)^2 = j_z(x)^2 - j_z(0)^2,$$

namely

$$\frac{dj_z(x)}{dx} = j_z(x)^2 - j_z(0)^2 + j_\perp(0)^2,$$

$$\frac{d\ln j_\perp(x)}{dx} = j_z(x).$$

In Fig. 8 it is shown the scaling flow of the exchange constants $j_z(x)$ and $j_\perp(x)$ as $x \to \infty$. Since the sign of $J_\perp$ can be always changed rotating by an angle π the impurity spin around the z-axis, we have just drawn the flow with $J_\perp > 0$, $J_z > 0$ and $J_z < 0$ implying antiferro- and ferro-magnetic Kondo exchange, respectively. We notice that for ferromagnetic exchange $J_z < 0$ with $|J_z| \geq J_\perp$, the couplings flow to a line of fixed points, where perturbation theory become valid. In particular for isotropic couplings $J_z = J_\perp = J < 0$, the value at $x = \ln W/T \gg 1$ is

$$j(T) = \frac{j}{1 - j\ln W/T} \simeq -\frac{1}{\ln W/T} \to 0. \tag{69}$$

This suggests that the impurity asymptotically decouples from the conduction electrons. For instance the impurity magnetic susceptibility, which can be calculated perturbatively

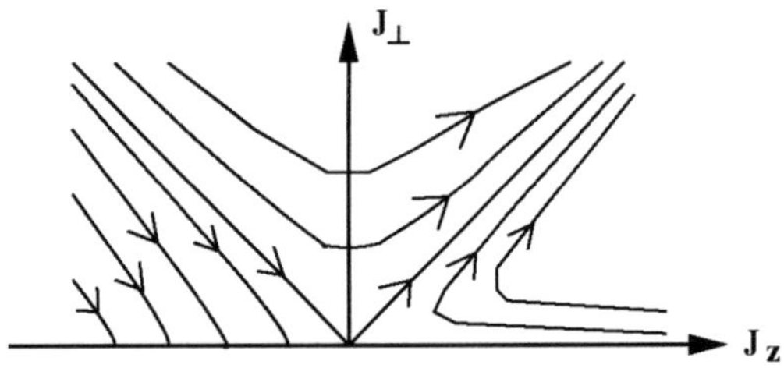

FIGURE 8. Scaling flow of the Kondo exchange constants as the cut-off is sent to zero.

in this case, behaves at low temperatures as

$$\chi(T) \sim \frac{1}{T \ln W/T},$$

almost the Curie-like behavior of a free spin.

For antiferromagnetic $J_z > 0$, as well as for $J_z < 0$ but $|J_z| < J_\perp$, the exchange constants flow to strong-coupling $j_z \simeq j_\perp \to +\infty$. This implies that the electrons very close to the chemical potential feel a very strong antiferromagnetic exchange with the impurity spin. It is therefore reasonable to assume that one has to diagonalize the Kondo spin-exchange first, Eq. (61), and only then treats what is left within perturbation theory. Taking this point of view, the first step leads to assume that the electronic state c_0, see (41), get strongly bound into a spin-singlet with the impurity spin. This singlet state has no internal degrees of freedom, hence it can act only as a scalar potential for the left-over electronic states orthogonal to c_0. The question is what should be the scattering phase shift induced by this spin-singlet object. Since the impurity traps just one electron, from the Friedel sum rule we expect

$$\delta_\uparrow + \delta_\downarrow = \pi.$$

Since the trapping occurs into a singlet state, which is spin-isotropic, $\delta_\uparrow = \delta_\downarrow$, hence

$$\delta_\uparrow = \delta_\downarrow = \frac{\pi}{2}.$$

In other words, the phase shift is indeed the same as we found at very small U, which supports the hypothesis that perturbation theory does not break down. In reality we do know that this is actually the case thanks to exact solutions obtained either by the Wilson numerical renormalization [28] group or by Bethe Ansatz [29, 30].

3.3. The impurity model from weak to strong repulsion and the DMFT results

Since the weak coupling regime $U \ll \Gamma$ is adiabatically connected to the strong coupling one, $U \gg \Gamma$, let us discuss more in detail the perturbative expression of the impurity Green's function Eq. (51) for very small ε, i.e.

$$\mathscr{G}(\varepsilon+i0^{+}) \simeq \frac{Z}{\varepsilon+iZ\Gamma}. \tag{70}$$

The low energy DOS is still a lorentian with a narrower width $T_K = Z\Gamma$ and total spectral weight Z. The rest $1-Z$ of the spectral weigth should be concentrated into some higher-energy part. The low-energy peak, commonly called Abrikosov-Suhl or Kondo resonance, has to survive even for very large U, although its spectral weigth would become very small, $Z \ll 1$.

On the other hand we have seen that, for very large U, when the mapping onto the Kondo model is valid, most of the spectral weight should be concentrated into two features at energies $\pm U/2$, the so-called Hubbard side-bands, see Fig. 6. Therefore the obvious conclusion, supported by numerical results, is that, for $U \gg \Gamma$, the impurity spectral function has three distinct features: two Hubbard side-bands centered at $\pm U/2$ and a very narrow Kondo resonance close to the chemical potential, see Fig. 9. This narrow resonance shrinks as U increases and disappears only if $U \to \infty$.

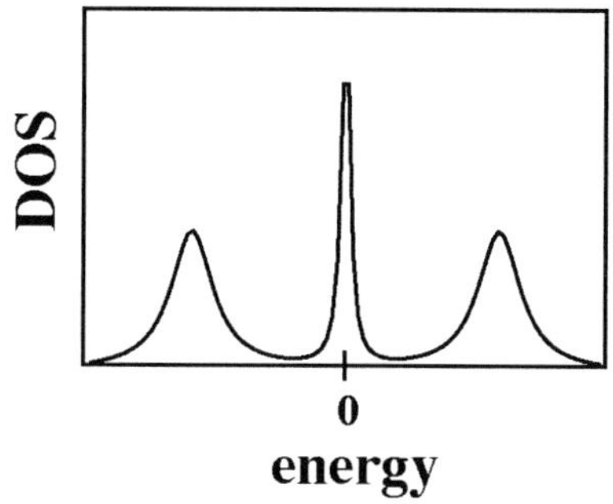

FIGURE 9. The impurity DOS for large U. The central peak is the Kondo resonance, while the other two are the Hubbard bands.

3.3.1. DMFT scenario for the Mott transition

Within DMFT, a single-band Hubbard model in the Bethe lattice can be mapped onto the single-orbital Anderson impurity model (26) provided that the hybridization function and the impurity Green's function satisfy the self-consistency relation

$$\Delta(z) = t^2 \mathscr{G}(z). \tag{71}$$

The role of self-consistency is crucial. Indeed, as U increases we know that impurity spectral weight is transferred from the central resonance into the Hubbard side-bands and, at the meantime, the central peak shrinks. As a result of self-consistency, the low-frequency component of the hybridization width decreases like the spectral weight of the Kondo resonance. This cooperative effect leads to a vanishing low-frequency hybridization, [6] which implies a vanishing T_K for the impurity model, at a finite value of $U = U_c$, while in the impurity model without self-consistency $T_K \to 0$ only if $U \to \infty$. Once $T_k = 0$, only the Hubbard side-bands remain, separated by an energy gap of order U. Since the impurity spectral function correponds, within the DMFT mapping, to the local spectral function of the lattice model, the vanishing of T_K corresponds to the opening of a gap in the single-particle spectral function of the lattice model, namely to the Mott transition.

Notice that, since the impurity self-energy $\Sigma(i\varepsilon_n)$ is also the lattice self-energy then, in the metallic phase prior to the Mott transition, the low-energy lattice Green's function is, through (50),

$$G(\mathbf{k}, i\varepsilon_n) = \frac{1}{i\varepsilon_n - \varepsilon_{\mathbf{k}} - \Sigma(i\varepsilon_n)} \simeq \frac{Z}{i\varepsilon_n - Z\,\varepsilon_{\mathbf{k}}}, \tag{72}$$

and describes low-energy coherent quasiparticles with residue Z, namely effective Fermi temperature Z times smaller than the non-interacting one, $T_F^* = Z\,T_F$, and renormalized bandwidth ZW, i.e. renormalized mass $m_* = m/Z$. In other words, until the effective impurity model displays the Kondo resonance, the lattice model describes a metal with coherent quasiparticles. The Mott transition, being signaled by $Z \to 0$, corresponds to $T_F^* \to 0$ and, at the same time, $m_* \to \infty$. This behavior is commonly referred to as the Brinkman-Rice scenario [31] for the Mott transition.

The last very important thing to point out is that the Mott transition corresponds to an impurity model with very small $T_K \to 0$, namely well inside the regime in which the mapping onto a Kondo model is valid.

4. BRIEF INTRODUCTION TO THE WILSON NUMERICAL RENORMALIZATION GROUP AND TO CONFORMAL FIELD THEORY

A very powerful way to study impurity models is by combining together the Wilson numerical renormalization group (NRG) [28, 32, 33] and Conformal Field Theory (CFT) [34, 35, 36]. The former allows to determine in a very efficient way the low energy spectrum of an impurity model which, unless in simple cases, requires the full machinery of CFT to be interpreted. In the following we are going to briefly discuss both techniques.

4.1. Reduction to a one dimensional model

Let us consider for simplicity a single-orbital AIM, the extension to more orbitals being straightforward,

$$\begin{aligned} \mathcal{H}_{AIM} &= \sum_{\mathbf{k}\sigma} \varepsilon_{\mathbf{k}}\, c^\dagger_{\mathbf{k}\sigma} c_{\mathbf{k}\sigma} + \left(V_{\mathbf{k}}\, c^\dagger_{\mathbf{k}\sigma} d_\sigma + H.c.\right) + \mathcal{H}_{int}\left[d^\dagger, d\right] \\ &= \mathcal{H}_0 + \mathcal{H}_{hyb} + \mathcal{H}_{int}. \end{aligned} \tag{73}$$

We already observed that he impurity is actually coupled only to a particular combination of conduction electrons:

$$c_0^\dagger = \frac{1}{V}\sum_{\mathbf{k}} V_{\mathbf{k}}\, c_{\mathbf{k}}^\dagger, \tag{74}$$

where we have droped for simplicity the spin index and by definition

$$V^2 = \sum_{\mathbf{k}} |V_{\mathbf{k}}|^2.$$

Let us construct the single-particle state

$$|c_0\rangle = c_0^\dagger\,|0\rangle.$$

This is not an eigenstate of the conduction electron Hamiltonian $\mathscr{H}_0$. Indeed we can always define a new state $|c_1\rangle$ orthogonal to $|c_0\rangle$ through

$$t_0\,|c_1\rangle \equiv \left(\mathscr{H}_0 - \varepsilon_0\right)|c_0\rangle,$$

where

$$\varepsilon_0 = \langle c_0|\,\mathscr{H}_0\,|c_0\rangle,$$

and

$$t_0^2 = \left(\langle c_0|\mathscr{H}_0|c_1\rangle\right)^2 = \langle c_0|\left(\mathscr{H}_0 - \varepsilon_0\right)^2|c_0\rangle.$$

Analogously, at the next step we introduce another state $|c_2\rangle$ orthogonal to the previous two through

$$t_1\,|c_2\rangle \equiv \left(\mathscr{H}_0 - \varepsilon_1\right)|c_1\rangle - t_0\,|c_0\rangle,$$

where

$$\varepsilon_1 = \langle c_1|\,\mathscr{H}_0\,|c_1\rangle,$$

and

$$t_1^2 = \left(\langle c_1|\mathscr{H}_0|c_2\rangle\right)^2 = \langle c_1|\left(\mathscr{H}_0 - \varepsilon_1\right)^2|c_1\rangle - t_0^2.$$

At the n-th iteration of this Lanczos algorithm, we obtain a state $|c_{n+1}\rangle$, orthogonal to the n previous ones, by

$$t_n\,|c_{n+1}\rangle \equiv \left(\mathscr{H}_0 - \varepsilon_n\right)|c_n\rangle - t_{n-1}\,|c_{n-1}\rangle,$$

where

$$\varepsilon_n = \langle c_n|\,\mathscr{H}_0\,|c_n\rangle,$$

and

$$t_n^2 = \left(\langle c_n|\mathscr{H}_0|c_{n+1}\rangle\right)^2 = \langle c_n|\left(\mathscr{H}_0 - \varepsilon_n\right)^2|c_n\rangle - t_{n-1}^2.$$

In this way we can rewrite the Hamiltonian (73) like

$$\mathscr{H}_{AIM} = \mathscr{H}_{int}\left[d^\dagger, d\right] + V\sum_\sigma \left(c^\dagger_{0\sigma} d_\sigma + H.c.\right) + \sum_{n\geq 0} \varepsilon_n\, c^\dagger_{n\sigma} c_{n\sigma} + t_n\left(c^\dagger_{n\sigma} c_{n+1\sigma} + H.c.\right), \tag{75}$$

which describes a semi-infinite chain with the impurity at the edge. Notice that, if the original model has particle-hole symmetry, which implies that for any **k** there is a **p** such that $V_{\mathbf{k}} = V^*_{\mathbf{p}}$ and $\varepsilon_{\mathbf{k}} = -\varepsilon_{\mathbf{p}}$, then, for any $n \geq 0$, $\varepsilon_n = 0$.

In the most common situations in which the conduction electron DOS and the hybridization function are smooth, the hopping matrix elements t_n converge fastly to a constant value n-independent.

As an example, let us consider a conduction electron DOS $\rho(\varepsilon) = \theta\left(1-\varepsilon^2\right)/2$, constant within the interval $-1 \leq \varepsilon \leq 1$ (in units of half-bandwidth) and zero outside. The chemical potential is $\mu = 0$. In addition we assume that the hybridization matrix elements V_ε are equal for any ε. In this case the conduction electron Hamiltonian is

$$\mathscr{H}_0 = \int_{-1}^{1} d\varepsilon\, \varepsilon\, c^\dagger_\varepsilon c_\varepsilon,$$

and

$$c^\dagger_0 = \sqrt{\frac{1}{2}}\int_{-1}^{1} d\varepsilon\, c^\dagger_\varepsilon.$$

If we write

$$c^\dagger_n = \sqrt{\frac{2n+1}{2}}\int_{-1}^{1} d\varepsilon\, P_n(\varepsilon)\, c^\dagger_\varepsilon,$$

with $P_0(\varepsilon) = 1$, and assume

$$\int_{-1}^{1} d\varepsilon\, P_n(\varepsilon)\, P_m(\varepsilon) = \delta_{nm}\frac{2}{2n+1},$$

the recursion equation becomes

$$\sqrt{2n+1}\,\varepsilon\, P_n(\varepsilon) = t_n\sqrt{2n+3}\,P_{n+1}(\varepsilon) + t_{n-1}\sqrt{2n-1}\,P_{n-1}(\varepsilon),$$

which, multiplying both sides by $\sqrt{2n+1}$, is

$$(2n+1)\,\varepsilon\, P_n(\varepsilon) = t_n\sqrt{(2n+3)(2n+1)}\,P_{n+1}(\varepsilon) + t_{n-1}\sqrt{(2n+1)(2n-1)}\,P_{n-1}(\varepsilon).$$

We realize that the above is the recurrence formula for the Legendre polynomials if

$$t_n = \frac{n+1}{\sqrt{(2n+3)(2n+1)}} \simeq \frac{1}{2}\left(1+\frac{1}{8n^2}\right),$$

which indeed converges soon to $1/2$.

Therefore, instead of studying the full model (75), one can equally well consider a simple tight-binding Hamiltonian with constant hopping on a semi-infinite chain. The advantage is that a lot of powerful techniques are available in one dimension, like Conformal Field Theory, which we discuss later.

4.2. The Wilson Numerical Renormalization group

Let us consider again the model with a constant DOS and matrix elements $V_\varepsilon = V$. We divide the energy interval $0 \leq \varepsilon \leq 1$ into sub-intervals $I_n^{(+)}$, $n \geq 0$, defined in the following way

$$\varepsilon \in I_n^{(+)} \qquad \text{means} \qquad \frac{1}{\Lambda^{n+1}} \leq \varepsilon \leq \frac{1}{\Lambda^n},$$

with $\Lambda > 1$. The same is done for $-1 \leq \varepsilon \leq 0$, in which case

$$\varepsilon \in I_n^{(-)} \qquad \text{means} \qquad -\frac{1}{\Lambda^n} \leq \varepsilon \leq -\frac{1}{\Lambda^{n+1}}.$$

In other words we have performed a logarithmic mesh of the energies. Let us focus on a single interval $I_n^{(+)}$, whose length is

$$L_n = \Lambda^{-n}\left(1 - \Lambda^{-1}\right).$$

In this finite interval we can introduce a complete basis of wave-functions in the energy space, defined as

$$\phi_{n,l}^{(+)}(\varepsilon) = \sqrt{\frac{1}{L_n}}\, e^{i\omega_n l\,\varepsilon},$$

where

$$\omega_n = \frac{2\pi}{L_n},$$

and l is an integer. The fermionic operators are

$$a_{n,l}^\dagger = \int_{I_n^{(+)}} d\varepsilon\, \phi_{n,l}^{(+)}(\varepsilon)\, c_\varepsilon^\dagger. \tag{76}$$

Analogously, within the interval $I_n^{(-)}$ of negative energies, we define the wave-functions

$$\phi_{n,l}^{(-)}(\varepsilon) = \sqrt{\frac{1}{L_n}}\, e^{-i\omega_n l\,\varepsilon},$$

and the corresponding operators

$$b_{n,l}^\dagger = \int_{I_n^{(-)}} d\varepsilon\, \phi_{n,l}^{(-)}(\varepsilon)\, c_\varepsilon^\dagger. \tag{77}$$

The conduction electron Hamiltonian in the new basis becomes

$$\mathcal{H}_0 = \int_{-1}^{1} d\varepsilon\, \varepsilon\, c_\varepsilon^\dagger c_\varepsilon = \sum_n \sum_{ll'} E_{n;ll'}\left(a_{n,l}^\dagger a_{n,l'} - b_{n,l}^\dagger b_{n,l'}\right), \tag{78}$$

where

$$E_{n;ll'} = \int_{I_n^{(+)}} d\varepsilon\, \varepsilon\, \phi_{n,l}^{(+)}(\varepsilon)^*\, \phi_{n,l'}^{(+)}(\varepsilon)$$

$$
\begin{aligned}
&= \int_{I_n^{(+)}} d\varepsilon \,(\varepsilon - E_n)\, \phi_{n,l}^{(+)}(\varepsilon)^* \phi_{n,l'}^{(+)}(\varepsilon) + E_n \,\delta_{ll'} \\
&= \Delta E_{n;ll'} + E_n \,\delta_{ll'}.
\end{aligned}
$$

In the previous formula, if we take E_n as the average energy within the interval, namely,

$$
E_n = \frac{1}{2\Lambda^n}\left(1 + \frac{1}{\Lambda}\right), \tag{79}
$$

then the off-diagonal matrix elements $\Delta E_{n;ll'}$ vanish in the limit $\Lambda \to 1$.

The conduction electron operator which is coupled to the impurity becomes in the new basis

$$
f_0^\dagger = \sqrt{\frac{1}{2}} \int_{-1}^{1} d\varepsilon\, c_\varepsilon^\dagger = \sum_n \sqrt{\frac{L_n}{2}} \left(a_{n,0}^\dagger + b_{n,0}^\dagger\right), \tag{80}
$$

namely involves only the $l = 0$ component within each interval. Therefore, neglecting the off-diagonal elements $\Delta E_{n;ll'}$, which is rigorously true only for $\Lambda \to 1$, the $l = 0$ sector decouples from the $l \neq 0$, which remains untouched by the impurity, and one needs to solve only the following Hamiltonian, dropping the label $l = 0$ and inserting back the spin index,

$$
\begin{aligned}
\mathscr{H}_{AIM} &= \sum_{n\sigma} E_n \left(a_{n\sigma}^\dagger a_{n\sigma} - b_{n\sigma}^\dagger b_{n\sigma}\right) \\
&\quad + V \sum_{n\sigma} \sqrt{\frac{L_n}{2}} \left[\left(a_{n\sigma}^\dagger + b_{n\sigma}^\dagger\right) d_\sigma^\dagger + H.c.\right] + \mathscr{H}_{int}\left[d^\dagger, d\right].
\end{aligned} \tag{81}
$$

Starting by this Hamiltonian with a logaritmic discretization of the energy, we can apply the previously discussed Lanczos algorithm to obtain the tight-binding Hamiltonian

$$
\mathscr{H}_{AIM} = \sum_{n\geq 0} \sum_\sigma t_n \left(f_{n\sigma}^\dagger f_{n+1\sigma} + H.c.\right) + V \sum_\sigma \left(f_{0\sigma}^\dagger d_\sigma + H.c.\right) + \mathscr{H}_{int}\left[d^\dagger, d\right]. \tag{82}
$$

Wilson [28] showed that the hopping matrix elements have the expression

$$
t_n = \frac{1}{2}\left(1 + \Lambda^{-1}\right) \Lambda^{-n/2} \frac{1 - \Lambda^{-n-1}}{\sqrt{\left(1 - \Lambda^{-2n-3}\right)\left(1 - \Lambda^{-2n-1}\right)}} = \frac{1}{2}\left(1 + \Lambda^{-1}\right) \Lambda^{-n/2} \zeta_n, \tag{83}
$$

where ζ_n fastly approaches 1, so that in the following we take it to be strictly unity. We notice that, thanks to the logarithmic mesh, the hopping decays exponentially with the chain-length, unlike the uniform mesh in which the hopping saturates to a constant value.

Let us define a set of Hamiltonians

$$
\mathscr{H}_N = \Lambda^{(N-1)/2} \left\{ \sum_{n=0}^{N-1} \sum_\sigma t_n \left(f_{n\sigma}^\dagger f_{n+1\sigma} + H.c.\right) + V \sum_\sigma \left(f_{0\sigma}^\dagger d_\sigma + H.c.\right) + \mathscr{H}_{int}\left[d^\dagger, d\right] \right\}. \tag{84}
$$

The pre-factor has the effect of making the lowest hopping $\Lambda^{(N-1)/2}\,t_{N-1}$ of order one. Clearly

$$\mathscr{H}_{AIM} = \lim_{N\to\infty} \Lambda^{-(N-1)/2}\,\mathscr{H}_N.$$

These Hamiltonians satisfy the recursion formula:

$$\mathscr{H}_{N+1} = \Lambda^{1/2}\,\mathscr{H}_N + \frac{1}{2}\left(1+\Lambda^{-1}\right)\sum_{\sigma}\left(f^{\dagger}_{N\sigma}f_{N+1\sigma} + H.c.\right). \tag{85}$$

We know that the Kondo temperature is exponentially small at large U/Γ. Therefore, in order to describe correctly the low temperature physics, one would need to go to an exponentially small $T \ll T_K$. In this case, finite-size effects in any physical observable should become negligible only when the typical level spacing Δ becomes smaller than T. In the uniform mesh, $\Delta \sim \pi/N$, where N is the chain length, hence we would need an exponentially large size to access the low temperature properties. On the contrary, thanks to the logarithmic mesh, a chain with $N \gg \ln(1/T)/\ln\Lambda$ is already sufficient. However it is practically impossible to diagonalize completely the Hamiltonian but for the first few N's. We notice that, after the rescaling by the factor $\Lambda^{1/2}$, typical Λ being $3\div 5$, the spectrum of $\mathscr{H}_N$ generally contains a bunch of eigenstates with eigenvalues ≤ 1, separated by a gap from states with higher energies ≥ 10. The hopping to site $N+1$ will have negligible effects on the high energy states, so that one can truncate the Hilbert space of the chain with N-site, by keeping only the low energy eigenstates, combine them with the states of the site $N+1$ and diagonalize $\mathscr{H}_{N+1}$ in this reduced Hilbert space. This procedure is iterated to larger chain sizes, until the recursion formula (85) flows to a fixed point Hamiltonian $\mathscr{H}_*$. Indeed (85) describes a renormalization group (RG) transformation, which can be shortly written as

$$\mathscr{H}_{N+1} = T\left[\mathscr{H}_N\right], \tag{86}$$

In reality it is more appropriate to define the RG transformation within chains of the same parity, namely

$$\mathscr{H}_N = T^2\left[\mathscr{H}_{N-2}\right]. \tag{87}$$

As N increases the RG transformation flows to a fixed point defined by

$$\mathscr{H}_* = T^2\left[\mathscr{H}_*\right]. \tag{88}$$

By analysing the spectrum of the fixed point one should be able to identify $\mathscr{H}_*$. Once this is done, one may perturb $\mathscr{H}_* \to \mathscr{H}_* + \delta\mathscr{H}_*$, with $\delta\mathscr{H}_*$ limited to the first few sites, and by iterating again the RG procedure, one can determine all scaling operators and their eigenvalues. [28] In most common cases this programme can be actually carried out, but there are situations in which the model flows to non trivial fixed points which require a more sophisticated analysis.

4.2.1. The simplest example: the single-channel Kondo effect

Just as an example, let us discuss how the Numerical Renormalization Group (NRG) works for a single-orbital AIM in the large-U Kondo-regime. [28] At the beginning it is convenient to consider the model in the absence of the impurity, which, as we discussed, is equivalent to a tight-binding model on a semi-infinite chain. At particle-hole symmetry, the number of electrons is equal to the number of sites. For a finite chain of N sites, the appropriate Block waves with open boundaries at site $r = 1$ and $r = N$ are of the form $\sin kr$ with

$$k = \frac{\pi}{N+1} n,$$

with $n > 0$ an integer. The single-particle spectrum near the Fermi energy can be linearized, leading to

$$\varepsilon_k \simeq \varepsilon_F + v_F k = \varepsilon_F + v_F \frac{\pi}{N+1} n,$$

and consists of equally spaced levels, with spacing $\Delta = v_F \pi/(N+1)$. The same property holds also for the full multi-particle spectrum. Since the number of electrons is equal to the number of sites, the chains with an even number of sites is obtained by filling with two electrons the lowest $N/2$ levels. The ground state is therefore non-degenerate and is followed by equally spaced excited states. If the number of sites is odd, the ground state is obtained by doubly occupying the lowest $(N-1)/2$ states, and occupying with a single electron the $(N+1)/2$ level. Since this can be done in two different ways, according to the spin of the electron, the ground state has spin 1/2, namely is doubly degenerate. If, instead of fixing the number of electrons, we keep the chemical potential fixed such that, on average, the number of electrons is equal to the number of sites, then, in the case of even chains, the chemical potential lies in the middle of two consecutive single-particle levels, while, for odd chains, it coincides with a single-particle level. In the latter case, the ground state turns out to be fourfold degenerate, since there are four equally probable states available on the level right at the chemical potential, empty, singly occupied by a spin up electron, singly occupied by a spin down electron, or doubly occupied.

We know that the Kondo effect is identified by a phase shift $\delta_\sigma = \pi/2$ for each spin-channel σ. By the Friedel's sum rule the variation of the electron number at fixed chemical potential is

$$\Delta N_{els} = \frac{1}{\pi}\left(\delta_\uparrow + \delta_\downarrow\right) = 1.$$

This implies that the ground state of a chain with N sites has now to accomodate $N+1$ electrons, hence that an even chain will eventually develop a fourfold degenerate state, while an odd chain a non-degenerate one. In other words the NRG flow for even N,

$$\mathcal{H}_N = T^2\left[\mathcal{H}_{N-2}\right], \tag{89}$$

will start from the spectrum of an even chain and flow to the spectrum of an odd chain. Viceversa for odd N's. This flow is the clear signal of a $\pi/2$ phase-shift typical of the Kondo effect.

In conclusion, the fixed point Hamiltonian still describes a tight-binding semi-infinite chain with equally spaced levels, the impurity being absorbed by the conduction sea into a simple change of boundary conditions. For what concerns the leading scaling operators, the only perturbing terms compatible with particle-hole symmetry are a variation of the first hopping matrix element, $t_0 \to t_0 + \delta t_0$, and a local interaction term $\delta U\, f^\dagger_{0\uparrow} f^\dagger_{0\downarrow} f_{0\downarrow} f_{0\uparrow}$.

4.3. Conformal Field Theory applied to impurity models

We have previously shown that an impurity model in any dimension is, for what it concerns the low energy properties, effectively equivalent to a model of electrons hopping with a constant matrix element on a semi-infinite chain, with the impurity sitting at the edge. We also know that one-dimensional gapless fermions have very peculiar features. In particular, the continuum of particle-hole excitations in one-dimension is exhausted by collective acoustic modes, the zero sounds. This property, which holds even if the electrons are interacting, provided the interaction does not open a gap in the spectrum, is the basis of the so-called **Bosonization technique**, [37] which allows to map the gapless fermionic model into a critical quantum field theory, which corresponds to a $1+1$ dimensional Conformal Field Theory (CFT). [36] In reality, if the electrons are interacting, the proper mapping is into a set of CFT's that reflect the overall symmetry of the Hamiltonian, which is generally lower if interaction is present. For instance, in a single band spin-isotropic model, one has to introduce a CFT that takes care of the charge degrees of freedom, and a non-abelian CFT to describe the $SU(2)$ spin sector, the well-known spin-charge separation which occurs in one dimension. This procedure, which is called *conformal embedding*, can be rigorously carried out by comparing, in the absence of interaction, the partition functions in the different representations.

In the specific example of impurity models, the conduction electrons are assumed to be non-interacting, hence by construction the fermionic model is massless and should be representable in terms of a CFT. Indeed, since the chain is semi-infinite, it actually maps into a CFT with an appropriate conformally invariant boundary condition (BC) at the edge. [38] A single impurity obviously can not induce any bulk gap so that, even in its presence, the model should still correspond to a CFT with the same number of massless degrees of freedom, namely the same central charge c. [36] Moreover, the impurity can not even change the sound velocity, in this non interacting case simply the Fermi velocity, since the latter is also a bulk property. The only thing that may change are the BC's, which still have to be conformally invariant. The change of boundary conditions has two effects. First it modifies the fine structure of the low-energy spectrum, namely the sequence and degeneracy of levels at finite but long chain-length. In addition, it may change the scaling properties of boundary operators. Both effects can be computed exactly by CFT once the new boudary conditions are known, and compared directly with NRG results.

This procedure encounters two main difficulties. The first is that the actual symmetry in the presence of the impurity is generally lower than the large symmetry of the free conduction bath. This implies that one has to identify the proper conformal embedding, which in some cases may not be simple to uncover. The second difficulty is to determine the conformally invariant BC's that are compatible with the embedding. This is accomplished by the so-called *fusion hypothesis*, according to which conformally invariant BC's can be obtained by *fusing* the spectrum of a known BC with the primary fields, which are the scaling operators of the theory. [38, 34, 35, 39] By *fusion* one means the product of two primary fields, which yields a combination of primary fields according to some specific rules, called *fusion rules*. In principle, if we were able to classify all allowed BC's, by comparing their CFT spectra with the NRG results we could identify which BC is actually realized.

We do not aim to give here a detailed introduction to CFT, but just want to show in some non-trivial cases how it works. If readers desire to acquire a deeper knowledge, a useful text-book is *Conformal Field Theory*, by P. Di Francesco, P. Mathieu and D. Senechal, [36] hereafter denoted as DMS book.

4.3.1. Some simple CFT results

We know that the first step of Bosonization in one-dimension is the linearization of the free-electron spectrum around the Fermi momentum. [37] This linearization is not expected to affect the low energy behavior provided the perturbations are weak compared to the band-width. Therefore let us consider, instead of a tight-binding model, free spinless Dirac fermions, which have indeed a linear spectrum, on a chain of length L with anti-periodic boundary conditions. We will consider Dirac fermions moving only in one direction, namely with a single chirality, because this is the case relevant to a semi-infinite chain. Indeed, as we have seen, the wave-functions with negative momenta of a semi-infinite chain are not indepenendent from those with positive momenta.

The single-particle wave-functions for a chiral Dirac fermion are plane waves with momentum

$$k = \frac{\pi}{L}(2n-1),$$

being n integer. The Hamiltonian in momentum space reads

$$\mathscr{H} = v_F \sum_k k c_k^\dagger c_k, \tag{90}$$

where v_F has to be identified with the Fermi velocity of the original tight-binding model. Let us define for positive k

$$\begin{aligned} a_k &= c_k, \\ b_k &= c_{-k}^\dagger, \end{aligned}$$

so that, apart from an actually infinite constant, the Hamiltonian becomes

$$\mathscr{H} = v_F \sum_{k>0} k \left(a_k^\dagger a_k + b_k^\dagger b_k \right). \tag{91}$$

The partition function at temperature T is simply

$$Z_{Dirac} = \prod_{k>0} \left[1 + \exp(-\beta v_F k) \right]^2 = \prod_{n\geq 1} \left(1 + q^{n-1/2} \right), \tag{92}$$

where conventionally q is defined as

$$q = \exp\left(-\beta \frac{2\pi}{L} \right) \equiv e^{2\pi i \tau}.$$

One can show that

$$Z_{Dirac}(q) = \frac{\theta_3(q)}{\phi(q)}, \tag{93}$$

where θ_3 is the third Jacobi theta-function

$$\theta_3(q) = \sum_{n=-\infty}^{\infty} q^{n^2/2},$$

and ϕ the Euler function

$$\phi(q) = \prod_{n\geq 1} (1-q^n).$$

On the other hand, we know by Bosonization that, if we define for positive $p = 2\pi n/L$ the operators

$$b_p = -i\sqrt{\frac{2\pi}{pL}}\rho(p) = -i\sqrt{\frac{2\pi}{pL}}\sum_k c_k^\dagger c_{k+p}, \qquad b_p^\dagger = i\sqrt{\frac{2\pi}{pL}}\rho(-p),$$

they satisfy bosonic commutation relations. In addition their equation of motion can be reproduced by the Hamiltonian

$$\mathscr{H} = \frac{\pi v_F}{L}\sum_p \rho(p)\rho(-p) = v_F \sum_{p>0} p b_p^\dagger b_p + \frac{\pi v_F}{L}\Delta N^2, \tag{94}$$

where we assumed that $\rho(p=0)$ is the variation ΔN of the electron number with respect to a reference value. The partition function of this bosonic model is the product of the bosonic term

$$\prod_{p>0} [1-\exp(-\beta v_F p)]^{-1} = \prod_{n>0} (1-q^n)^{-1} = \phi(q)^{-1}$$

plus the contribution of the ΔN which is simply, assuming an infinite reference number,

$$\sum_{n=-\infty}^{\infty} \exp\left(-\beta \frac{v_F \pi}{L} n^2\right) = \sum_{n=-\infty}^{\infty} q^{n^2/2} = \theta_3(q).$$

We immediately recognize that the bosonic partition function does coincide with the fermionic one.

We can proceed further on, and consider spinful Dirac fermions. Obviously the partition function is the square of Z_{Dirac} in Eq. (93). However, we would like to consider a perturbation that only preserves independently the spin $SU(2)$ symmetry and the charge isospin $SU(2)$, defined through the generators $\mathbf{I}$, which are the $q = 0$ components of the so-called isospin current operators

$$I_z(q) = \frac{1}{2}\sum_{k\sigma}\left(c_{k\sigma}^\dagger c_{k+q\sigma} - \delta_{q0}\right),$$

$$\begin{aligned} I^+(q) &= \sum_k c^\dagger_{k\uparrow} c^\dagger_{-k-q\downarrow}, \\ I^-(q) &= \left(I^+\right)^\dagger. \end{aligned}$$

The spin $SU(2)$ current operators are instead

$$\mathbf{S}(q) = \frac{1}{2} \sum_{k\alpha\beta} c^\dagger_{k\alpha}\, \sigma_{\alpha\beta}\, c_{k+q\beta}, \tag{95}$$

where σ are the Pauli matrices. In real space these current operators, $J_a = I_a, S_a$, satisfy the commutation relations

$$\left[J_a(x), J_b(y)\right] = i\varepsilon_{abc}\, \delta(x-y) J_c(x) - ik\frac{1}{4\pi}\delta_{ab}\frac{\partial\delta(x-y)}{\partial x},$$

with $k=1$. For generic $k \geq 1$, the above commutation relations identify an $SU(2)_k$ CFT, [36] where the label k may be regarded as the number of channels which are used to build up the generators. In general, an $SU(2)_k$ CFT has primary fields $\phi^{(k)}_{2j}$ with spin quantum numbers j, such that $2j = 0, 1, \ldots, k$. Their scaling dimension is $x_j = j(j+1)/(k+2)$. The product of two primary fields with spin j and j' yields all primary fields with spin bewteen $|j-j'|$ and $min(k-j-j', j+j')$. The Hilbert space of the theory is obtained by applying the primary fields on the reference vacuum state and, from this ancestor state, by generating all descendent states applying the current operators with $q<0$. This is what is called a *conformal tower.* [36] The energy difference between the descendent states and their ancestor is an integer multiple of the fundamental level spacing. The character $\chi^{(k)}_{2j}$ represents the contribution to the partition function of the conformal tower generated by the primary field $\phi^{(k)}_{2j}$. [36]

This construction may look abstruse but actually has a simple physical interpretation. Let us consider again a single spinful fermion, $k=1$. Let us further assume that on average the number of electrons is equal to the number of sites, and the latter is even. In this case the ground state is obtained by filling with two electrons of opposite spin all single-particle states below the chemical potential, which lies in the middle of two consecutive single-particle levels, which, we recall, are separated by the fundamental spacing Δ, from now on our energy unit. With this definition, the Hilbert space can be constructed as follows. One can start from the vacuum and act on it with particle-hole excitations, namely with $I_z(q)$ or $\mathbf{J}(q)$ with $q<0$. In addition one can apply the operators $I^+(q)$ or $I^-(q)$, again with $q<0$, to change by an even multiple the number of electrons, and then consider all particle-hole excitations on top of these states. In this way we obtain all states which have even number of electrons, like the vacuum state. This is nothing but the conformal tower obtained by the ancestor fields $\chi^{(1)}_0$ in both the charge and spin $SU(2)_1$ sectors, which should contribute to the partition function with the product of characters $\left(\chi^{(1)}_0\right)_{charge}\left(\chi^{(1)}_0\right)_{spin}$.

The rest of the Hilbert space includes all states with odd number of electrons. Since a single electron carries isospin and spin 1/2, all these states have half-odd integer values of I_z and S_z. One realizes that they can all be obtained by applying the product of the isospin and spin primary fields $\phi^{(1)}_{1\,charge} \times \phi^{(1)}_{1\,spin}$, which is nothing but the single electron operator, and construct out of it all descendent states. Their contribution to the partition function should then be $\left(\chi_1^{(1)}\right)_{charge}\left(\chi_1^{(1)}\right)_{spin}$. The expression of the $SU(2)_k$ characters, cfr. DMS book at page 586, is

$$\chi_l^{(k)}(q) = \frac{1}{\eta(q)^3} \sum_{n=-\infty}^{\infty} \left[2n(k+2)+l+1\right] q^{(2n(k+2)+l+1)^2/4(k+2)},$$

where

$$\eta(q) = q^{1/24}\phi(q),$$

is the Dedekind function. In the specific case of $k=1$,

$$\begin{aligned} \chi_0^{(1)}(q) &= \sqrt{\frac{\theta_3(q)^2+\theta_4(q)^2}{2\eta(q)^2}}, \\ \chi_1^{(1)}(q) &= \sqrt{\frac{\theta_3(q)^2-\theta_4(q)^2}{2\eta(q)^2}}, \end{aligned}$$

where

$$\theta_4(q) = \sum_{n=-\infty}^{\infty} (-1)^n q^{n^2/2},$$

is the fourth Jacobi theta-function. Hence we find that

$$\left(\chi_0^{(1)}\right)_{charge}\left(\chi_0^{(1)}\right)_{spin} + \left(\chi_1^{(1)}\right)_{charge}\left(\chi_1^{(1)}\right)_{spin} = \frac{\theta_3(q)^2}{\eta(q)^2} = q^{-1/12} Z_{Dirac}(q)^2, \quad (96)$$

which, apart from the vacuum polarization contribution $q^{-1/12}$, is exactly the partition function of two species of Dirac fermions. The spectrum, and correspondingly the partition function, can be represented as in Table 1. In that Table we identify each conformal tower by the quantum numbers of the primary fields which generate the ancestor states. x is the energy in units of the fundamental level spacing Δ of the ancestor state with respect to the chemical potential. The descendent levels of an ancestor have energies $x+n$, with n a positive integer. Notice that an important consequence of conformal invariance is that the energy of each state in units of Δ coincide with the scaling dimension of the operator which, applied to the vacuum, yields that state. In

TABLE 1. The spectrum of spinful electrons when the ground state contains an even number of particles

I	S	x
0	0	0
1/2	1/2	1/2

particular the single electron excitation has energy $x = 1/2$, 1/4 from the spin $S = 1/2$ and 1/4 from the isospin $I = 1/2$ primary fields, because the chemical potentials lies in the middle between two consecutive levels.

Following the same reasoning, one can easily show that the table of the energy spectrum in the case of odd chains at half-filling is the one of Table 2. As discussed

TABLE 2. The spectrum of spinful electrons when the ground state contains an odd number of particles

I	S	$x - 1/4$
$1/2$	0	0
0	$1/2$	0

before the ground state is indeed fourfold degenerate, the chemical potential coinciding with a single-particle level.

Following the *fusion hypothesis*, [35, 34] the Kondo fixed point should be obtained by fusing the spectrum in the absence of the impurity with a primary field of the embedded theory. Since, in the large U-limit, the Kondo exchange only involves the spin $SU(2)_1$ degrees of freedom, and since, apart from the identity $S = 0$, the only primary field has $S = 1/2$, we would predict that, upon fusion with the latter, the Kondo fixed point should be reproduced. This is actually the case. Indeed, if we fuse the spectrum of even chains with the spin field $\phi_1^{(1)}$ we do get the spectrum of odd chains, and viceversa. CFT also tells us that, in order to identify the local scaling operators, we have to double fuse the starting spectrum. Each row of the resulting table corresponds to a local scaling operator, whose dimension is the sum of the dimensions of each primary field involved. Descendent operators have dimensions which increase by unity. An operator is relevant if its dimension is < 1, marginal if it is exactly 1, and irrelevant otherwise. In the single-channel example, if, for instance, we double fuse the even chain spectrum, we obtain it back. This implies that the lowest dimension operators allowed by symmetry are scalar products of current operators with dimension 2, hence they are irrelevant.

The two channel model. Since it will be important for what follows, let us conclude by considering the case of two channels of spinful fermions. The partition function is the square of the partition function (96) of a single channel, and can be written for even chains as

$$Z = \sum_{n_1,n_2=0,1} \left(\chi_{n_1}^{(1)} \chi_{n_2}^{(1)} \right)_{charge} \left(\chi_{n_1}^{(1)} \chi_{n_2}^{(1)} \right)_{spin}, \tag{97}$$

where n_1 refers to channel 1 and n_2 to channel 2. For odd chains one readily finds that

$$Z = \sum_{n_1,n_2=0,1} \left(\chi_{n_1}^{(1)} \chi_{n_2}^{(1)} \right)_{charge} \left(\chi_{1-n_1}^{(1)} \chi_{1-n_2}^{(1)} \right)_{spin}. \tag{98}$$

These expressions manifestly show that the free two-channel conduction electrons are invariant under independently spin or isospin $SU(2)$ transformations for each channel, namely under a large symmetry $SU(2) \times SU(2) \times SU(2) \times SU(2)$.

Let us suppose that the impurity couples only to the spin-current operators in such a way that only the overall SU(2) symmetry is preserved. Therefore, while the charge degrees of freedom can still be represented by two $SU(2)_1$ CFT's, the appropriate confomal embedding for the spin sectors should involve an $SU(2)_2$ CFT, since the total spin current is made up of two channels, times the coset CFT, namely

$$SU(2)_1 \times SU(2)_1 \rightarrow SU(2)_2 \times \frac{SU(2)_1 \times SU(2)_1}{SU(2)_2}.$$

Since the central charge is conserved and each $SU(2)_k$ has a central charge $3k/(k+2)$, the coset theory should have $c = 1/2$, namely the central charge of an Ising CFT. This can be proved rigorously by the character decomposition.

The Ising CFT has three primary fields, the identity I, with dimension 0, the energy field ε, with dimension 1/2, and the spin field σ with dimension 1/16. The fusion rules are [36]

$$I \times I = I, \ \varepsilon \times \varepsilon = \varepsilon, \ \sigma \times \sigma = I + \varepsilon, \ I \times \varepsilon = \varepsilon, \ I \times \sigma = \sigma, \ \varepsilon \times \sigma = \sigma.$$

The characters χ_x^I, where x is the dimension of the primary field, are explicitly given by (all functions are assumed to depend on the variable q, even when not indicated)

$$\begin{aligned}
\chi_0^I &= \frac{1}{2}\left[\sqrt{\frac{\theta_3}{\eta}} + \sqrt{\frac{\theta_4}{\eta}}\right], \\
\chi_{1/2}^I &= \frac{1}{2}\left[\sqrt{\frac{\theta_3}{\eta}} - \sqrt{\frac{\theta_4}{\eta}}\right], \\
\chi_{1/16}^I &= \sqrt{\frac{1}{2}}\sqrt{\frac{\theta_2}{\eta}},
\end{aligned}$$

where the second Jaboi function is defined by

$$\theta_2 = \sum_{n=-\infty}^{\infty} q^{(n-1/2)^2/2}.$$

One can show that the product of characters of two $SU(2)_1$ CFTs, $\chi_{2j}^{(1)}\chi_{2j'}^{(1)}$, can be related to the product of characters of an $SU(2)_2$ and an Ising CFTs, $\chi_{2l}^{(2)}\chi_x^I$, by

$$\begin{aligned}
\chi_0^{(1)}\chi_0^{(1)} &= \chi_0^{(2)}\chi_0^I + \chi_2^{(2)}\chi_{1/2}^I, \\
\chi_0^{(1)}\chi_1^{(1)} &= \chi_1^{(1)}\chi_0^{(1)} = \chi_1^{(2)}\chi_{1/16}^I, \\
\chi_1^{(1)}\chi_1^{(1)} &= \chi_0^{(2)}\chi_{1/2}^I + \chi_2^{(2)}\chi_0^I.
\end{aligned}$$

By means of the above decomposition one can write the tables of the spectra for even and odd chains, as given in Table 3. By these tables we can easily realize the spectrum of even chains can be turned into that of odd chains, and viceversa, either by fusion with the $SU(2)_2$ primary field of spin 1, or by fusion with the Ising field ε.

TABLE 3. The spectra of two channels on even chains, right table, and odd chains, left table. I_1 and $_2$ are the values of the isospins of each channel, S the value of the total spin and *Ising* refers to the Ising primary fields.

I_1	I_2	S	*Ising*	x
0	0	0	0	0
0	0	1	1/2	1
1	0	1/2	1/16	1/2
0	1	1/2	1/16	1/2
1	1	0	1/2	1
1	1	1	0	1

I_1	I_2	S	*Ising*	$x-1/2$
0	0	0	1/2	0
0	0	1	0	0
1	0	1/2	1/16	0
0	1	1/2	1/16	0
1	1	0	0	0
1	1	1	1/2	1

5. THE FATE OF THE KONDO EFFECT IN IMPURITY CLUSTERS AND SPECULATIONS ON CLUSTER DMFT

In the previous sections we have discussed quite in detail the physics of a single-orbital Anderson impurity model in connection with DMFT. We have seen that already the simple single-site formulation of DMFT provides interesting results about the Mott transition. Most notably, we have shown that DMFT predicts this transition to occur by the gradual narrowing and disappearing of a quasiparticle peak coexisting with well preformed Mott-Hubbard side-bands. This behavior implies for instance that the effective impurity model used to solve the lattice problem is well inside the Kondo regime before the MIT, with a Kondo temperature approaching zero right at the transition.

In the case of the single-site DMFT all the physics was exhausted by knowing whether, after self-consistency, in the effective impurity model Kondo effect takes place or does not. If the self-consistent hybridization function is finite around the chemical potential, the impurity is Kondo-screened, which translates into a Landau-Fermi liquid behavior of the lattice model. On the contrary, if the self-consistent hybridization function develops a gap at the chemical potential, the impurity stays unscreened and the lattice model is insulating. However, an insulator with unscreened local moments is a poor description of a realistic Mott insulator, which may be appropriate only at high temperature. At low temperature, inter-site exchange processes come always into play and provide a screening mechanism within the insulating phase. These processes are not well described within the single-site DMFT, which motivates its extension to cluster-DMFT. As we discussed in the Introduction, we do expect that new physics emerges when one moves to cluster DMFT in order to account for short-range spatial correlations, otherwise missed within single-site DMFT. How does this new physics translates in the corresponding impurity clusters? Conversely, how the physics of impurity clusters affects the behavior of lattice models in cluster-DMFT? In this Section we are going to discuss briefly some attempts we have been doing along this direction in the past years, that refer to the works in Refs. [40, 41, 42, 43, 44, 45]. The whole project originated from a DMFT calculation on a three-band model for doped fullerenes, [40, 41] that showed the existence of an unexpected s-wave superconducting dome adjacent the Mott insulating phase. This behavior was argued to be related to the physics of the impurity model onto which the lattice model maps within DMFT, namely a three-orbital impurity model with an effective inverted exchange of Jahn-Teller origin that favors low-spin impurity configurations – the single-orbital analogous of J discussed in the Introduction.

This speculation was later pursued by studying a simpler two-orbital impurity model with an inverted exchange, [42, 43] and its lattice counterpart. [44] These analyses have revealed a close connection between the behavior of non-trivial impurity model without any DMFT self-consistency and the actual behavior of the corresponding lattice models after DMFT self-consistency, which we recently extended to more involved impurity clusters. [45] In the following we aim to present the physical idea that guided these works and some results obtained.

Let us consider therefore a typical cluster of impurities that would represent a lattice model within cluster-DMFT. Each impurity, besides being hybridized with the conduction bath, is also coupled to the other impurities, generally by a single-particle hopping. Since we still expect the Mott transition to occur when the impurities are well inside the Kondo regime, we can equally well consider a cluster of impurity-spins that are Kondo coupled by an exhange J_K to a conduction bath and, in addition, coupled among themselves by an exchange J, generally antiferromagnetic. In this situation new possibilities arise. If $J_K \gg J$, conventional Kondo screening occurs and the lattice model should be metallic with well defined Landau's quasiparticles. On the contrary, if $J \gg J_K$, the impurities strongly bound together into an overall singlet configuration, if their number is even, in which case no Kondo screening is needed anymore, or, if their number is odd, into a spin-1/2 configuration, in which case a partial Kondo screening can still take place. We will argue that, in both situations, the impurity behavior is not at all trivial. For instance, the impurity spectral function contains, besides the Hubbard side-bands, a low energy incoherent feature which displays either a deep or a singularity right at the chemical potential. When full self-consistency is carried out, two possibilities may arise. The first and boring one is that the insulating phase occupies the whole $J > J_K$ region. This would mean that the incoherent feature is not self-consistent in the DMFT scheme and would likely imply a first order phase transition from the metal to the Mott insulator. The second and more intriguing scenario is that the incoherent feature survives DMFT self-consistency, leading to a strange metallic behavior near the MIT. If this were the case, then we should also expect a transition in the metallic phase from a conventional Fermi-liquid metal, at $J_K \gg J$, into a less conventional metal at $J \gg J_K$.

One may immediately realize that this is the *impurity* analogue of what we discussed in the Introduction about a realistic MIT. Actually, the competition we envisaged between the effective quasiparticle Fermi temperature T_F^* and the energy scale J, that identifies the strength of the terms which quench the residual entropy of the ideal Mott insulator, has its counterpart into the competition between the Kondo temperature and the inter-impurity exchange in impurity clusters. Indeed, the Kondo energy gain is related to the impurity coherently tunneling among all degenerate configurations, and it is greater the more degenerate the impurity ground state is. The inter-impurity exchange instead splits this degeneracy, hence competes against Kondo screening. It turns out that this competition may lead to an unstable fixed point, which separates the Kondo screened phase from the unscreened one. The occurrence of this critical point might actually be the signal that, within cluster-DMFT, the Fermi-liquid metal does encounter an instability before the MIT takes place, as we discussed previously.

Let us speculate a bit more on this hypothesis. Let us suppose, as we will demonstrate in several examples, that, in an impurity cluster, the Kondo screened phase is indeed separated from the unscreened one by an unstable fixed point. The latter will be necessarily characterized by several instability channels, namely by several relevant operators of dimension smaller than one. In turns, this would correspond to singular, if the dimension is smaller than 1/2, or otherwise strongly enhanced impurity contribution to the susceptibilities in the corresponding channels, which also means strongly enhanced irreducible vertices at the impurity site. As we have learned, the lattice irreducible vertices in DMFT are local and coincide with the impurity ones. Therefore, if these vertices are inserted into the Bethe-Salpeter equation for the lattice susceptibilities, it is reasonable to expect that the impurity instability will turn into a true bulk instability of the lattice model, leading to a symmetry breaking in the dominant channel. [42] This is the channel in which the enhanced irreducible vertex and the Green's function bubble cooperate at most to a bulk instability. Obviously, this instability will affect a whole region around the impurity unstable fixed point, leading to a symmetry broken phase before the MIT takes place. In most cases, however, the cluster of impurities contains by construction small terms which are nevertheless relevant at the unstable fixed point. As a result, the phase transition through the unstable fixed point would turn into a crossover. Yet, if this crossover is sharp, we should still expect that an instability arises when DMFT self-consistency is carried out. [43, 45]

At this stage what we have discussed so far is merely a conjecture but at least suggests that, just like in the single-site DMFT, the knowledge of the physical behavior of impurity clusters *per se* might be of great help to interpret cluster-DMFT results.

5.1. The impurity dimer

The first and simpler impurity cluster that one can discuss is a dimer. In particular we consider the Hamiltonian

$$\begin{aligned}\mathcal{H} &= \sum_{\mathbf{k}\sigma}\sum_{a=1}^{2} \varepsilon_{\mathbf{k}}\, c^{\dagger}_{\mathbf{k}a\sigma}c_{\mathbf{k}a\sigma} + \sum_{\mathbf{k}\sigma}\sum_{a=1}^{2}\left(V_{\mathbf{k}}\, c^{\dagger}_{\mathbf{k}a\sigma}d_{a\sigma} + H.c.\right)\\ &\quad + \frac{U}{2}\sum_{a=1}^{2}\left(n_a-1\right)^2 + V\left(n_1-1\right)\left(n_2-1\right) + J\mathbf{S}_1\cdot\mathbf{S}_2. \end{aligned} \tag{99}$$

In this expression

$$n_a = \sum_{\sigma} d^{\dagger}_{a\sigma}d_{a\sigma}, \qquad \mathbf{S}_a = \frac{1}{2}\sum_{\alpha\beta} d^{\dagger}_{a\alpha}\,\sigma_{\alpha\beta}\, d_{a\beta},$$

where σ are the Pauli matrices. The Hamiltonian describes two impurities, each one hybridized with its own bath, and coupled together by a charge repulsion term $V > 0$ and by an antiferromagnetic exchange $J > 0$. For simplicity we assume the two conduction baths degenerate and particle-hole symmetric. This implies that both the orbital and the particle-hole symmetries remain, even in the presence of the impurities.

5.1.1. The isolated impurities

Let us start by considering the case of an isolated dimer, i.e. $V_{\mathbf{k}} = 0$. As we will assume always a large U, the ground state of the two impurities in the absence of hybridization lies in the two-electron sector, which contains six states. It is convenient to introduce pseudo-spin-1/2 operators even in the orbital sector, so that

$$T_z = \frac{1}{2}\sum_{\sigma}(n_1 - n_2), \qquad T^+ = \sum_{\sigma} d^\dagger_{1\sigma} d_{2\sigma}, \qquad T^- = \sum_{\sigma} d^\dagger_{2\sigma} d_{1\sigma}.$$

Among the six two-particle states, there is a spin-triplet with quantum numbers $S = 1$ and $T = 0$, and energy

$$E(S=1, T=0) = V + \frac{J}{4}.$$

The other three states are spin-singlets but form an orbital-triplet, $T = 1$. In particular the states with different $T_z = -1, 0, 1$ are

$$\begin{aligned}
|T_z = -1\rangle &= d^\dagger_{2\uparrow} d^\dagger_{2\downarrow} |0\rangle, \\
|T_z = 0\rangle &= \sqrt{\frac{1}{2}}\left(d^\dagger_{1\uparrow} d^\dagger_{2\downarrow} + d^\dagger_{2\uparrow} d^\dagger_{1\downarrow}\right)|0\rangle, \\
|T_z = +1\rangle &= d^\dagger_{1\uparrow} d^\dagger_{1\downarrow} |0\rangle.
\end{aligned}$$

However the interaction lowers the pseudo-spin $SU(2)$ down to $O(2)$, so that the two states with $T = 1$ but $T_z = \pm 1$ are still degenerate, with energy

$$E(S=0, T=1, T_z=\pm 1) = U,$$

but the state with $T_z = 0$ has a different energy

$$E(S=0, T=1, T_z=0) = V - \frac{3}{4}J.$$

Provided $V < U + 3J/4$, which is a reasonable assumption, the lowest energy state of the isolated dimer is the spin-singlet with $T_z = 0$, namely the intersite singlet. The next lowest energy state is the doublet of singlets if $V > U - J/4$, and the spin-triplet otherwise.

Since an incoming pair of electrons has the same quantum numbers of the two electron states, we can define by symmetry three particle-particle scattering vertices, in the triplet channel, Γ^1, in the singlet channels with $T_z = \pm 1$, $\Gamma^0_{\pm}$, and finally in the singlet channel with $T_z = 0$, Γ^0_0. Their bare values are

$$\begin{aligned}
\Gamma^1 &= V + \tfrac{1}{4}J, \\
\Gamma^0_{\pm} &= U, \\
\Gamma^0_0 &= V - \tfrac{3}{4}J.
\end{aligned} \tag{100}$$

Notice that the bare Γ^0_0 is attractive only if $V < 3J/4$.

5.1.2. The breakdown of perturbation theory

In the opposite limit of very weak U, J and V, one can do perturbation theory on the non-interacting state, which simply represents two uncoupled resonant level models. As we know, each bath acquires a phase shift $\pi/2$ per spin. Moreover we also know that, at weak coupling, this result does not change. Rather, we have shown that, provided perturbation theory does not break down, the $\pi/2$ phase shift persists up to $U \to \infty$ in the model with particle-hole symmetry. However, as U increases, the width of the resonance, which is also the Kondo temperature, diminishes and becomes exponentially small for large U. The question is whether perturbation theory actually works for any U or it breaks down at same critical value.

We notice that, if T_K becomes much smaller then the energy gap which separates the ground state of the dimer from the first excited one, the impurities lock into the $T_z = 0$ spin-singlet state. Since this state is non-degenerate, the dimer becomes invisible to the conduction electrons, no Kondo effect is needed anymore. This implies that for very large U, hence T_K very small, the actual phase shift of the conduction channels should be $\delta = 0$. This immediately suggests that something has to occur as U increases. In particular, since $\delta = \pi/2$ and $\delta = 0$ are the only values compatible with particle-hole symmetry, the two extreme cases can not be joined continuously. In other words, a critical point should separate the $\delta = \pi/2$ Kondo screened phase from the $\delta = 0$ unscreened one. Before presenting the NRG plus CFT results, let us use the results of the previous sections to show that this breakdown is actually dramatic for what concerns the impurity.

We have shown that the conduction electron phase shift is given by

$$\delta(\varepsilon) = \Im m \ln \mathscr{G}(\varepsilon + i0^+), \tag{101}$$

where

$$\mathscr{G}(i\varepsilon_n)^{-1} = i\varepsilon_n + i\Gamma \,\mathrm{sign}(\varepsilon_n) - \Sigma(i\varepsilon_n), \tag{102}$$

is the impurity Green's function. If we denote as

$$\Sigma'(\varepsilon) = \Re e \mathscr{G}(\varepsilon + i0^+), \qquad \Sigma''(\varepsilon) = -\Im m \mathscr{G}(\varepsilon + i0^+),$$

then

$$\delta(\varepsilon) = \tan^{-1}\left(\frac{\Gamma + \Sigma''(\varepsilon)}{-\varepsilon + \Sigma'(\varepsilon)}\right). \tag{103}$$

It is clear from this expression that the only way to obtain a $\delta(\varepsilon \to 0) = 0$ is through a $|\Sigma'(\varepsilon \to 0)| \to \infty$. This shows that a phase-shift $\delta = 0$ is not at all innocuous for the impurity self-energy, whose real part diverges at the chemical potential. Moreover, since the local Landau-Fermi liquid picture should still hold even if $\delta = 0$, simply because the impurity dimer has disappeared at low energy, we should expect that $\delta(\varepsilon) \sim \varepsilon$, which implies that $\Sigma'(\varepsilon) \sim 1/\varepsilon$. Finally we notice that the impurity DOS

$$\rho(\varepsilon) = \frac{1}{\pi} \frac{\Gamma + \Sigma''(\varepsilon)}{\left[\varepsilon - \Sigma'(\varepsilon)\right]^2 + \left[\Gamma + \Sigma''(\varepsilon)\right]^2} \sim \varepsilon^2, \tag{104}$$

namely it vanishes quadratically at the chemical potential. Therefore, moving from $\delta = \pi/2$ to $\delta \to 0$, the impurity DOS goes from a Kondo-resonance behavior into a pseudo-gapped one.

5.1.3. The dimer phase diagram

The impurity model (99) at $V = 0$ was originally studied using NRG by Jones and Varma in the late 80's. [46, 47, 48] Their results were later interpreted with CFT by Affleck and Ludwig [49], see also Ref. [50] The same model but with $V = U$ was studied with NRG in Refs. [42, 43] Indeed, irrespectively of the value of V, provided the lowest energy state is the $T_z = 0$ singlet, the phase diagram remains the same, and is sketched in Fig. 10. As expected there are two stable phases, one Kondo screened below a critical

J / U = constant << 1

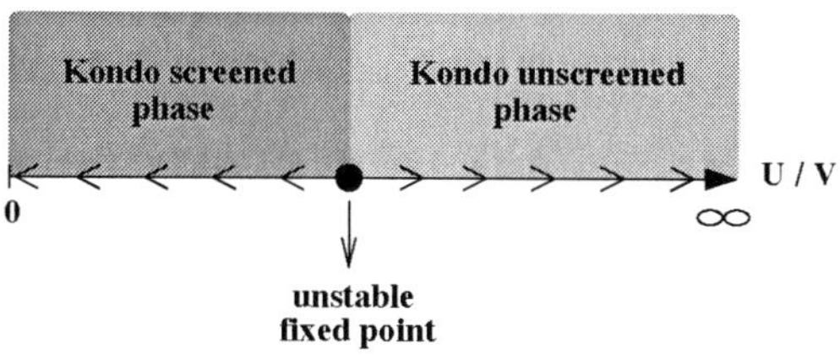

FIGURE 10. Phase diagram of (99) as function of U

U, and one unscreened above, separated by an unstable fixed point.

CFT analysis. Affleck and Ludwig realized [49] that the NRG spectrum of this fixed point can be reproduced within CFT by fusion of the spectra in Table 3 with the Ising primary field σ of dimension 1/16. The outcome of this fusion is shown in Table 4. In the same Table we also list the scaling operators and their dimensions

TABLE 4. Left: Spectrum at the unstable fixed point of the model (99). Right: Scaling operators with their dimension Δ.

I_1	I_2	S	*Ising*	$x - 1/16$
0	0	0	1/16	0
1	0	1/2	0	3/8
0	1	1/2	0	3/8
0	0	1	1/16	1/2
1	1	0	1/16	1/2
1	0	1/2	1/2	7/8
0	1	1/2	1/2	7/8
1	1	1	1/16	1

I_1	I_2	S	*Ising*	Δ
0	0	0	0	0
0	0	0	1/2	1/2
0	0	1	0	1/2
1	1	0	0	1/2
1	0	1/2	1/16	1/2
0	1	1/2	1/16	1/2
0	0	1	1/2	1
1	1	0	1/2	1
1	1	1	0	1
1	1	1	1/2	3/2

Δ. Apart from the identity, the only operator allowed by symmetry is the second, $(I_1, I_2, S, Ising) = (0, 0, 0, \varepsilon)$, with dimension 1/2. This is the operator which moves away from the fixed point. For instance, if at fixed J the transition occurs when $T_K = T_K^*$, then a change δJ will move away from the fixed point and introduce an energy scale

$T_- \sim (\delta J)^{1/(1-\Delta)} = (\delta J)^2$. In order to identify the other relevant operators, we notice that, if $\mathbf{S}_a(x)$ are the current operators of each channel, then $\mathbf{S}(x) = \mathbf{S}_1(x) + \mathbf{S}_2(x)$ is the current of the $SU(2)_2$ CFT, and

$$\mathbf{M}(x) = \mathbf{S}_1(x) - \mathbf{S}_2(x) \sim \phi_2^{(2)}\, \varepsilon, \tag{105}$$

as can be easily understood by checking that both sides have the same dimension 1. This is the third operator in the table. Since it carries spin and is odd under $1 \leftrightarrow 2$, it can not appear in the theory, not only it but also all its descendents. Notice that this operator is nothing but the staggered magnetization.

The fourth operator has also dimension 1/2. It is a spin singlet with $I_1 = I_2 = 1$, hence includes both an hopping term among the baths, equivalently among the impurities, namely

$$T = \sum_\sigma \mathrm{e}^{i\phi}\, d^\dagger_{1\sigma} d_{2\sigma} + H.c., \tag{106}$$

and the particle-particle singlet operator

$$\Delta_{SC} = \mathrm{e}^{i\phi} \left(d^\dagger_{1\uparrow} d^\dagger_{2\downarrow} + d^\dagger_{2\uparrow} d^\dagger_{1\downarrow} \right) + H.c.. \tag{107}$$

Again this operator is forbidden by the $O(2)$ orbital symmetry or by particle conservation, hence neither its descendents can appear. The other dimension-1/2 operators correspond to single particle excitation, which does not conserve charge. All the others are marginal or irrelevant.

The relevant operators $\mathbf{M}$, T and Δ_{SC} have a diverging susceptibility [50]

$$\chi(T) \sim \frac{1}{T_K} \ln\left(\frac{T_K}{T}\right).$$

Away from the fixed point this singularity is cut-off by T_-.

Additional informations can be gained through the so-called modular S-matrix, which is formally defined as follows. If $q = \exp(2i\pi\tau)$ and $\widetilde{q} = \exp(2i\pi/\tau)$, then the characters χ_x, where x label primary fields, satisfy

$$\chi_x(\widetilde{q}) = \sum_y S_x^y \chi_y(q). \tag{108}$$

The matrix elements S_x^y identify the unitary modular S-matrix, which is in general real. Suppose we start from a spectrum in which the ground state belongs to a conformal tower identified by the identity I, and assume that the actual boundary conditions are obtained by fusion with a primary field A. It has been shown in the context of boundary CFT [38, 51] that the ground state degeneracy g_A and residual entropy S_A with the new boundary conditions are

$$g_A = \frac{S_A^I}{S_I^I}, \qquad S_A = \ln g_A. \tag{109}$$

Next, suppose we want to calculate in the semi-infinite chain the correlation function between an incoming and an outgoing operator, say a primary field B. This correlation function will have obviously the same asymptotic behavior in the new boundary A as in the reference one I. What changes is an overall prefactor, which is the scattering S-matrix $S_B(A|I)$ of the primary field B. It turns out that

$$S_B(A|I) = \frac{S_A^B}{S_A^I}\frac{S_I^I}{S_I^B}. \tag{110}$$

In particular, if B is the primary field which enters the decomposition of the single-particle operator, in the two-channel example is the Ising field σ, then the scattering matrix corresponds to the single-particle S-matrix at zero frequency.

In the case under study, the fixed point is obtained by fusion in the Ising sector whose modular S-matrix is, using the convention that it acts on the vector (I, ε, σ), see DMS book at page 363,

$$S = \frac{1}{2}\begin{pmatrix} 1 & 1 & \sqrt{2} \\ 1 & -1 & -\sqrt{2} \\ \sqrt{2} & -\sqrt{2} & 0 \end{pmatrix}.$$

One readily finds that the fixed point has a residual entropy $\ln\sqrt{2}$ and a vanishing single-particle S-matrix

$$S(0) = \frac{S_\sigma^\sigma}{S_0^\sigma}\frac{S_0^0}{S_0^\sigma} = 0. \tag{111}$$

In turns this implies, from the relation

$$S(0) = 1 - 2\frac{\rho(0)}{\rho_0(0)},$$

that the impurity DOS at the chemical potential is exactly half of its value in the Kondo screened phase.

NRG results. It is clear that NRG is essential as input of any CFT analysis. In addition NRG provides dynamical informations which are not accessible to CFT, as for instance the whole energy dependence of the impurity spectral function. In Fig. 11 we show the DOS of the impurity across the fixed point of the model (99) with $V = U$. [43] In the top panel of that Figure we draw the DOSs well inside the Kondo screened phase and the unscreened one on a large energy range, where the Hubbard side-bands are well visible. We notice the Kondo resonance in the screened regime as well as the narrow pseudo-gap in the unscreened one. Yet, in the unscreened regime, even if pseudo-gaped, the DOS has spectral weight within the Hubbard side-bands. In the bottom panel, the low energy part is shown across the fixed point. The DOS has two main features. There is a broad resonance which is smooth across the transition. On top of that, in the screened phase there is a narrower peak which shrinks and vanishes at the transition, where only the broad resonance survives. Above the transition, in the unscreened phase, the narrow

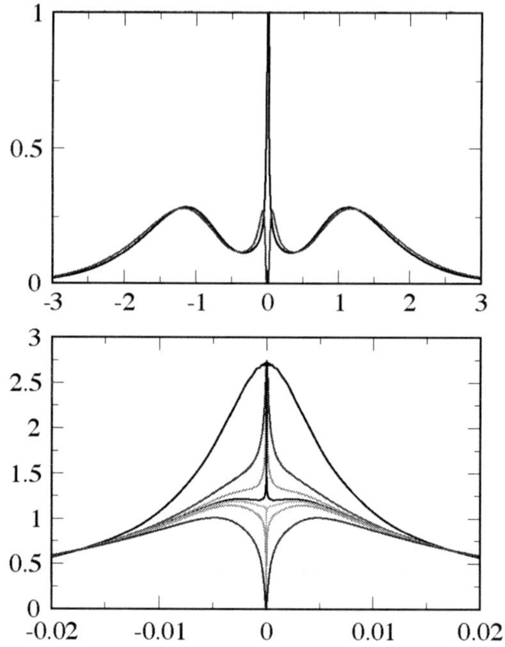

FIGURE 11. Impurity DOS across the unstable fixed point. In the top panel two DOS are shown on the full energy range, with well visible Hubbard-side bands. The bottom panel is a zoom of the low energy part. (From Ref. [43])

resonance turns into a narrow pseudogap within the broad resonance. Since we know, and it is also confirmed by NRG, that the fixed-point DOS at the chemical potential is half of its value in the Kondo regime, we guessed [43] the following ansatz for the low-energy DOS:

$$\rho_{\pm}(\varepsilon) = \frac{\rho_0}{2}\left(\frac{T_+^2}{\varepsilon^2+T_+^2} \pm \frac{T_-^2}{\varepsilon^2+T_-^2}\right), \tag{112}$$

where $\rho_0 = 1/(\pi\Gamma)$ is the DOS of the resonant level model in the absence of interaction, and the $+$ refers to the screened phase and the $-$ to the unscreened one. T_+ is the width of the broad resonance, while T_- is that one of the narrow resonance which turns into a narrow pseudogap. T_- has to be identified as the energy scale which controls the deviations δJ from the fixed point, namely $T_- \sim (\delta J)^2$. The model DOS (112) has all features we need. At the chemical potential is equal to the non-interacting value ρ_0 in the screened phase, is half of it at the fixed point, and vanishes quadratically in the unscreened regime. The model DOS (112) corresponds to a model Green's function at low energy

$$\mathcal{G}(i\varepsilon) = \frac{1}{2\Gamma}\left(\frac{T_+}{i\varepsilon + iT_+\,\mathrm{sign}(\varepsilon)} + \frac{T_-}{i\varepsilon + iT_-\,\mathrm{sign}(\varepsilon)}\right), \tag{113}$$

as well as to a model self-energy

$$\Sigma(i\varepsilon) = \mathscr{G}_0(i\varepsilon)^{-1} - \mathscr{G}(i\varepsilon)^{-1} = i\varepsilon + i\Gamma\,\mathrm{sign}(\varepsilon) - \mathscr{G}(i\varepsilon)^{-1}. \tag{114}$$

If we use (114) to fit the actual NRG results, we find an excellent agreement with the fitting parameters T_+ and T_- shown in Fig. 12. [43] Indeed T_+ is smooth across the transition, and can be identified as the fixed point value of $J = J_*$, while T_- vanishes quadratically, in accordance with CFT.

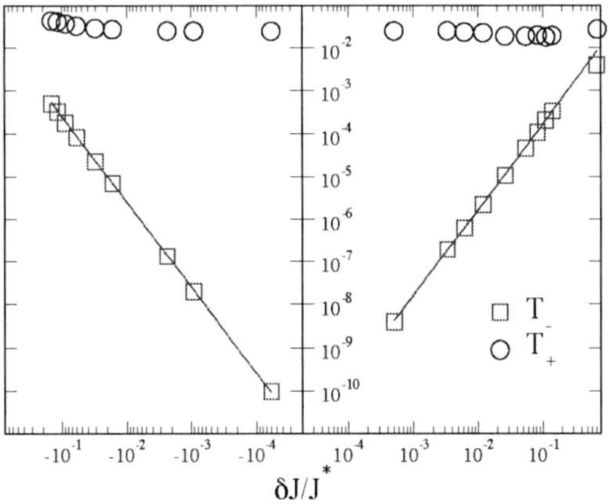

FIGURE 12. Fit values of T_+ and T_- across the fixed point. (From Ref. [43])

In Fig. 13 we draw the impurity self-energy $\Im m\Sigma(i\varepsilon)$, which is purely imaginary in Matsubara frequencies because of particle-hole symmetry, as function of ε in the screened phase, at the fixed point and in the unscreened regime. In the Kondo screened phase

$$\Sigma(i\varepsilon) \sim i\varepsilon\left(1 - \frac{1}{Z}\right),$$

with

$$Z = \frac{2}{\Gamma}\frac{T_+T_-}{T_+ + T_-}, \tag{115}$$

in accordance with perturbation theory. Notice that the non-interacting limit corresponds to $T_+ = T_- = \Gamma$. As the fixed point is approached, $T_- \to 0$ and $Z \to 0$, too. At the fixed point the self-energy is finite at zero frequency

$$\Sigma(i\varepsilon \to 0) = -i\Gamma,$$

like the self-energy of a disordered metal. Finally, in the unscreened phase

$$\Sigma(i\varepsilon) \sim -i\frac{2\Gamma}{\varepsilon}\frac{T_+T_-}{T_+ - T_-}.$$

Although the impurity self-energy in the unscreened phase is singular, yet the phase should be describable in terms of a local Fermi liquid theory in the Nozières sense. [52]

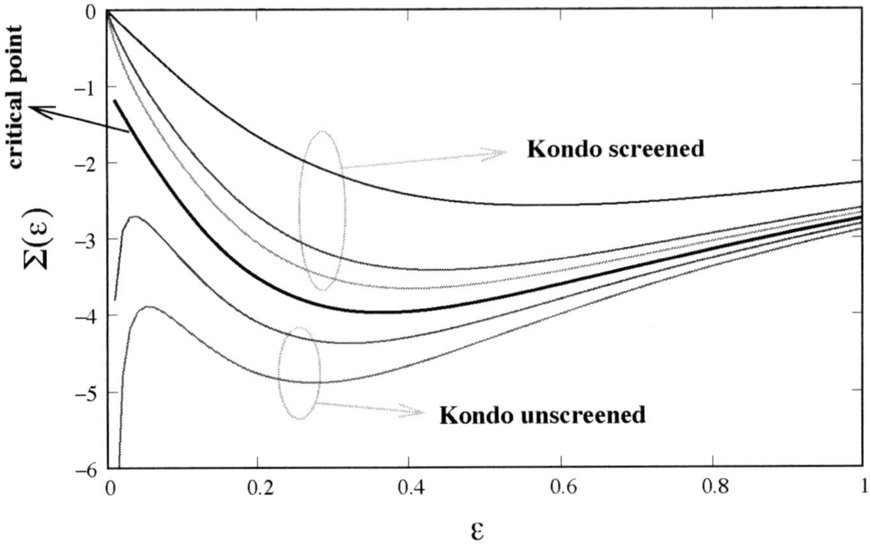

FIGURE 13. Behavior of the self-energy in the different phases.

Indeed the quasiparticle DOS as given by Eq. (178) is perfectly finite even in the unscreened phase

$$\rho_a^* = \int_{-\infty}^{\infty} \frac{d\varepsilon}{\pi} \frac{\partial f(\varepsilon)}{\partial \varepsilon} \Im m \left\{ \mathcal{G}_a(\varepsilon + i\delta) \left[1 - \left(\frac{\partial \Delta_a(i\varepsilon)}{\partial i\varepsilon} \right)_{i\varepsilon \to \varepsilon + i\delta} - \left(\frac{\partial \Sigma_a(i\varepsilon)}{\partial i\varepsilon} \right)_{i\varepsilon \to \varepsilon + i\delta} \right] \right\} = \frac{1}{\pi} \left(\frac{1}{T_+} + \frac{1}{T_-} \right), \tag{116}$$

since the singularity of the self-energy is compensated by the vanishing DOS. This allows to define an effective $Z = \rho_0 / \rho_*$, eventhough, strictly speaking, Z is not determined in the unscreened phase. Notice that ρ_*, defined as above, diverges at the fixed point, implying that a local Fermi liquid description is not possible there.

Moreover, since in the stable screened and unscreened phase a Fermi liquid description is legittimate, one can extract the Landau parameters discussed in C. This is done by calculating with NRG the Wilson ratios of the three conserved quantities, namely the charge (C), the spin (S) and the total T_z (T_z), and, through (189) one obtains the Landau parameters in the corresponding particle-hole channels, A_C, A_S and A_{T_z}, respectively. These in turns are related to the scattering amplitudes in the particle-particle channels,

$$A^1 = Z^2 \rho_* \Gamma^1, \qquad A^0_\pm = Z^2 \rho_* \Gamma^0_\pm, \qquad A^0_0 = Z^2 \rho_* \Gamma^0_0,$$

through

$$A_C = \frac{1}{4}\left(6A^1 + 4A^0_\pm + 2A^0_0 \right),$$

$$A_S = \frac{1}{4}\left(2A^1 - 4A^0_{\pm} - 2A^0_0\right),$$
$$A_{T_z} = \frac{1}{4}\left(-6A^1 + 4A^0_{\pm} - 2A^0_0\right).$$

Since the parameters on the left hand side are known by NRG, one can determine the the scattering amplitudes in the particle-particle channels. In Fig. 14 these amplitudes are drawn for the model (99) with $V = U$ and both J positive and negative. Notice that,

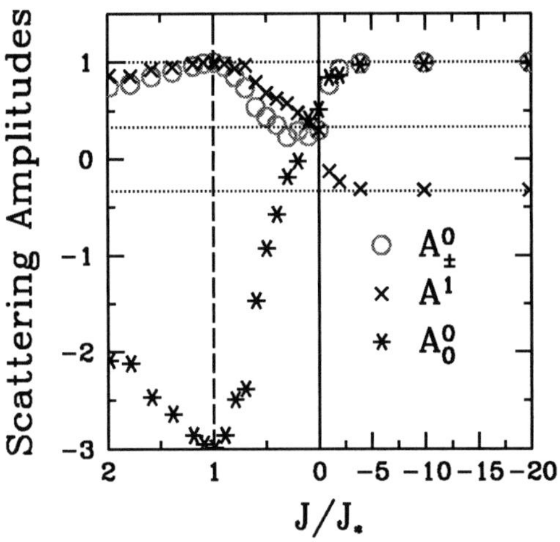

FIGURE 14. Scattering amplitudes in the particle-particle channels. (From Ref. [43])

approaching the fixed point from both sides of the transition, $A^0_0 \to -3$, even though its bare value, see Eq. (100), is $\rho_0 (U - 3J/4) \gg 0$ since in that calculation $U/J = 2 \times 10^2$ and $\rho_0 U = 6$. In other words, approaching the fixed point the scattering amplitude in the $S = 0$ $T_z = 0$ Cooper channel from repulsive turns strongly attractive.

5.1.4. The phase diagram away from particle-hole symmetry

One may realize that to conjecture the existence of the fixed point we used crucially the particle-hole symmetry to exclude a continuous cross-over from $\delta = \pi/2$ to $\delta = 0$. On the other hand, CFT predicts that ordinary particle-hole symmetry breaking is not a relevant perturbation, hence that the fixed point should survive even when it is broken. In the model (99) with $V = U$, we included particle-hole symmetry breaking by

$$\frac{U}{2}\sum_{a=1}^{2} (n_a - 1)^2 + V\,(n_1 - 1)\,(n_2 - 1) = \frac{U}{2}\left(n_1 + n_2 - 2\right)^2 \to \frac{U}{2}\left(n_1 + n_2 - 2 + \nu\right)^2.$$

The role of $\nu > 0$ is to lower the average occupancy of the impurity below 2. However, even in the presence of this perturbation, we still find the same unstable fixed point.

The phase diagram as function of J and the average impurity occupancy n, with all other parameters fixed, is sketched in Fig 15. [43] The critical line which separates the two stable phases moves at larger J's as we move away from particle-hole symmetry, and, for $n \to 1$, merges into a mixed-valence critical point which occurs when $J \simeq U$. The behavior of the DOS across the transition moving the parameter v is also shown

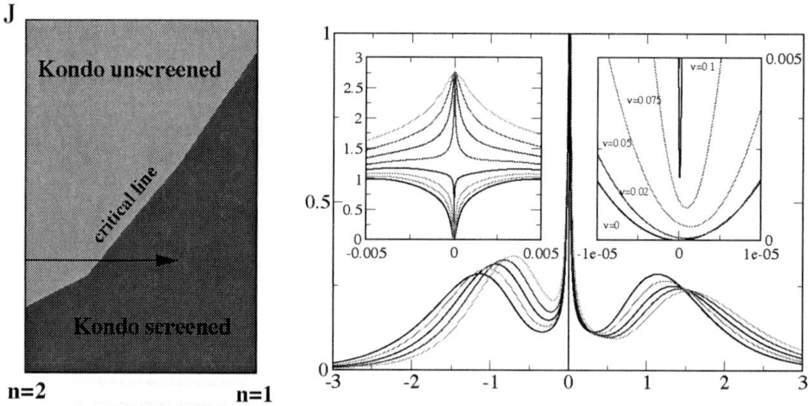

FIGURE 15. Left panel: the phase diagram of the model (99) with $V = U$ as function of the average impurity occupancy. Right panel: the impurity DOS along the path drawn in the phase diagram. (From Ref. [43])

in Fig 15. We notice that, as soon as we move away from particle-hole symmetry, the pseudo-gap becomes asymmetric and is gradually filled until, above a critical v, it turns into a Kondo resonance.

The NRG spectrum shows that across the transition there is still a $\pi/2$ jump of the phase shifts, even though on both phases the phase shift is in between 0 and $\pi/2$. Namely, if we assume in the unscreened phase a phase shift δ_-, in the screened phase $\delta_+ = \pi/2 + \delta_-$. In addition we know that the scattering S-matrix is

$$\Re e\, S_\pm = \cos 2\delta_\pm = 1 - 2\frac{\rho_\pm}{\rho_0},$$

from which we obtain

$$\rho_+ - \rho_- = \rho_0 \cos 2\delta_- .$$

From the definition of the phase-shifts in terms of the Green's function, we guessed a model DOS

$$\rho_\pm(\varepsilon) = \frac{\rho_0}{2}\left[\frac{T_+^2 + \mu_\pm^2}{(\varepsilon + \mu_\pm)^2 + T_+^2} \pm \cos 2\delta_- \frac{T_-^2}{\varepsilon^2 + T_-^2}\right], \tag{117}$$

with $\mu_\pm = \pm T_+ \sin 2\delta_-$ which fits well the numerical results. We notice from (117) that the second lorentian is always pinned at the chemical potential, so that the filling of the pseudo-gap occurs because the broad resonance is not anymore symmetric.

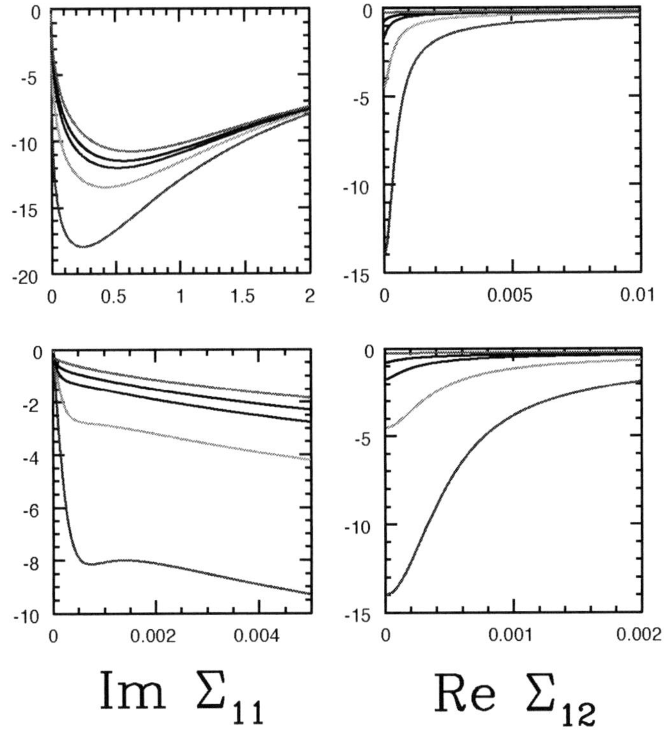

FIGURE 16. Left panels: diagonal self-energy on different energy ranges. Right panels: off-diagonal self-energy. (From Ref. [45])

5.1.5. A more realistic dimer model

We have seen that the unstable fixed point of the dimer model (99) is destabilized by a direct hopping between the two impuritites. This is however the most realistic situation, and would actually correspond to the impurity model onto which for instance an Hubbard model should be mapped within cluster DMFT. Therefore one may wander how much of the critical behavior survives if, instead of (99) we considered the impurity model

$$\begin{aligned} \mathscr{H} &= \sum_{\mathbf{k}\sigma}\sum_{a=1}^{2} \varepsilon_{\mathbf{k}}\, c^\dagger_{\mathbf{k}\,a\sigma} c_{\mathbf{k}\,a\sigma} + \sum_{\mathbf{k}\sigma}\sum_{a=1}^{2}\left(V_{\mathbf{k}}\, c^\dagger_{\mathbf{k}\,a\sigma} d_{a\sigma} + H.c.\right) \\ &+ \frac{U}{2}\sum_{a=1}^{2}\left(n_a - 1\right)^2 - t\sum_{\sigma}\left(d^\dagger_{1\sigma} d_{2\sigma} + H.c.\right). \end{aligned} \tag{118}$$

As we have mentioned several times, the approach to the Mott transition corresponds to an impurity model deep inside the Kondo regime, namely with $U \gg V_{\mathbf{k}}, t$. In this limit, (118) can be mapped by a Schrieffer-Wolff transformation onto the Kondo limit of (99), with $V = 0$. The Kondo exchange J_K is the same in both models, the impurity exchange is $J = 4t_{\perp}^2/U$, However, at next order in $1/U$, a local direct hybridization V_x among the baths is generated, with $V_x \sim J_k t_{\perp}/U$. Although small, V_x is a relevant perturbation which makes the unstable fixed point of model (99) unaccessible in model (118). Yet, since $J \gg V_x$, one may expect that a quantum critical region, but not a true quantum critical point, is still well visible.

This is indeed the case. Notice that, in the presence of the hopping, the impurity self-energy has a diagonal part, $\Sigma_{11} = \Sigma_{22}$, which we find is purely imaginary in Matsubara frequencies, and an off diagonal one, $\Sigma_{12} = \Sigma_{21}^*$, which turns out to be real. In Fig. 16 we draw the impurity self energy $\Sigma(i\varepsilon)$ as function of ε with $U = 8$, $t = 0.05$, hence $J = 1.25 \times 10^{-3}$, and $\Gamma = 0.5 \div 0.3$ from the top curve to the bottom one, namely $J_K = 0.08 \div 0.05$, in units of half the conduction bandwidth. [43] There are several interesting things to notice. The first is that a quite sharp cross-over from an "unscreened" phase, with a large $\Im m\Sigma_{11}$ to a "screened" one, with a small $\Im m\Sigma_{11}$, is visible. However, even if in the "unscreened" phase the $\Im m\Sigma_{11}$ seems to be diverging coming from high energy, at very low energy it upturns into a conventional Fermi-liquid behavior, with $\Im m\Sigma_{11} \propto -\varepsilon$, although with a very large slope. Also interesting is the behavior of $\Re e\Sigma_{12}$. Indeed, in spite of the fact that the off-diagonal term introduced in the Hamiltonian, $t = 0.05$, is very small, the off-diagonal self-energy becomes very large at small energy, so to compensate the big slope of the diagonal one, inversely proportional to Z. This behavior indicates that, although the fixed point is not accessible, still the cross-over between the two extreme regimes is sharp. This is further confirmed by the plot of the diagonal DOS in the same range of Γ's, shown in Fig 17. We see that the DOS moves from a Kondo-like behavior into a pseudo-gaped one quite sharply.

5.2. The dimer model versus DMFT

In summary, we have shown that a dimer of Anderson impurities includes generally two regimes. One is dominated by the Kondo effect, while in the other the inter-impurity coupling prevails. Consequently, in the former the impurity DOS have a Kondo resonance, but in the latter the DOS have a deep or even a pseudo-gap. In the highly symmetric model (99), which possesses an orbital $O(2)$ symmetry, these two regimes are separated by a true critical point. Right at the critical point there are singular susceptibilities in three relevant channels, namely the staggered magnetization, Eq. (107), the channel hybridization, Eq. (106), and the inter-channel single Cooper pair, see Eq. (105). In the more realistic model (118), which inludes by construction the relevant interchannel hybridization, the two extreme regimes are separated by a crossover, which however turns out to be quite sharp. This suggests that, although none of the above mentioned susceptibilities ever diverges, yet there should be a finite region where they are strongly enhanced, manifestation of the avoided criticality. Can we make any prediction out of these results about a lattice model which maps by DMFT just on the same class of im-

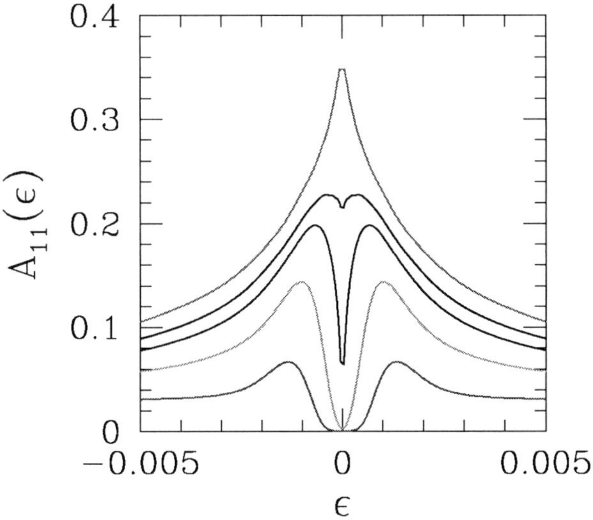

FIGURE 17. Diagonal component of the DOS. (From Ref. [43])

purity dimers.

Let us just concentrate on a lattice model which has been actually worked out by DMFT, [44] and corresponds to the lattice generalization of (99) with $U = V$, namely with an Hamiltonian

$$\mathscr{H} = -t \sum_{<\mathbf{RR'}>\,\sigma} \sum_{a=1}^{2} \left(c^{\dagger}_{\mathbf{R}a\sigma} c^{\dagger}_{\mathbf{R'}a\sigma} + H.c. \right) + \frac{U}{2} \sum_{\mathbf{R}} (n_{\mathbf{R}} - 2)^2 + J \sum_{\mathbf{R}} \mathbf{S}_{\mathbf{R}1} \cdot \mathbf{S}_{\mathbf{R}2}, \quad (119)$$

which describes a two-band Hubbard model with an inverted exchange splitting favoring an inter-orbital singlet state on each site. Within the original single-site DMFT, (119) maps onto (99), which we know has the right symmetry to encounter the unstable fixed point. The behavior of the Kondo temperature of the impurity after DMFT self-consistency has been carried out can be depicted as in Fig. 18. At half-filling, T_K is decreasing with U until, at a critical value, it should vanish, signalling the Mott transition. However, before that happens, T_K necessarily becomes of order J which corresponds to the location of the impurity critical point. Analogously, if we start at half-filling from the Mott insulator and dope it, T_K from its insulating value $T_K = 0$, will gradually increase with doping, until, at a critical doping, it will cross the value $T_K \simeq J$ of the critical point, which we know can be crossed also varying the impurity average occupancy. In both cases, the impurity model encounters its critical point before the MIT, and, from what we previously mentioned, one should expect, around this point, a region with symmetry breaking within one of the instability channels of the impurity.

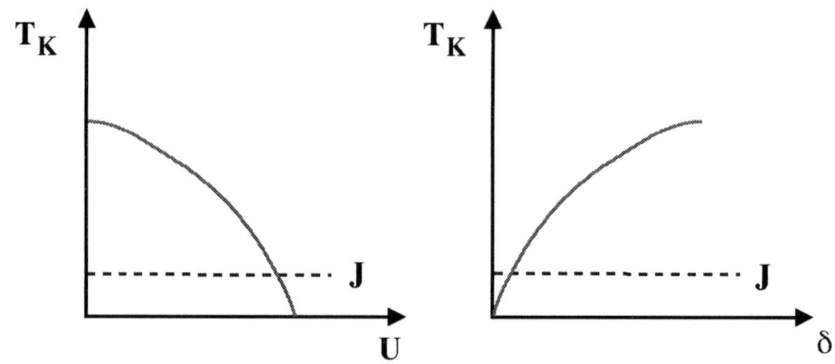

FIGURE 18. Behavior of the DMFT self-consistent Kondo temperature as compared to the inverted exchange J

The staggered magnetization (105) would likely correspond to an antiferromagnetic ordering with opposite staggered magnetization in the two orbitals, while (106) to an inter-orbital dimerization, presumably also staggered. Both these instabilities occur in the particle-hole channel. The last one, (107), occurs in the particle-particle inter-orbital singlet channel and would correspond to s-wave superconductivity. Since the irreducible vertices in all the three channels are equally singular, which of these instabilities really occurs mainly depends on the Green's function bubbles

$$T\sum_n \sum_{\mathbf{k}} G(i\varepsilon_n + i0^+, \mathbf{k}+\mathbf{Q})\, G(i\varepsilon_n, \mathbf{k}),$$

in the particle-hole channels, where $\mathbf{Q}$ is the modulation wave-vector, as opposed to

$$T\sum_n \sum_{\mathbf{k}} G(i\varepsilon_n, \mathbf{k})\, G(-i\varepsilon_n, -\mathbf{k}),$$

in the Cooper one. While the latter is always singular in a Fermi-liquid phase, the former is singular only if nesting occurs at some $\mathbf{Q}$, or if a van Hove singularity exists at the chemical potential, in which case $\mathbf{Q} = 0$. If we exclude both these events, which can always be done within DMFT by simply preventing particle-hole instabilities in the self-consistency procedure, only superconductivity might emerge. However, within model (119), we should not expect any room for a superconducting instability, especially near the MIT. In fact, the "bare" scattering amplitude in the inter-orbital singlet Cooper channel,

$$A_0^0 = U - 3J/4, \tag{120}$$

is always repulsive near the MIT, assuming $J \ll W \simeq U$. This naive expectation is actually wrong. In Fig. 19 we draw the anomalous superconducting self-energy at zero-frequency and half-filling as function of U/W for different J's. [44] For large J/W, as U increases, the anomalous self-energy decreases, in agreement with (120), until a first order phase transition from a superconductor to a Mott insulator takes place. For intermediate couplings, the first order transition remains but the off-diagonal self-energy starts to behave non-monotonically, namely after a first decrease, it increases again.

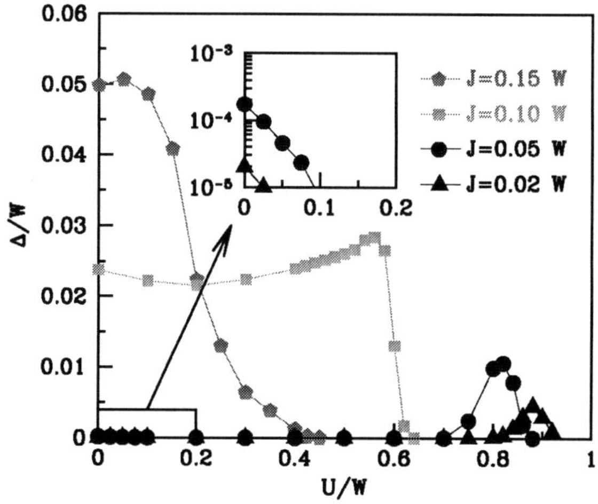

FIGURE 19. Anomalous self-energy at zero frequency as function of U/W for different J's. (From Ref. [44])

More interesting and surprising is the weak J/W behavior. Here the off-diagonal self-energy decreases until, roughly when $U = 3J/4$, see (120), superconductivity disappears into a normal metal behavior. However, just before the MIT, superconductivity reappears with an anomalous self-energy reaching values that are orders of magnitude larger than within the BCS $U = 0$ domain. This behavior might suggest, as we speculated, that this superconducting region before the MIT is just the response of the lattice model to the impurity critical behavior. Indeed, if we prevent superconductivity and calculate numerically the self-energy of the metastable metallic solution, which could be identified as the hypothetical normal phase, we find a behavior very similar to that one of the impurity. By fitting the calculated self-energy with the model one, Eq. (114), we can even extract the values of T_- and T_+ as shown in Fig. 20. [44] Several things are worth noticing. First, the Landa-Fermi liquid behavior breaks down and turns, before the MIT, into a pseudo-gaped phase. Next, the maximum of the anomalous self-energy occurs right at the point $T_- = 0$, as expected. Finally, the MIT is not accompanied by a vanishing energy scale, as in the Brinkmann-Rice scenario, [31] but occurs when $T_+ = T_-$ within the pseudo-gaped regime. Notice that, the pseudo-gaped phase is semi-metallic, with vanishing Drude weight. On the other hand, this "normal" phase is unstable to superconductivity, which does have a finite Drude weight which vanishes only at the MIT. This implies that, in the pseudo-gaped regime, the on-set of superconductivity is accompanied by an increase of Drude weight, unlike what happens

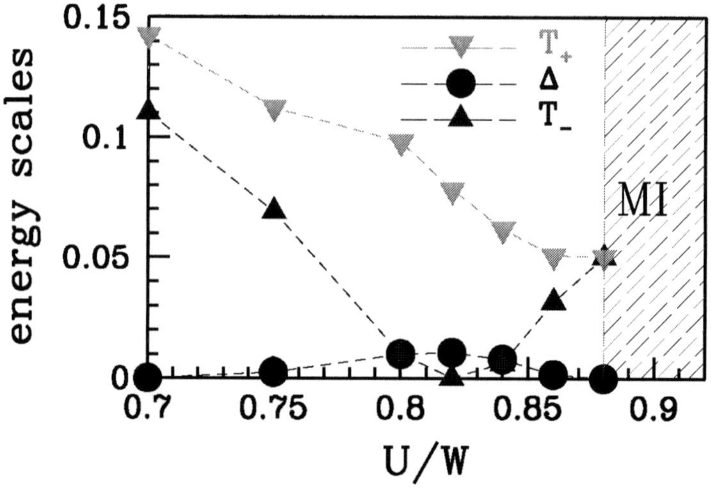

FIGURE 20. Fit values of T_+ and T_- of the normal phase. Also shown is the off-diagonal self-energy. (From Ref. [44])

in the BCS regime. This increase of Drude weight, has a counterpart in the impurity model. At the fixed point, the impurity model has a finite entropy, implying that the conduction bath is unable to fully screen the impurity degrees of freedom. If one allows for a bath where anyone of the relevant symmetries is broken, new screening channels open up, that are able to quench also the residual fixed point entropy. The fixed point therefore is washed out, but screening energy is gained. Translated in the lattice model, this screening energy gain does correspond to band energy gain.

In conclusion, at least in model (119), the physics of the underneath impurity model translates nicely into the behavior of the lattice model even after full DMFT-self-consistency. This is further supported by the full phase diagram as function of U/W and doping away from half-filling, shown in Fig. 21. [44]

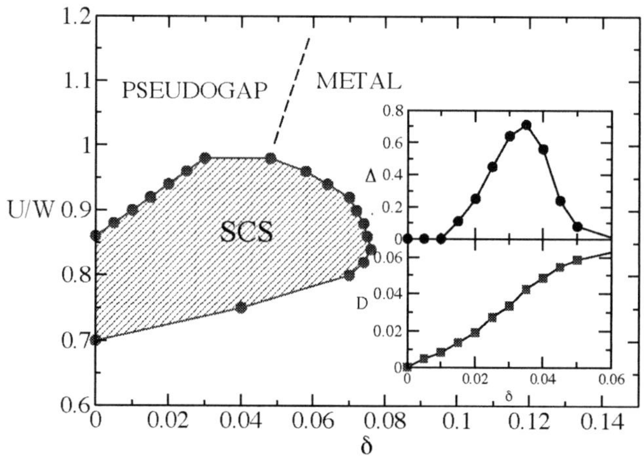

FIGURE 21. Phase diagram of (119) with $J/W = 0.02$ as function of U/W and doping $\delta = 2 - n$. (From Ref. [44])

ACKNOWLEDGMENTS

I am pleased to acknowledge all my collaborators who have contributed to the research activity I have presented in these lecture notes: M. Ferrero, L. De Leo, M. Capone, A.F. Ho, P. Lecheminant, G.E. Santoro, C. Castellani and E. Tosatti.

A. BRIEF INTRODUCTION TO SCATTERING THEORY

We give here a brief introduction to scattering theory to put in evidence why the existence of local moments in a metal for $T \ll T_F$ is so puzzling.

Let us consider an impurity imbedded in a normal metal. We only take into account the valence band, described by the Hamiltonian

$$\mathscr{H}_0 = \sum_{\mathbf{k},\sigma} \varepsilon_{\mathbf{k}}\, c^\dagger_{\mathbf{k}\sigma} c_{\mathbf{k}\sigma}. \tag{121}$$

The scattering potential provided by the impurity has the general form

$$\mathscr{V} = \sum_{\sigma} \sum_{\mathbf{k},\mathbf{p}} V_{\mathbf{kp}}\, c^\dagger_{\mathbf{k}\sigma} c_{\mathbf{p}\sigma}, \tag{122}$$

where $V_{\mathbf{kp}}$ are the matrix elements of the impurtity potential onto the valence band Block waves.

The single-particle Green's function in complex frequency for the full Hamiltonian $\mathscr{H} = \mathscr{H}_0 + \mathscr{V}$ can be formally written as the matrix

$$\hat{G}(z) = \frac{1}{z - \hat{H}}. \tag{123}$$

Analogously the unperturbed one is

$$\hat{G}_0(z) = \frac{1}{z - \hat{H}_0}. \tag{124}$$

The above operators have the following interpretation. If we consider for instance the inverse of the Green's function $\hat{G}$, namely

$$\hat{G}(z)^{-1} = z - \hat{H},$$

in the basis set of the valence band Block waves, this is the matrix

$$\left(\hat{G}(z)^{-1}\right)_{\mathbf{k}\sigma,\mathbf{p}\sigma'} = \delta_{\sigma\sigma'}\,\left[\delta_{\mathbf{kp}}\,z - \delta_{\mathbf{kp}}\,\varepsilon_{\mathbf{k}} - V_{\mathbf{kp}}\right], \tag{125}$$

hence the Green's function is the inverse of the above matrix. Suppose that we have diagonalized the full Hamiltonian and got the eigenvalues ε_a's. In the diagonal basis also the Green's function is diagonal with matrix elements

$$G(z)_{ab} = \delta_{ab}\,\frac{1}{z - \varepsilon_a}.$$

Therefore we readily find that

$$-\frac{1}{\pi}\lim_{\delta\to+0}\Im m\,\mathrm{Tr}\left(\hat{G}(z = \omega + i\delta)\right) = -\frac{1}{\pi}\lim_{\delta\to+0}\Im m\sum_{a\sigma}\frac{1}{\omega - \varepsilon_a + i\delta}$$
$$= \sum_{a\sigma}\delta(\omega - \varepsilon_a) \equiv \rho(\omega),$$

where ω is a real frequency and $\rho(\omega)$ is the density of states (DOS). Since the trace is invariant under unitary transformations, hence also under the transformation which diagonalizes the Hamiltonian, it is generally true that

$$-\frac{1}{\pi}\lim_{\delta\to+0}\Im m\,\mathrm{Tr}\left[\hat{G}(z = \omega + i\delta)\right] = \rho(\omega). \tag{126}$$

Analogously

$$-\frac{1}{\pi}\lim_{\delta\to+0}\Im m\,\mathrm{Tr}\left[\hat{G}_0(z = \omega + i\delta)\right] = \rho_0(\omega) = \sum_{\mathbf{k}\sigma}\delta(\varepsilon_{\mathbf{k}} - \omega), \tag{127}$$

gives the DOS of the host metal.

Let us formally write the full Green's function as

$$\begin{aligned}\hat{G} &= \frac{1}{z - \hat{H}_0 - \hat{V}} = \frac{1}{\hat{G}_0^{-1} - \hat{V}}\\ &= \frac{1}{\hat{1} - \hat{G}_0\hat{V}}\,\hat{G}_0 = \hat{G}_0 + \hat{G}_0\hat{V}\,\left[\hat{1} - \hat{G}_0\hat{V}\right]^{-1}\hat{G}_0\end{aligned}$$

$$\equiv \hat{G}_0 + \hat{G}_0 \hat{T} \hat{G}_0, \tag{128}$$

which provides the definition of the so-called T-matrix, namely

$$\hat{T}(z) = \hat{V} \left[\hat{1} - \hat{G}_0 \hat{V}\right]^{-1}. \tag{129}$$

From now on we will assume $z = \omega + i\delta$ with δ a positive infinitesimal number, so we will not explicitly indicate the $\lim_{\delta\to+0}$. We notice that

$$\frac{\partial}{\partial z} \ln \hat{G}(z) = -\hat{G}(z),$$

so that

$$\rho(\omega) = \frac{1}{\pi}\frac{\partial}{\partial z}\left\{\Im m \mathrm{Tr} \ln \hat{G}(z)\right\}. \tag{130}$$

We are interested in the variation of the DOS induced by the impurity, which, by making use of (128) and (129), is given by

$$\begin{aligned} \Delta\rho(\omega) &= \rho(\omega) - \rho_0(\omega) \\ &= \frac{1}{\pi}\frac{\partial}{\partial z}\left\{\Im m \mathrm{Tr}\ln\left[\hat{G}(z)\,\hat{G}_0(z)^{-1}\right]\right\} \\ &= \frac{1}{\pi}\frac{\partial}{\partial z}\left\{\Im m \mathrm{Tr}\ln\left[\frac{1}{\hat{1} - \hat{G}_0(z)\hat{V}}\right]\right\} \\ &= \frac{1}{\pi}\frac{\partial}{\partial z}\left\{\Im m \mathrm{Tr}\ln \hat{V}^{-1}\hat{T}(z)\right\}. \end{aligned} \tag{131}$$

We define the matrix of the scattering phase shifts

$$\hat{\delta}(z) = \Im m \ln \hat{V}^{-1}\hat{T}(z) = \mathrm{Arg}\left(\hat{V}^{-1}\hat{T}(z)\right), \tag{132}$$

through which

$$\Delta\rho(\omega) = \frac{1}{\pi}\frac{\partial}{\partial\omega}\mathrm{Tr}\hat{\delta}(\omega). \tag{133}$$

We notice that for $|z| \to \infty$, $\hat{G}_0(z) \to 1/z \to 0$, hence $\hat{T}(z) \to \hat{V}$ and $\hat{\delta}(z) \to 0$.

The variation of the total number of electrons, ΔN_{els}, induced by the impurity at fixed chemical potential μ is therefore

$$\Delta N_{els} = \int_{-\infty}^{\mu} d\omega\, \Delta\rho(\omega) = \frac{1}{\pi}\mathrm{Tr}\hat{\delta}(\mu), \tag{134}$$

which is the so-called Friedel sum rule.

Let us go back to the definition of the T-matrix (129). One readily finds that its inverse is given by

$$\hat{T}(z)^{-1} = \left[\hat{V}^{-1} - \hat{G}_0(z)\right].$$

Since the Hamiltonian is hermitean, then

$$\left[\hat{T}(z)^{-1}\right]^{\dagger} = \left[\hat{V}^{-1} - \hat{G}_0(z^*)\right],$$

so that

$$\left[\hat{T}(z)^{-1}\right]^{\dagger} - \hat{T}(z)^{-1} = \left[\hat{G}_0(z) - \hat{G}_0(z^*)\right].$$

Therefore

$$\hat{T}(z)^{\dagger}\left\{\left[\hat{T}(z)^{-1}\right]^{\dagger} - \hat{T}(z)^{-1}\right\}\hat{T}(z) = \hat{T}(z) - \hat{T}(z)^{\dagger}$$
$$= \hat{T}(z)^{\dagger}\left[\hat{G}_0(z) - \hat{G}_0(z^*)\right]\hat{T}(z). \tag{135}$$

Since

$$\hat{G}_0(z) - \hat{G}_0(z^*) = -2\pi i\delta\left(\omega - \hat{H}_0\right),$$

the Eq. (135) implies the following identity

$$T_{\mathbf{kp}}(\omega) - T^{\dagger}_{\mathbf{kp}}(\omega) = -2\pi i\sum_{\mathbf{q}} T^{\dagger}_{\mathbf{kq}}(\omega)\,\delta(\omega - \varepsilon_{\mathbf{q}})\,T_{\mathbf{qp}}(\omega), \tag{136}$$

which is the so-called *optical theorem*. It also shows that the imaginary part of the T-matrix is finite only within the conduction band.

Let us now analyse the on-shell T-matrix

$$T_{\mathbf{kp}}(\varepsilon),$$

where $\varepsilon_{\mathbf{k}} = \varepsilon_{\mathbf{p}} = \varepsilon$. We can rewrite the on-shell optical theorem as follows

$$-2\pi i\,\delta\left(\varepsilon_{\mathbf{k}} - \varepsilon_{\mathbf{p}}\right)\left[T_{\mathbf{kp}}(\varepsilon) - T^{\dagger}_{\mathbf{kp}}(\varepsilon)\right]$$
$$= (-2\pi i)^2\,\delta\left(\varepsilon_{\mathbf{k}} - \varepsilon_{\mathbf{p}}\right)\sum_{\mathbf{q}} T^{\dagger}_{\mathbf{kq}}(\varepsilon)\,\delta(\varepsilon_{\mathbf{k}} - \varepsilon_{\mathbf{q}})\,T_{\mathbf{qp}}(\varepsilon).$$

The above equation implies that, if we introduce the so-called on-shell S-matrix through

$$S_{\mathbf{kp}}(\varepsilon) = \delta_{\mathbf{kp}} - 2\pi i\,\delta\left(\varepsilon_{\mathbf{k}} - \varepsilon_{\mathbf{p}}\right)T_{\mathbf{kp}}(\varepsilon), \tag{137}$$

where, as before, $\varepsilon_{\mathbf{k}} = \varepsilon_{\mathbf{p}} = \varepsilon$, then it follows that the S-matrix is unitary, *i.e*

$$\hat{S}\hat{S}^{\dagger} = \hat{I}. \tag{138}$$

$S_{\mathbf{kp}}(\varepsilon)$ is the transition amplitude that an electron in state $\mathbf{k}$ scatters elastically into state $\mathbf{p}$. Since it is unitary, it follows that only elastic scattering survives.

A.0.1. One dimensional example

The simplest example of scattering theory is provided by free electrons moving in one dimension and scettering off an impurity potential. For sake of simplicity we assume that the impurity sits at the origin, $x = 0$, and that the scattering potential $V(x)$ is short rage and parity-invariant, $V(x) = V(-x)$. Therefore the eigenvectors can be classified into even and odd with respect to $x \to -x$. An even eigenvector for vary large $|x|$ can be taken to be of the form

$$\psi_{k,e}(x) = \begin{cases} x \gg 0 & \cos(kx + \delta_e) \\ x \ll 0 & \cos(kx - \delta_e) \end{cases}, \tag{139}$$

while the degenerate odd eigenvector

$$\psi_{k,o}(x) = \begin{cases} x \gg 0 & \sin(kx + \delta_o) \\ x \ll 0 & \sin(kx - \delta_o) \end{cases}. \tag{140}$$

If we define incoming, $\psi_{in,e(o)}$ and outgoing, $\psi_{out,e(o)}$, waves with given parity through

$$\begin{aligned} \psi_{in,e(o)}(x) &= \begin{cases} x \gg 0 & \pm e^{-ikx} \\ x \ll 0 & e^{ikx} \end{cases} \\ \psi_{out,e(o)}(x) &= \begin{cases} x \gg 0 & e^{ikx} \\ x \ll 0 & \pm e^{-ikx} \end{cases}, \end{aligned}$$

where the $\pm$ refers to even/odd, it follows by (139) and (140) that, in the scattering process,

$$\psi_{in,e(o)}(x) \to e^{2i\delta_{e(o)}}\, \psi_{out,e(o)}(x) \equiv S_{e(o)}\, \psi_{out,e(o)}(x), \tag{141}$$

that defines the scattering S-matrix. In addition one can take linear combinations of (139) and (140) and finds the two eigenstates

$$\begin{aligned} \psi_{<,e(o)}(x) &= \begin{cases} x \gg 0 & e^{-ikx} + r\, e^{ikx} \\ x \ll 0 & t\, e^{-ikx} \end{cases} \\ \psi_{>,e(o)}(x) &= \begin{cases} x \gg 0 & t\, e^{ikx} \\ x \ll 0 & e^{ikx} + r\, e^{-ikx} \end{cases}, \end{aligned}$$

with the reflection and transmission amplitudes given by

$$r = i\, e^{i(\delta_e + \delta_o)} \sin(\delta_e - \delta_o), \qquad t = e^{i(\delta_e + \delta_o)} \cos(\delta_e - \delta_o), \tag{142}$$

hence with total reflection and transmission

$$R = \sin^2(\delta_e - \delta_o), \qquad T = \cos^2(\delta_e - \delta_o). \tag{143}$$

Now let us impose boundary conditions far away from the impurity. For instance we can assume that the wavefunctions must vanish for $x = \pm L$. One readily finds that this imposes a quantization of momentum, so that, for the even channel

$$k_e(n) = \left(n + \frac{1}{2}\right)\frac{\pi}{L} - \frac{\delta_e}{L}, \qquad E(k_e) = E_e(n), \tag{144}$$

while, for the odd one,

$$k_o(n) = n\frac{\pi}{L} - \frac{\delta_o}{L}, \qquad E(k_o) = E_o(n). \tag{145}$$

In other words the scattering phase-shifts modify the level spectrum, or, conversely, that from the level spectrum one can determine the phase-shifts.

A.0.2. Three dimensional example

Let us consider the simpler case of a spherical Fermi surface, *i.e.* $\varepsilon_{\mathbf{k}} = \varepsilon_k$ depending only on the modulous of the wavevector. In addition we assume that the matrix elements $V_{\mathbf{kp}}$ only depend on the angle between the two wavevectors, $\theta_{\mathbf{kp}}$, hence can be expanded in Lagrange polynomials

$$V_{\mathbf{kp}} = \sum_l V_l\,(2l+1)\,P_l\left(\cos\theta_{\mathbf{kp}}\right). \tag{146}$$

Then also the T-matrix has a similar expansion

$$T_{\mathbf{kp}}(z) = \sum_l t_l(z)\,(2l+1)\,P_l\left(\cos\theta_{\mathbf{kp}}\right). \tag{147}$$

Since

$$\int \frac{d\Omega_{\mathbf{q}}}{4\pi} P_l\left(\cos\theta_{\mathbf{kq}}\right) P_{l'}\left(\cos\theta_{\mathbf{qp}}\right) = \frac{1}{2l+1}\,\delta_{ll'}\,P_l\left(\cos\theta_{\mathbf{kp}}\right),$$

the optical theorem transforms into

$$t_l\left(\omega\right) - t_l^*(\omega) = 2i\mathscr{I}m\,t_l\left(\omega\right) = -2\pi i\,\rho(\omega)\,\left|t_l\left(\omega\right)\right|^2, \tag{148}$$

where $\rho(\omega)$ is the bare conduction electron density of states per spin. Since

$$t_l(\omega) = |t_l(\omega)|\,\mathrm{e}^{i\delta_l(\omega)},$$

it follows that

$$t_l(\omega) = -\frac{1}{\pi\rho(\omega)}\,\sin\delta_l(\omega)\,\mathrm{e}^{i\delta_l(\omega)}, \tag{149}$$

and

$$S_l(\omega) = \mathrm{e}^{2i\delta_l(\omega)}. \tag{150}$$

If the concentration of impurities is very low, then the scattering rate suffered by the electrons is given by the Fermi-golden rule with the T-matrix, namely

$$\frac{1}{\tau_{\mathbf{k}}} = 2\pi n_i \sum_{\mathbf{q}} T^\dagger_{\mathbf{kq}}(\omega)\,\delta\left(\omega - \varepsilon_{\mathbf{q}}\right)\,T_{\mathbf{qk}}(\omega)\,\left(1 - \cos\theta_{\mathbf{kq}}\right),$$

where n_i is the impurity concentration and the last term is a geometric factor which guarantees that forward scattering, $\theta = 0$, does not contribute to the current dissipation. Since

$$(2l+1)\,z P_l(z) = (l+1)\,P_{l+1}(z) + l\,P_{l-1}(z),$$

we obtain

$$\frac{1}{n_i\,\tau_{\mathbf{k}}(\omega)} = 2\pi\,\rho(\omega)\sum_{ll'} t_l^*(\omega)\,t_{l'}(\omega)\int\frac{d\Omega_{\mathbf{q}}}{4\pi}(2l+1)\,P_l(\cos\theta_{\mathbf{kq}})$$
$$\left[(2l'+1)\,P_{l'}(\cos\theta_{\mathbf{kq}}) + (l'+1)\,P_{l'+1}(\cos\theta_{\mathbf{kq}}) + l'\,P_{l'-1}(\cos\theta_{\mathbf{kq}})\right]$$

$$= 2\pi \rho(\omega) \sum_l (2l+1) |t_l(\omega)|^2$$

$$+ l t_l^*(\omega) t_{l-1}(\omega) + (l+1) t_l^*(\omega) t_{l+1}(\omega), \qquad (151)$$

which, as expected, gives a scattering rate independent of momentum. If only scattering at fixed l occurs, then

$$\frac{1}{\tau(\omega)} = 2\pi n_i \rho(\omega) (2l+1) |t_l(\omega)|^2 = \frac{2}{\pi \rho(\omega)} (2l+1) \sin^2 \delta_l(\omega), \qquad (152)$$

thus leading to a zero-temperature residual resistivity proportional to $\sin^2 \delta_l(0)$.

A.1. General analysis of the phase-shifts

Let us keep assuming a spherically symmetric case and in addition only s-wave scattering. Then the phase shift is $\delta_l(\omega) = \delta_{l0}\, \delta(\omega)$ defined through the $l = 0$ component of the T-matrix

$$\delta(\omega) = \tan^{-1} \frac{\Im m\, t(\omega)}{\Re e\, t(\omega)}.$$

An important role is played by the frequencies at which the real part vanishes.

The first possibility is that this occurs outside the conduction band, either below or above. In this case we know that the imaginary part is zero. This implies that the phase-shift jumps by π at these energies so that the variation of the DOS is δ-like. In this case one speaks about bound states which appear outside the conduction band. Clearly this possibility can not explain a Curie-Weiss behavior for $T \ll T_F$. Indeed at very low temperature the bound state is either doubly occupied, if below the conduction band, or empty, if above, and one needs a temperature larger than the conduction bandwidth, hence larger that T_F, to release its spin entropy.

The other possibility is that the real part vanishes at a frequency ω_* within the conduction band where the imaginary part is non zero. Around ω_* and assuming a real part linearly vanishing and an imaginary part roughly constant we get

$$\delta(\omega) \simeq \tan^{-1} \frac{\Gamma}{\omega_* - \omega},$$

which implies a phase-shift $\delta_L(\omega_*) = \pm\pi/2$ and leads to a Lorentian DOS variation

$$\Delta\rho(\omega) \sim \frac{1}{\pi} \frac{\Gamma}{(\omega - \omega_*)^2 + \Gamma^2}. \qquad (153)$$

This is called a *resonance*. Clearly, in other for this resonance to contribute at $T \ll T_F$, ω_* should be very close to the chemical potential, in particular $|\omega_* - \mu| \ll T_K$, and, in addition, $\Gamma \simeq T_K$.

Let us therefore assume $\Gamma = T_K$ and for further simplicity that the resonance is right at the chemical potential, i.e. $\omega_* = 0$. The contribution of the resonant state to the magnetic susceptibility is then given by

$$\Delta\chi(T) = -\mu_B g \int d\varepsilon \, \frac{\partial f(\varepsilon)}{\partial \varepsilon} \frac{1}{\pi} \frac{T_K}{\varepsilon^2 + T_K^2}, \tag{154}$$

where

$$f(\varepsilon) = \left(1 + e^{\beta\varepsilon}\right)^{-1},$$

is the Fermi distribution. One readily finds that for $T \gg T_K$,

$$\Delta\chi \simeq \mu_B g \frac{1}{T},$$

while for $T \ll T_K$

$$\Delta\chi \simeq \mu_B g \frac{1}{\pi T_K},$$

similar to the observed behavior. In addition the resistivity would be given through (151) by ($\rho(\omega) \sim \rho(0) \equiv \rho_0$)

$$\begin{aligned} R(0) \quad &\propto \quad \frac{1}{\tau(0)} = n_i \frac{2}{\pi\rho_0} \sin^2\delta(0) \\ &= \quad n_i \frac{2}{\pi\rho_0}, \end{aligned} \tag{155}$$

again compatible with the almost unitary limit, $\delta(0) = \pi/2$, observed experimentally. Therefore the existence of a narrow resonance near the chemical potential with width exactly given by T_K would seem the natural explanation to what experiments find.

However this is evidently very strange, since the Kondo behavior is observed in many different host metals as well as for different magnetic impurities, hence it would be really surprising that in all these cases a resonance appears always pinned near the chemical potential.

Moreover this simple single-particle scenario fails to explain the entropy released above T_K. Indeed, for $T \gg T_K$ the resonance is effectively like an isolated level which can be empty, singly occupied with a spin up or down, or doubly occupied. Therefore its entropy should be $S = 2\ln 2$. On the other hand the experiments tells us that only spin and orbital degrees of freedom are released above T_K, which amounts in the above simplified model to an entropy $S = \ln 2$. In other words a resonance at the chemical potential at temperatures larger than its broadening is not at all the same as a local moment, since the former does have valence fluctuations which are absent in the latter.

Therefore, although the resonance scenario is suggestive, it is not at all the solution to the puzzle.

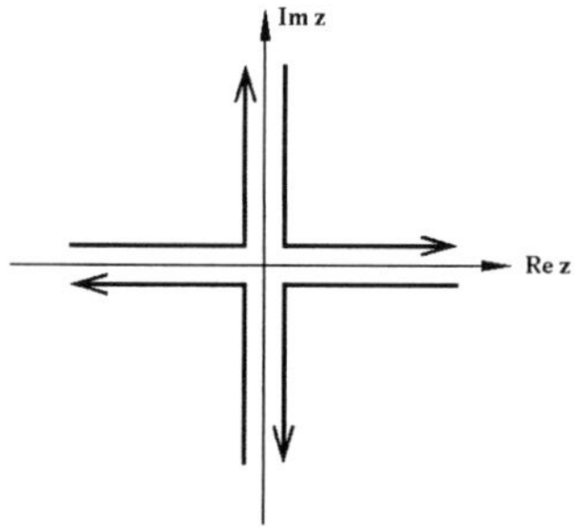

FIGURE 22. Integration contour.

B. THE RESONANT LEVEL MODEL

In this section we will derive some properties of the AIM in the absence of interation, which is also called a resonant level model.

B.1. Variation of the electron number

The variation of the electron number per spin due to the impurity is known to be related to the Green's functions through

$$\begin{aligned}\Delta N_{els} &= T\sum_n \left[\mathcal{G}(i\omega_n) + \sum_{\mathbf{k}} G_{\mathbf{kk}}(i\omega_n) - G^{(0)}_{\mathbf{kk}}(i\omega_n)\right] e^{i\omega_n 0^+} \\ &= T\sum_n \frac{1}{i\omega_n - \varepsilon_d - \Delta(i\omega_n)}\left[1 + \sum_{\mathbf{k}} |V_{\mathbf{k}}|^2 \left(\frac{1}{i\omega_n - \varepsilon_{\mathbf{k}}}\right)^2\right] \\ &= T\sum_n \frac{\partial}{\partial i\omega_n} \ln\left(i\omega_n - \varepsilon_d - \Delta(i\omega_n)\right), \qquad (156)\end{aligned}$$

Let us consider the contour drawn in Fig. 22. The function $S(z) = \ln(z - \varepsilon_d - \Delta(z))$ has a branch cut on the real axis and it is analytic everywhere else. Therefore

$$\oint \frac{dz}{2\pi i} f(z) \frac{\partial}{\partial z} S(z) = 0,$$

because in the region which excludes both axes the function is analytic. On the other hand

$$\begin{aligned}0 &= \oint \frac{dz}{2\pi i} f(z) \frac{\partial S(z)}{\partial z} = T\sum_n \frac{\partial}{\partial i\omega_n} \ln\left(i\omega_n - \varepsilon_d - \Delta(i\omega_n)\right) \\ &+ \int_{-\infty}^{+\infty} \frac{d\varepsilon}{2\pi i} f(\varepsilon) \frac{\partial}{\partial \varepsilon}\left[S(\varepsilon + i0^+) - S(\varepsilon + i0^+)\right]\end{aligned}$$

Therefore we finally get

$$\begin{aligned}
\Delta N_{els} &= -\int_{-\infty}^{+\infty} \frac{d\varepsilon}{2\pi i} f(\varepsilon) \frac{\partial}{\partial \varepsilon} \left[S(\varepsilon + i0^+) - S(\varepsilon - i0^+) \right] \\
&= \int_{-\infty}^{+\infty} \frac{d\varepsilon}{2\pi i} f(\varepsilon) \frac{\partial}{\partial \varepsilon} \ln \frac{\mathscr{G}(\varepsilon + i0^+)}{\mathscr{G}(\varepsilon - i0^+)} \\
&= \int_{-\infty}^{+\infty} \frac{d\varepsilon}{\pi} f(\varepsilon) \frac{\partial}{\partial \varepsilon} \delta(\varepsilon),
\end{aligned} \tag{157}$$

where by definition the phase-shift is

$$\delta(\varepsilon) = \frac{1}{2i} \ln \frac{\mathscr{G}(\varepsilon + i0^+)}{\mathscr{G}(\varepsilon - i0^+)}. \tag{158}$$

Assuming a constant hybridization width $\Gamma(\varepsilon) = \Gamma$, one finds that

$$\delta(\varepsilon) = \tan^{-1} \frac{\Gamma}{\varepsilon_d - \varepsilon},$$

hence, through Eq. (38),

$$\Delta N_{els} = \int_{-\infty}^{+\infty} d\varepsilon\, f(\varepsilon) \frac{1}{\pi} \frac{\Gamma}{(\varepsilon - \varepsilon_d)^2 + \Gamma^2} = N_d. \tag{159}$$

In other words, within the above assumption, the change in the number of electrons only comes from those electrons occupying the impurity level. If for instance $\varepsilon_{d\uparrow}$ is different from $\varepsilon_{d\downarrow}$, at zero temperature the number of electrons at the impurity level with spin σ is

$$N_{d\sigma} = \frac{1}{2} - \frac{1}{\pi} \tan^{-1} \frac{\varepsilon_{d\sigma}}{\Gamma}. \tag{160}$$

B.2. Energy variation

Let us also calculate the impurity contribution to the free energy. Suppose that we change $V_{\mathbf{k}} \to \lambda V_{\mathbf{k}}$. The free energy becomes a function of λ, $E(\lambda)$, which, by the Helmann-Feynmann theorem, satisfies

$$\begin{aligned}
\frac{\partial F(\lambda)}{\partial \lambda} &= \langle \frac{\partial \mathscr{H}}{\partial \lambda} \rangle \\
&= \sum_{\mathbf{k}\sigma} \langle \Psi_\lambda | V_{\mathbf{k}} c^\dagger_{\mathbf{k}\sigma} d_\sigma + H.c. | \Psi_\lambda \rangle \\
&= 2T \sum_n \sum_{\mathbf{k}} V_{\mathbf{k}}\, G_{d\mathbf{k}}(i\omega_n; \lambda) + V^*_{\mathbf{k}}\, G_{\mathbf{k}d}(i\omega_n; \lambda).
\end{aligned}$$

Here $G(i\omega_n; \lambda)$ are the Green's functions at a fixed λ. Specifically

$$G_{\mathbf{k}d}(z; \lambda) = \lambda \frac{V_{\mathbf{k}}}{z - \varepsilon_{\mathbf{k}}} \frac{1}{z - \varepsilon_d - \lambda^2 \Delta(z)}, \tag{161}$$

$$G_{d\mathbf{k}}(z;\lambda) = \lambda\,\frac{V_{\mathbf{k}}^*}{z-\varepsilon_{\mathbf{k}}}\,\frac{1}{z-\varepsilon_d-\lambda^2\Delta(z)}, \tag{162}$$

with $\Delta(z)$ defined in Eq. (30). Therefore

$$\begin{aligned}
\frac{\partial F(\lambda)}{\partial\lambda} &= 4T\sum_n\sum_{\mathbf{k}}\lambda\,\frac{|V_{\mathbf{k}}|^2}{i\omega_n-\varepsilon_{\mathbf{k}}}\,\frac{1}{i\omega_n-\varepsilon_d-\lambda^2\Delta(i\omega_n)} \\
&= 4T\sum_n\lambda\,\Delta(i\omega_n)\,\frac{1}{i\omega_n-\varepsilon_d-\lambda^2\Delta(i\omega_n)} \\
&= -2T\sum_n\frac{\partial}{\partial\lambda}\ln\left(i\omega_n-\varepsilon_d-\lambda^2\Delta(i\omega_n)\right),
\end{aligned}$$

so that the free energy variation due to the impurity is

$$F(\lambda)-F(0) = -2T\sum_n\ln\left(\frac{i\omega_n-\varepsilon_d-\lambda^2\Delta(i\omega_n)}{i\omega_n-\varepsilon_d}\right). \tag{163}$$

In addition the average value of the hybridization at fixed λ is

$$E_{hyb}(\lambda) = \sum_{\mathbf{k}\sigma}\langle\lambda V_{\mathbf{k}}\,c^\dagger_{\mathbf{k}\sigma}d_\sigma + H.c.\rangle = \lambda\,\frac{\partial F(\lambda)}{\partial\lambda}. \tag{164}$$

To calculate (163) we use the same integration contour as in Fig. 22. By noting that

$$\ln(\varepsilon-\varepsilon_d+i0^+)-\ln(\varepsilon-\varepsilon_d-i0^+) \equiv -2i\,\delta_0(\varepsilon) = -2\pi i\,\theta\,(\varepsilon-\varepsilon_d),$$

where $\delta_0(\varepsilon)$ is the phase-shift in the absence of hybridization, and that

$$F(0) = E_{bath}(0)+2\,\varepsilon_d\,f(\varepsilon) \equiv F_0+2\,\varepsilon_d\,f(\varepsilon),$$

the factor 2 coming from spin, we find that

$$\begin{aligned}
F(\lambda) &= F_0+2\,\varepsilon_d\,f(\varepsilon_d)+2\int\frac{d\varepsilon}{\pi}\,f(\varepsilon)\,[\delta_0(\varepsilon)-\delta(\varepsilon)] \\
&= F_0+2\int\frac{d\varepsilon}{\pi}\,\varepsilon\,f(\varepsilon)\,\frac{\partial\delta(\varepsilon)}{\partial\varepsilon}.
\end{aligned} \tag{165}$$

C. FERMI LIQUID THEORY OF THE ANDERSON IMPURITY MODEL

In this Appendix we show how to build up a Fermi liquid theory of a generic AIM closely following the original work by Mihaly and Zawadowskii [53] and Yoshimori and Zawadowskii [54].

Let us consider more generally a multi-orbital Anderson impurity model with Hamiltonian

$$\mathscr{H}_{AIM} = \sum_{\mathbf{k}a}\varepsilon_{\mathbf{k}a}\,c^\dagger_{\mathbf{k}a}c_{\mathbf{k}a}+\sum_{\mathbf{k}a}\left(V_{\mathbf{k}a}\,c^\dagger_{\mathbf{k}a}d_a+H.c.\right)$$

$$+\sum_a \varepsilon_a d_a^\dagger d_a + \mathscr{H}_{int}\left[d^\dagger, d\right]. \tag{166}$$

The Hamiltonian admits conserved quantities labelled by an index (i),

$$\mathscr{M}^{(i)} = \sum_{\mathbf{k}ab} c_{\mathbf{k}a}^\dagger M_{ab}^{(i)} c_{\mathbf{k}b} + \sum_{ab} d_a^\dagger M_{ab}^{(i)} d_b. \tag{167}$$

For convenience we adopt the normalization $\mathrm{Tr}\left(M^{(i)}\cdot M^{(i)}\right) = 1$. By means of the Luttinger-Ward functional one can prove that the variation $\Delta M^{(i)}$ of the average value $M^{(i)} = \langle \mathscr{M}^{(i)} \rangle$ associated with the presence of the impurity is given by

$$\Delta M^{(i)} = \sum_{ab} M_{ab}^{(i)} \oint \frac{dz}{2\pi i} f(z) \frac{\partial}{\partial z} \ln \mathscr{G}_{ba}(z) = \oint \frac{dz}{2\pi i} f(z) \frac{\partial}{\partial z} \mathrm{Tr}\left(M^{(i)} \ln \mathscr{G}(z)\right),$$

where the integration contour encloses clockwise the real axis, $f(z)$ is the Fermi distribution function in the complex plane and $\mathscr{G}(z)$ the matrix of the impurity single-particle Green's functions. Since the latters have a branch cut on the real axis, the above expression is also equal to

$$\Delta M^{(i)} = -\frac{1}{\pi}\int_{-\infty}^{\infty} d\varepsilon \frac{\partial f(\varepsilon)}{\partial \varepsilon} \Im m\, \mathrm{Tr}\left(M^{(i)} \ln \mathscr{G}(\varepsilon + i\delta)\right) \tag{168}$$

with δ an infinitesimal positive number. Let us consider one of the conserved operators (to simplify notations we will drop its index). We can always rotate the basis set in such a way to make this operator diagonal

$$\mathscr{M} = \sum_{\mathbf{k}a} M_a c_{\mathbf{k}a}^\dagger c_{\mathbf{k}a} + \sum_a M_a d_a^\dagger d_a, \tag{169}$$

leaving unchanged the non-interacting part of the Hamiltonian (166). In this representation, if we add to (166) a perturbation

$$\delta \mathscr{H}_{AIM} = -h \mathscr{M}, \tag{170}$$

the impurity Green's functions become

$$\begin{aligned}
\mathscr{G}_{ab}(z,h)^{-1} &= \delta_{ab} \mathscr{G}_a(z,h)^{-1} \\
&= \delta_{ab}\left[\left(\mathscr{G}_a^{(0)}(z,h)\right)^{-1} - \Sigma_a(z,h)\right] = z - \varepsilon_a + hM_a - \Delta_a(z,h) - \Sigma_a(z,h),
\end{aligned}$$

where

$$\Delta_a(z,h) = \sum_{\mathbf{k}} |V_{\mathbf{k}a}|^2 \frac{1}{z - \varepsilon_{\mathbf{k}a} + hM_a} \tag{171}$$

is the hybridization function in the presence of the external field h, as opposed to the one in its absence $\Delta_a(z,0) \equiv \Delta_a(z)$. We notice that

$$\left(\frac{\partial \Delta_a(z,h)}{\partial h}\right)_{h=0} = M_a \frac{\partial \Delta_a(z)}{\partial z},$$

from which it follows that, by defining $\mathscr{G}_a(z,h=0) \equiv \mathscr{G}_a(z)$,

$$\begin{aligned}
\left(\frac{\partial \ln \mathscr{G}_a(z,h)}{\partial h}\right)_{h=0} &= -\left(\frac{\partial \ln \mathscr{G}_a(z,h)^{-1}}{\partial h}\right)_{h=0} \\
&= -\mathscr{G}_a(z)\left[M_a\left(1-\frac{\partial \Delta_a(z)}{\partial z}\right)-\left(\frac{\partial \Sigma_a(z,h)}{\partial h}\right)_{h=0}\right].
\end{aligned}$$

In the presence of the field, also the average value of $\mathscr{M}$ changes, $M \to M(h)$, as well as its variation in the presence of the impurity $\Delta M \to \Delta M(h)$. From the above formulas it follows that the variation of the susceptibility $\Delta\chi$ is given by

$$\begin{aligned}
\Delta\chi &= \left(\frac{\partial \Delta M(h)}{\partial h}\right)_{h=0} = \sum_a \int_{-\infty}^{\infty} \frac{d\varepsilon}{\pi}\frac{\partial f(\varepsilon)}{\partial \varepsilon}\,\Im m\left\{M_a \mathscr{G}_a(\varepsilon+i\delta)\right. \\
&\times\left.\left[M_a\left(1-\left(\frac{\partial \Delta_a(z)}{\partial z}\right)_{z=\varepsilon+i\delta}\right)-\left(\frac{\partial \Sigma_a(\varepsilon+i\delta)}{\partial h}\right)_{h=0}\right]\right\}. \qquad (172)
\end{aligned}$$

On the other hand one can prove diagrammatically that

$$\begin{aligned}
\left(\frac{\partial \Sigma_a(i\omega_n)}{\partial h}\right)_{h=0} &= -\frac{1}{\beta}\sum_m\sum_b \Gamma_{a,b;b,a}(i\omega_n,i\varepsilon_m;i\varepsilon_m,i\omega_n)\,\mathscr{G}_b(i\varepsilon_m)^2 \\
&\times\left(\frac{\partial\left(\mathscr{G}_a^{(0)}(i\varepsilon_m,h)\right)^{-1}}{\partial h}\right)_{h=0} \\
&= -\frac{1}{\beta}\sum_m\sum_b \Gamma_{a,b;b,a}(i\omega_n,i\varepsilon_m;i\varepsilon_m,i\omega_n)\,\mathscr{G}_b(i\varepsilon_m)^2 M_b\left(1-\frac{\partial \Delta_b(i\varepsilon_m)}{\partial i\varepsilon_m}\right),
\end{aligned} \qquad (173)$$

where the interaction vertex is the reducible one.

Using the fact that $\mathscr{M}$ is conserved, the following Ward identity also holds

$$\begin{aligned}
&\left[\Sigma_a(i\varepsilon+i\omega)-\Sigma_a(i\varepsilon)\right]M_a = -\frac{1}{\beta}\sum_n\sum_b \Gamma_{a,b;b,a}(i\varepsilon+i\omega,i\varepsilon_n;i\varepsilon_n+i\omega,i\varepsilon) \\
&\times M_b\mathscr{G}_b(i\varepsilon_n+i\omega)\mathscr{G}_b(i\varepsilon_n)\left[i\omega-\Delta_b(i\varepsilon_n+i\omega)+\Delta_b(i\varepsilon_n)\right]. \qquad (174)
\end{aligned}$$

It follows that

$$\begin{aligned}
\frac{\partial \Sigma_a(i\varepsilon)}{\partial i\varepsilon}M_a &= -\frac{1}{\beta}\sum_n\sum_b \Gamma_{a,b;b,a}(i\varepsilon,i\varepsilon_n;i\varepsilon_n,i\varepsilon)M_b\mathscr{G}_b(i\varepsilon_n)^2 \\
&- \lim_{i\omega\to 0}\frac{1}{\beta}\sum_n\sum_b \Gamma_{a,b;b,a}(i\varepsilon+i\omega,i\varepsilon_n;i\varepsilon_n+i\omega,i\varepsilon)
\end{aligned}$$

$$\times M_b \mathscr{G}_b(i\varepsilon_n + i\omega)\mathscr{G}_b(i\varepsilon_n)\frac{\left[-\Delta_b(i\varepsilon_n+i\omega)+\Delta_b(i\varepsilon_n)\right]}{i\omega}$$

$$\begin{aligned}
&= -\frac{1}{\beta}\sum_n\sum_b \Gamma_{a,b;b,a}(i\varepsilon, i\varepsilon_n; i\varepsilon_n, i\varepsilon) M_b \mathscr{G}_b(i\varepsilon_n)^2 \left(1-\frac{\partial\Delta_b(i\varepsilon_n)}{\partial i\varepsilon_n}\right)\\
&+\int_{-\infty}^{\infty}\frac{d\varepsilon'}{2\pi}\frac{\partial f(\varepsilon')}{\partial\varepsilon'}\sum_b \Gamma_{a,b;b,a}(i\varepsilon,\varepsilon'-i\delta';\varepsilon'+i\delta',i\varepsilon)\\
&\times M_b \mathscr{G}_b(\varepsilon'+i\delta')\mathscr{G}_b(\varepsilon'-i\delta')\Im m\left[\Delta_b(\varepsilon'-i\delta')-\Delta_b(\varepsilon'+i\delta')\right]\\
&= -\frac{1}{\beta}\sum_n\sum_b \Gamma_{a,b;b,a}(i\varepsilon, i\varepsilon_n; i\varepsilon_n, i\varepsilon) M_b \mathscr{G}_b(i\varepsilon_n)^2 \left(1-\frac{\partial\Delta_b(i\varepsilon_n)}{\partial i\varepsilon_n}\right)\\
&+\int_{-\infty}^{\infty}\frac{d\varepsilon'}{\pi}\frac{\partial f(\varepsilon')}{\partial\varepsilon'}\sum_b \Gamma_{a,b;b,a}(i\varepsilon,\varepsilon'-i\delta';\varepsilon'+i\delta',i\varepsilon)\\
&\times M_b \Gamma_b(\varepsilon')\mathscr{G}_b(\varepsilon'+i\delta')\mathscr{G}_b(\varepsilon'-i\delta'),
\end{aligned} \tag{175}$$

where, by definition,

$$\Gamma_b(\varepsilon') = \frac{1}{2}\Im m\left(\Delta_b(\varepsilon'-i\delta')-\Delta_b(\varepsilon'+i\delta')\right)$$

is the non-interacting resonance-width.

By comparing (175) with (173) we find that

$$\begin{aligned}
\left(\frac{\partial\Sigma_a(i\varepsilon)}{\partial h}\right)_{h=0} &= \frac{\partial\Sigma_a(i\varepsilon)}{\partial i\varepsilon}M_a - \int_{-\infty}^{\infty}\frac{d\varepsilon'}{\pi}\frac{\partial f(\varepsilon')}{\partial\varepsilon'}\sum_b \Gamma_{a,b;b,a}(i\varepsilon,\varepsilon'-i\delta';\varepsilon'+i\delta',i\varepsilon)\\
&\times M_b \Gamma_b(\varepsilon')\mathscr{G}_b(\varepsilon'+i\delta')\mathscr{G}_b(\varepsilon'-i\delta').
\end{aligned} \tag{176}$$

Going back to the expression for the variation of the susceptibility, Eq. (172), and inserting (176), we obtain the following result

$$\begin{aligned}
\Delta\chi &= \sum_a \int_{-\infty}^{\infty}\frac{d\varepsilon}{\pi}\frac{\partial f(\varepsilon)}{\partial\varepsilon}\Im m\Bigg\{M_a\mathscr{G}_a(\varepsilon+i\delta)\\
&\times\Bigg[M_a\left(1-\left(\frac{\partial\Delta_a(z)}{\partial z}\right)_{z=\varepsilon+i\delta}-\left(\frac{\partial\Sigma_a(z)}{\partial z}\right)_{z=\varepsilon+i\delta}\right)\\
&+\int_{-\infty}^{\infty}\frac{d\varepsilon'}{\pi}\frac{\partial f(\varepsilon')}{\partial\varepsilon'}\sum_b \Gamma_{a,b;b,a}(\varepsilon,\varepsilon';\varepsilon',\varepsilon)\\
&\times M_b \Gamma_b(\varepsilon')\mathscr{G}_b(\varepsilon'+i\delta')\mathscr{G}_b(\varepsilon'-i\delta')\Bigg]\Bigg\}
\end{aligned} \tag{177}$$

in which we have assumed that the reducible vertex is analytic on the real axis.

Let us define the quantity

$$\rho_a^* = \int_{-\infty}^{\infty} \frac{d\varepsilon}{\pi} \frac{\partial f(\varepsilon)}{\partial \varepsilon} \Im m \left\{ \mathscr{G}_a(\varepsilon + i\delta) \left[1 - \left(\frac{\partial \Delta_a(i\varepsilon)}{\partial i\varepsilon} \right)_{i\varepsilon \to \varepsilon + i\delta} - \left(\frac{\partial \Sigma_a(i\varepsilon)}{\partial i\varepsilon} \right)_{i\varepsilon \to \varepsilon + i\delta} \right] \right\}, \tag{178}$$

which plays actually the role of the quasi-particle DOS at the chemical potential, as opposed to the particle DOS

$$\rho_a = \int_{-\infty}^{\infty} \frac{d\varepsilon}{\pi} \frac{\partial f(\varepsilon)}{\partial \varepsilon} \Im m \mathscr{G}_a(\varepsilon + i\delta) \equiv - \int_{-\infty}^{\infty} \frac{d\varepsilon}{\pi} \frac{\partial f(\varepsilon)}{\partial \varepsilon} \rho_a(\varepsilon). \tag{179}$$

Finally we notice that

$$\begin{aligned} \mathscr{G}_b(\varepsilon' + i\delta')\mathscr{G}_b(\varepsilon' - i\delta') &= \frac{1}{\varepsilon' - \varepsilon_b - \Delta_b(\varepsilon' + i\delta') - \Sigma_b(\varepsilon' + i\delta')} \\ &\quad \times \frac{1}{\varepsilon' - \varepsilon_b - \Delta_b(\varepsilon' - i\delta') - \Sigma_b(\varepsilon' - i\delta')} \\ &= \frac{1}{\Delta_b(\varepsilon' - i\delta') - \Delta_b(\varepsilon' + i\delta') + \Sigma_b(\varepsilon' - i\delta') - \Sigma_b(\varepsilon' + i\delta')} \\ &\quad \times \left(\mathscr{G}_b(\varepsilon' - i\delta') - \mathscr{G}_b(\varepsilon' + i\delta') \right) \\ &= \pi \rho_b(\varepsilon') \frac{1}{\Gamma_b(\varepsilon') + \gamma_b(\varepsilon')}, \end{aligned}$$

where we have defined

$$\gamma_b(\varepsilon') = \frac{1}{2} \Im m \left(\Sigma_b(\varepsilon' - i\delta') - \Sigma_b(\varepsilon' + i\delta') \right), \tag{180}$$

which is related to the quasiparticle decay rate. By means of the above expression and Eq. (178), we can rewrite (177) as

$$\begin{aligned} \Delta\chi &= \sum_a M_a^2 \rho_a^* + \sum_{ab} M_a M_b \int_{-\infty}^{\infty} \frac{d\varepsilon\, d\varepsilon'}{\pi} \frac{\partial f(\varepsilon)}{\partial \varepsilon} \frac{\partial f(\varepsilon')}{\partial \varepsilon'} \\ &\quad \Im m \left\{ \mathscr{G}_a(\varepsilon + i\delta) \Gamma_{a,b;b,a}(\varepsilon, \varepsilon'; \varepsilon', \varepsilon) \rho_b(\varepsilon') \frac{\Gamma_b(\varepsilon')}{\Gamma_b(\varepsilon') + \gamma_b(\varepsilon')} \right\} \\ &= \sum_a M_a^2 \rho_a^* - \sum_{ab} M_a M_b \int_{-\infty}^{\infty} d\varepsilon\, d\varepsilon' \frac{\partial f(\varepsilon)}{\partial \varepsilon} \frac{\partial f(\varepsilon')}{\partial \varepsilon'} \end{aligned}$$

$$\times \rho_a(\varepsilon)\,\rho_b(\varepsilon')\,\Gamma_{a,b;b,a}(\varepsilon,\varepsilon';\varepsilon',\varepsilon)\,\frac{\Gamma_b(\varepsilon')}{\Gamma_b(\varepsilon')+\gamma_b(\varepsilon')}, \tag{181}$$

where the last expression is obtained by assuming that, for small frequencies, the reducible vertex is real, hence that the only contribution to the imaginary part comes from

$$\Im m\,\mathcal{G}_a(\varepsilon+i\delta) = -\pi\,\rho_a(\varepsilon).$$

Eq. (181) has a broad validity, even in cases like DMFT, in which the hybridization functions may be strongly frequency-dependent.

In general the matrix elements of a conserved operators may be either diagonal or, if different states are degenerate, off-diagonal. In the former case only the diagonal elements of the scattering vertex, $\Gamma_{a,b;b,a}$, enter the expression of the susceptibility. More interesting is the case of degenerate levels, $\varepsilon_{\mathbf{k}a} = \varepsilon_{\mathbf{k}}$ and $\varepsilon_a = \varepsilon$ in (166). Under the assumption that the interaction does not spoil this degeneracy, $\rho_a = \rho$ and $\rho^*_a = \rho^*$. Since we assumed that $\mathrm{Tr}(M\cdot M) = 1$, in a generic basis in which M is not diagonal the susceptibility variation reads

$$\begin{aligned}\Delta\chi &= \rho^* - \sum_{abcd} M_{ba}M_{dc}\int_{-\infty}^{\infty} d\varepsilon\, d\varepsilon'\,\frac{\partial f(\varepsilon)}{\partial\varepsilon}\,\frac{\partial f(\varepsilon')}{\partial\varepsilon'}\rho(\varepsilon)\,\rho(\varepsilon')\,\Gamma_{a,c;d,b}(\varepsilon,\varepsilon';\varepsilon',\varepsilon)\\ &\quad\times\frac{\Gamma(\varepsilon')}{\Gamma(\varepsilon')+\gamma(\varepsilon')}\\ &\equiv \rho^*\left[1-A\right],\end{aligned} \tag{182}$$

with the Landau scattering parameters A defined through

$$\begin{aligned}A &\equiv \frac{1}{\rho^*}\sum_{abcd} M_{ba}M_{dc}\int_{-\infty}^{\infty} d\varepsilon\, d\varepsilon'\,\frac{\partial f(\varepsilon)}{\partial\varepsilon}\,\frac{\partial f(\varepsilon')}{\partial\varepsilon'}\rho(\varepsilon)\,\rho(\varepsilon')\,\Gamma_{a,c;d,b}(\varepsilon,\varepsilon';\varepsilon',\varepsilon)\\ &\quad\times\frac{\Gamma(\varepsilon')}{\Gamma(\varepsilon')+\gamma(\varepsilon')}.\end{aligned} \tag{183}$$

In most common impurity models the hybridization functions have a smooth frequency-dependence and, in addition, perturbation theory does not break down at any value of the interaction. In these cases we have already seen that, at small frequency,

$$\Sigma_a(i\varepsilon) \simeq i\varepsilon\left(1-\frac{1}{Z_a}\right),$$

where Z_a is the so-called quasiparticle residue, hence that

$$\rho^*_a = \frac{\rho_a}{Z_a},$$

and $\gamma_a(\varepsilon\to 0) = 0$. Therefore (181) simplifies notably:

$$\Delta\chi = \sum_a M_a^2\,\rho^*_a - \sum_{ab} M_a M_b\,\rho_a\,\rho_b\,\Gamma_{a,b;b,a}(0,0;0,0)$$

$$= \sum_a M_a^2 \rho_a^* - \sum_{ab} M_a M_b Z_a Z_b \rho_a^* \rho_b^* \Gamma_{a,b;b,a}(0,0;0,0). \quad (184)$$

When orbital degeneracy also occurs

$$\Delta\chi = \rho^* \left[1 - A\right], \quad (185)$$

when this time

$$A = Z^2 \rho^* \sum_{abcd} M_{ba} M_{dc} \Gamma_{a,c;d,b}(0,0;0,0), \quad (186)$$

which is the common definition of the Landau A-parameters in terms of the quasiparticle scattering amplitudes.

We conclude by noticing that the variation of the specific heat is, in the degenerate example,

$$\frac{\Delta C_v}{C_{V0}} = \frac{\rho^*}{\rho_0}, \quad (187)$$

where C_{V0} and ρ_0 are the specific heat and DOS, respectively, of the conduction electrons in the absence of the impurity. On the other hand, for any conserved quantity, $\mathscr{M}^{(i)}$, by Eq. (185) we find that

$$\frac{\Delta\chi^{(i)}}{\chi_0} = \frac{\rho^*}{\rho_0} \left[1 - A^{(i)}\right]. \quad (188)$$

In other words, we can define for each conserved quantity a corresponding Wilson ratio whose value

$$R^{(i)} = \frac{\Delta\chi^{(i)}}{\chi_0} \frac{C_{V0}}{\Delta C_v} = 1 - A^{(i)}, \quad (189)$$

is directly related to the Landau parameters.

C.1. Application to the single-orbital AIM

Let us apply these results to the single-orbital AIM. Two incoming quasi-particles at the same frequency can only be into a singlet state, hence there is a single scattering parameter

$$\Gamma = \Gamma_{\uparrow,\downarrow;\downarrow,\uparrow} = \Gamma_{\downarrow,\uparrow;\uparrow,\downarrow} = -\Gamma_{\uparrow,\downarrow;\uparrow,\downarrow} = -\Gamma_{\downarrow,\uparrow;\downarrow,\uparrow}. \quad (190)$$

There are only two conserved quantities, the total charge, with $M^c_{\sigma\sigma'} = \delta_{\sigma\sigma'}/\sqrt{2}$, and the total spin, with $\mathbf{M}^s = \sigma/\sqrt{2}$. Through (186) one finds that the charge, A^c, and spin, A^s, Landau parameters are

$$A^c = Z^2 \rho^* \Gamma = -A^s \equiv A. \quad (191)$$

We know that, for large U, the charge fluctuations of the impurity are suppressed, hence we can assume that

$$\chi^c = \rho^* \left[1 - A\right] = 0,$$

namely that $A = 1$. It follows that

$$\chi^s = \rho^* \left[1 + A\right] = 2\rho^*,$$

as well as

$$R^c = 0, \qquad R^s = 2.$$

REFERENCES

1. N. Mott, *Proc. Phys. Soc. (London)* **62**, 416 (1949).
2. N. Mott, *Metal Insulator Transition*, Taylor and Francis, London, 1990.
3. J. Hubbard, *Proceedings of the Royal Society of London. Series A, Mathematical and Physical Sciences* **276**, 238–257 (1963).
4. J. Hubbard, *Proceedings of the Royal Society of London. Series A, Mathematical and Physical Sciences* **277**, 237–259 (1964).
5. J. Hubbard, *Royal Society of London Proceedings Series A* **281**, 401–419 (1964).
6. A. Georges, G. Kotliar, W. Krauth, and M. J. Rozenberg, *Rev. Mod. Phys.* **68**, 13 (1996).
7. S. Sorella, *Phys. Rev. B* **71**, 241103 (2005).
8. M. Capello, F. Becca, M. Fabrizio, S. Sorella, and E. Tosatti, *Phys. Rev. Lett.* **94**, 026406 (2005).
9. S. Sorella, G. B. Martins, F. Becca, C. Gazza, L. Capriotti, A. Parola, and E. Dagotto, *Phys. Rev. Lett.* **88**, 117002 (2002).
10. G. Misguich, and C. Lhuillier, "Two-dimensional quantum antiferromagnets," in *Frustrated Spin Systems*, edited by H. Diep, World-Scientific, Singapore, 2005, p. 981, and references therein.
11. A. Millis, *Phys. Rev. B* **48**, 7183 (1993).
12. P. Coleman, C. Pepin, Q. Si, and R. Ramazashvili, *J. Phys.: Condens. Matter* **13**, 723 (2001).
13. Q. Si, *Advances in Solid State Physics* **44**, 253 (2004).
14. A. Georges, and G. Kotliar, *Phys. Rev. B* **45**, 6479–6483 (1992).
15. W. Metzner, and D. Vollhardt, *Phys. Rev. Lett.* **59**, 121–124 (1987).
16. W. Metzner, and D. Vollhardt, *Phys. Rev. B* **37**, 7382–7399 (1988).
17. A. I. Lichtenstein, and M. I. Katsnelson, *Phys. Rev. B* **62**, R9283–R9286 (2000).
18. D. Senechal, D. Perez, and M. Pioro-Ladriere, *Phys. Rev. Lett.* **84**, 522 (2000).
19. M. Potthoff, M. Aichhorn, and C. Dahnken, *Phys. Rev. Lett.* **91**, 206402 (2003).
20. G. Kotliar, S. Y. Savrasov, G. Pálsson, and G. Biroli, *Phys. Rev. Lett.* **87**, 186401 (2001).
21. T. Maier, M. Jarrell, T. Pruschke, and M. H. Hettler, *Rev. Mod. Phys.* **77**, 1027 (2005).
22. P. W. Anderson, *Phys. Rev.* **124**, 41–53 (1961).
23. A. Hewson, *The Kondo Problem to Heavy Fermions*, Cambridge University Press, 1997.
24. J. Friedel, *Nuovo Cimento* **7**, 287–311 (1958).
25. A. A. Abrikosov, *Physics* **2**, 5–20 (1965).
26. J. Kondo, *Progress of Theoretical Physics* **32**, 37–49 (1964).
27. P. W. Anderson, *Journal of Physics C: Solid State Physics* **3**, 2436–2441 (1970).
28. K. G. Wilson, *Rev. Mod. Phys.* **47**, 773 (1975).
29. N. Andrei, *Phys. Rev. Lett.* **45**, 379–382 (1980).
30. A. M. Tsvelik, and P. B. Wiegmann, *Advances in Physics* **32**, 453–713 (1983).
31. W. F. Brinkman, and T. M. Rice, *Phys. Rev. B* **2**, 4302–4304 (1970).
32. H. R. Krishnamurthy, J. W. Wilkins, and K. G. Wilson, *Phys. Rev. B* **21**, 1003 (1980).
33. H. R. Krishnamurthy, J. W. Wilkins, and K. G. Wilson, *Phys. Rev. B* **21**, 1044 (1980).
34. I. Affleck, *ACTA PHYS.POLON.B* **26**, 1869 (1995).
35. I. Affleck, and A. W. W. Ludwig, *Nucl. Phys.* **B352**, 849–862 (1991).
36. P. Di Francesco, P. Mathieu, and D. Sénéchal, *Conformal Field Theory*, Springer, 1996.
37. A. Gogolin, A. Nersesyan, and A. Tsvelik, *Bosonization and strongly correlated systems*, Cambridge University Press, 1998.
38. J. L. Cardy, *Nucl. Phys.* **B324**, 581 (1989).
39. I. Affleck, and A. W. W. Ludwig, *Nucl. Phys.* **B360**, 641–696 (1991).
40. M. Capone, M. Fabrizio, and E. Tosatti, *Physical Review Letters* **86**, 5361–5364 (2001).

41. M. Capone, M. Fabrizio, C. Castellani, and E. Tosatti, *Science* **296**, 2364 (2002).
42. M. Fabrizio, A. F. Ho, L. De Leo, and G. E. Santoro, *Phys. Rev. Lett.* **91**, 246402 (2003).
43. L. D. Leo, and M. Fabrizio, *Phys. Rev. B* **69**, 245114 (2004).
44. M. Capone, M. Fabrizio, C. Castellani, and E. Tosatti, *Phys. Rev. Lett.* **93**, 047001 (2004).
45. M. Ferrero, L. De Leo, P. Lecheminant, and M. Fabrizio, Strong correlations in a nutshell (2007), URL `http://www.citebase.org/abstract?id=oai:arXiv.org:cond-mat/07%02629`.
46. B. A. Jones, and C. M. Varma, *Phys. Rev. B* **40**, 324 (1989).
47. B. A. Jones, C. M. Varma, and J. W. Wilkins, *Phys. Rev. Lett.* **61**, 125 (1988).
48. B. A. Jones, and C. M. Varma, *Phys. Rev. B* **40**, 324 (1989).
49. I. Affleck, and A. W. W. Ludwig, *Phys. Rev. Lett.* **68**, 1046 (1992).
50. I. Affleck, A. W. W. Ludwig, and B. A. Jones, *Phys. Rev. B* **52**, 9528 (1995).
51. I. Affleck, and A. W. W. Ludwig, *Phys. Rev. Lett.* **67**, 161–164 (1991).
52. P. Nozières, *J. Low Temp. Phys.* **17**, 31 (1974).
53. L. Mihaly, and A. Zawadowski, *J. Phys. Lett. (France)* **39**, L483–6 (1978).
54. A. Yoshimori, and A. Zawadowski, *Journal of Physics C: Solid State Physics* **15**, 5241–5253 (1982).

Modelling and simulating strongly correlated fermions

Didier Poilblanc

Laboratoire de Physique Théorique
CNRS & Université Paul Sabatier, 118 rte de Narbonne, 31062 Toulouse, Cedex

Abstract. I shall briefly review here some aspects of the Exact Diagonalization technique and put it in perspective with other numerical techniques such as Density Matrix Renormalization Group (DMRG) or Quantum Monte Carlo methods (partially) covered in this AIP Series by Reinhard Noack [1] and Richard Scalettar [2] respectively. I will then illustrate this lecture in discussing some simple applications, mainly for one dimensional systems.

LANCZOS ALGORITHM

Algorithm

Many systems of Condensed Matter consist of fermions (electrons) moving on a lattice and experiencing strong repulsive interactions [3]. In such cases, the traditional perturbative methods to treat the electronic correlations often break done. Well controlled systems of ultra-cold atoms confined on optical lattices [4] also offer a rich variety of correlated lattice fermions (and bosons). In a pioneering work, Bonner and Fisher [5] revealed the exact diagonalisation (ED) method as a powerful tool to study the properties of one dimensional (1D) spin chains. Later on, it was extended to investigate two dimensional (2D) localized spin systems [6]. This work initiated a more extensive use of the method to investigate a wide variety of different systems such as strongly correlated lattice electrons (Hubbard, t–J models,...), mesoscopic systems [7], magneto-elastic systems [8], frustrated magnets [9] etc....

The exact diagonalisation method is widely based on the Lanczos algorithm [10] which I shall describe here. Complementary aspects can also be found in Refs. [8]. This algorithm is particularly suited to handle sparse matrices and there is, in fact, a wide variety of lattice models belonging to this class as we shall see later on.

Let us consider some lattice model corresponding to some Hamiltonian H with its symmetry group $\mathscr{G} = \{g\}$, namely $[H,g] = 0$. Let us also assume, for the moment, that irreducible representations of the symmetry group can be constructed. They consist of complete subsets of states $\mathscr{A}_l = \{|\alpha\rangle\}$ which are globally invariant under the application of the Hamiltonian H (we postpone to the next part of this section the explicit construction of these states). Clearly, H can be diagonalized in each of the subsets $\mathscr{A}_l$ independently. It can be written as a tridiagonal matrix in a new orthonormal basis set

CP918, *Lectures on the Physics of Strongly Correlated Systems XI: Eleventh Training Course*, edited by A. Avella and F. Mancini

$\{|\Phi_m\rangle\}$ defined as [11]

$$\begin{aligned} H|\Phi_1\rangle &= e_1|\Phi_1\rangle + b_2|\Phi_2\rangle, \\ &\vdots \\ H|\Phi_n\rangle &= e_n|\Phi_n\rangle + b_{n+1}|\Phi_{n+1}\rangle + b_n|\Phi_{n-1}\rangle \ , \end{aligned} \tag{1}$$

where the various coefficients and new basis states can be calculated recursively. The proof is as follows: let us suppose that the procedure has been applied until the step n, i.e. an orthonormal set of states $|\Phi_1\rangle, ..., |\Phi_n\rangle$ has been constructed. Assuming the knowledge of $e_1, ..., e_{n-1}$, $b_2, ..., b_n$, $|\Phi_{n-1}\rangle$ and $|\Phi_n\rangle$, one can then determine

$$e_n = \langle\Phi_n|H|\Phi_n\rangle. \tag{2}$$

Hence, the new state defined by

$$|\phi_{n+1}\rangle = H|\Phi_n\rangle - e_n|\Phi_n\rangle - b_n|\Phi_{n-1}\rangle, \tag{3}$$

is clearly orthogonal to $|\Phi_n\rangle$. Moreover, $\langle\Phi_{n-1}|\phi_{n+1}\rangle = \langle\Phi_{n-1}|H|\Phi_n\rangle - b_n$ which is also vanishing as can be seen by substituting the expression for $H|\Phi_{n-1}\rangle$. More generally, $|\phi_{n+1}\rangle$ is, in fact, orthogonal to all the previous states $|\Phi_p\rangle$, $p \leq n$ as can be shown recursively. Indeed, let us assume that, for $p < n$,

$$\forall i, \ \ p \leq i \leq n \ \ \langle\Phi_i|\phi_{n+1}\rangle = 0, \tag{4}$$

then $\langle\Phi_{p-1}|\phi_{n+1}\rangle = \langle\Phi_{p-1}|H|\Phi_n\rangle$ where we have used $\langle\Phi_{p-1}|\Phi_n\rangle = \langle\Phi_{p-1}|\Phi_{n-1}\rangle = 0$. Substituting the expression given by Eq. (2) for $H|\Phi_{p-1}\rangle$ leads to the expected result

$$\langle\Phi_{p-1}|\phi_{n+1}\rangle = 0. \tag{5}$$

The (positive) number b_{n+1} is simply defined as a normalization factor,

$$b_{n+1}^2 = \langle\phi_{n+1}|\phi_{n+1}\rangle, \tag{6}$$

i.e. $|\Phi_{n+1}\rangle = \frac{1}{b_{n+1}}|\phi_{n+1}\rangle$.

In principal, a zero vector will be generated after iterating the Hamiltonian a number of times corresponding to the size $\mathcal{N}_l$ of the Hilbert space. However, the number of iterations necessary to obtain the lowest eigenvalues and eigenvectors is much smaller. Typically, for $\mathcal{N}_l \sim 10^7$, the GS can be obtained with an accuracy better than 10^{-10}, by truncating the procedure after only $N_{it} \sim 100$ iterations and by diagonalising the resulting tri-diagonal matrix by using a standard library subroutine. However, for a given size $\mathcal{N}_l$ of the Hilbert space, the necessary number N_{it} of iterations might vary by a factor of 2 or 3 depending on the model Hamiltonian. In practice, the convergence is faster for models for which high energy configurations have been integrated out (e.g. t–J models in contrast to Hubbard models). Note however that, in some cases (models with strong finite range interaction), the energy vs N_{it} curve can exhibit steps before convergence to the true GS is achieved. Once space group symmetries have been implemented, the best choice for the initial state $|\Phi_1\rangle$ is a purely random vector. The GS is also easily obtained as a function of the states $|\Phi_n\rangle$. However, to express it in terms of the initial basis $\{|\alpha\rangle\}$ (as it is often useful) it becomes necessary to store temporarily the intermediate vectors $|\Phi_n\rangle$. This step is usually the most demanding in terms of disk space and/or mass storage. However, note that, at runtime, only three Lanczos vectors of size $\mathcal{N}_I$ need to be assigned in memory.

Full diagonalisation: In some special cases (spectrum statistics [12], thermodynamics [13], etc ...) the complete spectrum (or a least the lower part of it) is needed. This can also be performed by the Lanczos algorithm. In this case, usage of a more sophisticated standard library package (e.g. HARWELL) is preferable. Indeed, more involved tests become then necessary to eliminate the unphysical "ghost" levels appearing (always above the GS energy) after the diagonalisation of the tridiagonal matrix. However, the input for the library subroutine consists only on the set of states $|\Phi_n\rangle$ which have to be calculated separately (see below). Note also that full diagonalisation of small Hilbert spaces can also be performed by standard LAPACK routines.

Space group symmetries

Usually, some efforts have to be carried out in order to take full advantage of the symmetries of the Hamiltonian. Let me consider a model defined on a D-dimensional lattice describing interacting fermions as e.g. the simple Heisenberg (spin) model,

$$H_J = \sum_{\mathbf{x},\mathbf{y}} J_{\mathbf{xy}}\, \mathbf{S_x} \cdot \mathbf{S_y}, \tag{7}$$

where interaction (in this case the exchange coupling) is not necessarily restricted to NN sites but, nevertheless, is supposed to exhibit translation and point group symmetry. In other words, denoting the point group[14] by $\mathcal{G}_P = \{g_P\}$, we assume, e.g. in the case of Eq. (7),

$$\begin{aligned} J_{\mathbf{xy}} &= J(\mathbf{x}-\mathbf{y}) \\ &\text{and} \\ \forall g_P &\in \mathcal{G}_P,\ J(g_P(\mathbf{r})) = J(\mathbf{r}). \end{aligned} \tag{8}$$

Clearly, such properties are easily generalized to generic spin or fermionic Hamiltonians. In addition, we shall also assume, as in Eq. (7), spin rotational invariance (the total spin S is a good quantum number) or, at least, invariance of the Hamiltonian under a spin rotation around some quantization axis. Translation symmetry can be preserved on finite systems provided the geometry is that of a D-dimensional torus with periodic or twisted boundary conditions (BC). In 1D, this simply corresponds to closed rings. In 2D, examples are shown in Fig. 1 in the case of the square lattice. Such clusters can be seen as periodic approximants of the infinite lattice and are characterized by two translation vectors $\mathbf{T}_1$ and $\mathbf{T}_2$. In order to preserve also the 90^0-rotation symmetry of the square lattice ($\mathcal{G}_P = C_{4v}$), it is convenient to choose orthogonal translation vectors of equal lengths i.e. $\mathbf{T}_1 = (n,m)$ and $\mathbf{T}_2 = (-m,n)$. The number of sites in the cluster is then

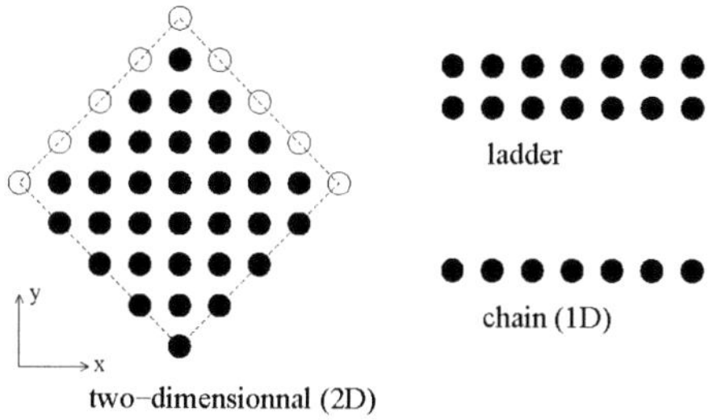

FIGURE 1. Typical clusters used in finite size studies of models defined on 1D and 2D square lattices.

given by $N = n^2 + m^2$ e.g. N=26 and N=32 correspond to the choices $(n,m) = (5,1)$ and $(n,m) = (4,4)$, respectively. Note that reflection symmetries can be lost e.g. the 26-site cluster of Fig. 1 corresponds to $\mathscr{G}_P = C_4$ instead of C_{4v} for the original infinite lattice. On the torus geometry, the full space group reads,

$$\mathscr{G} = \mathscr{G}_P \otimes \mathscr{T}, \tag{9}$$

where we denote by $\mathscr{T} = \{t_p\}$, $p = 1,...,N$, the translation group. It is clear that H_J is invariant under any $g \in \mathscr{G}$.

Construction of the Hilbert space

The motivation to take into account the Hamiltonian symmetries is two-fold. First, the Hamiltonian can be block diagonalised, each block corresponding to an irreducible representation of the symmetry group. Practically, the sizes of the various blocks $\mathscr{N}_l$ are much smaller than the size of the full Hilbert space (see e.g. Table 1), hence, minimizing the numerical effort to diagonalise the Hamiltonian matrix. Secondly, each irreducible representation of the symmetry group is characterized by quantum numbers (such as the momentum **k**) which can be connected directly to physical properties.

I shall consider, here, three examples; the Heisenberg defined by Eq. (7), the t–J and the (generalized) Hubbard models. The t-J model describes the motion of holes ($\emptyset_{\mathbf{x}}$) in a quantum antiferromagnet i.e.

$$\begin{aligned} H_{t-J} &= H_J + \mathscr{P} H_K \mathscr{P}, \\ H_K &= \sum_{\mathbf{x},\mathbf{y},s} t_{\mathbf{xy}}\, c^{\dagger}_{\mathbf{x},s} c_{\mathbf{y},s}\,. \end{aligned} \tag{10}$$

The Gutzwiller projector $\mathscr{P}$ restricts the on-site configurations to $P = 3$ possible states $\emptyset$ (i.e. no occupation or "hole"), $\uparrow$ (i.e. single electron with up spin) and $\downarrow$ (i.e. single

TABLE 1. Sizes of the full and reduced Hilbert spaces (corresponding to the ground-state) for various N-sites Heisenberg clusters (2D). T_N and I_2 stand for the translation group $\mathcal{T}$ (see text) and the spin inversion symmetry $S^Z_{\mathbf{x}} \to -S^Z_{\mathbf{x}}$, respectively.

$\sqrt{N}\times\sqrt{N}$	Full Hilbert space (fixed S_Z)	Symmetry group	Reduced Hilbert space
4×4	12 870	$T_{16}\otimes C_{4v}\otimes I_2$	107
$\sqrt{20}\times\sqrt{20}$	184 756	$T_{20}\otimes C_4\otimes I_2$	1 206
$\sqrt{32}\times\sqrt{32}$	601 080 390	$T_{32}\otimes C_{4v}\otimes I_2$	1 184 480
6×6	9 075 135 300	$T_{36}\otimes C_{4v}\otimes I_2$	15 804 956
$\sqrt{40}\times\sqrt{40}$	137 846 528 820	$T_{40}\otimes C_4\otimes I_2$	137 846 528 820 [15]

electron with down spin) (no doubly occupied sites). On the contrary, in Hubbard-like models,

$$H_U = H_K + \sum_{\mathbf{x},\mathbf{y}} V_{\mathbf{xy}} n_{\mathbf{x}} n_{\mathbf{y}}\,, \tag{11}$$

each lattice site can be in $P = 4$ possible states $\emptyset$, $\uparrow$, $\downarrow$ and $\uparrow\downarrow$. In order to minimize memory occupation, configurations can be stored in binary form. One, first, assumes an arbitrary labelling of the lattice sites from 1 to N and denotes a configuration in real space by

$$|c\rangle = |s_1, ..., s_i, ..., s_N\rangle\ . \tag{12}$$

For the Heisenberg model, the information s_i on each lattice site $\mathbf{x}_i$ can be stored on a single bit, e.g. $\uparrow$ and $\downarrow$ correspond to 1 and 0 respectively. On the contrary, for t–J or Hubbard-like models, 2 bits per site is required; one can, for example, code the 3 states $\emptyset$, $\uparrow$, $\downarrow$ of the t–J model by 00, 10, and 01 respectively. A spin configuration with up to 64 spins or an electronic configuration (t–J or Hubbard) on up to 32 sites can then be represented by a single 64-bits word. As an example, a $N = 4$ site configuration of the t–J model such as $|\uparrow,\emptyset,\uparrow,\downarrow\rangle$ is coded as $0_{64}...0_9 0_8 1_7 1_6 0_5 0_4 0_3 1_2 0_1$ (where the subscripts indicate the place of the bits) i.e. the integer "194". An integer $N(|c\rangle)$ can then be uniquely associated to each configuration $|c\rangle$ and formally written as,

$$N(|c\rangle) = \sum_1^N 2^{i-1}\sigma_i\,, \tag{13}$$

for spin-1/2 models ($\sigma_i = 0$, or 1) and

$$N(|c\rangle) = \sum_1^N 2^{2(i-1)}\sigma_i\,, \tag{14}$$

for (generalized) t–J models ($\sigma_i = 0$, 1 or 2).

For Hubbard-like models, spins up and down can be independently positioned on the lattice so that it becomes more convenient, for each word, to use the N first bits to store the information related to the down spins while the N following bits are used for the up spins (or vice versa). Within this convention, the previous configuration $|\uparrow,\emptyset,\uparrow,\downarrow\rangle$ is coded as $0_{64}...0_9 0_8 1_7 0_6 1_5 1_4 0_3 0_2 0_1$ where 1 (0) corresponds to the presence (absence) of a spin σ on the corresponding site.

TABLE 2. Sizes of the full and reduced Hilbert spaces corresponding to the ground-state (or to a momentum $\mathbf{K} = \mathbf{0}$ *) for various 2D N-sites t–J clusters with N_h holes.

$N_h / \sqrt{N} \times \sqrt{N}$	Full Hilbert space (fixed S_Z)	Symmetry group	Reduced Hilbert space
$1/4\times 4$	102 960	T_{16}	6 435
$1/\sqrt{26}\times\sqrt{26}$	135 207 800	T_{26}	5 200 300
$1/\sqrt{32}\times\sqrt{32}$	9 617 286 240	$T_{32}\otimes C_{4v}$	37 596 701*
$2/\sqrt{32}\times\sqrt{32}$	76 938 289 920	$T_{32}\otimes C_{4v}\otimes I_2$	150 295 402
$4/\sqrt{26}\times\sqrt{26}$	10 546 208 400	$T_{26}\otimes C_4\otimes I_2$	50 717 244

TABLE 3. Sizes of the full and reduced Hilbert spaces corresponding to the ground-state of the 1D Hubbard model at a density $n = N_e/N = 1/3$ and various system sizes.

N_e / N	Full Hilbert space (fixed S_Z)	Symmetry group	Reduced Hilbert space
$4/12$	4 356	T_{12}	366
$6/18$	665 856	T_{18}	36 996
$8/24$	112 911 876	T_{24}	4 704 846

For a system of N lattice sites, there is *a priori* P^N possible configurations. However, the number of electrons ($N_e = N_\uparrow + N_\downarrow$) is conserved and, due to spin rotation symmetry, it is possible to, first, restrict oneself to a fixed value of the component $S_Z = (N_\uparrow - N_\downarrow)/2$ of the total spin. If the minimal value is chosen, all eigenenergies of the various spin multiplets can be found. As seen from the numbers listed in the second column of Tables 1 and 2, the numbers of such configurations are much smaller than P^N although these sizes still grow exponentially fast with the system size.

At this point, it becomes useful to consider space group symmetry. Each irreducible representation can be characterized by a momentum

$$\mathbf{K} = \sum_\mu n_\mu \mathbf{K}_\mu \,, \tag{15}$$

where $\mathbf{K}_\mu$ are the reciprocal lattice vectors (e.g., in 2D, $\mathbf{K}_\mu = \frac{2\pi}{N}\mathbf{T}_\mu \wedge \mathbf{e}_z$) and n_μ are integers. For each value of $\mathbf{K}$, one then considers $\mathscr{G}^P_{\mathbf{K}}$, the so-called little group of $\mathbf{K}$ ($\mathscr{G}^P_{\mathbf{K}} \subset \mathscr{G}_P$), containing group elements g_P such that

$$g_P(\mathbf{K}) = \mathbf{K}. \tag{16}$$

The relevant subgroup of $\mathscr{G}$ to be considered is then

$$\mathscr{G}_{\mathbf{K}} = \mathscr{G}^P_{\mathbf{K}} \otimes \mathscr{T}. \tag{17}$$

For a given symmetry sector $l = (\mathbf{K}, \tau_{\mathbf{K}})$ ($\tau_{\mathbf{K}}$ is one of the irreducible representations of $\mathscr{G}^P_{\mathbf{K}}$) a "symmetric" state $|\alpha\rangle$ can then be constructed from a single configuration $|c\rangle$ as a linear combination which reads, up to a normalization factor,

$$|\alpha\rangle \equiv |\alpha\rangle\{|c\rangle\} = \sum_{g_P\in\mathscr{G}^P_{\mathbf{K}},\, t\in\mathscr{T}} e(\tau_{\mathbf{K}}, g_P)\exp(i\mathbf{K}\cdot\mathbf{T}_t)\,(g_P t)(|c\rangle), \tag{18}$$

where $e(\tau_{\mathbf{K}}, g_P)$ are the characters (tabulated) of the representation $\tau_{\mathbf{K}}$ and $\mathbf{T}_t$ are the translation vectors associated to the translations t. Since the procedure to construct the symmetric state is well defined, it is clear that one needs to keep only a single one of the related configurations $g_{Pt}(|c\rangle)$, this state being called "representative" of the symmetric state $|\alpha\rangle$ and denoted by $|r\rangle = R(|c\rangle)$. This naturally implies,

$$\forall g \in \mathscr{G}_{\mathbf{K}},\ R(g(|c\rangle)) = |r\rangle\,. \tag{19}$$

In other words, one retains only the configurations $|c\rangle$ which can not be related to each other by any space symmetry $g \in \mathscr{G}_{\mathbf{K}}$. The set of all the representatives (labelled from 1 to $\mathscr{N}_l$) defines then unambiguously the Hilbert space $\mathscr{A}_l = \{|\alpha\rangle\}$. Typically, the size of this symmetric subspace is reduced by a factor of card$(\mathscr{G}_{\mathbf{K}})$ compared to the size of the original basis $|c\rangle$. Note that, in some cases, there might exist some configurations $|c\rangle$ (to be eliminated) which do not give rise to any representative. This occurs when there is a subset $\mathscr{G}' \subset \mathscr{G}_{\mathbf{K}}$ such that

$$\begin{gathered}\forall g \in \mathscr{G}', \qquad g(|c\rangle) = |c\rangle \\ \text{and} \\ \sum_{g\in\mathscr{G}'} e(\tau_{\mathbf{K}}, g_P(g)) \quad \exp(i\mathbf{K}\cdot\mathbf{T}(g)) = 0\,.\end{gathered} \tag{20}$$

The choice of the representative among the set of equivalent states (i.e. states related by some symmetry operations of $\mathscr{G}_{\mathbf{K}}$) is, in principle, arbitrary. However, as we shall see in the following, it is convenient to define it as the state $|c\rangle$ of a given class giving rise to the *smallest* integer $N(|c\rangle)$ i.e.

$$N(|r\rangle) = \min_{g\in\mathscr{G}_{\mathbf{K}}}\{N(g(|c\rangle))\} \tag{21}$$

Indeed, for Hubbard-like models, one can construct the configurations by a tensorial product of the up and down spin parts

$$|c\rangle = |c(\uparrow)\rangle \otimes |c(\downarrow)\rangle\,. \tag{22}$$

and $N(|c\rangle)$ is of the form

$$N(|c\rangle) = N'(|c(\uparrow)\rangle) \times 2^N + N'(|c(\downarrow)\rangle)\,, \tag{23}$$

where $N'(|c(\sigma)\rangle)$ corresponds to the binary coding ($N'(|c(\sigma)\rangle) < 2^N$)) of the σ-spin part of the configuration, as explained above (see Eq. (13)). Therefore, the minimization of $N(g(|c\rangle))$ over $g \in \mathscr{G}_{\mathbf{K}}$ can be done in two steps. First, one generates all possible configurations for the up spins (this usually involves a limited number of states) and only representatives $|r(\uparrow)\rangle$ are kept. At this stage, one needs to keep track of the subsets of symmetries $\mathscr{E}_{\mathbf{K}}[r(\uparrow)\rangle]$ of $\mathscr{G}_{\mathbf{K}}$ leaving these representatives invariant. In a second step, one constructs the full set of configurations as a tensorial product of the form $|r(\uparrow)\rangle \otimes |c(\downarrow)\rangle$. The remaining symmetries of $\mathscr{E}_{\mathbf{K}}[|r(\uparrow)\rangle]$ are then applied to the spin $\downarrow$ part and one only retains the configurations $|c(\downarrow)\rangle$ such that,

$$\forall g' \in \mathscr{E}_{\mathbf{K}}[r(\uparrow)\rangle]\ N'(|c(\downarrow)\rangle) \leq N'[g'(|c(\downarrow)\rangle)]\,. \tag{24}$$

The Hilbert space is then formally defined by all the symmetric states $|\alpha\rangle\{|r\rangle\}$ (see Eq.(18)). However, one only needs to store the binary codes $N(|r\rangle)$ as well as the normalizations of the related symmetric states. The normalization of (18) requires some caution. In some cases, there might exist more than a single (i.e the identity operator $\mathbf{1}$) group element g' of $\mathscr{E}_{\mathbf{K}}[|r(\uparrow)\rangle]$ which keeps $|r\rangle$ invariant. Then,

$$\mathscr{F}_{\mathbf{K}}[|r\rangle] = \{g' \in \mathscr{E}_{\mathbf{K}}(|r(\uparrow)\rangle); N'[g'(|c(\downarrow)\rangle)] = N'(|c(\downarrow)\rangle)\} \tag{25}$$

defines the subgroup of such symmetry operations. The sum over the group elements in Eq. (18) should, in fact, be restricted to the elements of the coset $\mathscr{G}_{\mathbf{K}}/\mathscr{F}_{\mathbf{K}}[|r\rangle]$ and the appropriate normalization factor is then given by

$$n(|r\rangle) = [\frac{card\{\mathscr{F}_{\mathbf{K}}[|r\rangle]\}}{card\{\mathscr{G}_{\mathbf{K}}\}}]^{1/2} . \tag{26}$$

The two integers $N(|r\rangle)$ and $n(|r\rangle)$ corresponding to the binary code of a representative and to the normalization of the related symmetric state, respectively, can be combined and stored in a single 64-bits computer word.

In the case of the generic Heisenberg or t–J models, the previous two-step procedure can, in most cases, also be done but it is more involved and I shall only indicate here the spirit of the method. More technical details for spin-1/2 models can be found in Refs. [9] and [16]. The decomposition between up and down spins is replaced here by an (appropriate) partition of the lattice sites into two subsets A and B. The computer word (integer) (13) or (14) coding each configuration contains then two parts, each part corresponding to one subset of the lattice sites, $|c(A)\rangle$ and $|c(B)\rangle$. The partition is chosen in a way such that the group $\mathscr{G}_{\mathbf{K}}$ can be decomposed as

$$\mathscr{G}_{\mathbf{K}} = \mathscr{G}^S_{\mathbf{K}} + \mathscr{S}(\mathscr{G}^S_{\mathbf{K}}), \tag{27}$$

where $\mathscr{S}$ is a group symmetry which fulfils $\mathscr{S}^2 = \mathbf{1}$, the subgroup $\mathscr{G}^S_{\mathbf{K}}$ leaves the two subsets of lattice sites *globally* invariant and $\mathscr{S}(\mathscr{G}^S_{\mathbf{K}})$ is the coset of $\mathscr{S}$ relative to $\mathscr{G}^S_{\mathbf{K}}$. The decomposition (27) is not always unique. For 2D clusters such as the $\sqrt{26} \times \sqrt{26}$ lattice of Fig. 1 ($4p+2$ sites, C_4 point group symmetry) a convenient choice for $\mathscr{S}$ is the translation

$$\mathscr{S} = t(\mathbf{T}_0) \text{ where } \mathbf{T}_0 = \frac{1}{2}(\mathbf{T}_1 + \mathbf{T}_2) \tag{28}$$

and the A–B decomposition just corresponds to the well known checkerboard decomposition. For clusters like the $\sqrt{32} \times \sqrt{32}$ lattice of Fig. 1 (C_{4v} symmetry) a reflection symmetry is more appropriate for $\mathscr{S}$. Note that, in the case of 1D rings, (27) is only possible if the number of sites is of the form $N = 4p+2$ or for very special values of the momentum K (like $K = 0$ or $K = \pi$). Then, the previous procedure can be extended to this case by writing the configurations as

$$|c\rangle = |c(A)\rangle \otimes |c(B)\rangle \tag{29}$$

and by (i) applying the subgroup $\mathscr{G}^S_{\mathbf{K}}$ on $|c(A)\rangle$ to generate its corresponding representative $|r(A)\rangle$ and then (ii) applying all the symmetries of $\mathscr{E}_{\mathbf{K}}[|r(A)\rangle]$ on the part $|c(B)\rangle$. The action of the remaining symmetry $\mathscr{S}$ is considered at last; if the lattice sites are labelled in such a way that

$$\mathbf{x}_{i+N/2} = \mathscr{S}(\mathbf{x}_i), \tag{30}$$

for $i \leq N/2$, the application of $\mathscr{S}$ can be implemented as a simple permutation of the two sub-words of $N(|c\rangle)$.

Construction of the Hamiltonian matrix

We turn now to the implementation of the basic operation $|\Phi_n\rangle \rightarrow H|\Phi_n\rangle$ appearing in Eq. (2) which is always specific to the model Hamiltonian H and constitutes the central part of the Lanczos code. Since the states $|\Phi_n\rangle$ are expressed in terms of the symmetric states $|\alpha\rangle$ of (18) the Hamiltonian matrix has to be computed in this basis. For this purpose, it is only necessary to apply H on the set of representatives $|r_\gamma\rangle$ (labelled from 1 to $\mathscr{N}_l$). In general, each configuration $|r\rangle$ leads to a small number β_{max} (at most equal to the number of terms in H) of generated states,

$$H|r_\gamma\rangle \propto \sum_{\beta=1}^{\beta_{max}} (-1)^{\theta_{\gamma,\beta}} |c_{\gamma,\beta}\rangle , \tag{31}$$

where different signs $(-1)^{\theta_{\gamma,\beta}}$ might arise (in the case of fermion models) from fermionic commutation relations. The matrix is then very sparse. Note that the full Hamiltonian can always be split in a small number of separate terms so that the amplitude of the matrix elements in (31) is just a constant and hence does not need to be stored.

At this point, it becomes necessary to determine the representatives (in binary form) of the various generated states on the right hand side of (31) by applying to them all the symmetries of $\mathscr{G}_\mathbf{K}$. To achieve this, the choice of Eq. (23) for the binary coding of the configurations is very convenient. It is, indeed, a simple way to take advantage of the natural decomposition of the generated states,

$$|c_{\gamma,\beta}\rangle = |c_{\gamma,\beta}(\uparrow)\rangle \otimes |c_{\gamma,\beta}(\downarrow)\rangle . \tag{32}$$

Although, we restrict ourselves here, for sake of simplicity, to the Hubbard model, the following procedure can be straightforwardly generalized to the case of Heisenberg or t–J Hamiltonians by using the form (29) for the configurations $|c_{\gamma,\beta}\rangle$. The calculation of the representative

$$|r_{\gamma,\beta}\rangle_f = R\{|c_{\gamma,\beta}(\uparrow)\rangle \otimes |c_{\gamma,\beta}(\downarrow)\rangle\} \tag{33}$$

can be done in two steps. First, one applies all the symmetries of $\mathscr{G}_\mathbf{K}$ to $|c_{\gamma,\beta}(\uparrow)\rangle$. Since this procedure has to be repeated a large number of times, the function

$$R:\ |c(\uparrow)\rangle \longmapsto |r(\uparrow)\rangle . \tag{34}$$

can be, in fact, tabulated, prior to the actual calculation of the matrix elements. This is made possible since the number of possible states $|c(\uparrow)\rangle$ remains, in general quite modest. This procedure enormously speeds up the calculation of the representatives and justifies the choice of Eq. (23). Note that one also needs, in this preliminary calculation, to store, for each configuration $|c(\uparrow)\rangle$, the corresponding ensembles,

$$\mathscr{R}_{\mathbf{K}}[(|c(\uparrow)\rangle] = \{g \in \mathscr{G}_{\mathbf{K}}; g(|c(\uparrow)\rangle) = |r(\uparrow)\rangle\}. \tag{35}$$

This also requires limited storage since, in most cases, $\mathscr{R}_{\mathbf{K}}[|c(\uparrow)\rangle]$ contains just a single element. In a second step, it is sufficient to apply only the remaining symmetries of $\mathscr{R}_{\mathbf{K}}[|c_{\gamma,\beta}(\uparrow)\rangle]$ to the $|c_{\gamma,\beta}(\downarrow)\rangle$ part. A standard hashing table [17] is then used to find the positions of the representatives in the list. The connectivity matrix connecting the labels of the initial set of representatives to the labels of the new set of generated states is then stored on disk.

Note that the phases related to commutations of fermion operators and/or to the characters of the symmetry operations involved in the transformation of the generated states to their representatives have also to be kept. These phases have the general form,

$$\lambda_{\gamma,\beta} = (-1)^{\theta_{\gamma,\beta}} e(\tau_{\mathbf{K}}, g_P(g^*_{\gamma,\beta})) \exp(i\mathbf{K}\cdot\mathbf{T}(g^*_{\gamma,\beta})), \tag{36}$$

where $g^*_{\gamma,\beta}$ is a group element of $\mathscr{R}_{\mathbf{K}}[|c_{\gamma,\beta}(\uparrow)\rangle]$ such that

$$|r_{\gamma,\beta}\rangle_f = g^*_{\gamma,\beta}(|c_{\gamma,\beta}\rangle), \tag{37}$$

which depends on γ and β. It is easy to show that, if there exists more than a single group element which fulfils (37) then, all of them lead to the same phase factor (36).

Since the number of possible different phases given by (36) is quite small, it is possible to store the $\lambda_{\gamma,\beta}$ (in some convenient integer form) together with $N\{|r_{\gamma,\beta}\rangle_f\}$ on a small number of computer bits. For an Hilbert space of 10^8 (hundred millions) representatives with typically an average of ~ 50 images per state, the occupation of the disk is of the order of 5 Gw i.e. 40 Gb. This can even be reduced by a factor of two by using each computer word to store the information corresponding to two images instead of a single one. Once it has been generated, the Hamiltonian matrix is cut into several pieces (typically of the order of 10 to 100 Mw) and the various parts are successively read from the disk in order to calculate $H|\Phi_n\rangle$. The best performances are obtained when the calculation using the nth part of the matrix and the access to the disk to read the (n+1)th part are simultaneous. Note that the Lanczos algorithm as it has been described above is well adapted to be implemented on a vector supercomputer (e.g. on a NEC SX8) where parallisation on a single node can also be achieved.

How to play with boundary conditions

Finite size effects might be quite large. However, the deviations of the various physical quantities from their values in the thermodynamic limit is generally directly related to the system size N in some systematic way. When the size dependence of the finite size corrections is known or can be extracted from the set of numerical data, a systematic finite size scaling analysis can be performed to get, in most cases, very accurate results as shown, later, in the case of 1D systems.

Finite size corrections can also, by themselves, give valuable physical information on the physical nature of the electronic system. To illustrate this aspect, let us, for example, consider the behavior of the GS energy under a twist in the boundary conditions, which can be physically interpreted as a magnetic flux threading the ring (1D) or the torus (2D or 3D). The local curvature of the GS energy with respect to the twist gives directly the charge [18] stiffness and the spin stiffness [19] in metallic and localized spin systems, respectively. Note that, in the case of the spin systems, the two different spin species ($\uparrow$ or $\downarrow$) have to feel opposite fluxes so that the spin current (instead of the charge current which is absent here) is coupled to the "electromagnetic" field. In addition to the determination of the stiffness of the system, the investigation of the complete GS energy vs flux curve in metallic systems is a very useful tool to characterize the exact nature of the carriers. In particular, the existence of pairs in correlated 1D [20, 21, 22, 23], ladders [24, 25] or 2D [26, 27, 28] systems can be revealed by such a study.

Let us assume, for simplicity, that the translation vectors are of the form $\mathbf{T}_\mu = (0,...,0,L_\mu,0,...,0)$, i.e. the torus is not tilted compared to the underlying lattice. A flux Φ (in unit of the flux quantum) through one of the hole of the d-dimensional torus, or equivalently a twist in the boundary conditions along some direction $\mathbf{e}_\mu$ can be introduced by the substitution,

$$t_{\mathbf{xy}}\, c^\dagger_{\mathbf{x},s} c_{\mathbf{y},s} \rightarrow t_{\mathbf{xy}}\, c^\dagger_{\mathbf{x},s} c_{\mathbf{y},s} \exp\left(2i\pi \frac{\Phi}{L_\mu}(\mathbf{x}-\mathbf{y})\cdot \mathbf{e}_\mu\right). \tag{38}$$

Other gauge choices are possible, but (38) is the most convenient since it preserves translation symmetry so that the whole analysis described before can be used.

To be more specific, I now consider the simple example of a Hubbard model (11) with on-site interaction $U n_{\mathbf{x},\uparrow} n_{\mathbf{x},\downarrow}$ (i.e. assuming $V_{\mathbf{xy}} = (U/2)\delta(\mathbf{x}-\mathbf{y})$) on a 1D ring). It is known that repulsive and attractive models belong to different universality classes; while the $U > 0$ model is an ordinary Luttinger liquid [29, 30, 31] (see later), the $U < 0$ model has a spin gap associated to the existence of on-site singlet pairs and belongs to the Luther-Emery universality class [32]. This last feature implies that the GS energy vs flux has two (almost) degenerate minima at $\Phi = 0$ and $\Phi = 1/2$. In other words, if $U < 0$, the relevant carriers are pairs carrying a charge $2e$ and, hence, flux quantization in terms of half of a flux quantum is expected in the ring. The repulsive case might also exhibit a second minimum. However, it is usually much less pronounced and a finite size scaling of the energy difference between the two minima reveals that this feature disappears in the thermodynamic limit (see e.g. Ref. [23] for a similar analysis).

Dynamical properties

Continued fraction expansion

In contrast to the Quantum Monte Carlo (QMC) technique (which requires analytical continuation) the ED method provides a more straightforward way to calculate dynamical correlations directly on the real frequency (or time) axis [33]. I shall here describe the method as well as practical applications to physical quantities. Time dependent correlation functions such as

$$C(t) = \langle\Psi_0|A(t)A^\dagger(0)|\Psi_0\rangle \,, \tag{39}$$

where $A(t) = e^{iHt}Ae^{-iHt}$ is a time-dependent operator, are of physical importance to investigate dynamical properties of correlated systems. It is often more convenient to consider its Fourier transform,

$$\tilde{C}(z) = \langle\Psi_0|A\frac{1}{z-H+E_0}A^\dagger|\Psi_0\rangle \,, \tag{40}$$

since many measurable frequency-dependent response function can be written as a spectral function ,

$$I(\omega) = -\frac{1}{\pi}\lim_{\varepsilon\to 0}\mathrm{Im}\,\tilde{\mathrm{C}}(\omega + \mathrm{i}\varepsilon) \,. \tag{41}$$

The single particle spectral function (seen in angular resolved photo-emission spectroscopy), the dynamical spin structure factor (seen inelastic neutron scattering experiments) or the optical conductivity (seen in optics) are such examples.

The Lanczos method provides a very simple way to calculate *exactly* (40) on finite clusters [33]. As noted above, this should be contrasted to the Max Ent method used in QMC. One proceeds in two steps; first, the GS $|\Psi_0\rangle$ is calculated by the Lanczos algorithm as described in Sec. II and the operator $A^\dagger$ is applied to it to construct a new orthogonal state,

$$|\tilde{\Phi}_1\rangle = \frac{1}{(\langle\Psi_0|AA^\dagger|\Psi_0\rangle)^{1/2}}A^\dagger|\Psi_0\rangle \,. \tag{42}$$

Note that the matrix elements of the operator $A^\dagger$ can be calculated in the same way as described in Sec. II.D for the Hamiltonian. However, $A^\dagger$ usually connects different symmetry sectors so that both Hilbert subspaces (Sec. II.C.) have to be constructed first. Using (42), (40) takes the form,

$$\tilde{C}(z) = \langle\Psi_0|AA^\dagger|\Psi_0\rangle\langle\tilde{\Phi}_1|(z'-H)^{-1}|\tilde{\Phi}_1\rangle \,, \tag{43}$$

where $z' = z + E_0$.

In a second step, a second Lanczos iterative procedure (2), up to order M, is performed starting with (42) as an initial state. According to (2) one gets,

$$z' - H = \begin{matrix} z'-\tilde{e}_1 & -\tilde{b}_2 & \dots & 0 \\ -\tilde{b}_2 & \ddots & \ddots & \vdots \\ \vdots & \ddots & \ddots & -\tilde{b}_M \\ 0 & \dots & -\tilde{b}_M & z'-\tilde{e}_M \end{matrix} \tag{44}$$

where the matrix is expressed in the new basis $\{|\tilde{\Phi}_n\rangle\}$ obtained from (2) with the choice $|\Phi_1\rangle = |\tilde{\Phi}_1\rangle$. Straightforwardly, (43) gives

$$\tilde{C}(z) = \langle\Psi_0|AA^\dagger|\Psi_0\rangle\,\frac{D_2}{D_1} \,, \tag{45}$$

where D_n is defined as a determinant $D_n = \det\Delta_n$ and Δ_n is the $(M-n+1)\times(M-n+1)$ matrix,

$$\Delta_n = \begin{vmatrix} z' - \tilde{e}_n & -\tilde{b}_{n+1} & \dots & 0 \\ -\tilde{b}_{n+1} & \ddots & \ddots & \vdots \\ \vdots & \ddots & \ddots & -\tilde{b}_M \\ 0 & \dots & -\tilde{b}_M & z' - \tilde{e}_M \end{vmatrix} \tag{46}$$

The various determinants D_n (for a given z) can be calculated recursively by using

$$D_n = (z' - \tilde{e}_n)D_{n+1} - \tilde{b}_{n+1}^2 D_{n+2} \ , \tag{47}$$

for $1 \leq n \leq M-2$. This formula is of practical importance and gives directly the spectral weight (41) from the knowledge of the tridiagonal matrix (44). It is important to notice that the intermediate states $\{|\tilde{\Phi}_n\rangle\}$ do not need to be stored to calculate the final results (41).

For completeness, let me also mention that $\tilde{C}(z)$ can be written as a continued fraction expansion as it is clear from combining (45) and (47),

$$\tilde{C}(z) = \cfrac{\langle\Psi_0|AA^\dagger|\Psi_0\rangle}{z + E_0 - \tilde{e}_1 - \cfrac{\tilde{b}_2^2}{z + E_0 - \tilde{e}_2 - \cfrac{\tilde{b}_3^2}{z + E_0 - \tilde{e}_3 - \dots}}} \tag{48}$$

This new form explains naturally why the procedure can be truncated after N_{it} iterations (typically N_{it} is of order 200) when a good convergence is reached.

The physical interpretation of the spectral function is clear. Indeed, using a complete set of eigenvectors, it can be re-written as

$$I(\omega) = \sum_m |\langle\Psi_m|A^\dagger|\Psi_0\rangle|^2 \delta(\omega - E_m + E_0) \ . \tag{49}$$

The poles and the associated weights then determine the dynamics of the operator. In practice, the operator $A^\dagger$ has some symmetry properties and is characterized by well defined quantum number. This means that, due to selection rules, the set of eigenstates $|\Psi_m\rangle$ is often restricted to an irreducible representation of the symmetry group. The power of the continued fraction expansion is that it does not require the actual calculation of the eigenstates and the eigenvectors. Note that z must contain a small but non-zero imaginary part i.e. $z = \omega + i\varepsilon$ to avoid true divergences at frequencies corresponding to the poles of the spectral function.

Examples of dynamical correlation functions

The interest of the investigation of dynamical correlations is two-fold. Fist, in the vicinity of phase transitions, one might expect some operators to exhibit low energy

TABLE 4. Examples of dynamical correlation functions accessible in experiment. The related operator A involved in the correlator (see Eq. 40) as well as the commonly used notations are also indicated for each case.

Correlation functions	Notations	Operators	Experiments
Spectral function	$A(\mathbf{k},\omega)$	$A = c_{\mathbf{k},\sigma}$	Photo-emission
Spin structure factor	$S_{zz}(\mathbf{q},\omega)$	$A = S^z_{\mathbf{q}}$	Inelastic neutron scattering
Optical conductivity	$\sigma_{xx}(\omega)$	$A = j_x$	Infrared reflectivity
Two magnons	$R(\omega)$	$A = \sum_{\mathbf{k}} R_{\mathbf{k}} \mathbf{S}_{\mathbf{k}} \cdot \mathbf{S}_{-\mathbf{k}}$	Raman scattering

fluctuations. The calculation of dynamical correlations e.g. in the frustrated Heisenberg model [34] has given further insights on the transition towards a quantum disordered state. Similarly, in a search for time reversal symmetry breaking, chiral fluctuations have been investigated in the t-J model [35]. The second motivation for the calculation of dynamical correlations comes from experiment. Indeed, various experiments (see Table 4) provide a direct measure of such correlation functions [36]. Results of Lanczos calculations can be directly compared to angular resolved photo-emission in the Mott insulator [37] and the doped phase [38, 39], inelastic neutron scattering [36], infrared reflectivity [40, 26, 41] and Raman scattering [42] experiments, to cite only a few.

COMPARISON WITH OTHER NUMERICAL METHODS

Density Matrix Renormalization Group

A numerical Renormalization Group (RG) was successfully implemented by Wilson [43] for the Kondo problem in 1975. Later, in 1992, White [44] invented a real space RG devoted to extended (or lattice) correlated systems. The major breakthrough was to introduce the density matrix to select the relevant low energy states. Since complete reviews on the Density Matrix Renormalization technique are available in the literature [45, 1] I will only briefly sketch the outline of the method.

The procedure is based on a gradual build up of a system while keeping the number of degrees of freedom (i.e. the number of states) fixed. It can be summarized as follows:

- 1. Start with block B and construct "super-block" B-site$_1$-site$_2$-B,
- 2. Diagonalize super-block Hamiltonian to get $\Psi(i_1,i_2,i_3,i_4)$,
- 3. Calculate the *reduced density matrix*,
  $\rho(i_1,i_2,i'_1,i'_2) = \sum_{i_3,i_4} \Psi(i_1,i_2,i_3,i_4)\Psi(i'_1,i'_2,i_3,i_4)$
  and keep m eigenvectors of *maximum* weights,
- 4. Re-write all operators in the new truncated basis:
  B $\Rightarrow$ B'=B-site $\Rightarrow$ step 1.

In principle, the efficiency (i.e. the possibility of keeping a small number m of states carrying a large fraction of the reduced density matrix weight) of the method is much higher for low dimensional systems due to a simpler form of the effective interaction between blocks. In that case, very large systems with, typically, up to 200 sites can easily be handled. In addition, time dependent or dynamical correlations can also be

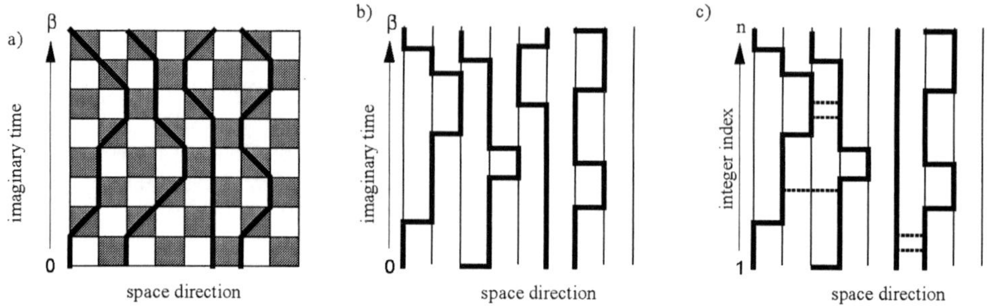

FIGURE 2. Comparison between vaious world-line representation: (a) discrete time; (b) continuous time; (c) SSE representation. Reproduced from Ref. [46]

computed with DMRG [45]. Note that, in contrast to the ED technique, open boundary conditions are much more favorable for DMRG. Example of DMRG calculations for 1D Hubbard chains will be given in the following.

Quantum Monte Carlo Methods

The Quantum Monte Carlo (QMC) technique has made tremendous progress in the last few years and I will only sketch briefly basic concepts. Interested readers can e.g. refer to Ref. [46] for details.

For classical systems the Monte Carlo algorithm relies upon an iterative stochastic procedure in configuration space $\{i\}$. A major breakthrough was realized in 1953 when Metropolis and collaborators [47] proposed a simple algorithm to sample the probability distribution $p(i)$ using the transition probability $P(i \to j) = \min[1, \frac{p(j)}{p(i)}]$.

The generalization of Monte Carlo ideas to quantum systems was then made possible thanks to the Trotter-Susuki formula [48] of the partition function,

$$\begin{aligned} Z &= Tr[\exp(-\frac{\beta}{M}(H_1+H_2))]^M \\ &= Tr\ [\exp(-\frac{\beta}{M}H_1)\exp(-\frac{\beta}{M}H_2)]^M + O(1/M^2) \qquad (50) \\ &= \Sigma\ \langle\Psi_1|e^{-\frac{\beta}{M}H_1}|\Psi_2\rangle...\langle\Psi_{2M}|e^{-\frac{\beta}{M}H_2}|\Psi_1\rangle, \qquad (51) \end{aligned}$$

assuming a decomposition of the Hamiltonian into $H_1 + H_2$ associated to e.g. different types of links. This representation enables then to treat e.g. quantum spins in *d dimensions* as a classical problem in *d+1 dimensions*. For example, in the case of a spin chain, H_1 and H_2 can be chosen on the even and odd bonds, respectively, leading to the so-called "world-line" representation shown in Fig. 2(a) (the lines refer to the trajectory of let's say the up spins in imaginary time).

Recently, it was shown that the continuous time limit [49] could be taken, $\Delta\tau \Rightarrow 0$, $M \Rightarrow \infty$, hence eliminating systematic errors due to the discretization of the imaginary time axis (see Fig. 2(b)). In the Stochastic Series Expansion (SSE) representation [50] a Taylor expansion of Z is performed,

$$
\begin{aligned}
Z = & \; Tr \;\; \exp(-\beta H) = \sum_n \frac{\beta^n}{n!} Tr(-H)^n \\
= \sum_n \frac{\beta^n}{n!} & \;\; \Sigma \;\; \langle \Psi_1 | - H_{b_1} | \Psi_2 > \ldots \\
& \ldots \;\; < \Psi_n | - H_{b_n} | \Psi_1 >,
\end{aligned}
\tag{52}
$$

leading to the world-line representation of Fig. 2(c).

In contrast to ED and DMRG, QMC techniques suffer from the so-called "minus sign problem" for frustrated spin systems and systems where fermionic commutation rules play a role (except in 1D fermionic systems where exchange of fermions only occurs through the boundaries). However, e.g. in the case of unfrustrated quantum spins models, QMC enables to treat very large systems (of the order of thousand spins) down to very low temperatures. In fact, such achievements have been made possible recently thanks to the developpement of non-local updates such as “loop” or “worm” algorithms (see Refs. [46] for details).

We finish this short summary of QMC techniques by briefly mentioning the Determinental QMC technique designed for fermions in dimension higher than 1[51, 2] based on a Hubbard-Stratonovitch transformation to decouple the interaction term. To summarize: (i) one first uses the Trotter-Suzuki formula to decouple the kinetic and interaction terms, (ii) one decouples the interaction term using

$$
e^{-\Delta\tau U n_{i,\uparrow} n_{i,\downarrow}} \propto \sum_{s=\pm 1} e^{-\Delta\tau s_{i,l} \lambda (n_{i,\uparrow} - n_{i,\downarrow})}\,. \tag{53}
$$

where $s_{i,l}$ is the Ising variable at site i and imaginary time slice l and (iii) integrates out fermionic variables so that the partition function can be written as,

$$
Z = \sum_{s=\pm 1} \det M^+(s) \det M^-(s), \tag{54}
$$

where $M^{\pm}(s)$ are matrices (of the dimension of the system size) depending on the Ising variables $\{s\}$. A Monte Carlo sampling of the sum over the Ising configurations can then be realized.

Partial conclusions

The Lanczos exact diagonalization technique is a versatile and unbiased technique which can treat frustrated and fermionic systems, long range interactions, etc... It also gives access to static and dynamical correlations as well as to the precise symmetries and quantum numbers of the GS and of the low energy excited states, a useful information to investigate e.g. quantum phase transitions. Unfortunately, it is limited to small clusters where finite size corrections have to be taken into account. Larger systems can sometimes be handled using the Contractor Renormalization procedure [52] to construct effective Hamiltonians.

The QMC methods are (mostly) finite-T methods which can reach large systems [46]. The so-called Max-Ent technique is useful for dynamics. In order to beat "critical slowing down" more involved updates (such as the loop algorithms) can be used. However, when frustration or doping ($n \neq 1/2$) is present the "minus sign problem" prevents to reach low temperatures.

The DMRG technique is (mostly) a T=0 technique [45, 1] also capable to reach large systems. New developments for excited states and/or time-dependent quantities are now available. DMRG is very accurate in 1D (or quasi-1D) (frustration & doping *is* possible) but it has not been generalized really satisfactorily in 2D yet.

In the next two Sections, I will present simple dedactic exemples to illustrate the use of ED and DMRG techniques. For simplicity, these exemples are chosen in the area of 1D correlated systems [31].

APPLICATION NO. 1: FINITE SIZE SCALING ANALYSIS FOR 1D FERMIONS

The XXZ chain

It is instructive to start this Section by considering first a simple practical case, namely the so-called XXZ chain,

$$H_{\mathrm{XXZ}} = \frac{J}{2}\sum_i \{(S_i^+ S_j^- + S_i^- S_j^+) + J_z S_i^z S_j^z\}. \tag{55}$$

As we shall see further on, the XXZ chain can be mapped, using a Jordan-Wigner transformation, onto a chain of spinless fermions with NN repulsion. The SU(2)-symmetric Heisenberg point $J_z = J$ was solved by Bethe in 1931 [53], so-called Bethe-Ansatz, and generalized by Luther and Peschel in 1975 [54] for arbitrary spin anisotropy. This model constitutes then a simple "laboratory" to "test" our numerical techniques and the field-theoretic ideas briefly sketched below.

At this stage, in order to construct an appropriate field theory(for details see [31]), it is convenient to introduce transformations of the spin operator in the *long wavelength* limit, using conjugate bosonic field (the so-called "bosonisation" procedure) Φ and Π,

$$S^+(x) = \frac{e^{-i\theta(x)}}{\sqrt{2\pi a}}(\exp(-i\pi x/a) + \cos 2\Phi(x)), \tag{56}$$

$$S^z(x) = -\frac{1}{\pi}\partial_x\Phi(x) + \exp(i\pi x/a)\frac{\cos 2\Phi(x)}{a\pi}, \tag{57}$$

where $\theta = \pi \int \Pi dx$. Physically, the θ and Φ angles define the orientation of the spin at position x. Inserting such expressions in the Hamiltonian, one obtains a Sine-Gordon field-theory model,

$$H_{\mathrm{SG}} = \int \frac{dx}{2\pi}(uK(\pi\Pi)^2 + \frac{u}{K}(\partial_x\Phi)^2) + g\int dx \cos 4\Phi, \tag{58}$$

where the prefactors uK and $\frac{u}{K}$ have the physical meaning of stiffness and inverse compressibility of an elastic chain characteristic of a "Luttinger liquid" (LL), the paradigm of the 1D electron gas. The last term stands for the interactions of strength g. In principle, the connection with the original Hamiltonian (55) can be obtained in perturbation in J_z, e.g., $g = -\frac{2aJ_z}{(2\pi a)^2}$, $K = 1 - \frac{2J_z}{\pi J} + ...$ and $u/aJ = 1 + \frac{2J_z}{\pi J} + ...$ Fortunately, for arbitrary anisotropy, K and u can be obtained with a high accuracy by the numerical procedure explained below. Determining K_ρ is of central importance, since K_ρ governs the asymptotic algebraic behavior of the correlations functions,

$$\langle S^z(x)S^z(0)\rangle \sim (-1)^x \frac{1}{x^{2K}}, \tag{59}$$

$$\langle S^+(x)S^-(0)\rangle \sim (-1)^x \frac{1}{x^{1/2K}}. \tag{60}$$

In the XX limit corresponding to $g = 0$ one gets $K = 1$. On the other hand, in the Heisenberg case $J_z = J$, the SU(2) invariance immediately implies $K = 1/2$.

Before describing the numerical procedure, it is interesting to first explore the stability of the LL w.r.t. g using a Renormalization Group (RG) analysis [55]. Under a change of scale $l = \ln(L) \to l + dl$ one gets,

$$\frac{dK}{dl} = -g^2, \tag{61}$$

$$\frac{dg}{dl} = (2-4K)g + O(g^2), \tag{62}$$

equivalent to the Kosterlitz-Thouless (KT) equations for the 2D classical XY model (superfluid transition). Note that the linear term of the second equation vanishes for $K = K^* = \frac{1}{2}$. Therefore, the LL is stable (unstable) when $K \geq 1/2$ ($K < 1/2$). By borrowing the following expression from the Bethe-Ansatz,

$$K = \frac{\pi}{2(\pi - \arccos(J_z/J))}, \tag{63}$$

one can immediately deduce that,

- for $|J_z/J| \leq 1$: a Luttinger liquid phase occurs,
- for $|J_z/J| > 1$: an Ising (AF or F) gapped phase is stable.

Note that for e.g. $J_z/J > 1$, the last term of the Sine-Gordon Hamiltonian (58) is minimized for $2\Phi = 0$ and π which, according to (57), correspond indeed to the two-fold Ising GS.

General relations

Finite size scaling analysis are usually necessary to complement the Bosonisation and RG results (see above) in order to extract reliable and *quantitative* physical results. Here, we shall consider the more general framework of e.g. a 1D extended Hubbard model so that we now deal with both spin *and* charge degrees of freedom. Since the model (11) belongs generically to the LL universality class its low energy spin and charge collective modes have linear dispersion relations characterized by spin and charge velocities u_ρ and u_σ [29, 30, 31] (new labels "ρ" and "σ" refer now to charge and spin degrees of freedom). The density of state (or equivalently the occupation number) exhibits a singularity of the form [29],

$$N(\omega) \sim |\omega|^\alpha , \tag{64}$$

where α is a non-universal exponent. Remarkably enough, the exponents characterizing the long distance behavior of all the various correlation functions depend only on a single parameter K_ρ similar to the parameter K encountered in the XXZ chain. The parameters α and K_ρ are simply related by

$$\alpha = \frac{1}{4}(K_\rho + \frac{1}{K_\rho} - 2) . \tag{65}$$

As the XXZ chain, the on-site Hubbard model is integrable [56] and one can then rely on the Bethe-Ansatz equations to calculate the various parameters of the model[57]. However, the generic model (11) is not integrable and ED is a very efficient technique to compute the LL parameters. The method summarized here was first applied to quantum spin chains [58] and, later on, to the 1D t-J model [59], the extended Hubbard model [60, 23] and models of spinless fermions [61] with extended interactions. Its remarkable accuracy is due to the existence of conformal invariance identities [62] which relate the LL parameters to simple physical quantities easily computed numerically by the Lanczos method. This avoids the hard task of a direct calculation of the long distance behavior of equal time correlation functions and, hence, greatly simplifies and improves the accuracy the characterization of the LL. For instance the Drude weight πD (D is the charge stiffness) and the charge velocity u_ρ are related by [57]

$$\pi D = 2 u_\rho K_\rho . \tag{66}$$

The quantities u_ρ and D can be directly obtained on finite systems. The charge stiffness weight for a single chain is given by the Kohn formula [18],

$$D = \frac{\partial^2 (E_0/L)}{\partial \phi^2} , \tag{67}$$

where L is the chain length and $\phi = 2\pi \frac{\Phi}{L}$. The charge velocity can be extracted from the difference in energy of the two singlet symmetry sectors $k = 0$ and $k = 2\pi/L$ for closed shell configurations [63]. K_ρ is then obtained from Eq. (66).

In practice, the charge stiffness can be computed from the local curvature at a minimum of the $E_0(\phi)$ curve. Typically, a scaling behavior of the form,

$$D(L) = D(\infty) + \frac{A}{L^2} + \mathcal{O}(1/L^2) \tag{68}$$

is usually very well satisfied. Interestingly enough, two different sets of boundary conditions (typically periodic and anti-periodic) lead generally to coefficients A in (68) of different signs so that a very accurate determination of $D(\infty)$ can be made. The charge velocity u_ρ follows the same scaling as (68) and hence the numerical value of α can be easily computed. More extensive studies of the quarter filled t-U-V model can be found in Refs. [60, 23].

Note that the conformal invariance picture implies the simple identity,

$$\frac{\pi}{2}\frac{u_\rho}{K_\rho} = \frac{1}{\kappa}, \tag{69}$$

relating the two previous quantities to the inverse compressibility $\kappa^{-1} = \frac{1}{2}\frac{\partial^2(E_0/L)}{\partial n^2}$. The above relation enables one to check the consistency of the LL picture by numerically investigating the compressibility [64]. Alternatively, (69) can also be used as a test identity to investigate whether an electronic system belongs to the LL (or Luther-Emery) class. Such an analysis was done, for example, in the case of the extended Hubbard chain [60] or the t–J ladder [65, 25].

A second consistency check can be obtained from the finite size scaling of the GS energy [62],

$$\frac{E_0(L)}{L} = e_\infty - \frac{\pi(u_\rho + u_\sigma)}{6L^2}c + \mathcal{O}(1/L^2), \tag{70}$$

where the central charge c is given by the symmetry of the model (here $c = 1$) and the spin velocity u_σ can be calculated independently. Such an analysis was carried out e.g. for the t-U-V model [23].

In principle, corrections to scaling laws can also be predicted from the knowledge of anomalous dimensions. Including logarithmic corrections in the finite size scaling analysis is sometimes necessary. The S=3/2 quantum spin chain is, for example, a case where a more extended analysis including the logarithmic corrections was shown to be essential [66].

The case of the spinless fermion chain

As a practical example we now consider the case of a chain of spinless fermions with repulsive interaction V at half-filling ($n = 1/2$). In fact, starting from the above XXZ chain, one can use the Jordan-Wigner transformation to introduce a fermionic representation of the spin algebra as follows,

$$S_i^- = (-1)^i c_i \exp(i\pi \sum_{j=1}^{i-1} n_j), \tag{71}$$

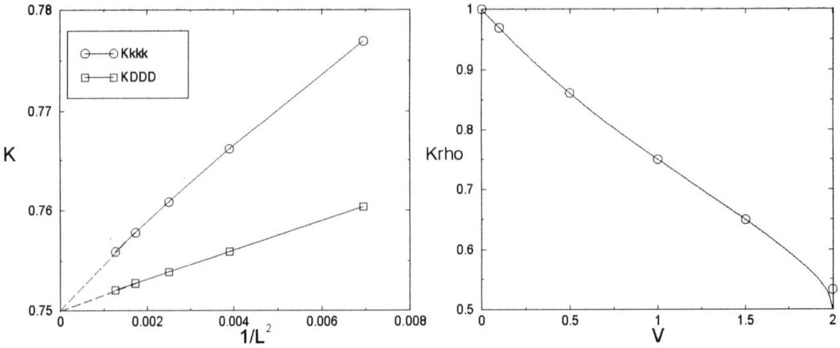

FIGURE 3. LL exponent K of the spinless fermion t-V chain (solvable by Bethe-Ansatz) obtained by ED on periodic rings; (a) finite size scaling using different conformal invariance relations using the compressibility (circles) or the Drude weight (squares); (b) extrapolated value vs V/t compared to the BA solution. S. Capponi, Thesis 1999 (Toulouse).

to map the XXZ chain into the spinless fermion chain (t-V model),

$$H_{\text{t-V}} = -t\sum_i (c_{i+1}^\dagger c_i + H.C.) \tag{72}$$

$$+ \; V(n_i - \frac{1}{2})(n_{i+1} - \frac{1}{2}), \tag{73}$$

where $V = J_z$ and $t = J/2$.

Bethe's equations solved by Haldane [67] provide exact expressions quite useful to test the numerical procedure. Phase separation (ferromagnet in the XXZ framework) and a Metal-Insulator transition (antiferromagnet in the XXZ framework) occur for $V/t < -2$ and $V/t > 2$ (KT transition) while the Luttinger liquid is stable in the range $-2 \le V/t \le 2$. In terms of λ defined by $\cos 2\lambda = -\frac{V}{2t}$ simple expressions can be obtained for the charge velocity, the charge stiffness, the exponent K, etc...

$$u = \frac{\pi \sin 2\lambda}{\pi - 2\lambda}, \tag{74}$$

$$D = \frac{\pi^2 \sin 2\lambda}{4\lambda(\pi - 2\lambda)}, \tag{75}$$

$$K = \frac{D}{u} = \pi/(4\lambda). \tag{76}$$

An example of numerical calculation of K is shown in Figs. 3. The finite size scaling of K shown in Fig. 3(a) for $V/t = 1$ follows very closely the expected $1/L^2$ law and extrapolates to the exact value $3/4$ (for $n = 1/2$ and $V/t = 1$). The extrapolated values vs V/t shown in Fig. 3(b) agree quite well with the exact BA curve, except very near $V/t = 2$ at the onset of the Insulating phase. More extensive ED studies of the extended spinless fermion model (where no exact solution is available) can be found e.g. in Ref. [61].

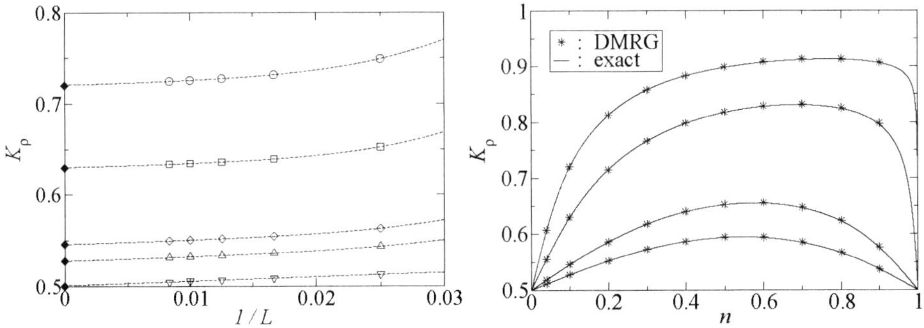

FIGURE 4. LL exponent K_ρ of the Hubbard chain (solvable by Bethe-Ansatz) obtained by DMRG on opened chains; (a) finite size scaling; (b) extrapolated value vs n for $U = 1, 2, 6, 10, \infty$ (top to bottom) compared to the exact BA (full lines); reproduced from S. Ejima et al., Europhys. Lett. **70**, 492 (2005).

The case of the Hubbard chain

We now turn to the case of the Hubbard chain which bears an additional complexity brought by the spin degrees of freedom complicating the determination of the charge velocity [63]. In a recent DMRG study [68] performed on open chains (preventing the direct calculation of the Drude weight by the above Kohn's method), the simple behavior of the charge structure factor in the long wavelength limit ($q \to 0$),

$$S(\mathbf{q}) \sim \frac{K_\rho}{2\pi} q, \tag{77}$$

has been used to determined K_ρ very accurately. Results are reproduced in Fig. 4 showing $1/L$ finite size corrections (in contrast to $1/L^2$ for periodic chains) (Fig. 3(a)) and excellent agreements with BA (Fig. 4(b)).

About Metal-Insulator transition and charge localization in 1D

Finite size scaling is also a powerful tool to investigate quantum phase transitions. The extended Hubbard model (11) on a ring with extended interactions provides a simple example of a metal-insulator transition [60, 30]. At commensurate densities, for example $n = 1/3$, strong interactions extended to intermediate distances can localize the particles in a charge density wave (CDW) state. This occurs when Umklapp scattering becomes a relevant perturbation [30]. This is also the case for the spinless fermion model at $n = 1/2$ and $V/t > 2$. The insulator phase can be characterized by a vanishing of the charge stiffness. This can be evidenced numerically by analyzing the finite size behavior of D. Typically, the $1/L^2$ behavior is no longer fulfilled and the data suggest $D(\infty) = 0$. The formation of a (pinned) CDW can also be detected by the formation of a gap in the single excitation spectrum. This means that the *single* particle gap

$$\Delta_{c,1} = E_0^{N_e+1} + E_0^{N_e-1} - 2E_0^{N_e} \tag{78}$$

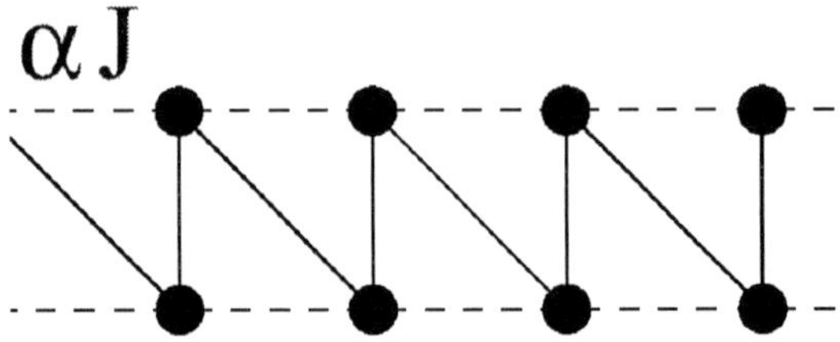

FIGURE 5. The frustrated Heisenberg chain.

(where $N_e = nN$ is the number of electrons) scales to a finite value in the thermodynamic limit.

APPLICATION NO. 2: DIMERIZED SPIN CHAINS.

The frustrated chain

As a second simple instructive example to illustrate the ED and DMRG techniques, I will deal in this Section with the case of the dimerized spin chain. Complementary aspects can also be found in Ref. [8]. Some spin-1/2 spin chains like $CuGeO_3$ might undergo at low temperature a transition towards a spontaneously dimerized state schematically depicted in Fig. 5. To some extents, the simplest approach to understand this phenomena consists in adding to the Heisenberg chain model a next-nearest neighbor (NNN) magnetic frustrating interaction J_2,

$$H_{\rm F} = J\sum_i \mathbf{S}_i \cdot \mathbf{S}_{i+1} + J_2 \sum_i \mathbf{S}_i \cdot \mathbf{S}_{i+2}\,, \tag{79}$$

where the amount of frustration can be defined by a single parameter $\alpha = J_2/J$.

For large enough frustration, one expects a spontaneous dimerization. Indeed, for $\alpha = 1/2$, as shown by Majumdar and Gosh [70] (MG), the GS is doubly degenerate and consists in a product of singlets on all the even (or all the odd) bonds, a so-called Valence bond Crystal (VBC). This can be easily proven by re-writing the MG Hamiltonian,

$$H_{MG} = \sum_i \mathbf{S}_i \cdot \mathbf{S}_{i+1} + \frac{1}{2}\mathbf{S}_i \cdot \mathbf{S}_{i+2}\,, \tag{80}$$

as

$$H_{MG} = \frac{1}{4}\sum_i (\mathbf{S}_i + \mathbf{S}_{i+1} + \mathbf{S}_{i+2})^2 - \frac{3}{4}\sum_i (\mathbf{S}_i)^2\,, \tag{81}$$

so that a spin 1/2 is expected on each "triangle" $(i, i+1, i+2)$ leading to an energy $E_{MG} = -\frac{1}{2}S(S+1) = -\frac{3}{8}$.

In the Bosonization scheme frustration leads to a term,

$$H_{\text{pert}} = (\alpha - \alpha_c)\cos(4\Phi), \tag{82}$$

which becomes a relevant perturbation as soon as $\alpha > \alpha_c$ leading to two solutions $2\Phi = \pm\pi/2$. A numerical DMRG investigation carried out with up to 200 sites by White and Affleck [71] revealed indeed a Kosterlitz-Thouless transition at $\alpha_c \simeq 0.2412$ and a dimerized phase characterized by a spin gap whose behavior agrees with the KT prediction,

$$\Delta^{01} \propto \exp(-\text{Cst}/(\alpha - \alpha_c)). \tag{83}$$

The "spin-Peierls" model

In the spin-Peierls mechanism the lattice (phonons) is also expected to play an important role through e.g. a magneto-elastic effect. In other-word, a deformation of a (3D) lattice following (or in some cases inducing) the dimerization discussed above will produce a dimerization of the NN AF exchange couplings [69]. This effect can be described by the following Hamiltonian:

$$H_{\text{SP}} = \sum_i (1 + (-1)^i \delta)\mathbf{S}_i \cdot \mathbf{S}_{i+1} + \alpha J \sum_i \mathbf{S}_i \cdot \mathbf{S}_{i+2}, \tag{84}$$

where a fixed lattice dimerization δ (in reduced units) has been assumed. The extra *explicit* dimerization $H_D = J\delta\sum_i(-1)^i\mathbf{S}_i \cdot \mathbf{S}_{i+1}$ translates in the Bosonisation language into,

$$H_D = c\delta \int \frac{dx}{2\pi} \sin(2\Phi), \tag{85}$$

an always relevant perturbation which selects a unique GS, defined by $2\Phi = -\pi/2$, among the two MG-like GS.

The phase diagram of the model is shown in Fig. 6. Note that a MG line separates two regions characterized by (short-ranged) commensurate and incommensurate correlations.

Excitation spectrum

Interestingly enough, the dimerized spin chain exhibits exotic spin-1/2 topological excitations named solitons. In the bosonization picture, the latter correspond to kinks (or antikinks) separating $2\Phi = \pi/2$ and $2\Phi = -\pi/2$ GS's. They are similar to domain walls in ferromagnetic systems. As soon as $\delta \neq 0$, two such excitations experience a *confining* force, i.e. a force that grows with separating x like $\sim \delta x$. Indeed, the region with let's say $2\Phi = \pi/2$ between two $2\Phi = -\pi/2$ GS's and delimited by the soliton and the antsoliton corresponds to a local excitation of the system with energy density δ. Therefore, elementary excitations are unbreakable soliton-antisoliton (singlet or triplet) pairs: any *gedanken* experiment consisting of pulling off the two components will result

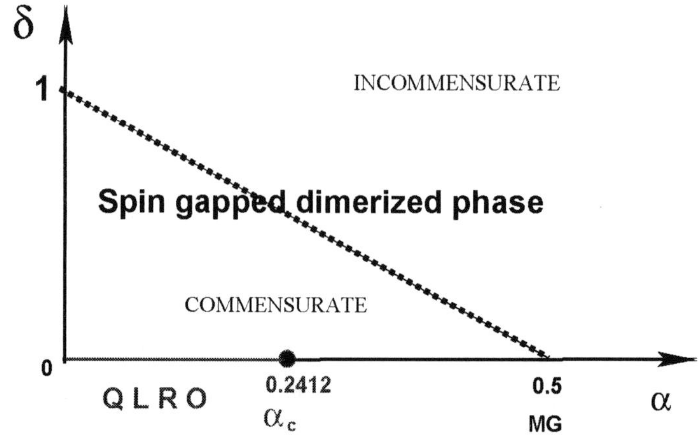

FIGURE 6. Phase diagram of the spin-Peierls model vs frustration and *explicit* dimerization. QLRO stands for "quasi-long range order" characterizing the critical AF phase.

in a spontaneous generation of new soliton-antisoliton ($s\bar{s}$) pairs above a given energy threshold (the spin gap). By analogy with a particle in a linear potential,

$$-\frac{\partial^2\Phi}{\partial x^2} + c\delta x\Phi = E\Phi, \tag{86}$$

one expects an energy spectrum similar to Fig. 7: at low energy the spectrum is composed of discrete levels corresponding to soliton-antisoliton pairs. By definition the spin gap Δ^{01} is the smallest energy to create a soliton-antisoliton pair. By a simple change of variables $x \to \delta y$ one gets,

$$(c\delta)^{2/3}(-\frac{\partial^2\Phi}{\partial y^2} + cy\Phi) = E\Phi, \tag{87}$$

implying that the spin gap behaves as,

$$\Delta^{01}(\delta) - \Delta^{01}(0) \sim \delta^{2/3}. \tag{88}$$

The energy $2\Delta^{01}$ corresponds to the onset of a two-soliton-antisoliton-pairs ($s\bar{s} - s\bar{s}$) continuum.

This simple scenario can be tested by numerical simulations. Fig. 8 shows results obtained on 28-site and 32-site rings. The low energy part of the spectrum (left) consists of discrete levels with finite dispersions (the soliton-antisoliton pairs are mobile). The finite spectral weight associated to these levels (right) suggests that (at least) the low energy level should be observable in Inelastic Neutron Scattering (INS) experiments.

Note that, if open chains are used like in DMRG, solitons get bound to one of the edge of the chainas shown in Fig. 9. Note that the number of nodes increases regularly with the excitation energy.

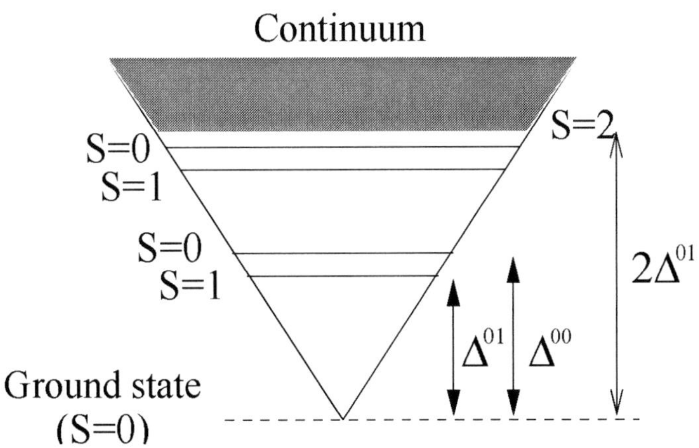

FIGURE 7. Schematic low energy spectrum of the dimerized spin chain based on an analogy with a particle in a linear potential. Δ^{01} stands for the spin gap.

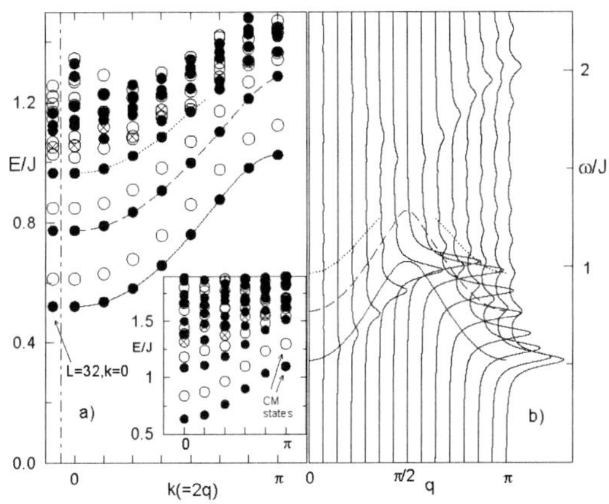

FIGURE 8. Left: low energy spectrum vs momentum of the dimerized spin chain for $\alpha = 0.5$ and $\delta = 0.1$ computed by ED on a 28-site periodic ring. $k = 0$ data are also shown for a 32-site ring. Right: associated spectral weight (INS). Reproduced from Ref. [72].

Acknowledgements

I am very happy to thank all my colleagues with whom I had the great plaisure to discuss about numerical techniques for strongly correlated systems over the years. Since it would be too long to list them all here, let me just list here my close collaborators of the Strongly Correlated Group in Toulouse, Fabien Alet, Sylvain Capponi, Matthieu Mambrini and Pierre Pujol.

The material covered by the "Lectures notes for XI Training Course on Strongly Correlated Systems" is broader than the present proceedings and can be found (in-

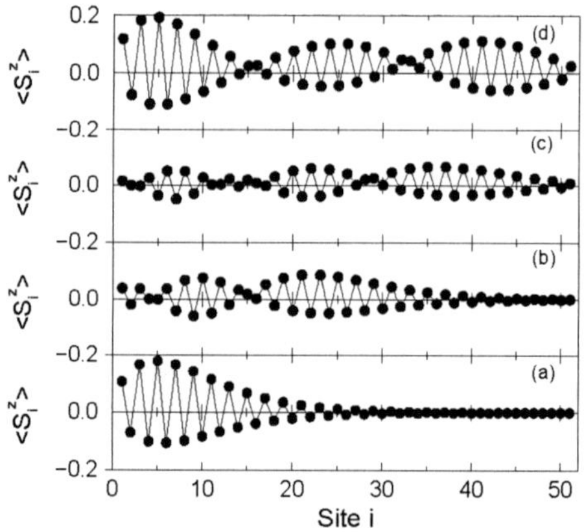

FIGURE 9. Lowest energy solitonic wavefunctions (local spin polarization) vs distance from the edge of an open chain for $\alpha = 0.5$ and $\delta = 0.1$. Increasing energy is considered from bottom to top. DMRG data reproduced from Ref. [73].

cluding simple sample FORTRAN Codes) on my Web page at "http://www.lpt.ups-tlse.fr/poilblanc".

REFERENCES

1. R.M. Noack, in *"Lectures on The Physics of Highly Correlated Electron Systems IX"*, AIP Conf. Proceedings Vol. **789** (2004).
2. R. Scalettar, this volume.
3. For a general introduction on correlated electrons see e.g. P. Fulde, *"Electron correlations in molecules and solids"*, Springer Series in Solid State Sciences, Vol. 114, 2nd ed., Springer-Verlag (1993).
4. M. Greiner et al., Nature **415**, 39 (2002).
5. J. C. Bonner and M. E. Fisher, Phys. Rev. **135**, 640 (1964).
6. J. Oitmaa and D. D. Betts, Can. J. Phys. **56**, 897 (1978).
7. For calculations of interacting electrons on 1D disordered mesoscopic rings, see e.g. M. Abraham and R. Berkovits, Phys. Rev. Lett. **70**, 1509 (1993); G. Bouzerar, D. Poilblanc and G. Montambaux, Phys. Rev. B **49**, 8258 (1994).
8. N. Laflorencie and D. Poilblanc, *Quantum Magnetism*, "Lecture notes in Physics", Vol. 645, p.227, Ed. U. Schollwöck et al., Spinger (2004).
9. H.J. Schulz, T. Ziman and D. Poilblanc, in *"Magnetic Systems with Competing Interactions"*, p120-160, Ed. H.T. Diep, World-Scientific, Singapore (1994); G. Misguich and C. Lhuillier, in *"Frustrated Spin Systems"*, p. 981, Ed. H.T. Diep, World-Scientific, Singapore (2005).
10. C. Lanczos, J. Res. Natl. Bur. Stand. **45**, 255 (1950).
11. A modified Lanczos method has also been introduced by E. Dagotto and A. Moreo, Phys. Rev. D **31**, 865 (1985) and E. Gagliano, E. Dagotto, A. Moreo and F. Alcaraz, Phys. Rev. B **34**, 1677 (1986).
12. G. Montambaux, D. Poilblanc, J. Bellissard and C. Sire Phys. Rev. Lett. **70**, 497 (1993); D. Poilblanc, T. Ziman, J. Bellissard, F. Mila and G. Montambaux, Europhys. Lett., **22** (7), 537 (1993).
13. E. Gagliano, S. Bacci, and E. Dagotto, Phys. Rev. B **44**, 285 (1991); J. Jaklic and P. Prelovsek, Phys. Rev. B **49**, 5065 (1994); *ibid.* Phys. Rev. B **50**, 7129 (1994); *ibid.* Phys. Rev. Lett. **74** 3411 (1995).
14. All operations of the point group will be defined with respect to an (arbitrary) lattice site.
15. A. Läuchli, private communication (2007).

16. H. J. Schulz, T. A. Ziman and D. Poilblanc, J. Phys. I France **6** (1996) 675–703.
17. For standard hashing algorithms see e.g. *"Numerical Recipes in Fortran: The Art of Scientific Computing"*, 2nd ed., p. 320, W. H. Press, S. A. Teukolsky, W. T. Vetterling, B. P. Flannery (Cambridge University Press, 1992).
18. W. Kohn, Phys. Rev. **133**, A 171 (1964).
19. T. Einarsson and H. J. Schulz, Phys. Rev. B **51**, 6151 (1995).
20. R. Fye, M. Martins, D. J. Scalapino, H. Wagner and W. Hanke, Phys. Rev. B**44**, 6909 (1991); M. Martins and R. Fye, J. Stat. Phys. **64**, 271 (1991);
21. C. A. Stafford, A. J. Millis and B. S. Shastry, Phys. Rev. B **43**, 13660 (1991).
22. For 3-bands Hubbard models see e.g. A. Sudbo, C. M. Varma, T. Giamarchi, E. B. Stechel and R. T. Scalettar, Phys. Rev. Lett. **72**, 3292 (1994).
23. K. Penc and F. Mila, Phys. Rev. B **49**, 9670 (1994).
24. C. A. Hayward, D. Poilblanc, R. M. Noack, D. J. Scalapino and W. Hanke, Phys. Rev. Lett. **75**, 926 (1995).
25. For a review about ladder materials see e.g. E. Dagotto and T. M. Rice, Science **671**, 619 (1996).
26. D. Poilblanc, Phys. Rev. B **44**, 9562 (1991).
27. F. F. Assaad, W. Hanke and D. J. Scalapino, Phys. Rev. Lett. **71**, 1915 (1993); *ibid.* Phys. Rev. B **50**, 12 835 (1994).
28. In doped Quantum Dimer Models, coupling of holes to gauge fields leads also to an elementary flux quantum involving $Q = 2e$; see e.g. A. Ralko, F. Mila and D. Poilblanc, cond-mat (2007).
29. The Luttinger liquid fixed point is generic of one dimensionnal interacting systems. Its low energy excitations are collective spin and charge modes and are bosonic in nature. See e.g. J. Solyom, Adv. Phys. **28**, 209 (1979); F.D.M. Haldane, J. Phys. C **14** 2585 (1981).
30. H.J. Schulz, *"Strongly Correlated Electronic Metals"*, p. 187, The Los Alamos Symposium – 1993, Ed. K.S. Bedell, Z. Wang, D.E. Meltzer, A.V. Balatsky and E. Abrahams, Addison-Wesley Publishing Company (1994).
31. T. Giamarchi, *"Quantum Physics in One Dimension"*, The International Series of Monographs on Physics, Oxford Science Publications, Clarendon Press Oxford (2004); T. Giamarchi, in *"Lectures on The Physics of Highly Correlated Electron Systems IX"*, AIP Conf. Proceedings Vol. 789 (2004).
32. A. Luther and V. J. Emery, Phys. Rev. Lett. **33**, 589 (1974).
33. R. Haydock, V. Heine and M.J. Kelly, J. Phys. C**8**, 2591 (1975); E. R. Gagliano and C. A. Balseiro, Phys. Rev. Lett. **59**, 2999 (1987).
34. D. Poilblanc, E. Gagliano, S. Bacci, E. Dagotto, Phys. Rev. B **43** 10970 (1991); See also D. Poilblanc and E. Dagotto, Phys. Rev. B **45**, 10111 (1992).
35. D. Poilblanc, E. Dagotto, J. Riera, Phys. Rev. B **43**, 7899 (1991).
36. Overviews of ED studies of the 2D t–J model can be found in E. Dagotto, A. Moreo, F. Ortolani, D. Poilblanc, and J. Riera, Phys. Rev. B **45**, 10741 (1992); P. Horsch, Int. J. Mod. Phys. B **6**, 589 (1992); E. Dagotto, Rev. Mod. Physics **66**, 763 (1994).
37. For ED results on the single hole problem see e.g. K. J. von Szczepanski, P. Horsch, W. Stephan, and M. Ziegler, Phys. Rev. B **41**, 2017 (1990); P. Prelovsek, I. Sega, and J. Bonca, Phys. Rev. B **42**, 10706 (1990); D. Poilblanc, H. Schulz, T. Ziman and E. Dagotto, Phys. Rev. B **47**, 14267 (1993) and references therein; R. Eder and Y. Ohta, Phys. Rev. B **50**, 10043 (1994); P. W. Leung and R. J. Gooding, Phys. Rev. B **52**, R15711 (1995); P. Béran, D. Poilblanc and R.B. Laughlin, Nucl. Phys. B **473**, 707-720 (1996).
38. For calculations of the spectral function of the t–J model at finite doping see e.g. Y. Hasegawa and D. Poilblanc, Phys. Rev. B**41**, 9555 (1990); D. Poilblanc and E. Dagotto, Phys. Rev. B**42**, 4861 (1990). W. Stephan and P. Horsch, Phys. Rev. Lett. **66**, 2258 (1991); E. Dagotto, F. Ortolani and D.J. Scalapino, Phys. Rev. B. **46**, 3183, (1992).
39. The photoemission spectra of the 3-band Hubbard model has been calculated by H. Eskes and G. A. Sawatsky, Phys. Rev. B **43**, 119 (1991).
40. A. Moreo and E. Dagotto, Phys. Rev. B **42**, 4786 (1990); W. Stephan and P. Horsch, Phys. Rev. B **42**, 8736 (1990); D. Poilblanc and E. Dagotto, Phys. Rev. B **44** 466 (1991).
41. For ladders with a spin gap, a pseudogap is expected in the optical conductivity. See e.g. C. A. Hayward, D. Poilblanc and D. J. Scalapino, Phys. Rev. B **53**, R8863 (1996).
42. The "Loudon-Fleury" two magnons Raman spectrum has been investigated by E. Dagotto and D. Poilblanc, Phys. Rev. B **42**, 7940 (1990); A.W. Sandvik, S. Capponi, D. Poilblanc, and E. Dagotto

Phys. Rev. B **57**, 8478 (1998).
43. K.G. Wilson, Rev. Mod. Physics **47**, 773 (1975).
44. S.R. White, Phys. Rev. Lett. **69**, 2863 (1992).
45. U. Schollwöck, Rev. Mod. Physics **77**, No. 1 (2005).
46. M. Troyer et al., "Non-local updates for QMC simulations", p.156 in "The Monte Carlo Method in the Physical Sciences", Ed. J. Gubernatis, AIP Conf. Proc., Vol. 690 (2003).
47. N. Metropolis, A.R. Rosenbluth, M.N. Rosenbluth, A.H. Teller & E. Teller, J. of Chem. Phys. **21**, 1087 (1953).
48. H. Trotter, Proc. Amer. Math. Soc. **10**, 545 (1959); M. Susuki, Prog. Theor. Phys. **56**, 1454 (1976).
49. N.V. Prokof'ev, B.V. Svistunov and I.S. Tupitsyn, Pis'ma v Zh. Eks. Teor. Fiz., **64**, 853 (1996).
50. A.W. Sandvik and J. Kurkijärvi, Phys. Rev. B **43**, 5950 (1991).
51. J.E. Hirsch, Phys. Rev. Lett. **51**, 1900 (1983); J.E. Hirsch, Phys. Rev. B **31**, 4403 (1985).
52. A. Aeurbach, in "Effective Models for Low-dimensional Strongly Correlated Systems", p. 1, Eds. G.G. Batrouni and D. Poilblanc, AIP Conference Proceedings, Vol. **816** (2006); S. Capponi, *ibid*, p. 16.
53. H.A. Bethe, Z. Phys. **49** 205 (1931).
54. A. Luther and I. Peschel Phys. Rev. B **12**, 3908 (1975).
55. J.V. José, L.P. Kadanoff, S. Kirkpatrick, and D.R. Nelson, Phys. Rev. B **16**, 1217 (1977).
56. E. H. Lieb and F. Y. Wu, Phys. Rev. Lett. **20**, 1445 (1968).
57. H. J. Schulz, Phys. Rev. Lett. **64**, 2831 (1990). See also H. J. Schulz, *"Correlated Electron Systems"*, p. 199, ed. V. J. Emery (World scientific, Singapore, 1993); J. Voit, Rep. Prog. Phys. **58**, 977 (1995) and references therein.
58. H. J. Schulz and T. A. Ziman, Phys. Rev. B **33**, 6545 (1986).
59. Masao Ogata, M. U. Luchini, S. Sorella and F. F. Assaad, Phys. Rev. Lett. **66**, 2388 (1991).
60. F. Mila and X. Zotos, Europhys. Lett. **24**, 133 (1993).
61. S. Capponi, D. Poilblanc and F. Mila, Phys. Rev. B **54**, 17547 (1996).
62. See e.g. H. Frahm and V. E. Korepin, Phys. Rev. B **42**, 10553 (1990) and references therein.
63. In general, in the presence of spin degrees of freedom, the singlet $k = \frac{2\pi}{L}$ excitation corresponding to the collective charge mode is *not* the first excited state in this symmetry sector. In that case, one has to follow the correct state by varying the model parameters starting from a Bethe-Ansatz integrable case where a precise assignement can be done (see Ref. [59]).
64. The inverse compressibility $\kappa^{-1} = L\frac{\partial\mu}{\partial N_e}$ can be estimated from the behavior of the chemical potential $\mu(N_e) = \frac{1}{2}(E_0^{N_e+1} - E_0^{N_e-1})$. However, special care is needed since shell effects often produce plateaux in $\mu(N_e)$ which complicate the analysis.
65. C. A. Hayward and D. Poilblanc, Phys. Rev. B **53**, 11721 (1996).
66. T. Ziman and H. J. Schulz, Phys. Rev. Lett. **59**, 140 (1987); For a more general discussion see e.g. I. Affleck, D. Gepner, H.J. Schulz and T. Ziman, J. Phys. A **22**, 5291 (1989).
67. F.D.M. Haldane Phys. Rev. Lett. **45**, 1358 (1980).
68. S. Ejima et al., Europhys. Lett. **70**, 492 (2005). C. K. Majumdar and D. K. Ghosh, J. Math. Phys. **10**, 1399 (1969).
69. The static deformation approach has also been applied to 1D spin-Peierls systems. See e.g. J. Riera and A. Dobry, Phys. Rev. B **51**, 16098 (1995).
70. C. K. Majumdar and D. K. Ghosh, J. Math. Phys. **10**, 1399 (1969).
71. S.W. White and I.A. Affleck, Phys. Rev. B **54**, 9862 (1996).
72. E. S. Sørensen, I. Affleck, D. Augier, and D. Poilblanc, Phys. Rev. B. **58**, 14701 (1998).
73. D. Augier, E. Sørensen, J. Riera, and D. Poilblanc, Phys. Rev. B **60**, 1075 (1999).

Numerical Studies of Disordered Tight-Binding Hamiltonians

R.T. Scalettar

Physics Department, University of California, Davis, CA 95616

Abstract. These are notes used for a set of lectures delivered at the Vietri summer school on Condensed Matter Physics in Fall 2006. They concern the general problem of the interplay of interactions and disorder in two dimensional electronic systems, as realized in the specific context of Quantum Monte Carlo simulations of the Anderson-Hubbard Hamiltonian. I wish to thank the organizers of this school for their hospitality during my visit, and their work in general in providing this educational opportunity for students over the years. It is a pleasure also to acknowledge the collaborators together with whom I have learned much of the physics and numerics presented in these notes: Zhaojun Bai, Andrew Baldwin, George Batrouni, Karim Bouadim, Wenbin Chen, Peter Denteneer, Fred Hébert, Norman Paris, Matt Schram, Nandini Trivedi, Martin Ulmke, Ichitaro Yamazaki and Gergely Zimanyi. This work was supported by the National Science Foundation (NSF-DMR-0312261 and NSF–ITR–0313390), and China Special Funds for Major State Basic Research Projects under contract 2005CB321700.

I. INTRODUCTION

A. Contents of these notes

These notes discuss the physics of electrons which move in a disordered potential and which also interact with each other. This is an enormous field, one of the most interesting in condensed matter physics, both for theory and experiment. In order to make it tractable, the focus will be on a specific model- the Hubbard Hamiltonian with random bond and site energies, although a brief foray into an interesting case when the hopping is non-Hermitian will be taken. A particular computational method, "determinant Quantum Monte Carlo" [1], will be described in detail and will provide the bulk of the non-trivial numerical results.

Section II. begins by examining the physics of the Hubbard Hamiltonian in the limit of no disorder, and presents its strong, weak, and mean field treatments. The exact diagonalization method is also described. Section III. will summarize some of the central features of the Anderson localization transition, that is, the case with disorder and no interactions. A generalization, the Hatano-Nelson model, in which the left and right hopping amplitudes are unequal, and hence the Hamiltonian is not Hermitian provides an interesting way to look at the localization transition. Section IV. outlines the determinant Quantum Monte Carlo (DQMC) method. The emphasis will be on presenting the 'pseudocode' for DQMC, relegating a more complete derivation to Appendix C. Two further appendices give additional background: Appendix A briefly reviews the formalism of second quantization, for those students for whom it is not so familiar. Quantum Monte

CP918, *Lectures on the Physics of Strongly Correlated Systems XI: Eleventh Training Course,* edited by A. Avella and F. Mancini

Carlo involves an understanding of classical monte carlo, a topic which is summarized in Appendix B.

After having considered the effects of randomness and interactions separately, and introduced DQMC, Section V. puts things together. Results of simulations of the "Anderson-Hubbard Hamiltonian", including evidence that, in two dimensions, interactions can drive a state that is localized by randomness into a metallic phase form the first topic. A Zeeman field, which spin polarizes the electrons and thereby reduces their effective interaction, can return the system to insulating behavior. Evidence is presented that particle-hole symmetry, and whether the disorder directly competes with the formation of magnetic moments, plays a crucial role in these transitions. Appendix D contains a review of aspects of this particle-hole mapping. This section concludes with a discussion of the band insulator to metal transition in which it is shown that an insulating phase caused by a periodic potential which doubles the unit cell can, like a disorder-induced Anderson insulator, also be made metallic by interactions.

Since this is a paper prepared for a summer school, these notes are interspersed with exercises. Some are rather simple- a few lines of analytic derivation might serve to complete them. Others go down the road to being small research topics, and involve writing a nontrivial program. The lecture notes contain substantially more material than can be covered in a week, especially when the Exercises are attempted. They will provide additional avenues for exploration after this summer school is completed, or topics for the more advanced student who wishes to proceed a bit further or more rapidly.

This introduction closes with a very brief review of the history of some of these problems. For details concerning the two dimensional MIT and the Anderson-Hubbard Hamiltonian, the student is encouraged to look at the various reviews of the field [2, 3]. An exceptionally nice introduction to the correlated electron problem and to the Hubbard model, is contained in the book by Fazekas[4]. Further summaries of DQMC may be found in a variety of articles and summer school notes, including [5, 6, 7, 8].

B. Brief Overview of Physics

When electrons are confined to two dimensions in a disordered environment, common understanding until relatively recently was that the electronic states would always be localized and the system would therefore be an insulator. This idea is based on the scaling theory of localization [9] for non-interacting electrons and corroborated by subsequent studies using renormalization group (RG) methods [10, 11]. The scaling theory highlights the importance of the number of spatial dimensions and demonstrates that while in three dimensions for non-interacting electrons there exists a transition from a metal to an Anderson insulator upon increasing the amount of disorder, a similar metal–insulator transition (MIT) is not possible in two dimensions.

The inclusion of interactions into the theory has been problematic, certainly when both disorder and interactions are strong and perturbative approaches break down. Following the scaling theory the effect of weak interactions in the presence of weak disorder was studied by diagrammatic techniques and found to increase the tendency to localize [12]. Subsequent perturbative RG calculations, including both interactions and disorder,

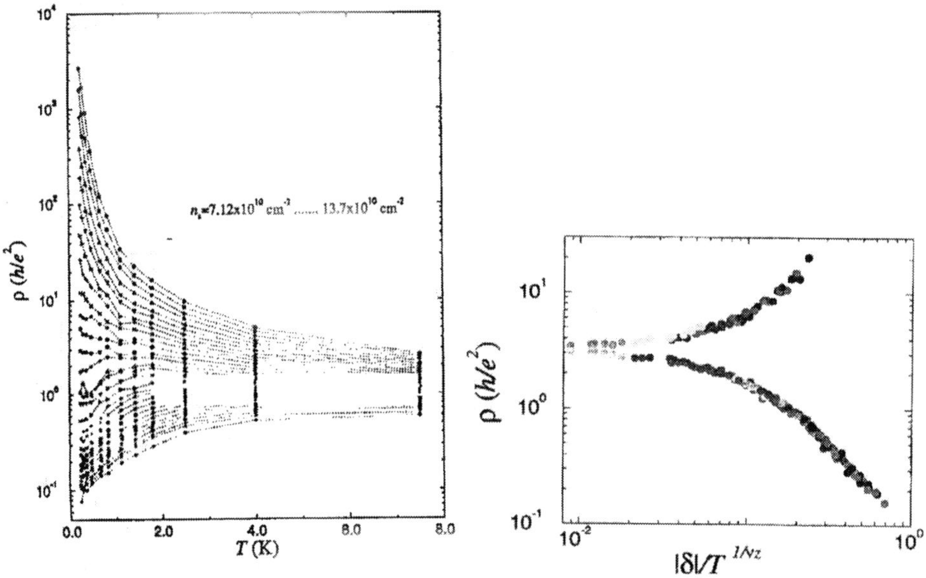

FIGURE 1. Left: Resistivity ρ versus temperature T for samples with a varying density of electrons n. As the density increases, the system undergoes a transition from insulating to metallic. Right: Similar data for a transition tuned by disorder strength δ. Horizontal axis is an appropriately scaled form. A large family of curves collapses to just two trajectories, one for metallic disorder strengths, and one for insulating ones, suggesting a quantum phase transition.

found indications of metallic behavior, but also, for the case without a magnetic field or magnetic impurities, found runaway flows to strong coupling outside the controlled perturbative regime, and hence were not conclusive [13, 14, 15]. The results of such approaches therefore did not change the widely held opinion that, in the absence of a magnetic field coupling to orbital motion, or magnetic impurities, the MIT does not occur in two dimensions.

In the middle of the 1990's, the situation changed dramatically with transport experiments on effectively 2D electron systems in silicon MOSFETs which provided surprising evidence that a MIT can occur in 2D [16, 17, 18]. In these experiments the temperature dependence of the resistivity ρ changes from that typical of an insulator (increase of ρ upon lowering T) at lower density to that typical of a conductor (decrease of ρ upon lowering T) as the density is increased above a critical value. (See Fig. 1, left panel.) The fact that the data can be scaled onto two curves (one for the metal, one for the insulator) is seen as evidence for a quantum phase transition with carrier density n as the tuning parameter. (See Fig. 1, right panel.) The possibility of such a transition has stimulated a large number of further experimental [19, 20, 21, 22] and also theoretical investigations [23, 24, 25, 26], including proposals that a superconducting state is involved [27, 28]. Explanations in terms of trapping of electrons at impurities, i.e. not requiring a quantum

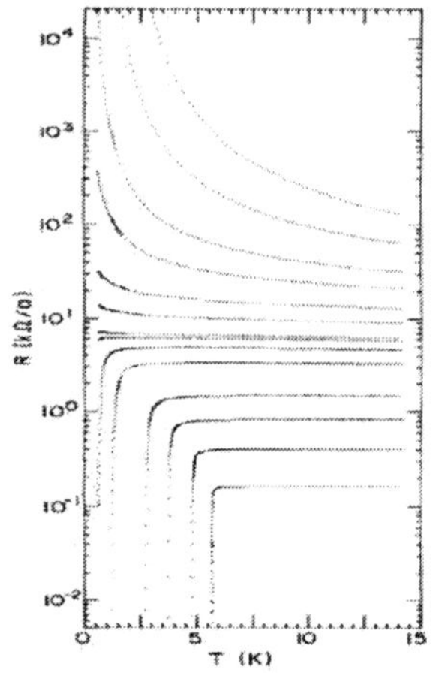

Source: Haviland etal, [38]

FIGURE 2. A superconductor to insulator transition is well established in two dimensional thin films. Here results for the resistance $R(T)$ of Pb films of varying thickness are shown. The transition can also be controlled with degree of disorder, and magnetic field.

phase transition have also been put forward [29, 30].

A common thread in many of the different explanations of the 2D MIT is the importance of electron-electron interactions. Indeed, one of the most central questions motivated by the experiments is whether correlations can enhance the conductivity of a 2D disordered electron system, and possibly lead to a conducting phase [31, 32]. In order to address this issue theoretically, one approach is to identify an appropriate model and, then, work out its properties. This will be the strategy here.

To this end, these lectures will begin by providing some background on the "Anderson-Hubbard" Hamiltonian, which is one of the most simple models that incorporates both disorder and interactions. Several methods are discussed which help develop an initial insight into the physics of this Hamiltonian, but the focus will be on Determinant Quantum Monte Carlo (DQMC). The notes conclude with a detailed description of the MIT studied with this model and method. DQMC has previously been applied extensively to the Hubbard model without disorder. [1, 6, 33, 34, 35, 36] While the strengths of DQMC in addressing this problem are considerable: disorder and interaction can be varied in a controlled (ie exact) way and strong interaction is treatable, it is important also to acknowledge its serious limitations: the size of the lattice is finite (a few hundred sites) and low temperatures are often not accessible due to the sign problem.

There are many analogies between the MIT and the superconducting-insulator transition (SIT). As for the MIT, one of the most interesting issues again focuses on the behavior in two dimensions. A transition from superconductor to insulator is observed as the degree of disorder or the magnetic field is changed, but the fundamental question is not whether a superconducting phase can exist (it surely can!), but rather why the value of resistance at the transition appears to take on a universal value. Interestingly, the connection between the MIT and SIT can be brought out in the context of the Hubbard Hamiltonian through a particle-hole transformation which changes the sign of the interaction between the electrons from repulsive to attractive, since the latter case allows for superconducting phases. The interested reader can examine reviews such as [37, 38, 39, 40, 41, 42] or, for applications of DQMC, to [43, 44].

C. Guide to Exercises

These notes attempt to provide interesting problems for students of different levels of familiarity with numerical work and with the Hubbard model. The most advanced/ambitious students might wish to construct a determinant Quantum Monte Carlo code. Though not phrased in the form of exercises, Section IV. writes down all the required equations and shows some simple results. Appendix C provides a derivation. It is, however, unlikely such a task could be fully accomplished in the evening sessions of this summer school. A less time consuming objective for a student who wants a 'research' code is to try a mean field or exact diagonalization program. These are discussed in Section II., and also in the material presented by other lecturers. The exercises provide many options to those wishing to develop further their knowledge of the material, without writing a major code. Here is a rough guide to where the different problems can be found.

Section II: Exercises extending aspects of the weak ($U = 0$) and strong ($t = 0$) coupling limits of the Hubbard Hamiltonian, for example, determining the tight binding dispersion relations and density of states of the periodic Anderson model, the three band Emery model, or the Hubbard model on triangular and honeycomb lattices.
Section III: Exercises involving the Hatano-Nelson model.
Appendix A: Exercises involving the manipulation of creation and destruction operators and the Hubbard Hamiltonian.
Appendix B: Exercises involving the formal foundations of *classical* monte carlo (detailed balance, the Metropolis algorithm, the relation of monte carlo to molecular dynamics and the Langevin equation, etc).
Appendix D: Exercises involving particle-hole symmetry and the relation between the attractive and repulsive Hubbard Hamiltonians.
Tutorials: The tutorials contain tasks involving using the DQMC code provided in the summer school.

II. THE TRANSLATIONALLY INVARIANT HUBBARD MODEL

The Hubbard Hamiltonian, in the absence of disorder, is,

$$H = -t \sum_{\langle \mathbf{j},\mathbf{l}\rangle\sigma} c^{\dagger}_{\mathbf{j}\sigma} c_{\mathbf{l}\sigma} + U \sum_{\mathbf{j}} n_{\mathbf{j}\uparrow} n_{\mathbf{j}\downarrow} - \mu \sum_{\mathbf{j}} (n_{\mathbf{j}\uparrow} + n_{\mathbf{j}\downarrow}). \tag{1}$$

Here $c_{\mathbf{j}\sigma}^{\dagger}(c_{\mathbf{j}\sigma})$ are creation(destruction) operators for electrons on site $\mathbf{j}$ with spin σ. (Their properties are reviewed in Appendix A.) The first term of H is the kinetic energy since it describes the destruction of an electron of spin σ on site $\mathbf{l}$ and its creation on site $\mathbf{j}$ (or *vice-versa*). The symbol $\langle \mathbf{j}, \mathbf{l} \rangle$ indicates that hopping is allowed only between specified pairs of sites (usually the near neighbors). These notes mostly concern square lattices in two dimensions. The second term is the interaction energy: a doubly occupied site ($n_{\mathbf{j}\uparrow} = n_{\mathbf{j}\downarrow} = 1$) adds an energy U to the state. The final term is a chemical potential which controls the filling. The situation where the filling is one electron per site is referred to as 'half-filling,' since the lattice contains half as many electrons as the maximum number (two per site). The value $\mu = U/2$ results in half-filling for any choice of hopping t or temperature T, if the lattice is bipartite and the hopping only connects sites on the two independent sublattices. Studies of the Hubbard model often focus on the half-filled case because it exhibits many interesting phenomena (Mott insulating behavior, antiferromagnetic order, etc.)

The Hubbard Hamiltonian offers a way to get qualitative insight into how the interactions between electrons can give rise to insulating, magnetic, and even novel superconducting effects in a solid. It was written down in the early 1960's and initially applied to understanding the behavior of the transition metal monoxides like FeO, NiO, CoO. These compounds are antiferromagnetic insulators, but are predicted to be metallic by "electronic structure" methods which treat strong interactions less carefully.

Over the intervening years, the Hubbard model has been applied to the understanding of many systems, from 'heavy fermion' systems in the 1980's, to high temperature superconductors in the 1990's. In the last several years the 'boson-Hubbard' and fermion Hubbard models have been invoked to study the physics of ultra-cold atoms. Indeed, it is an amazing feature of the Hamiltonian that, despite its simplicity, it exhibits behavior relevant to many of the most subtle and beautiful properties of solid state (and now atomic) systems.

The Hubbard model has been studied by the full range of analytic techniques developed by condensed matter theorists, from simple mean field approaches to field theoretic methods employing Feynman diagrams, expansions in the degeneracy of the number of 'flavors' (spin, orbital angular momentum), the Bethe Ansatz, etc. It has also been extensively attacked with numerical methods like diagonalization and variational and path-integral Quantum Monte Carlo (QMC). In these lectures, the main tool will be determinant QMC, but the non-interacting and zero hopping limits, and static mean field theory, will be described to develop some simple pictures of the physics.

A. Noninteracting Limit ($U = 0$)

In Appendix A, two, equivalent, ways to solve the Hubbard Hamiltonian at $U = 0$ are presented. The first is based on explicitly constructing the matrix for $\hat{H}$ in the single particle sector. The second considers a change to momentum creation and destruction operators,

$$c_{\mathbf{k}\sigma}^{\dagger} = \frac{1}{\sqrt{N}} \sum_{\mathbf{l}} e^{i\mathbf{k}\cdot\mathbf{l}} c_{\mathbf{l}\sigma}^{\dagger}. \tag{2}$$

Appendix A provides many details concerning this change of basis such as the preservation of the anticommutation relations etc. The key result is that the kinetic energy term in the Hamiltonian can be rewritten as

$$H = \sum_{\mathbf{k}\sigma} (\varepsilon_{\mathbf{k}} - \mu) c^{\dagger}_{\mathbf{k}\sigma} c_{\mathbf{k}\sigma} \tag{3}$$

where $\varepsilon_{\mathbf{k}}$ depends on the specific lattice geometry, as do the discrete allowed values of the momentum $\mathbf{k}$. For a one dimensional lattice of length N with near neighbor hopping, $\varepsilon_k = -2t\cos k$ and $k = k_n = 2\pi n/N$ with $n = 1, 2, 3, \ldots, N$. This choice of momenta reflects the use of periodic boundary conditions, which minimize finite size effects, an important consideration in numerical work which tends to be on lattices of limited size. The use of momentum space in Eq. 3 provides a much more simple representation than the original real-space expression of Eq. 1 because it is diagonal. That is, the Hamiltonian only counts the number of electrons in each momentum state without converting electrons from one momentum to another. A useful analogy is of course with normal modes in classical physics which can be excited without setting other modes into motion. Just as the normal mode construction in classical physics is possible only for quadratic potential energies, the diagonalization of a tight binding Hamiltonian can be done only when it is quadratic in the fermion creation and destruction operators.

Because the case of the two dimensional square lattice is of interest physically for the cuprate superconductors, and because it forms the bulk of the applications discussed in Section V., it is useful to review a few of its properties. The dispersion relation is $\varepsilon_{\mathbf{k}} = -2t(\cos k_x + \cos k_y)$. This dispersion relation has a number of interesting and important features. First, its Fermi surface, ie the contours of constant energy, is 'nested' at half-filling. That is, a particular wave-vector, $\mathbf{k} = (\pi, \pi)$ connects extended lengths of the Fermi surface. See Fig. 3. A number of response functions, for example the non-interacting magnetic susceptibility $\chi^0(\mathbf{q})$, involve energy denominators of the form $\varepsilon_{\mathbf{k}+\mathbf{q}} - \varepsilon_{\mathbf{k}}$ and hence become large at (π, π) when nesting occurs.

A second interesting feature of the square lattice dispersion is that its density of states,

$$N(E) = \frac{1}{N} \sum_{\mathbf{k}} \delta(E - \varepsilon_{\mathbf{k}}). \tag{4}$$

is singular at half-filling, $E = 0$. Possible implications of both nesting and van-Hove singularities have been discussed in the context of theories of high temperature superconductivity.

The average energy of the $U = 0$ Hubbard model is,

$$\langle E \rangle (T) = \frac{1}{N} \sum_{\mathbf{k}} \varepsilon_{\mathbf{k}} (1 + e^{+\beta(\varepsilon_{\mathbf{k}} - \mu)})^{-1}. \tag{5}$$

Fig. 4 shows the energy and specific heat, $\langle E \rangle (T)$ and $C(T) = d\langle E \rangle / dT$, as functions of temperature. As expected, $C(T)$ has a peak at a temperature T set by the hopping t. The following exercises explore several refinements of the $U = 0$ Hubbard Hamiltonian.

Exercise 1: Consider the one-dimensional Hubbard Hamiltonian with a staggered site energy $V = A \sum_l (-1)^l n_l$. Solve the $U = 0$ limit by going to momentum space. Show that you get two energy bands separated by a gap.

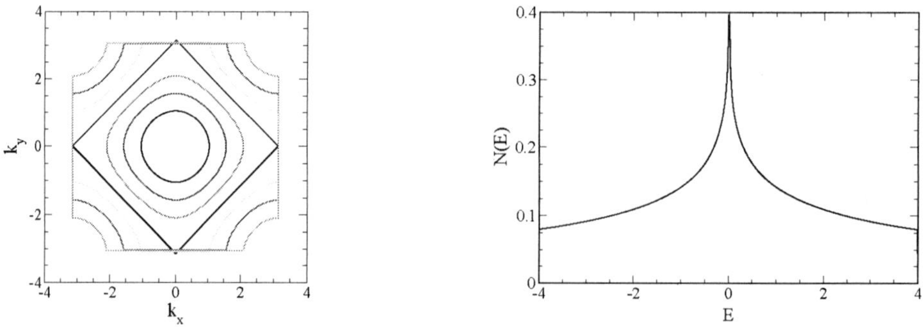

FIGURE 3. Left: Surfaces of constant energy E for the square lattice dispersion relation. Note that for $E = 0$ the topology is a rotated square and that $\varepsilon_{\mathbf{k}+(\pi,\pi)} = \varepsilon_{\mathbf{k}}$ for points $\mathbf{k}$ on the surface. Right: The density of states. Note the divergence ('van Hove singularity') at $E = 0$.

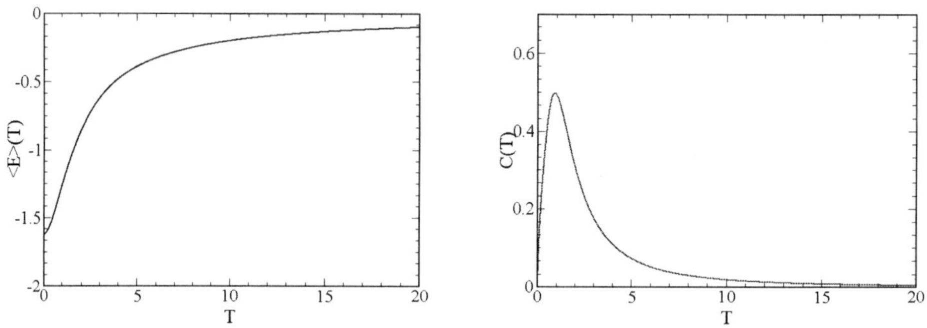

FIGURE 4. Left: Energy $\langle E\rangle(T)$ of the square lattice Hubbard Hamiltonian with $t = 1$, $U = 0$, and density $\rho = 1$ (half-filling). Right: Specific heat $C(T)$.

Exercise 2: Compute (numerically) the density of states $N(E)$ of the Hubbard model on a two dimensional honeycomb lattice. You will need to determine the dispersion relation and how the periodic boundary conditions restrict the allowed $\mathbf{k}$ values. You should find that $N(E)$ vanishes linearly at $E = 0$. The system is said to be a *semi-metal* there. Appendix A contains exercises which explore the dispersion relation and density of states for a number of other tight binding Hamiltonians.

Exercise 3: Compute (numerically) the density of states $N(E)$ of the Hubbard model on a two dimensional triangular lattice. Again, you will need to determine the dispersion relation and how the periodic boundary conditions restrict the allowed $\mathbf{k}$ values. Unlike the preceding cases, you will find that $N(E) \neq N(-E)$. This is a consequence of the fact that the model is not "particle-hole" symmetric on a triangular lattice. This will be

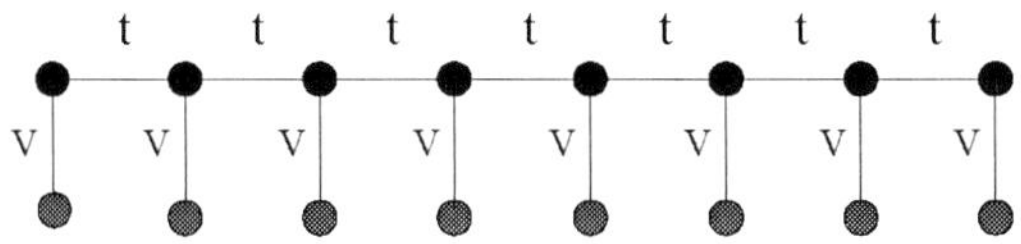

FIGURE 5. Lattice connections of a one-dimensional periodic Anderson model. Sites of the top ('conduction') chain are connected to one another by hopping parameter t. Sites of the lower ('localized') chain are not connected to each other, but hybridize with a hopping parameter V with the sites in the top chain.

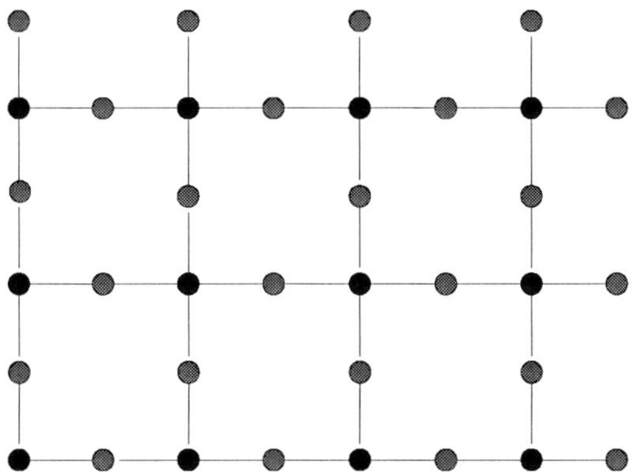

FIGURE 6. Lattice connections of a three band Emery model of the CuO_2 sheets of the cuprate superconductors. The Cu atoms lie in a square array with intervening O atoms. Pairs of Cu and O atoms are connected by hopping t. Sometimes an additional hopping t' between oxygen atoms is also included. The O sites are also usually given a site energy ε_{pd} relative to the Cu sites.

discussed further in Sections V.C. and Appendix D below.

Exercise 4: Almost as widely studied as the Hubbard Hamiltonian is the 'Periodic Anderson Model' (PAM). It consists first of a set of 'conduction' sites (orbitals) whose near neighbors are connected with hopping 't'. These might be in a one-d chain, or a two-d square lattice arrangement for example. A second set of 'localized' sites (orbitals) are disconnected from each other but hybridize with their conduction partners with amplitude 'V', as illustrated in Fig. 5. Compute $E(k)$ and $N(E)$ for this model, in $d = 1$. Show there is a band gap when $V \neq 0$. If one adds a Hubbard U on the localized orbitals one gets the Periodic Anderson model.

Exercise 5: A three-band Hubbard model introduced by Emery and widely studied for high temperature superconductivity has the geometry shown in Fig. 6. One set of atoms forms a square array while a second set, with higher on-site energy, sits at the midpoints of the bonds between them. Compute $E(k)$ and $N(E)$ for this model.

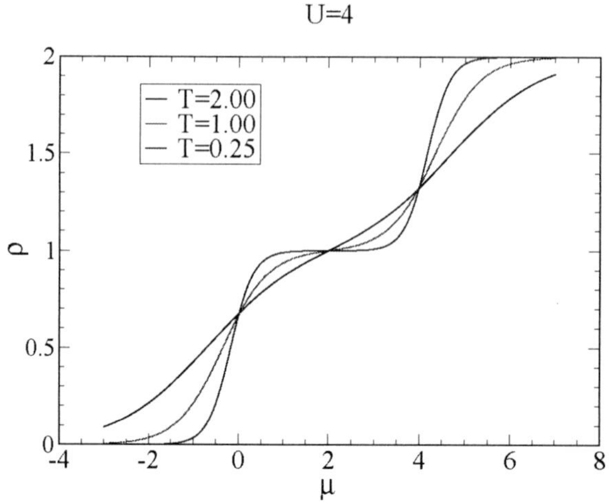

FIGURE 7. Density ρ as a function of chemical potential for the single site ($t = 0$) Hubbard model at $U = 4$ for three different temperatures. At low T a 'Mott plateau' develops.

B. Strong Coupling Limit (t=0)

Having looked at some of features of the noninteracting Hubbard Hamiltonian in various geometries, a first insight into the role of interactions is obtained by considering a single site. That is, by setting $t = 0$ in the Hamiltonian. This situation is easily solved. There are four possibilities corresponding to the site being empty, having a single electron (either spin up or spin down) or being doubly occupied. Each of the states $|0\rangle, |\uparrow\rangle, |\downarrow\rangle, |\uparrow\downarrow\rangle$ is an eigenstate of H with eigenvalues $0, -\mu, -\mu, U - 2\mu$ respectively. The partition function is

$$Z = \sum_{\alpha} \langle \alpha | e^{-\beta H} | \alpha \rangle = 1 + 2e^{\beta\mu} + e^{2\beta\mu - \beta U}, \tag{6}$$

and the energy is,

$$E = \langle H + \mu n \rangle = Z^{-1} \sum_{\alpha} \langle \alpha | H e^{-\beta H} | \alpha \rangle = U\, e^{2\beta\mu - \beta U}\, (1 + 2e^{\beta\mu} + e^{2\beta\mu - \beta U})^{-1} \tag{7}$$

The occupation is given by,

$$\rho = \langle n \rangle = 2\,(e^{\beta\mu} + e^{2\beta\mu - \beta U})\,(1 + 2e^{\beta\mu} + e^{2\beta\mu - \beta U})^{-1} \tag{8}$$

Fig. 7 is a plot of ρ vs. μ for $U = 4$ and $T = 2, 1$ and $T = 0.25$ and exhibits one of the fundamental features of the Hubbard model, namely the "Mott insulating gap". How is this understood? At $T = 0$ the chemical potential $\mu = \partial E / \partial \rho$ measures how much the energy changes when the density changes, ie. the cost to add a particle. In a noninteracting system described by a set of energy levels, with levels filled up to some 'Fermi energy' E_F, the cost to add a particle is the next energy level just above the last occupied level, that is, $\mu = E_F$. The jump in μ at $\rho = 1$ arises from the interactions: Consider a nearly empty lattice and ask the energy cost to add an electron. This cost need not involve U because empty sites are abundant. When one gets to half-filling, however, suddenly the cost to add an electron jumps by U since inevitably an added electron must sit on top of an electron which is already there. This sudden jump in the cost to add a particle is referred to as the "Mott gap." Besides the plateau in $\rho(\mu)$, Mott and band gaps are characterized by a density of states (spectral function) which vanishes at the Fermi surface. Examples of this are given in Section V.

Similar jumps in μ also occur in the context of band theory, where a gap between two bands likewise causes μ to change discontinuously. In both cases, the jump in μ indicates the existence of a gapped, insulating phase. However, band and Mott insulators are very different in other ways. 'Anderson insulators', arising from disorder, differ from both by having a finite $N(E_F)$, as discussed in the next section.

Half-filling $\rho = 1$ occurs when $\mu = U/2$. Because half-filling is so often studied, it is convenient to write the Hubbard Hamiltonian as,

$$H = -t \sum_{\langle \mathbf{j},\mathbf{l} \rangle \sigma} c^{\dagger}_{\mathbf{j}\sigma} c_{\mathbf{l}\sigma} + U \sum_{\mathbf{j}} (n_{\mathbf{j}\uparrow} - \frac{1}{2})(n_{\mathbf{j}\downarrow} - \frac{1}{2}) - \mu \sum_{\mathbf{j}} (n_{\mathbf{j}\uparrow} + n_{\mathbf{j}\downarrow}) \tag{9}$$

This just corresponds to a shift in the chemical potential μ by $U/2$. With this convention, half-filling always occurs at $\mu = 0$ for any value of t, T, U on a bipartite lattice. To emphasize, the properties of this 'new' model are identical to the old one, if one compares them at the same density. The chemical potentials used to get those densities are simply offset.

Exercise 6: Write expressions for Z, E and ρ at $t = 0$ with this new convention for the interaction term in the Hubbard model. You should notice that they are a bit more symmetric looking at $\mu = 0$.

A fundamental physical quantity in the Hubbard model is the 'local moment',

$$\langle m_{\mathbf{j}}^2 \rangle = \langle (n_{\mathbf{j}\uparrow} - n_{\mathbf{j}\downarrow})^2 \rangle. \tag{10}$$

The local moment is zero if the site is either empty, $||0\rangle$, or has two oppositely pointed spins, $||\uparrow\downarrow\rangle$, but takes the value one if the site has a single electron, $||\uparrow\rangle$ or $||\downarrow\rangle$. Fig. 8 is a plot of the local moment as a function of U at half-filling for fixed $T = 0.5$. As U increases, the local moment rises from its uncorrelated value $\langle m_{\mathbf{j}}^2 \rangle = 0.5$ which reflects a uniform mixture of empty, singly occupied, and doubly occupied sites, to the value $\langle m_{\mathbf{j}}^2 \rangle = 1.0$ which reflects the presence only of singly occupied sites. This is a first indication of the tendency towards magnetism in the Hubbard model. An obvious question is whether, when these moments form on individual sites, there is a mechanism for developing correlations between them. This will be one of the central issues addressed in these notes. In Fig. 9 we show the energy $\langle E \rangle$ and the specific heat $C = d\langle E \rangle / dT$ for the half-filled $t = 0$ Hubbard Hamiltonian, as a function of T for $U = 8$. $C(T)$ has a peak at $T \sim U/3$ associated with the suppression of double occupation.

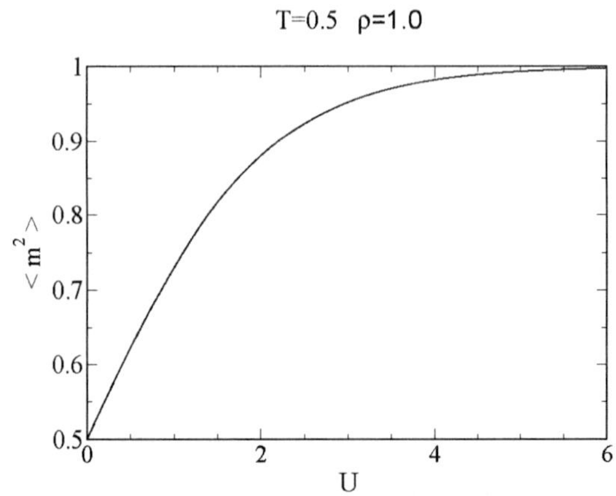

FIGURE 8. Local moment $\langle m^2 \rangle$ as a function of U for the single site ($t = 0$) Hubbard model at half-filling $\rho = 1$ and temperature $T = 0.5$. As U increases, a perfect moment develops.

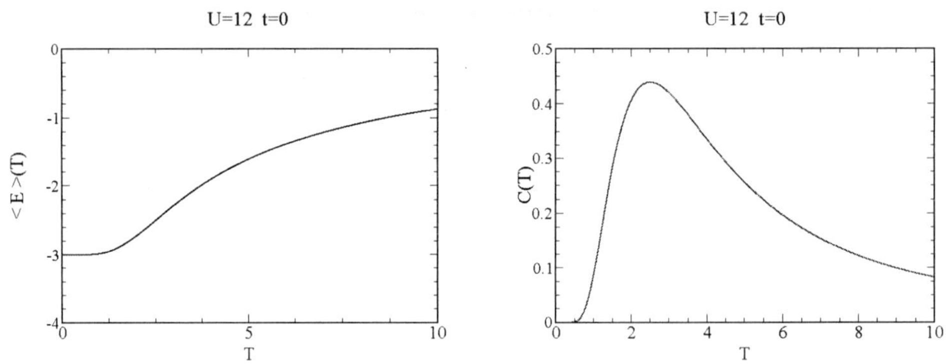

FIGURE 9. Energy (left) and specific heat (right) as functions of temperature T for the single site Hubbard model with $U = 12$.

<u>Exercise 7:</u> Show the local moment is related to the 'double occupancy' $d_{\mathbf{j}} = \langle n_{\mathbf{j}\uparrow} n_{\mathbf{j}\downarrow} \rangle$ by

$$\langle m_{\mathbf{j}}^2 \rangle = \langle n_{\mathbf{j}\uparrow} + n_{\mathbf{j}\downarrow} \rangle - 2d_{\mathbf{j}}. \tag{11}$$

At half-filling, the relation between the moment and the double occupancy becomes $\langle m_{\mathbf{j}}^2 \rangle = 1 - 2d_{\mathbf{j}}$. Interpret the evolution of the local moment between the two limits $\langle m_{\mathbf{j}}^2 \rangle = \frac{1}{2}$ and $\langle m_{\mathbf{j}}^2 \rangle = 1$ in terms of the behavior of the double occupancy $d_{\mathbf{j}}$.

Having understood the limit of strictly zero hopping, $t = 0$, it is natural to consider perturbation theory in t. Indeed, this is a very fruitful approach and connects the half-filled Hubbard Hamiltonian with the spin-$\frac{1}{2}$ Heisenberg model. A discussion of this is deferred until Section IIB. There, a solution of the two site Hubbard Hamiltonian by exact diagonalization is presented which can be examined in the limit of small, but nonzero, t/U.

C. Green's Functions for the Hubbard Hamiltonian

As mentioned, the Hubbard Hamiltonian has been extensively studied via diagrammatic Green's function methods. These lectures do not discuss that approach, but it may be useful to students who do know that technology to make contact with the material discussed so far by computing the Green's function in the weak and strong coupling limits. The one particle Greens function,

$$\begin{aligned} G_{\mathbf{jn}}(\tau) &= \langle c_{\mathbf{j}}(\tau) c_{\mathbf{n}}^{\dagger}(0) \rangle \\ c_{\mathbf{l}}(\tau) &= e^{H\tau} c_{\mathbf{l}}(0) e^{-H\tau}, \end{aligned} \tag{12}$$

is a fundamental quantity in understanding the many body physics of interacting electron systems. Its momentum space and frequency transform, the spectral function $A(\mathbf{k},\omega)$, yields the angle-resolved photoemission spectrum. From $G_{\mathbf{jn}}(\tau)$, the two particle Greens functions which yield the charge and spin susceptibilities can also be obtained. In the limit of no interactions, $G_{\mathbf{jn}}(\tau)$ can be computed analytically.

<u>Exercise 8:</u> Show that at $U = 0$,

$$c_{\mathbf{k}}(\tau) = e^{H\tau} c_{\mathbf{k}}(0) e^{-H\tau} = e^{-\varepsilon_{\mathbf{k}}\tau} c_{\mathbf{k}}(0) \tag{13}$$

You should do this two ways: First show that both expressions give the same result on the two states $\|0\rangle$ and $\|1\rangle$. Next, prove the result using $\partial \hat{O}(\tau) \partial \tau = [\hat{H}, \hat{O}(\tau)]$ which immediately follows from the definition $\hat{O}(\tau) = e^{\hat{H}\tau} \hat{O}(0) e^{-\hat{H}\tau}$.

<u>Exercise 9:</u> Show that at $U = 0$,

$$G_{\mathbf{jn}}(\tau) = \frac{1}{N} \sum_{\mathbf{k}} e^{i\mathbf{k}\cdot(\mathbf{n}-\mathbf{j})} (1 - f_{\mathbf{k}}) e^{-\varepsilon_{\mathbf{k}}\tau}. \tag{14}$$

G is a function of the difference $\mathbf{n}-\mathbf{j}$, as you would expect for a translationally invariant Hamiltonian.

<u>Exercise 10:</u> Write a program to evaluate $G_{\mathbf{jn}}(\tau)$ numerically at $U = 0$ for a $d = 1$ chain and a $d = 2$ square lattice. This is a useful calculation in checking a DQMC code.

Actually, the definition of G is a bit more subtle. In many-body theory one considers the so-called 'time ordered' Green's function, $G_{\mathbf{k}}(\tau) = -\langle \mathscr{T} c_{\mathbf{j}}(\tau) c_{\mathbf{n}}(0) \rangle$ where the time ordering operator T,

$$\begin{aligned} \mathscr{T} c_{\mathbf{j}}(\tau) c_{\mathbf{n}}(0) &= c_{\mathbf{j}}(\tau) c_{\mathbf{n}}(0) & \tau > 0 \\ \mathscr{T} c_{\mathbf{j}}(\tau) c_{\mathbf{n}}(0) &= -c_{\mathbf{n}}(0) c_{\mathbf{j}}(\tau) & \tau < 0 \end{aligned} \tag{15}$$

This definition of G and associated formalism opens the door into the huge world of diagrammatic perturbation theory and its application to the Hubbard model. The next few Exercises take you down the road just a little way.

Exercise 11: Prove that the time-ordered Greens function obeys $G(\tau+\beta) = -G(\tau)$ for $-\beta < \tau < 0$.

Exercise 12: Using the preceding exercise, argue that the Fourier transform of G given by,

$$G(\tau) = \sum_n G(i\omega_n) e^{-i\omega_n \tau} \tag{16}$$

involves the 'Matsubara frequencies' $\omega_n = \pi(2n+1)/\beta$.

Exercise 13: Prove the inversion relation

$$G(i\omega_n) = \int_0^\beta \frac{d\tau}{\beta} G(\tau) e^{i\omega_n \tau} \tag{17}$$

Exercise 14: Show that the Greens function in momentum space is given by,

$$\begin{aligned} G_{\mathbf{k}}(\tau) &= -e^{-\varepsilon_{\mathbf{k}}\tau}(1-f_{\mathbf{k}}) & 0 < \tau < \beta \\ G_{\mathbf{k}}(\tau) &= e^{-\varepsilon_{\mathbf{k}}\tau} f_{\mathbf{k}} & -\beta < \tau < 0 \end{aligned} \tag{18}$$

and hence that,

$$G_{\mathbf{k}}(i\omega_n) = \frac{1}{i\omega_n - \varepsilon_{\mathbf{k}}}. \tag{19}$$

Exercise 15: Provide an alternate demonstration of this result by considering $\partial/\partial\tau$ of the definition of the time ordered Greens function written in the form

$$G_{\mathbf{k}}(\tau) = \langle c_{\mathbf{k}}(\tau) c_{\mathbf{k}}(0)\rangle\, \theta(\tau) - \langle c_{\mathbf{k}}(0) c_{\mathbf{k}}(\tau)\rangle\, \theta(-\tau). \tag{20}$$

Be careful to take the appropriate derivatives of the step functions! Then Fourier transform both sides and solve for $G_{\mathbf{k}}(i\omega_n)$.

The approach used in the last exercise is the basis of the 'equation of motion' method for computing G. One starts with the definition of G, takes a time derivative, evaluates the resulting commutators of H with $c_{\mathbf{k}}$ and then Fourier transforms. If the Hamiltonian is quadratic in the fermion operators, then the set of equations closes, even if the different fermion operators mix. Otherwise, one can introduce an approximate truncation to terminate the hierarchy.

Exercise 16: The ambitious student should use the above procedure to evaluate $G_{\mathbf{k}}(i\omega_n)$ and $G_d(i\omega_n)$ for

$$H = \sum_{\mathbf{k}} \varepsilon_{\mathbf{k}} c^\dagger_{\mathbf{k}} c_{\mathbf{k}} + V \sum_{\mathbf{k}} (c^\dagger_{\mathbf{k}} d + d^\dagger c_{\mathbf{k}}) + \varepsilon_d d^\dagger d \tag{21}$$

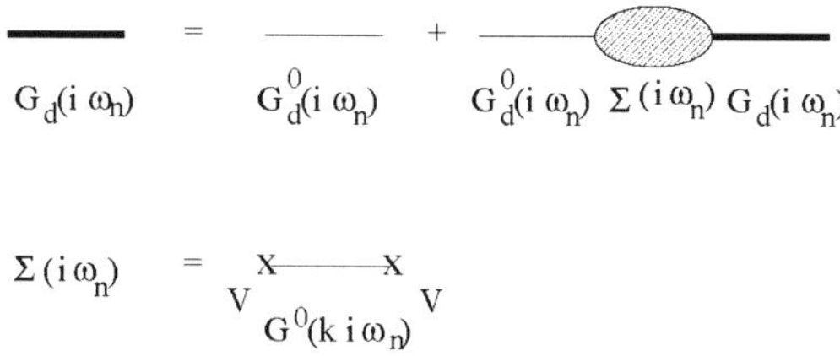

FIGURE 10. Dyson equation and self-energy for the localized (d) electron for the Anderson impurity model.

which describes the mixing of a *single* impurity orbital (labeled by 'd') with a band of conduction electrons (labeled by '**k**'). This is a model akin to the PAM mentioned earlier and is referred to as the Anderson impurity model (AIM). You will need to write the two definitions of $G_{\mathbf{k}}(\tau)$ and $G_d(\tau)$, take their τ derivatives, and Fourier transform. You'll end up with two equations in two unknowns (the two Greens functions). Solving, your result for G_d should be

$$G_d(i\omega_n) = \frac{1}{i\omega_n - \varepsilon_d - V^2 \sum_{\mathbf{k}} \frac{1}{i\omega_n - \varepsilon_{\mathbf{k}}}} \tag{22}$$

Exercise 17: The Greens function for $U = 0$ offers the jumping-off place for perturbative studies of the Hubbard, and related, Hamiltonians. If you have some familiarity with those approaches, you can attempt to rederive the Greens function for the AIM by evaluating the diagrammatic form of the self-energy, pictured in Fig. 10, by inserting the forms of the $U = 0$ Greens function and performing the necessary summations. Plugging into the Dyson Equation

$$\begin{aligned} G(k,i\omega_n) &= G_0(k,i\omega_n) + G_0(k,i\omega_n)\Sigma(k,i\omega_n)G(k,i\omega_n) \\ G^{-1}(k,i\omega_n) &= G_0^{-1}(k,i\omega_n) + \Sigma(k,i\omega_n) \end{aligned} \tag{23}$$

should reproduce the equation of motion result for the Greens function. In this case, the perturbative Feynman diagram method, like the equation of motion method, provides an exact expression (ie correct to all orders in V). This provides a simple illustration of the method, but it should be kept in mind that this simple result is highly atypical.

It is also instructive to look at the Green's function for a single site, that is, the $t = 0$ Hubbard model. Previously the Hilbert space for this problem was written down and the partition function and various equal time quantities were obtained. Now consider the calculation of

$$G_\uparrow(\tau) = \langle c_\uparrow(\tau) c_\uparrow^\dagger(0) \rangle. \tag{24}$$

Only the states $|00\rangle$ and $|01\rangle$ contribute to the expectation value since the creation operator for up electrons needs to see an empty up state. The action of the sequence of operators on the states is, $|00\rangle$:

$$c_\uparrow(\tau)c_\uparrow^\dagger(0)\,|00\rangle \;=\; e^{H\tau}c_\uparrow(0)e^{-H\tau}c_\uparrow^\dagger(0)\,|00\rangle = e^{H\tau}c_\uparrow(0)e^{-H\tau}\,|10\rangle$$

$$= \quad e^{H\tau} c_\uparrow(0) e^{+U\tau/4} \,|\, 1 0 \rangle = e^{H\tau} e^{+U\tau/4} \,|\, 0 0 \rangle = e^{+U\tau/2} \,|\, 0 0 \rangle \quad (25)$$

and similarly for $|\, 0 1 \rangle$.

Exercise 18: Complete the calculation begun above to show that,

$$G_\uparrow(\tau) = \frac{e^{+\beta U/4} e^{-\tau U/2} + e^{-\beta U/4} e^{\tau U/2}}{2 e^{\beta U/4} + 2 e^{-\beta U/4}}. \quad (26)$$

Exercise 19: The Green's function is related to the spectral density $A(\omega)$ by the relation,

$$G(\tau) = \int_{-\infty}^{+\infty} A(\omega) \frac{e^{-\omega\tau}}{e^{-\beta\omega} + 1} d\omega. \quad (27)$$

Show that if you plug in

$$A(\omega) = \frac{1}{2} \left(\delta(\omega - U/2) + \delta(\omega + U/2) \right) \quad (28)$$

and do the integral you get the correct $G(\tau)$. The spectral function of the single site Hubbard model consists of two delta function peaks separated by U (the Mott gap).

As discussed earlier (Fig. 7 and below Eq. 8), the Mott transition is associated with the development of a plateau in ρ vs μ at $\rho = 1$. The connection to insulating behavior is simple: In a dilute lattice electrons can move without double occupation. However, for an electron to move in a half-filled lattice in which each site is singly occupied, double occupation must occur. This costs an energy U. It is plausible to imagine that if U is very large, the electrons will not want to move at all, and one will have an "Mott" insulator. The structure of $A(\omega)$, evaluated in the $t = 0$ limit above, which consistes of two delta functions separated by U, provides another illustration of the opening of a gap and Mott insulating phenomenon. Though the Mott gap is similar to the way that the cost to add an electron jumps if there is a gap separating single particle energy bands, it is worth noting that this analogy goes only so far, and the Mott gap differs in very fundamental ways from band gaps.

Exercise 20: It is also interesting to compute G and A when there is a local site energy or chemical potential present. Work with the Hamiltonian $H = U(n_\uparrow - \frac{1}{2})(n_\downarrow - \frac{1}{2}) - \mu(n_\uparrow + n_\downarrow)$. Show that,

$$\begin{aligned} A(\omega) &= a_+ \; \delta(\omega - U/2 + \mu) + a_- \; \delta(\omega + U/2 + \mu). \\ a_+ &= \left(e^{\beta U/2} e^{\beta\mu} + e^{2\beta\mu} \right) / \left(1 + 2 e^{\beta U/2} e^{\beta\mu} + e^{2\beta\mu} \right) \\ a_- &= \left(1 + e^{\beta U/2} e^{\beta\mu} \right) / \left(1 + 2 e^{\beta U/2} e^{\beta\mu} + e^{2\beta\mu} \right) \end{aligned}$$

by evaluating $G(\tau)$ explicitly from its definition. μ now enters both the partition function Z and the imaginary time propagation. The coefficients a_+ and a_- are obtained by equating this expression for $G(\tau)$ with what you get from plugging $A(\omega)$ into the formula relating G and A. Notice that at $\mu = 0$, $a_+ = a_- = \frac{1}{2}$, and that, regardless of what μ, β and μ are, $a_+ + a_- = 1$. Interestingly, the spectral function is not just shifted by μ as one might have expected. The peak heights are also changed.

We have discussed the Hubbard model at $U = 0$ and at $t = 0$. In the noninteracting case, going to momentum creation and destruction operators diagonalized H. The single site limit was more easy in the sense that the interaction term is already diagonal in the original site operators. It is sometimes useful to consider the representation of the interaction term in momentum space, as done in the following exercise.

Exercise 21: Consider the form of the interaction term in momentum space. Substitute the equation which relates real and momentum space operators for each of the four real space creation operators in the interaction term of the Hubbard model. As with the hopping term, the sum over sites leads to momentum conservation and reduces the four momentum sums to three. Show that the result is,

$$\frac{U}{N} \sum_{\mathbf{k}_1,\mathbf{k}_2,\mathbf{k}_3} c^{\dagger}_{\mathbf{k}_1+\mathbf{k}_2-\mathbf{k}_3\uparrow} c^{\dagger}_{\mathbf{k}_3\downarrow} c_{\mathbf{k}_2\uparrow} c_{\mathbf{k}_1\downarrow}. \tag{29}$$

The physical content of this form is that an up and down electron of momentum $\mathbf{k}_2$ and $\mathbf{k}_1$ scatter and emerge with momenta $\mathbf{k}_3$ and $\mathbf{k}_1 + \mathbf{k}_2 - \mathbf{k}_3$, the same total momentum as initially. One can rewrite the sum over the three momentum variables in the more appealing form,

$$\frac{U}{N} \sum_{\mathbf{k}_1,\mathbf{k}_2,\mathbf{q}} c^{\dagger}_{\mathbf{k}_2+\mathbf{q}\uparrow} c^{\dagger}_{\mathbf{k}_1-\mathbf{q}\downarrow} c_{\mathbf{k}_2\uparrow} c_{\mathbf{k}_1\downarrow}. \tag{30}$$

Here $\mathbf{q}$ is seen to be the momentum exchanged in the collision of the two electrons of initial momenta $\mathbf{k}_1$ and $\mathbf{k}_2$.

D. Exact Diagonalization

After considering the limits $t = 0$ and $U = 0$, it is natural to examine the full problem on a small cluster of sites. However, before discussing diagonalization of the Hubbard model, it is useful to examine the Heisenberg model, since it describes the physics of the Hubbard model when U is large. Consider a set of quantum mechanical spins, with spin-1/2, localized on a set of sites i, and interacting on neighboring sites $\langle ij \rangle$ through inter-atomic exchange,

$$H = J \sum_{\langle ij \rangle} \vec{S}_i \cdot \vec{S}_j. \tag{31}$$

For two sites,

$$H = J\vec{S}_1 \cdot \vec{S}_2 = J\,[S_{1,x}S_{2,x} + S_{1,y}S_{2,y} + S_{1,z}S_{2,z}] = J\,[\frac{1}{2}(S_{1,+}S_{2,-} + S_{1,-}S_{2,+}) + S_{1,z}S_{2,z}] \tag{32}$$

There are four states in the Hilbert space, since each z component can take on one of two values. Denote these by $|S_{1,z}S_{2,z}\rangle = |++\rangle, |--\rangle, |+-\rangle, |-+\rangle$.

Compute the eigenvalues of the two site Heisenberg model by writing,

$$H = J\vec{S}_1 \cdot \vec{S}_2 = \frac{1}{2}J\,[(S_1 + S_2)^2 - S_1^2 - S_2^2]. \tag{33}$$

Replace S_1^2 and S_2^2 with $j(j+1) = \frac{1}{2}(\frac{1}{2}+1) = \frac{3}{4}$, since these operators give the same value on all states in the Hilbert space. S_1+S_2 can take the two possible values $S_1+S_2 = 0,1$ when two spin-1/2 angular momenta are added. The choice $S_1+S_2 = 1$ is three-fold degenerate, corresponding to $S_{1z}+S_{2z} = -1,0,1$. Finally, the eigenvalues of H are $\frac{1}{2}J[1(1+1)-\frac{3}{4}-\frac{3}{4}] = +\frac{1}{4}J$ (three-fold degenerate) and $\frac{1}{2}J[0(0+1)-\frac{3}{4}-\frac{3}{4}] = -\frac{3}{4}J$ (nondegenerate).

The 'brute force' method of constructing the matrix of H and diagonalizing of course yields the same result.

$$
\begin{aligned}
H|++\rangle &= +\frac{1}{4}J|++\rangle \\
H|--\rangle &= +\frac{1}{4}J|--\rangle \\
H|+-\rangle &= -\frac{1}{4}J|+-\rangle + \frac{1}{2}J|-+\rangle \\
H|-+\rangle &= -\frac{1}{4}J|-+\rangle + \frac{1}{2}J|+-\rangle.
\end{aligned} \tag{34}
$$

Thus the matrix for H (with the basis vectors in the order listed above) is

$$
H = \frac{1}{4}J \begin{pmatrix} 1 & 0 & 0 & 0 \\ 0 & 1 & 0 & 0 \\ 0 & 0 & -1 & 2 \\ 0 & 0 & 2 & -1 \end{pmatrix}. \tag{35}
$$

Two of the basis vectors, $|++\rangle$ and $|--\rangle$ are already eigenstates of H with eigenvalue $\frac{1}{4}J$. These are the states with $S_1+S_2 = 1$ and $S_1^z+S_2^z = 1,-1$. form the symmetric and antisymmetric linear combinations of the other two to get the remaining eigenvectors: $\frac{1}{\sqrt{2}}(|+-\rangle+|-+\rangle)$ and $\frac{1}{\sqrt{2}}(|+-\rangle-|-+\rangle)$. These have eigenvalues $\frac{1}{4}J$ and $-\frac{3}{4}J$ respectively. Confirming the earlier analysis with the method using the square of the total spin, there are three eigenvectors of eigenvalue $\frac{1}{4}J$, and a single eigenvector of eigenvalue $-\frac{3}{4}J$. The 'singlet' is the ground state for $J>0$ (antiferromagnetic coupling).

It is amusing that the four site Heisenberg model is also amenable to an analysis through examining the squares of the angular momentum.

$$
H = J[\vec{S}_1\cdot\vec{S}_2+\vec{S}_2\cdot\vec{S}_3+\vec{S}_3\cdot\vec{S}_4+\vec{S}_4\cdot\vec{S}_1] = \frac{1}{2}J[(\vec{S}_1+\vec{S}_2+\vec{S}_3+\vec{S}_4)^2-(\vec{S}_1+\vec{S}_3)^2-(\vec{S}_2+\vec{S}_4)^2] \tag{36}
$$

The possible values of S_1+S_3 are $0,1$ and likewise for S_2+S_4. When $S_1+S_3 = 0$ and $S_2+S_4 = 0$ are combined, a single total spin $S_1+S_3+S_2+S_4 = 0$ state results. Putting together $S_1+S_3 = 1$ and $S_2+S_4 = 0$ yields three $S_1+S_3+S_2+S_4 = 1$ states, as does combining $S_1+S_3 = 0$ and $S_2+S_4 = 1$. Adding $S_1+S_3 = 1$ and $S_2+S_4 = 1$ yields one $S_1+S_3+S_2+S_4 = 0$ state, three $S_1+S_3+S_2+S_4 = 1$ states, and five $S_1+S_3+S_2+S_4 = 2$ states.

In the sector formed by putting together $S_1+S_3 = 0$ and $S_2+S_4 = 0$, the eigenvalue is $\frac{1}{2}J[0(0+1)-0(0+1)-0(0+1)] = 0$. In the sector formed by putting together

$S_1+S_3=1$ and $S_2+S_4=0$, the (three) eigenvalues are $\frac{1}{2}J[1(1+1)-1(1+1)-0(0+1)]=0$. The same holds for the sector formed by putting together $S_1+S_3=0$ and $S_2+S_4=1$. In the sector formed by putting together $S_1+S_3=1$ and $S_2+S_4=1$, the eigenvalues are $\frac{1}{2}J[2(2+1)-1(1+1)-1(1+1)]=J$ for the five total spin 2 states, $\frac{1}{2}J[1(1+1)-1(1+1)-1(1+1)]=-J$ for the three total spin 1 states, and $\frac{1}{2}J[0(0+1)-1(1+1)-1(1+1)]=-2J$ for the spin 0 state.

Exercise 22: Solve the four spin "J_1-J_2" Heisenberg in which the bonds along the sides of the square are J_1 and additional bonds of strength J_2 are introduced along the diagonal.

Exercise 23: Show there is a level crossing as the relative size of J_1 and J_2 changes. What is the physics in the limits of J_1 dominant and J_2 dominant? You might speculate that in the thermodynamic limit this small size level crossing might evolve into a phase transition.

With this localized spin model background in hand, the two site Hubbard model, whose Hilbert space has dimension sixteen, can be diagonalized and its strong coupling limit understood. Consider the largest subspace, the sector with one spin up and one spin down electron. Denote by $|\uparrow\downarrow\ \ 0\rangle, |0\ \ \uparrow\downarrow\rangle, |\uparrow\ \ \downarrow\rangle$, and $|\uparrow\ \ \downarrow\rangle$, the state with a spin up electron and a spin down electron on site 1, and site 2 empty; the state with a spin up electron and a spin down electron on site 2, and site 1 empty; the state with a spin up on site 1 and a spin down electron on site 2; and the state with a spin up on site 2 and a spin down electron on site 1, respectively.

It is easy to see that the associated matrix for the Hamiltonian is

$$H=\begin{pmatrix} U & 0 & -t & -t \\ 0 & U & -t & -t \\ -t & -t & 0 & 0 \\ -t & -t & 0 & 0 \end{pmatrix} \tag{37}$$

It is immediately obvious that two of the eigenvalues of H are 0 and U. Two others are $\frac{1}{2}(U\pm\sqrt{U^2+16t^2})$. If the last two eigenvalues are expanded in t/U, the result is $-4t^2/U$ and $U+4t^2/U$, and the energy scale $J=4t^2/U$ arises. The significance of J will be discussed further below.

The Heisenberg model emerges as a limit of the Hubbard model when t/U is small. To see this, note that the sector of the two site Hubbard model with one up and one down spin, considered above, can be expanded to include all sectors with two electrons, by adding the states $|\uparrow\ \ \uparrow\rangle$, and $|\downarrow\ \ \downarrow\rangle$. These states are eigenstates of H with eigenvalue 0. All together, the two electron space of the two site Hubbard model has four 'small' eigenvalues 0,0,0, and $\frac{1}{2}(U-\sqrt{U^2+16t^2})\approx-4t^2/U$ and two 'large' ones U and $\frac{1}{2}(U+\sqrt{U^2+16t^2})$. The large eigenvalues are associated with eigenvectors whose components have significant mixtures of the states with doubly occupied sites. The existence of the two groups of states whose eigenvalues are separated by U is a reflection of the 'upper and lower Hubbard bands'. The 'Mott-Hubbard' gap in the spectrum gives rise to a metal-insulator transition. The two site Heisenberg model had three eigenvalues $J/4$ and one $-3J/4$. Apart from a trivial shift in energies, this is the same spec-

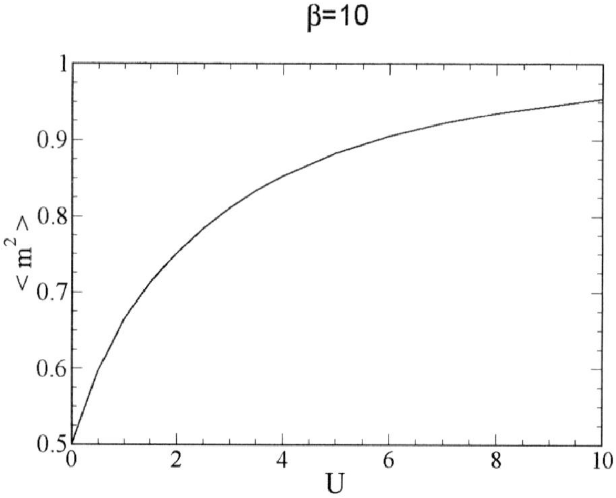

FIGURE 11. Local moment of the 2x2 Hubbard model as a function of U. Here $t = 1$, $\rho = 1$. The temperature $T = 0.1$ is low enough so that the suppression of the moment at small U is due to quantum, rather than thermal, fluctuations.

trum as that of the small eigenvalue sector of the Hubbard model, with the identification $J = 4t^2/U$.

Fig. 8 showed how the moment $\langle m^2 \rangle$ of the single site Hubbard model ($t = 0$) develops as the interaction U is increased at constant temperature T and $\rho = 1$. When U is small, thermal fluctuations reduce the moment to its uncorrelated value $\langle m^2 \rangle = 0.5$. Quantum fluctuations similarly reduce the moment, as is apparent in Fig. 11.

Exercise 24: Write a program to diagonalize the four site Hubbard model. Compute E and $\langle m^2 \rangle$.

The four site Hubbard model matrices are small enough that they can be constructed by hand. However, for larger lattices, one clearly wants to write a program which will generate the matrix elements automatically. Indeed, the routines entering such a code are needed also to evaluate expectation values of physically interesting quantities. This will be covered in lectures of other speakers. Here it is noted that there are two crucial elements to such a program: A first routine which computes the Hilbert state number given the occupations of the individual sites and a second routine which does the converse, namely computes the individual site occupation, for a given state number. With these available, the Hamiltonian can be constucted by looping over all states i, computing the occupation numbers using the second routine, then applying the kinetic energy to change the occupations and using the first routine to figure out the associated state number j. H_{ij} is then set to $-t$ (or $+t$) depending on the number of anticommutations involved, see Appendix A. The diagonal entry H_{ii} is a trivial function of the occupations.

Here is some pseudocode that builds the required arrays for a four site system.

```
state=0
loop:  n1=0,1
loop:  n2=0,1
loop:  n3=0,1
loop:  n4=0,1
       if (n1+n2+n3+n4=npart)
           occ1(state)=n1
           occ2(state)=n2
           occ3(state)=n3
           occ4(state)=n4
           state=state+1
           binary=n1+2*n2+4*n3+8*n4
           getstate(binary)=state
       endif
```

This code loops over the occupations of the different sites. (Here for simplicity four sites are chosen.) A check is performed to see if the total occupation equals the desired one: npart. If it does, arrays occ1, occ2, ... are used to store the site occupations, and the counter recording the number of states is incremented. These arrays allow us to look up the occupation on any site given the state. At the same time, the array getstate contains the state corresponding to a given set of site occupations. To access it, one first constructs the single number 'binary' from the occupations. This part is a little sloppy, because the array getstate is considerably larger (dimension 2^4, or more generally 2^N where N is the number of sites) than necessary. (If npart=2 the Hilbert space is only dimension 6, not 16.)

A few final comments: First, the action of the kinetic energy term in the Hubbard model is very similar to that of the exchange term $S_{+i}S_{-j} + S_{+j}S_{-i}$ of the Heisenberg model. So the same ideas/codes that work for the Heisenberg case for labeling the states and constructing the Hamiltonian can be carried over to the Hubbard case (and vice-versa). One very important difference, though, is the existence of minus signs which arise from the anticommutation of the fermion operators on different sites. In one dimension, these minus signs can be eliminated for near-neighbor hopping with a choice of appropriate boundary conditions. Indeed, this is one way of understanding the existence of exact maps between fermions and quantum spins, like the Jordan-Wigner transformation, in 1-d. In 2-d, it is no longer possible to eliminate the minus signs.

Indeed, in one dimension, for the half-filled Hubbard model, it is sometimes useful to alternate periodic (pbc) and antiperiodic boundary conditions (apbc) for lattices of length $4n$ and $4n$+2. (Even chain lengths are usually chosen, otherwise the antiferromagnetic order would be frustrated.) The reason is that at half-filling, lattices of length $4n$ have an even number of electrons of each spin species. Thus when an electron hops across the ends of the chain, ie between sites 1 and n, it passes an *odd* number of the remaining electrons of the same spin. This introduces a minus sign in the end-to-end hopping t. For lattice lengths $4n$+2, clearly this minus sign does not occur. Employing apbc for length $4n$ eliminates the minus sign. With the pbc/apbc alternation, measured quantities monotonically approach the large spatial size limit rather than oscillating above and below the asymptotic value. A more general trick to reducing finite size effects is boundary condition averaging [46, 47].

The obvious drawback of the exact diagonalization approach is the system size it is able to handle. Even using symmetries of the Hamiltonian, the matrix to be diagonalized grows exponentially with the number of sites. Typically, diagonalization methods are limited to a few tens of lattice sites, where the exact number is determined by the values of the degrees of freedom at each site (and the effort the programmer is willing to make!). To explore the systematic effects of the restriction to finite size it is useful to be able to compare results on lattices of different extent. One trick to facilitate this is to consider non-standard tilings. For example, rather than considering only 2x2, 3x3, 4x4, ... lattices, one can tile the 2-d square lattice with other configurations such as groups of eight sites,

$$\begin{array}{cccc} & A & B & \\ C & D & E & F \\ & G & H & \end{array} \tag{38}$$

as follows:

$$\begin{array}{cccccccc} & & & A & B & & & \\ & & C & D & E & F & & \\ & A & B & G & H & A & B & \\ C & D & E & F & C & D & E & F \\ & G & H & A & B & G & H & \\ & & C & D & E & F & & \\ & & & G & H & & & \end{array} \tag{39}$$

This construction corresponds to dividing the 2-d lattice into $\sqrt{8}$ by $\sqrt{8}$ squares whose axes are rotated with respect to the grid lines defining the array of sites. Notice that the natural connections which implement periodic boundary conditions can be seen from the picture. For example, site B has neighbors C (above), A (left), G (right) and E (below).

A further reason for looking at these oddly shaped lattices in two dimensions has to do with special symmetries of the 4x4 lattice which is frequently studied in 2-d exact diagonalization. It is easy to verify that the 4x4 lattice has the same connectivity as the 2x2x2x2 lattice in four dimensions. That the 4x4 lattice has unusual properties can be illustrated by the observation that correlation functions between observables separated by $\hat{x}+\hat{y}$ are identical to those separated by $2\hat{x}$ or $2\hat{y}$!

E. Mean Field Theory

So far, the solutions to the Hubbard model in the simple limits of no hopping ($t = 0$), no interactions ($U = 0$), and very small system sizes (two sites) have been described. The next goal is to study ferromagnetism using mean field theory. Before doing this, it is useful to describe Stoner's argument for magnetism, which will later be demonstrated to give quantitatively equivalent results to the (more general) mean field approach.

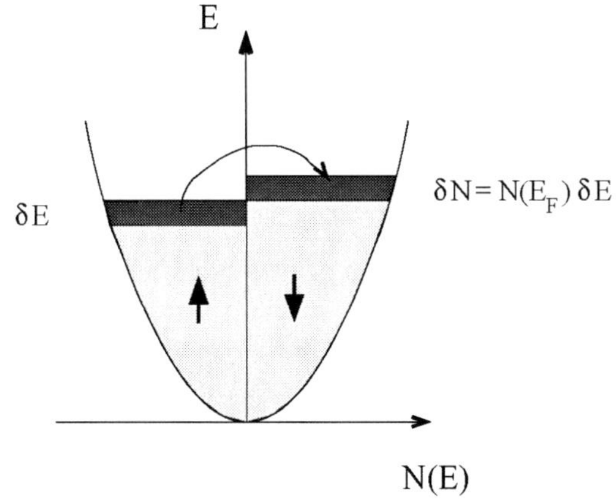

FIGURE 12. Qualitative picture of Stoner construction. δN down spin electrons are converted to up spin electrons, at a cost in kinetic energy $\delta K = +\delta n \delta E$. If this is less than the potential energy lowering, a ferromagnetic state is favored.

Stoner developed a very simple picture of ferromagnetism based on the competition between the kinetic energy cost of making the up and down spin electron numbers different and the associated potential energy gain. The basic idea is the following: Because of the Pauli principle, the way to occupy a given set of energy levels with the lowest energy is to start filling from the bottom and put two electrons, one of each spin, in each level. Otherwise, if you make the numbers of up and down electrons unequal, and don't fill each level with two electrons, you have to occupy higher energies. However, if you make the number of up and down electrons unequal, you can reduce the potential energy. In the limit of complete spin polarization where there are no electrons of one spin species, the potential energy is zero in the Hubbard model.

Perhaps this competition can lead to non-trivial effects. Consider a system with density of states $N(E)$ and both up and down spin electrons filling the energy levels up to the same 'Fermi level' E_F. The density of up and down electrons is equal. Call it n. Let's compute the change in energy which results from a reduction in the density of down spin electrons by δn and at the same time an increase in the number of up spin electrons by δn. The potential energy changes by,

$$\delta P = U(n+\delta n)(n-\delta n) - Un^2 = -U(\delta n)^2.$$

If an additional δn electrons are shifted into the up group, then energy levels above the original E_F are occupied. See Fig. 12. The definition of the density of states as the number of levels at an energy E, that is $N(E) = dN/dE$, implies $\delta n = N(E_F)\delta E$. This describes the range of energies above E_F filled in terms of δn. Likewise, levels below E_F that used to be occupied by down spin electrons are emptied. The net result of this process is to shift δn electrons up in energy by an amount δE. The change in the kinetic energy is then,

$$\delta K = +\delta n \delta E = +\frac{1}{N(E_F)}(\delta n)^2.$$

Putting these two expressions together,

$$\delta E = \delta P + \delta K = (-U + \frac{1}{N(E_F)})(\delta n)^2 = (-UN(E_F)+1)\frac{(\delta n)^2}{N(E_F)}.$$

If $UN(E_F) > 1$ the total energy change $\delta E < 0$, so it is favorable to have the up and down electron densities different and hence to have ferromagnetism. This is called the Stoner criterion. It tells us that magnetism is associated with large electron interactions and density of states. As described later, this simple calculation yields results in precise agreement with mean field theory.

What is mean field theory? As commented in an earlier section, a Hamiltonian which is quadratic in the fermion creation and destruction operators, $H = \sum_{\mathbf{i},\mathbf{j}} c_{\mathbf{i}}^\dagger h_{\mathbf{i},\mathbf{j}} c_{\mathbf{j}}$, can be solved by diagonalizing the matrix h. Mean field theory is a method which produces such a quadratic Hamiltonian from a model like the Hubbard model which has quartic terms $Uc_\uparrow^\dagger c_\uparrow c_\downarrow^\dagger c_\downarrow$ involving four fermion creation and destruction operators. The approach begins by expressing the number operators as an average value plus a deviation from the average:

$$\begin{aligned} n_{\mathbf{i}\uparrow} &= \langle n_{\mathbf{i}\uparrow}\rangle + (n_{\mathbf{i}\uparrow} - \langle n_{\mathbf{i}\uparrow}\rangle) \\ n_{\mathbf{i}\downarrow} &= \langle n_{\mathbf{i}\downarrow}\rangle + (n_{\mathbf{i}\downarrow} - \langle n_{\mathbf{i}\downarrow}\rangle). \end{aligned} \tag{40}$$

Substituting these expressions into the Hubbard interaction term, and dropping the 'small' term which is the product of the two deviations from the average yields,

$$\begin{aligned} n_{\mathbf{i}\uparrow} n_{\mathbf{i}\downarrow} &= [\langle n_{\mathbf{i}\uparrow}\rangle + (n_{\mathbf{i}\uparrow} - \langle n_{\mathbf{i}\uparrow}\rangle)][\langle n_{\mathbf{i}\downarrow}\rangle + (n_{\mathbf{i}\downarrow} - \langle n_{\mathbf{i}\downarrow}\rangle)] \\ &\approx \langle n_{\mathbf{i}\uparrow}\rangle\langle n_{\mathbf{i}\downarrow}\rangle + \langle n_{\mathbf{i}\downarrow}\rangle(n_{\mathbf{i}\uparrow} - \langle n_{\mathbf{i}\uparrow}\rangle) + \langle n_{\mathbf{i}\uparrow}\rangle(n_{\mathbf{i}\downarrow} - \langle n_{\mathbf{i}\downarrow}\rangle)] \\ &= n_{\mathbf{i}\uparrow}\langle n_{\mathbf{i}\downarrow}\rangle + n_{\mathbf{i}\downarrow}\langle n_{\mathbf{i}\uparrow}\rangle) - \langle n_{\mathbf{i}\uparrow}\rangle\langle n_{\mathbf{i}\downarrow}\rangle. \end{aligned} \tag{41}$$

The interpretation of this expression is simple. The up spin electrons interact with the average density of down spin electrons, and similarly the down spin electrons interact with the average density of up spin electrons. These two terms overcount the original single interaction term, so the product of the average densities is subtracted off.

Within this mean field replacement, the Hubbard Hamiltonian is now quadratic, and takes the form (in one dimension)

$$H = -t\sum_{l\sigma}(c_{l\sigma}^\dagger c_{l+1\sigma} + c_{l+1\sigma}^\dagger c_{l\sigma}) + n_{\mathbf{i}\uparrow}\langle n_{\mathbf{i}\downarrow}\rangle + n_{\mathbf{i}\downarrow}\langle n_{\mathbf{i}\uparrow}\rangle) - \langle n_{\mathbf{i}\uparrow}\rangle\langle n_{\mathbf{i}\downarrow}\rangle. \tag{42}$$

Since H is quadratic, its solution is a matter of diagonalizing an appropriate matrix. Specifically, for the case of ferromagnetism, one imagines that the average occupation is independent of spatial site **i** but allowed to be different for the two spin species. That is, $\langle n_{\mathbf{i}\uparrow}\rangle = n+m$ and $\langle n_{\mathbf{i}\downarrow}\rangle = n-m$. The goal is to calculate the energy E for fixed n

as a function of m and see whether the minimum is at $m = 0$ (paramagnetic state, no ferromagnetism) or $m \neq 0$ (ferromagnetism). Because the expectation values $\langle n_{i\uparrow} \rangle$ and $\langle n_{i\downarrow} \rangle$ have such a simple, site independent form, the energy levels can easily be written down. In $d = 1$ they are,

$$\begin{aligned} \varepsilon_{k\uparrow} &= U(n-m) - 2t\cos k \\ \varepsilon_{k\downarrow} &= U(n+m) - 2t\cos k. \end{aligned} \tag{43}$$

To determine the ground state phase diagram,

(1) Fix the lattice size, N, and choose a ratio U/t and particle number $N_{\rm tot}$.

(2) Loop over $N_\uparrow = 0, 1, 2, \ldots N_{\rm tot}$. setting $N_\downarrow = N_{\rm tot} - N_\uparrow$. Define the densities, $n_\uparrow = N_\uparrow / N$ and $n_\downarrow = N_\downarrow / N$.

(3) Fill up the lowest $N_\uparrow$ and $N_\downarrow$ energy levels, $\varepsilon_{k\uparrow} = -2t\cos k + U\langle n_\downarrow \rangle$ and $\varepsilon_{k\downarrow} = -2t\cos k + U\langle n_\uparrow \rangle$. The momentum values are $k = 2\pi/N\{-N/2+1, -N/2+2, \ldots N/2\}$.

(4) Normalize the energy accumulator to the number of sites and include the term $-U\langle n_\uparrow \rangle \langle n_\downarrow \rangle$. This gives the energy for the given $N_\uparrow$ and $N_\downarrow = N_{\rm tot} - N_\uparrow$. Make a list of them and see which is lowest.

(5) Repeat the calculation for different U and $N_{\rm tot}$ to get the phase diagram.

Simple (Fortran) code

```
implicit none
integer i,N,Nup,Ndn,Ntot
real*8 t,U,tpin,k,ekup,ekdn,denup,dendn
real*8 efup,efdn,eftot

write (6,*) 'N,Ntot,t,U'
read  (5,*)  N,Ntot,t,U
tpin=8.d0*datan(1.d0)/dfloat(N)
do 1000 Nup=0,Ntot,2

Ndn=Ntot-Nup
denup=dfloat(Nup)/dfloat(N)
dendn=dfloat(Ndn)/dfloat(N)

efup=0.d0
efdn=0.d0
do 200 i=-N/2+1,N/2
     k=tpin*dfloat(i)
     if (i.ge.-Nup/2+1.and.i.le.Nup/2) then
        ekup=-2.d0*t*dcos(k)+U*dendn
        efup=efup+ekup
     endif
     if (i.ge.-Ndn/2+1.and.i.le.Ndn/2) then
        ekdn=-2.d0*t*dcos(k)+U*denup
        efdn=efdn+ekdn
     endif
200 continue
eftot=(efup+efdn)/dfloat(N)-U*denup*dendn

write (36,990) Nup,Ndn,eftot
990 format(2i6,f16.6)
```

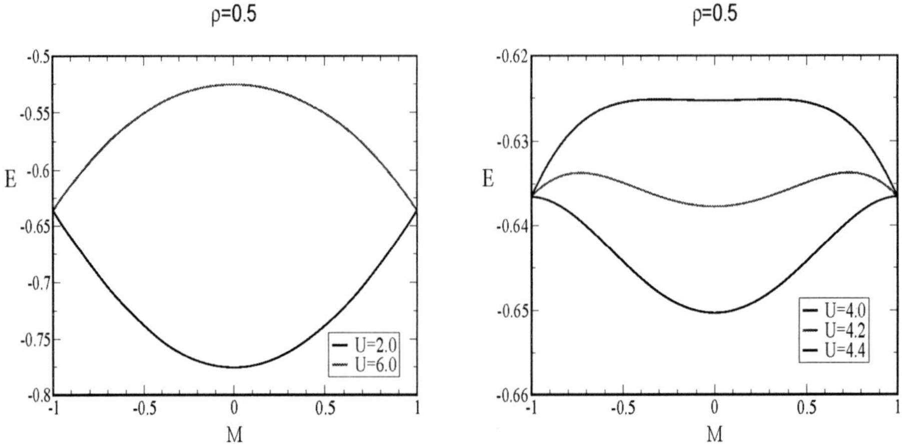

FIGURE 13. Left: Energy versus magnetization of the $d = 1$ Hubbard model at quarter filling, $\rho = 1/2$ (128 electrons on an $N = 256$ site lattice). A paramagnetic state, $M = 0$, is favored at $U = 2.0t$, and a ferromagnetic one, with full spin polarization $M = 1$, at $U = 6.0t$. Right: Same as left panel, except in the transition region $U \approx 4.0t$. The Stoner critierion gives $U_{\rm crit} = \sqrt{2}\pi t \approx 4.44t$.

```
1000 continue
end
```

Fig. 13(left) shows $E(M)$ for one quarter filling, $\rho = \rho_\uparrow + \rho_\downarrow = 1/2$. The magnetization $M = (\rho_\uparrow - \rho_\downarrow)/(\rho_\uparrow + \rho_\downarrow)$. At $U/t = 2$ the optimal energy is paramagnetic: the energy E is minimized at $M = 0$. For $U/t = 6$ the minima are fully spin polarized, $M = \pm 1$. Fig. 13(right) focuses on the transition region. When $U/t = 4.2$ the energies for large $|M|$ have started to turn down and are lower than intermediate M, though $E(M = 0)$ is still lowest. $U/t = 4.4$ has just gone ferromagnetic. Notice that the transition is *first order*. That is, as U/t increases the minimum jumps from $M = 0$ to $M = \pm 1$. Another possibility would have been a second order transition in which the minimum at $M = 0$ gradually shifts to larger M and partially polarized solutions are best for a range of U/t. An examination of other MFT solutions (like antiferromagnetic ones) is required to determine if this first order transition is 'real' or simply occurs because of the restriction to ferromagnetic solutions.

These MFT results are consistent with the Stoner Criterion for Ferromagnetism $UN(E_{\rm F}) > 1$. For the d=1 Hubbard model where $\varepsilon_k = -2t\cos(k)$,

$$N(E) = 2\sum_k \delta(E - \varepsilon_k) = 2\int \frac{dk}{2\pi}\delta(E - \varepsilon_k) = \frac{1}{\pi\sqrt{4t^2 - E^2}}. \tag{44}$$

The density ρ and the Fermi energy $E_{\rm F}$ are related by:

$$\rho = 2\int_{-2t}^{E_{\rm F}} dE\, N(E) = \frac{2}{\pi}\cos^{-1}(-\frac{E}{2t}). \tag{45}$$

This relation obeys the expected limits: $\rho = 0$ when $E_F = -2t$, $\rho = 1$ when $E_F = 0$, and $\rho = 2$ when $E_F = +2t$.

Putting these equations together, the density of states at E_F for a given filling is:

$$N(\rho) = \frac{1}{2\pi t}\frac{1}{\sin(\pi\rho/2)}. \tag{46}$$

For half-filling, $N(\rho = 1) = 1/2\pi t$ and hence $U_{\text{crit}} = 2\pi t$. For quarter-filling, $N(\rho = 1/2) = 1/\sqrt{2}\pi t$ and hence $U_{\text{crit}} = \sqrt{2}\pi t = 4.44t$. This is in good agreement with Fig. 13(right) where U_{crit} appears to be around $4.4t$. The slight difference between mean field theory and Stoner (Fig. 13 suggests U_{crit} a bit less than $4.4t$ while Stoner gives U_{crit} a bit more than $4.4t$) is a finite size effect. $N = 256$ was used in Fig. 13.

Exercise 25: Write a code to do mean field theory for the $d = 1$ Hubbard model and compute the critical U above which the ferromagnetic state is lower in energy than the paramagnetic one for $N_{tot} = 3N/4$. Verify that the critical U agrees with the Stoner criterion.

It is also interesting to work through the antiferromagnetic MFT solution in detail. The reason is the formal connection to many other problems in solid state physics. The most obvious is the opening of a gap in an energy band by a periodic potential $V(G)$ with wavevector G (e.g. see Ashcroft and Mermin). As discussed below, in the antiferromagnetic MFT solution, the up spin electrons move in a periodic potential resulting from the oscillating down spin density (and vice-versa). A gap is opened at $k = \pm\pi/2$. There is also a connection to simple phonon problems where one makes the masses or spring constants vary. Again, the single phonon dispersion curve for uniform masses and springs breaks into two branches, optic and acoustic.

An antiferromagnetic configuration allows a simple *spatial* dependence in which the densities alternate by m_s, the 'staggered magnetization': $n_{l\uparrow} = n + (-1)^l m_s$, and $n_{l\downarrow} = n - (-1)^l m_s$. The even sites have a surplus of up spin electron density: $n_{\text{even}\uparrow} = n + m_s$, and $n_{\text{even}\downarrow} = n - m_s$. The odd sites have a surplus of down spin electron density: $n_{\text{odd}\uparrow} = n - m_s$, and $n_{\text{odd}\downarrow} = n + m_s$. Note that the total number of up and down electrons in the whole system is the same, nN, and that each site has the same total density $2n$.

The form of the Hamiltonian in mean field theory is, $H = \sum_{j,l} c^\dagger_{j\sigma} h_\sigma(j,l) c_{l\sigma}$, where $h_\sigma(j,l)$ has $-t$ just above and below the main diagonal, with $h_\uparrow(l,l) = U(n - (-1)^l m_s)$, or $h_\downarrow(l,l) = U(n + (-1)^l m_s)$ along the diagonal. When h is nonzero, the eigenvectors of momentum k and $k + \pi$ are mixed and the eigenvalues are, $E(k) = \pm\sqrt{(-2t\cos k)^2 + (Um_s)^2} + Un$. Here k is now defined in a 'reduced zone', $k = 2\pi/N\{-N/4+1, -N/4+2, \ldots, +N/4\}$. The process for computing the energy of an antiferromagnetic configuration is basically the same as for ferromagnetism: the lowest energy levels are simply filled up. The difference is that here $N_\uparrow = N_\downarrow$ and one loops over different m_s.

Simple (Fortran) code

```
implicit none
integer i,N,Ntot,istag
real*8 t,U,tpin,k,ek,mstag
real*8 rho,Umstag,Urho
real*8 eaf,eaftot,lambdaminus
```

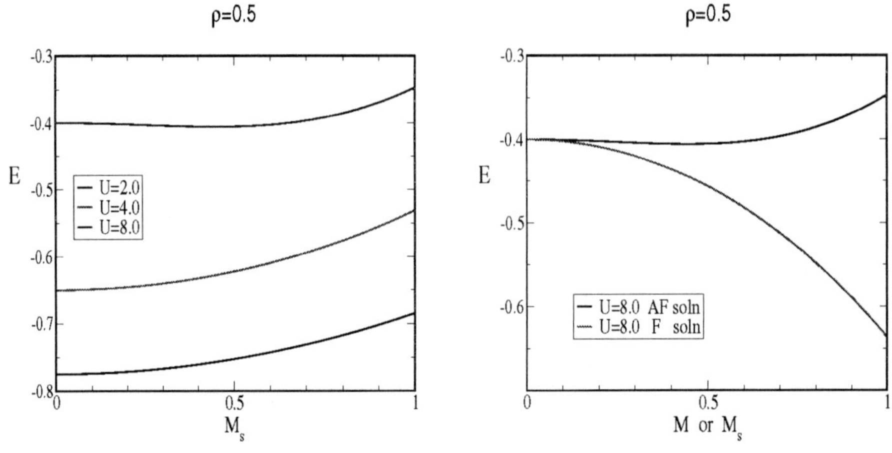

FIGURE 14. Left: Energy versus staggered magnetization of $d = 1$ Hubbard model at quarter filling, $\rho = \frac{1}{2}$ (128 electrons on an $N = 256$ site lattice). A paramagnetic state, $M_s = 0$, is favored at $U = 2.0t$ and $U = 4.0t$. Nonzero staggered magnetization with partial staggered spin polarization, $M_s = 21/48$, has lower energy than $M_s = 0$ at $U = 8.0t$. Right: However, the ferromagnetic solution has yet lower energy for this filling and interaction strength.

```
      write  (6,*)  'N,Ntot,t,U'
      read   (5,*)   N,Ntot,t,U
      write  (36,*)  Ntot/2+1

      tpin=8.d0*datan(1.d0)/dfloat(N)
      rho=dfloat(Ntot)/dfloat(N)
      Urho=U*rho/2.d0

      do 1000 istag=0,Ntot,2

      mstag=dfloat(istag)/dfloat(N)
      Umstag=U*mstag/2.d0

      eaftot=0.d0
      do 200 i=-Ntot/4+1,Ntot/4
           k=tpin*dfloat(i)
           ek = -2.d0*t*dcos(k)
           lambdaminus=-dsqrt(ek*ek+Umstag*Umstag)
           lambdaminus=lambdaminus+Urho
           eaftot=eaftot+lambdaminus
200    continue
       eaftot=2.d0*eaftot/dfloat(N)-U*(rho*rho-mstag*mstag)/4.d0

      write  (36,990)  istag,eaftot
990   format(i6,f16.6)

1000  continue

      end
```

Fig. 14(left) shows that antiferromagnetic solutions are not favored at quarter filling

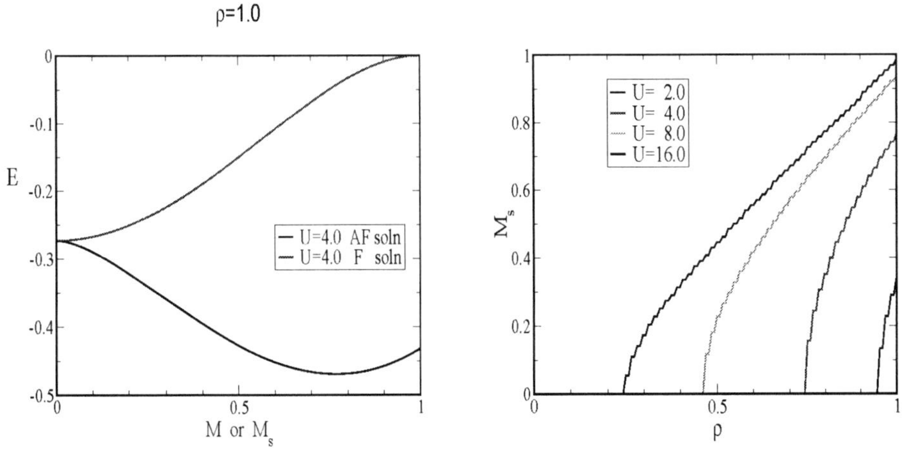

FIGURE 15. Left: Energy versus magnetization and staggered magnetization of $d = 1$ Hubbard model with $U = 4.0t$ and at half filling, $\rho = 1$ (256 electrons on an $N = 256$ site lattice). Nonzero staggered magnetization with partial staggered spin polarization, $M_s = 21/48$, has lower energy than $M_s = 0$. Unlike the quarter filled case of the right panel of Fig. 14 this antiferromagnetic solution is lower in energy than ferromagnetic ones. Right: Staggered magnetization which minimizes the energy, when restricted to the antiferromagnetic sector, as a function of filling, for different interaction strengths. In all cases, it is essential to check to see whether ferromagnetic solutions might be yet lower. Steps in the curves are a result of the finite lattice size.

and weak coupling $U = 2t, 4t$: The energy is minimized for $M_s = 0$. For $U = 8t$, E has its minimum at nonzero M_s. However, as seen in Fig. 14(right) the ferromagnetic energy is lower still.

Fig.15(left) shows the energy as a function of (ferromagnetic) magnetization m and (antiferromagnetic) staggered magnetization m_s for half-filling $\rho = 1$. While the ferromagnetic energy is minimized at $m = 0$, a lower value is obtained by a non-zero m_s. In Fig. 15(right) the staggered magnetization m_s which minimizes the energy is shown as a function of density ρ for different couplings. Let's consider the case of one quarter filling, that is, a density $n = n_\uparrow + n_\downarrow = 1/2$ electrons per site. For $U = 2$, m_s=0 has lowest energy. From the preceding analysis it is seen that $m = 0$ is the lowest of the ferromagnetic energies. Thus the Hubbard model at $U/t = 2$ and $\rho = 1/2$ is paramagnetic. It is a simple matter to construct the MFT phase diagram by considering all choices of filling and coupling strength. (Fig. 15(right) has results for a range of U values.) The complete diagram is shown in Fazekas [4] and also by Hirsch [36].

MFT codes can also be written in the grand canonical ensemble where one provides a chemical potential μ and then *computes* $N_\downarrow$ and $N_\uparrow$ by filling those levels which are below μ. An advantage of this method is that one easily work at finite temperature. by accumulating $\varepsilon_\sigma(k)$ times the Fermi function $1/[1 + e^{\beta(\varepsilon_\sigma(k)-\mu)}]$ for the energy and the Fermi function alone for the density. Grand canonical approaches are also convenient since they generalize better to considering complex, density patterns ('striped phases' etc). There is no longer an analytic form for the energy levels but it is a simple (and fast) matrix diagonalization problem to get them. The site and spin dependent densities can be computed self-consistently from the eigenvalues and eigenvectors, and iterated to convergence. (However, be careful that it is fairly common for this iteration to get stuck in metastable patterns.)

Exercise 26: Write a finite temperature MFT code for the $d = 1$ Hubbard model in the grand canonical ensemble. Show that at half-filling the Neel temperature for antiferromagnetism increases linearly with U at strong coupling. The correct result, as seen in the discussion of the Heisenberg model, is that $T_{\rm Neel} \propto t^2/U$.

In concluding the discussion of mean field theory, it should be emphasized that the approach, while very useful in yielding insight into the possible phases of the system, is a completely uncontrolled approximation. MFT overestimates the tendency for ordered phases, and can (and does) predict magnetic order where none occurs. Even if a particular phase transition is correctly predicted, the details of the transition (critical temperature, critical exponents, etc) are usually incorrect. These failures can be illustrated by noting that MFT gets the functional form of the Neel temperature wrong at strong coupling in the half-filled Hubabrd model, giving $T_N \propto U$ rather than the correct $T_N \propto t^2/U$ seen from the mapping to the Heisenberg model. MFT does not distinguish the moment formation and moment ordering temperatures. MFT is also privides a fundamentally misleading picture of the physics, since it implies the existence of sharp (non-decaying) single particle energy levels which are just shifted from their $U = 0$ values. Instead, interactions also introduce a finite lifetime, a phenomena which MFT misses. Phrased more formally, MFT, like $U = 0$ produces a spectral function consisting of delta-function peaks. One of the great achievements of 'Dynamical Mean Field Theory' is that it includes the finite lifetime broadening of energy levels.

III. DISORDER IN THE ABSENCE OF INTERACTIONS

A. The Anderson Transition

The development of insulating behavior of noninteracting electrons in a random potential is called 'Anderson localization.' To make contact with the material in these lectures, the Anderson transition can be examined as a particular instance of the Hubbard Hamiltonian at $U = 0$ with a spatially varying (random) chemical potential $-\mu \sum_{j\sigma} n_{j\sigma} \rightarrow -\sum_{j\sigma} \mu_j n_{j\sigma}$. Numerically, the Anderson transition then can be analyzed by diagonalizing matrices of dimension the number of spatial sites N (where of course the interest is in the thermodynamic limit, N large) which have the random chemical potential μ_j down the diagonal, and the hopping t in positions of the matrix corresponding to pairs of sites between which electrons can hop in the geometry in question.

The general picture is this: When randomness is turned on, the eigenstates with energies near the extremes of the density of states become localized, with those near the center remaining extended. As the randomness increases, the 'mobility edges' which demark the energies dividing localized from extended states move towards the center of the density of states. Given a fixed number of particles, when the mobility edge sweeps upward through E_F for increasing randomness, there is a metal to insulator transition. Alternately, the transition can be driven by varying particle number at fixed randomness, moving the Fermi level through the mobility edge. Ultimately, given sufficiently large randomness, it may be that all states are localized.

As for phase transitions in general, dimensionality plays a key role in whether the insulator-metal transition occurs. In one dimension, it is fairly easy to see that all states are localized by an arbitrarily small amount of randomness. Only the situation where there is zero disorder is metallic. The celebrated 'gang-of-four' paper [9] showed the much less trivial result that the same applies in two dimensions. As discussed previously,

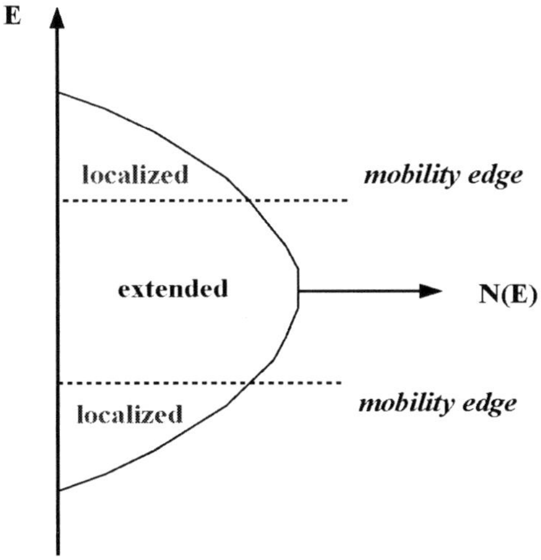

FIGURE 16. Disorder localizes states at the edges of the density of states. The mobility edge separates localized and extended states, and moves towards the band center as randomness increases.

this was part of the basis for the conjecture that, perhaps, even when interactions are included, there is no metallic phase in two dimensions. Only in three dimensions does the above picture of the coexistence of localized and delocalized states, separated by a mobility edge, really apply.

Approaches to the inclusion of correlations in the Anderson localization problem form a vast field. However, within the context of ideas developed in these notes, one can easily imagine a MFT treatment of the problem. Disorder can be very simply included in a grand canonical MFT code for the Hubbard model through an additional random chemical potential term along the diagonal. A simple physical picture that the insulator to metal transition originates in a screening (reduction) of the disorder. This occurs because the sites with low energy ε_i have the highest local density $\langle n_i \rangle$ and the sites with greatest ε_i have the lowest local density. ε_i becomes screened to $\tilde{\varepsilon}_i + U \langle n_i \rangle$, which has a smaller variance and thus represents a smoother energy landscape, in which the electrons might be more delocalized. Of course, in $d = 2$, since *any amount* of disorder localizes all the states, this metallic behavior cannot occur, and so the above picture is fatally flawed.

Exercise 27: Use a grand canonical MFT code to explore the localization-delocalization transition. Specifically, look at how the localization length and participation ratio (see below) of the eigenstates evolve as U is increased.

B. A Digression into Non-Hermiticity: The Hatano-Nelson Model

A non-Hermitian generalization of the Anderson Hamiltonian is the Hatano-Nelson model, which has been used to describe a number of phenomena, including the motion of flux lines in disordered type-II superconductors [48] and nuclear decay, dissipative systems, and quantum chromodynamics [49, 50, 51, 52, 53, 54, 55, 56, 57]. It is interesting here because it provides useful insight into the localization transition [48, 58],

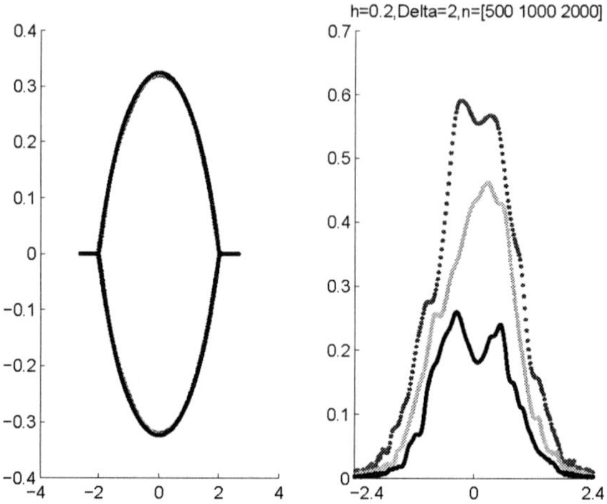

FIGURE 17. Left: Eigenspectrum of the Hatano-Nelson Hamiltonian with $h = 0.2$ and $\Delta = 2$. The eigenvalues are located on a ellipse centered at the origin, with two 'wings' extending out along the real axis. Right: Corresponding (normalized) participation ratios: $\mathscr{P}_n/N$. Numerical results were obtained for $N = 500, 1000, 2000$ sites, averaging over ten disorder realizations.

especially since, unlike the Anderson Hamiltonian, it allows for the existence of both localized and extended states in low dimension.

In one dimension, the Hatano-Nelson Hamiltonian is,

$$H = -\frac{t}{2}\sum_{l=1}^{N}(e^{h}c_{l+1}^{\dagger}c_l + e^{-h}c_l^{\dagger}c_{l+1}) + \sum_{l}\mu_l c_l^{\dagger}c_l \,. \tag{47}$$

The parameter h controls the asymmetry between the hopping amplitudes in the $+x$ and $-x$ directions. μ_l are a collection of random site energies which is chosen to have a uniform distribution on $[-\Delta/2, +\Delta/2]$. N is the number of lattice sites. The periodic boundary conditions which connect the two ends of the lattice have a fundamental influence on localization, as discussed further below. The choice of hopping parameter $t = 1$ sets the scale of energy,

When $\Delta = 0$, the eigenvalues of the Hatano-Nelson Hamiltonian are easily obtained,

$$\lambda(k) = \frac{1}{2}(e^{h}e^{ik} + e^{-h}e^{-ik}) \,. \tag{48}$$

Here $0 < k < 2\pi$. The eigenvalues lie on an ellipse in the complex plane, centered at the origin, with a semimajor axis of length $2\cosh h$ aligned with the real axis, and a semiminor axis of length $2\sinh h$ aligned with the imaginary axis. When disorder is turned on, $\Delta \neq 0$, lines of real eigenvalues extend outward from this ellipse. Fig. 17 shows the eigenspectrum for $h = 0.2$ and $\Delta = 2$.

A crucial observation to make is the relationship between the location of the eigenvalue: on the ellipse (complex) or the wings (real), and the nature of the associated eigenvector. Consider the eigenvalue problem in component form,

$$-\frac{t}{2}e^{h}\,\psi_{l+1} + \mu_l\,\psi_l - \frac{t}{2}e^{-h}\,\psi_{l-1} = \lambda\,\psi_l\,.$$

Performing the 'gauge transformation' $\psi_l = e^{-hl}\,\tilde{\psi}_l$ yields

$$-\frac{t}{2}\tilde{\psi}_{l+1} + \mu_l\,\tilde{\psi}_l - \frac{t}{2}\tilde{\psi}_{l-1} = \lambda\,\tilde{\psi}_l\,. \tag{49}$$

h has vanished from the equation and the eigenproblem of the (Hermitian) Anderson problem is recovered! Since the eigenstates are localized in $d = 1$ for the Anderson Hamiltonian, this analysis has seemingly proved the same to be true for the Hatano-Nelson model!

The flaw in this argument concerns the boundary conditions. The gauge transformation does not work if the ends of the $d = 1$ lattice are connected. This observation holds the key to connecting the topology of the eigenvalue phase diagram (complex ring with real wings) to whether the associated eigenvectors are extended or localized. Consider a localized eigenvector (and periodic boundary conditions). As described above, a gauge transformation which moves all the h factors away from the region of the lattice where the eigenvector is non-zero can be performed. For a localized eigenvector, the boundary conditions are irrelevant because by the time the edge of the lattice is reached, the eigenvector components are exponentially small. It is then plausible that the eigenvector will be described by the same properties as the Anderson Hamiltonian. Hence its eigenvalue will be real. The conclusion is that the eigenvectors whose eigenvalues lie on the real wings are localized.

Meanwhile, for delocalized eigenvectors, this reduction to the Anderson model fails, suggesting that the complex eigenvalues on the ring encircling the origin are delocalized. Fig. 17(right) shows the participation ratio,

$$\mathscr{P}_n = \Big(\sum_{l=1}^{N} |\psi_l^n|^4\Big)^{-1} \tag{50}$$

a measure of the number of components which are 'large' for the eigenvector ψ^n with components ψ_l^n. Parameters are the same as Fig. 17(left). This plot is consistent with the above picture. The number of sites participating in eigenvectors with real eigenvalues does not grow with lattice size, and hence are loalized. For complex eigenvalues the participation ratio grows roughly linearly with N. Of course, a more careful analysis of the finite size scaling is called for to make this completely compelling.

Perhaps not surprisingly for a problem concerning localization and transport, boundaries are crucial to the physics. The eigenvalues of the non-Hermitian matrix of the Hatano-Nelson Hamiltonian are *real* and identical to those of the Anderson Hamiltonian, when open boundary conditions are used, but differ dramatically otherwise. (Interestingly, even in the case when periodic boundary conditions are included, the gauge transformation allows all the factors of h to be accumulated in a single link.)

Quite a bit is known analytically about the Hatano-Nelson model. For example, if the disordered site energies are taken from the Cauchy distribution,

$$p(\mu_l) = \frac{1}{\pi}\frac{\gamma}{\gamma^2 + \mu_l^2}\,, \tag{51}$$

then the height of the ellipse (parallel to the imaginary axis), gets reduced by γ while the base of the ellipse (parallel to the real axis) remains fixed! Since the ellipse height in the imaginary direction is $\sinh h$ (with $t = 1$), the ellipse gets completely squashed down when γ exceeds a critical value $\gamma_c = \sinh h$. The eigenspectrum is completely real (all eigenvalues localized) for $\gamma > \gamma_c$. The $d = 1$ Hatano-Nelson model thereby mimics what happens in the Anderson Hamiltonian only in higher dimension.

IV. PRESCRIPTION FOR DETERMINANT QUANTUM MONTE CARLO

A. Basic Formalism

These notes now turn to the Determinant Quantum Monte Carlo (DQMC) method. Section V shows that considerable insight can be obtained concerning the interplay of disorder and interactions with this approach. A possible goal for students going through these notes is to write, or at least begin to write, a DQMC code. In fact, this is a rather ambitious target. To make it a bit more feasible, this discussion begins with the prescription for writing a DQMC program, literally by writing down some 'pseudo-code' for it in the same way as for diagonalization. This makes the discussion succinct and understandable in the 'what to do' sense. Of course, it is unsatisfactory since nothing is proven. Appendix C presents the background derivations.

By now the discussion of MFT has emphasized that the non-interacting electron problem, that is, one whose Hamiltonian is quadratic in the fermion creation and destruction operators, can be rather easily solved by diagonalizing an appropriate matrix. DQMC works by reducing the interacting problem into a non-interacting one! In order to do this, a classical Hubbard-Stratonovich field is introduced to decouple the interaction. The Monte Carlo then sums over different field configurations. Appendix C provides the details. For now, these notes simply describe how to construct the matrix used in the simulation.

Consider the part of $\hat{H}$ which includes the terms which are quadratic in the fermion creation and destruction operators, eg the hopping and site energies (chemical potential),

$$\hat{K} = \sum_\sigma \begin{pmatrix} c_{1\sigma}^\dagger & c_{2\sigma}^\dagger & \cdots \end{pmatrix} \begin{pmatrix} k_{11} & k_{12} & \cdots \\ k_{21} & k_{22} & \cdots \\ \vdots & \vdots & \ddots \end{pmatrix} \begin{pmatrix} c_{1\sigma} \\ c_{2\sigma} \\ \vdots \end{pmatrix} \tag{52}$$

k is defined to be the NxN matrix with elements $\Delta\tau\, k_{ij}$.

As an example, for a one dimensional Hubbard model with $N = 6$ sites, nearest neighbor hopping, no disorder, and periodic boundary conditions,

$$k = \Delta\tau \begin{pmatrix} -\mu & -t & 0 & 0 & 0 & -t \\ -t & -\mu & -t & 0 & 0 & 0 \\ 0 & -t & -\mu & -t & 0 & 0 \\ 0 & 0 & -t & -\mu & -t & 0 \\ 0 & 0 & 0 & -t & -\mu & -t \\ -t & 0 & 0 & 0 & -t & -\mu \end{pmatrix} \tag{53}$$

The matrix k will be one of the ingredients in constructing the probability used in the DQMC simulation.

For reasons made clear in Appendix C, in DQMC it is necessary to divide the inverse temperature β into L 'imaginary time intervals', $\beta = L\Delta\tau$. The length of each interval

$\Delta\tau$ should be chosen such that $t\,U\,(\Delta\tau)^2 < 1/10$. The interaction term U is then included via the 'Hubbard-Stratonovich field' $s(i,l)$ which has a spatial index $i = 1,2,...N$ and an 'imaginary ime index $l = 1,2,...L$ and takes on values $s(i,l) = \pm 1$. (This discrete version of the Hubbard-Stratonovich transformation introduced by Hirsch [36] is more efficient than a continuous one.) Again, the reader is referred to Appendix C for the details. The bottom line is that the simulation begins by filling the array $s(i,l)$ randomly with values ± 1. One then defines a set of L diagonal matrices, each of dimension N,

$$v_\uparrow(l) = \lambda \begin{pmatrix} s(1,l) & 0 & 0 & 0 & \cdots \\ 0 & s(2,l) & 0 & 0 & \cdots \\ 0 & 0 & s(3,l) & 0 & \cdots \\ 0 & 0 & 0 & s(4,l) & \cdots \\ \vdots & \vdots & \vdots & \vdots & \ddots \end{pmatrix} \tag{54}$$

Here the parameter λ is given by $\cosh\lambda = e^{U\Delta\tau/2}$. The matrices for the down spin electrons differ only by a sign: $v_\downarrow(l) = -v_\uparrow(l)$.

The next step is to compute the up and down 'Green's functions',

$$G_\sigma = [\,I + e^k\,e^{v_\sigma(1)}\,e^k\,e^{v_\sigma(2)}\,e^k\,e^{v_\sigma(3)} \ldots e^k\,e^{v_\sigma(L)}\,]^{-1} \tag{55}$$

Here I is the N dimensional identity matrix.

In order to sample the Hubbard-Stratonovich field configurations, suggest a change in $s(i,l) \to -s(i,l)$ on the first spatial site $i = 1$ of imaginary time slice $l = L$. Compute the quantities,

$$\begin{aligned} d_\uparrow &= 1 + (1 - [G_\uparrow]_{ii})\,(e^{-2\lambda\,s(i,l)} - 1) \\ d_\downarrow &= 1 + (1 - [G_\downarrow]_{ii})\,(e^{+2\lambda\,s(i,l)} - 1) \\ d &= d_\uparrow d_\downarrow\,. \end{aligned} \tag{56}$$

Next throw a uniformly distributed random number, $0 < r < 1$. If $r < d$, accept the update of the Hubbard Stratonovich field by setting $s(i,l) = -s(i,l)$.

If the move was accepted, the Green's functions, which depend on s (see Eqs. 54-55) will now be different. One could recompute G_σ from Eq. 55, using the new s. This will take a time which goes as N^3, since it involves a matrix inversion. There's a faster (order N^2) trick to get the new G_σ, which takes advantage of the fact that only one element in one of the $v_\sigma(l)$ has changed. Compute the vectors,

$$\begin{aligned} a_{k\uparrow} &= -(e^{-2\lambda\,s(i,l)} - 1)\,[G_\uparrow]_{ik} + \delta_{ki}\,(e^{-2\lambda\,s(i,l)} - 1) \\ a_{k\downarrow} &= -(e^{+2\lambda\,s(i,l)} - 1)\,[G_\downarrow]_{ik} + \delta_{ki}\,(e^{+2\lambda\,s(i,l)} - 1) \\ b_{j\uparrow} &= [G_\uparrow]_{ji}\,/\,(1 + c_{i\uparrow}) \\ b_{j\downarrow} &= [G_\downarrow]_{ji}\,/\,(1 + c_{i\downarrow}) \end{aligned} \tag{57}$$

Here δ_{ij} is the usual Kronecker δ. Remember that i,l is the fixed site (time slice) whose Hubbard-Stratonovich field is being updated. The free indices j,k run from 1 to N. Then the new G_σ are given by

$$[G_\sigma]_{jk} = [G_\sigma]_{jk} - b_{j\sigma}\,a_{k\sigma}. \tag{58}$$

The interaction energy matrix (Eq. 54) is updated if the move is accepted. After the new G_σ are computed, go to Hubbard-Statonovich field on the second spatial site $i = 2$ on imaginary time slice $l = L$ and suggest a change to it, and follow the procedure of Eq. 56 again to see if the move is accepted and Eqs. 57-58 to update G if it is. Continue this until all spatial sites of time slice $l = L$ are updated.

After all spatial sites i of imaginary time slice $l = L$ have been updated, change the Green's functions via,

$$G_\sigma = [e^k e^{v_\sigma(l)}]\, G_\sigma\, [e^k e^{v_\sigma(l)}]^{-1} \tag{59}$$

This 'wrapping' replaces the exponential of $v_\sigma(L)$ by that of $v_\sigma(L-1)$ at the end of the string of matrices in Eq. 55. The Hubbard-Stratonovich variables on imaginary time slice $l = L-1$ can now be updated following the procedures of Eqs. 56-58. When all spatial sites of imaginary time slice are finished, wrap the Greens functions using Eq. 59 again. Continue the process of Eqs. 56-59 until all imaginary time slices are updated.

After completing an entire set of updates to all the space-time points of the lattice, make measurements. For example, the density of electrons of spin σ on site i is given by,

$$\langle n_{i\sigma} \rangle = 1 - [G_\sigma]_{ii}. \tag{60}$$

The double occupancy rate on site i is

$$\langle n_{i\uparrow} n_{i\downarrow} \rangle = (\,1 - [G_\uparrow]_{ii}\,)\,(\,1 - [G_\downarrow]_{ii}\,) \tag{61}$$

The local moment on site i is,

$$\langle (n_{i\uparrow} - n_{i\downarrow})^2 \rangle = \langle n_{i\uparrow} + n_{i\downarrow} \rangle - 2 \langle n_{i\uparrow} n_{i\downarrow} \rangle. \tag{62}$$

The spin correlation function between moments on sites i and j, for $i \neq j$, is given by,

$$\begin{aligned} S_{+i} &= c^\dagger_{i\uparrow} c_{i\downarrow} \\ S_{-j} &= c^\dagger_{j\downarrow} c_{j\uparrow} \\ \langle S_{+i} S_{-j} \rangle &= -[G_\uparrow]_{ji}\, [G_\downarrow]_{ij} \end{aligned} \tag{63}$$

You can also measure pairing correlations, charge density wave correlations, etc.

A full simulation consists of performing the above updates of all the Hubbard-Stratonovich variables of the space-time lattice for several hundred 'equilibration' sweeps *without* making any measurements. This is followed by a few thousand 'measurement' sweeps in which you perform the update operations and also the measurements.

Note that there are a number of alternate Quantum Monte Carlo approaches to the Hubbard Hamiltonian closely related to that discussed here. These include a ground state projection method [34, 35], and approximate techniques which deal with the sign problem [68]. Dynamical mean field theory [69, 70, 71], especially when it employs the Hirsch-Fye QMC method [72] as its impurity solver, is a method of particular appeal since it not only provides a solution to tight-binding models (with the limit where the mometum dependence of the self-energy is neglected) but also offers a powerful methodology with which to put together electronic stucture and many body physics.

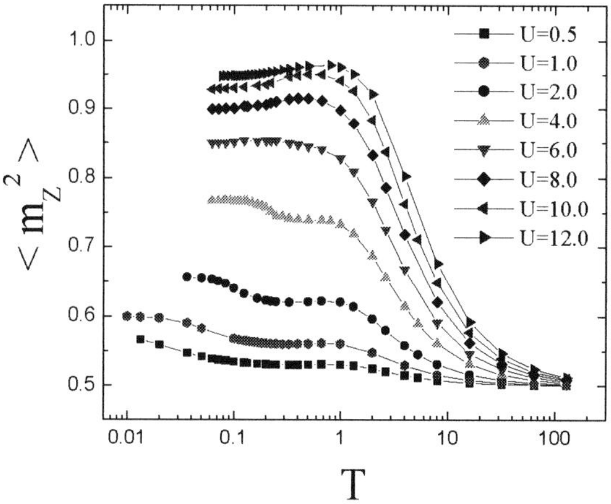

FIGURE 18. The local moment $\langle m_z^2 \rangle$ as a function of temperature for different interaction strengths U and lattice size 6x6. The lattice is half-filled.

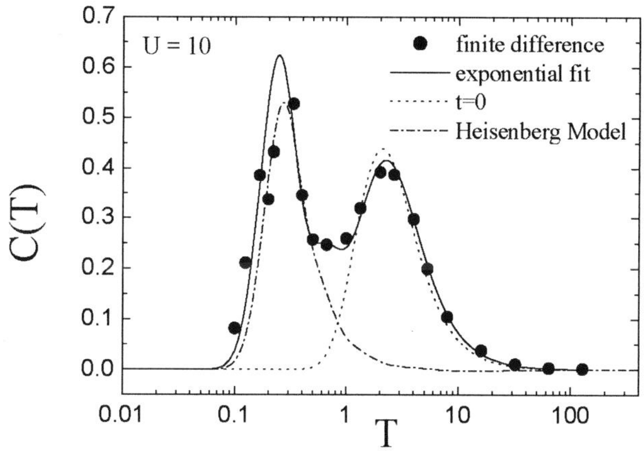

FIGURE 19. The specific heat of the Hubbard model for $U = 10t$. There is a high temperature peak which is fit well by considering a single site Hubbard model ($t = 0$) and a low temperature peak which agrees well with the Heisenberg model with $J = 4t^2/U = 0.4$.

B. Results: Local Moment, Spin Correlation, and Specific Heat

The evolution of the local moment (Eq. 62) at half-filling as the temperature is decreased is shown in Fig. 18 on a 6x6 lattice for different interaction strengths U. The

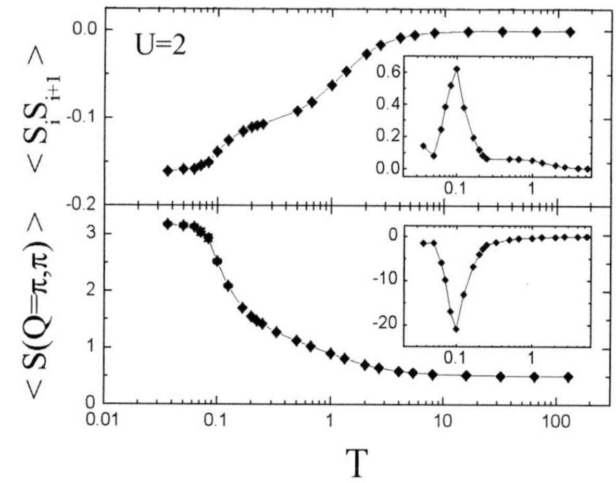

FIGURE 20. The near neighbor spin correlations and magnetic structure factor of the half-filled Hubbard model at $U = 2t$.

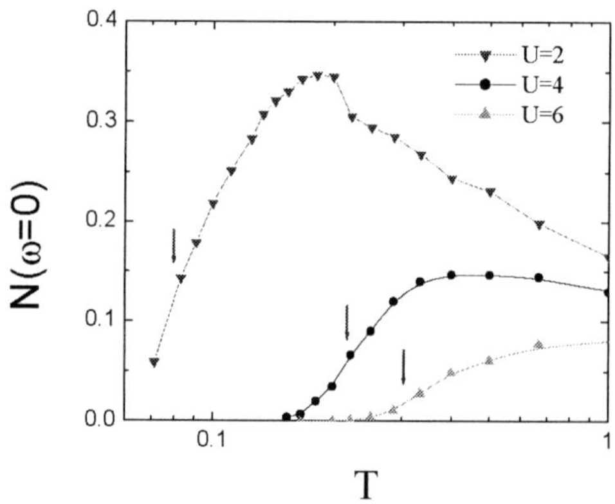

FIGURE 21. The density of states at $N(\omega = 0)$. As T is lowered, a Mott-Hubbard gap opens up. The half-filled Hubbard model is insulating.

local moment begins to develop from its uncorrelated value $1/2$ at a temperature set by U, and then saturates at low T. The local moment does not reach 1 at $T = 0$ because significant quantum fluctuations allow doubly occupied and empty sites to occur even in the ground state. (Compare with Fig. 8.) However, as U/t increases, these fluctuations are suppressed and the moment becomes better and better formed. The local moment also makes a further small adjustment at low T, which is due to the onset of magnetic order.

The specific heat $C(T)$ (Fig. 19) is determined by the energy. It shows an interesting two peak structure. The high temperature peak is associated with the formation of local moments, and the low temperature peak with their ordering. The Hubbard model maps onto the Heisenberg model at large U. This connection is emphasized in Fig. 19 which shows that the low temperature peak in the specific heat of the Hubbard model can be mapped onto that of the Heisenberg model with $J = 4t^2/U$.

The near-neighbor spin correlation is shown in teh top panel of Fig. 20. The bottom panel is the magnetic structure factor $S(\pi,\pi) = \frac{1}{N}\sum_{ij}\langle S_{z,i}S_{z,j}\rangle$ which sums the spin-spin correlations over the whole lattice. It is found that $S(\pi,\pi)$ grows linearly with N at low T, indicating that the correlations extend over the whole lattice. A finite size scaling analysis of these quantities can be used t o demonstrate the existence of long range order in the ground state of the half-filled Hubbard model on a square lattice [36]. Finally, Fig. 21 shows the density of states at $\omega = 0$ for the half-filled Hubbard model at different values of U. The suppression of $N(\omega = 0)$ at low T and large U is a signature of the presence of an insulating gap caused by the on-site repulsion.

V. DQMC FOR THE ANDERSON-HUBBARD HAMILTONIAN

These notes have now provided an overview of simple analytic approaches to the Hubbard model- weak and strong coupling and mean field theory. The basics of the DQMC method have also been established. This section presents the results of such DQMC simulations, specifically of the Anderson-Hubbard Hamiltonian, that is, studies of the interplay of disordered one-body potential and two-body interactions. The effect of this interplay and of a Zeeman field on magnetism, thermodynamics, and metal-insulator transitions will be shown. The final topic concerns the effect of interactions on the band insulating phase which arises from a periodic potential, as opposed to a random one. Journal references are given at the beginning of each subsection.

A. Interaction driven Anderson Insulator to Metal Transition

Journal Reference: "Conducting phase in the two-dimensional disordered Hubbard model", P.J.H. Denteneer, R.T. Scalettar, and N. Trivedi, Phys. Rev. Lett. **83**, 4610 (1999).

As discussed in Section I, the central question motivated by experiments on silicon metal-oxide semiconductor field effect transistors is whether electron-electron interactions enhance the conductivity of a 2D disordered electron system, and possibly lead to a conducting phase and a metal–insulator transition. This question can be addressed by studying the disordered Hubbard (or 'Anderson-Hubbard' model, which contains both relevant ingredients. Renormalization arguments suggest that the precise location of disorder is not important, since under changes of length scale randomness in one parameter introduces randomness into others. This is, in fact, not true if symmetries protect the spread of randomness. Initially the focus of the work here will be on bond randomness. It will be shown later that, indeed, different types of disorder can lead to different physics.

While the Hubbard model does not include the long range nature of the Coulomb repulsion, studying the simpler model of screened interactions is an important first step in answering the central qualitative question posed above. The Determinant Quantum Monte Carlo simulation techniques introduced in Section IV avoid the limitations of perturbative approaches (while of course being confronted with others). Recent studies using very different techniques have indicated that interactions may enhance conduc-

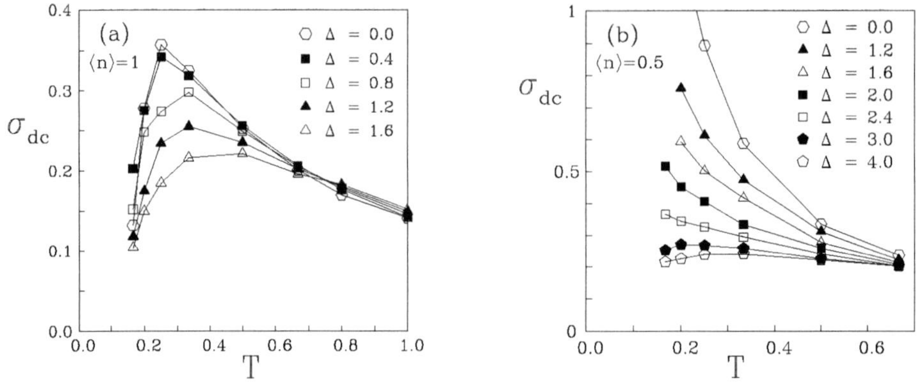

FIGURE 22. Conductivity $\sigma_{\rm dc}$ as a function of temperature T for various values of disorder strength Δ at $U = 4$ for (a) half-filling ($\langle n \rangle = 1$) and (b) $\langle n \rangle = 0.5$. Calculations are performed on an 8×8 square lattice; data points are averages over 4 realizations for a given disorder strength.

tivity: two interacting particles in a random potential experience a mutual delocalizing effect [31], and weak Coulomb interactions were found to increase the conductance of spinless electrons in (small) strongly disordered systems [32].

The quantity of immediate interest when studying a possible metal–insulator transition is the *conductivity* and especially its T-dependence. By the fluctuation–dissipation theorem σ_{dc} is related to the zero-frequency limit of the current-current correlation function. A complication of the QMC simulations is that the correlation functions are obtained as a function of *imaginary* time. To avoid a numerical analytic continuation procedure to obtain frequency-dependent quantities, which would require Monte Carlo data of high accuracy, one can employ an approximation that was used and tested in studies of the superconductor–insulator transition in the attractive Hubbard model [44]. This approximation is valid when the temperature is smaller than an appropriate energy scale in the problem. Additional checks and applicability to the present problem are discussed below. The approximation allows $\sigma_{\rm dc}$ to be computed directly from the wavevector $\mathbf{q}$- and imaginary time τ-dependent current-current correlation function $\Lambda_{xx}(\mathbf{q},\tau)$:

$$\sigma_{\rm dc} = \frac{\beta^2}{\pi}\Lambda_{xx}(\mathbf{q}=0,\tau=\beta/2)\,. \tag{64}$$

Here $\beta = 1/T$, $\Lambda_{xx}(\mathbf{q},\tau) = \langle j_x(\mathbf{q},\tau)\, j_x(-\mathbf{q},0)\rangle$, and $j_x(\mathbf{q},\tau)$ the $\mathbf{q},\tau$-dependent current in the x-direction, is the Fourier transform of $j_x(\ell) = i\sum_\sigma t_{\ell+\hat{x},\ell}(c^\dagger_{\ell+\hat{x},\sigma}c_{\ell\sigma} - c^\dagger_{\ell\sigma}c_{\ell+\hat{x},\sigma})$. (See also Ref. [73]).

As a test for the conductivity formula (Eq. 64) Fig. 22(a) presents results for $\sigma_{\rm dc}(T)$ at half-filling for $U = 4$ and various disorder strengths Δ. The behavior of the conductivity shows that as the temperature is lowered below a characteristic gap energy, the high T "metallic" behavior crosses over to the expected low T Mott insulating behavior for all Δ, thereby providing a reassuring check of the formula and of the numerics. Fig. 22(b)

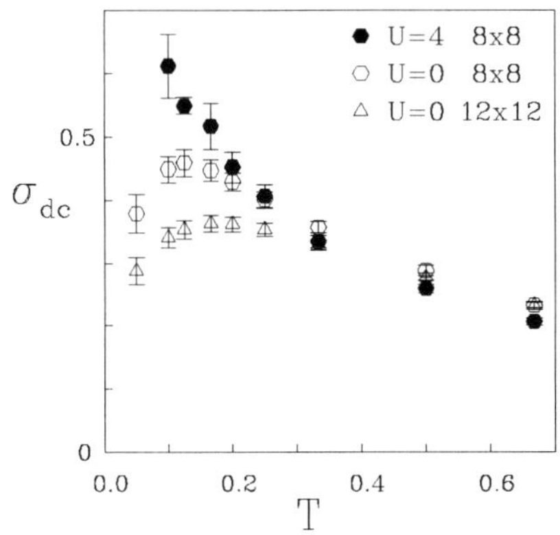

FIGURE 23. Conductivity σ_{dc} as a function of temperature T comparing $U = 4$ and $U = 0$ at $\langle n \rangle = 0.5$ and disorder strength $\Delta = 2.0$. Data points are averages over many realizations for this disorder strength (see text). Error bars are determined by the disorder averaging and not the Monte Carlo simulation.

shows $\sigma_{dc}(T)$ for a range of disorder strengths at density $\langle n \rangle = 0.5$ and $U = 4$. The figure displays a change from metallic behavior at low disorder to insulating behavior above a critical disorder strength, $\Delta_c \simeq 2.7$. If this persists to $T = 0$ and in the thermodynamic limit, it would describe a ground state metal–insulator transition driven by disorder.

In order to obtain a more precise understanding of the role of interactions on the conductivity, Fig. 23 compares the results of Fig. 22(b) with the disordered *non-interacting* σ_0 [74]. The comparison is made at strong enough disorder $\Delta = 2.0$ that the localization length is less than the lattice size and the non-interacting system is therefore insulating with $d\sigma_0/dT > 0$ at low T. Interactions are found to have a profound effect on the conductivity: in the high-temperature "metallic" region, interactions slightly reduce σ compared to the non-interacting σ_0 behavior. On the other hand in the low-temperature "insulating" region of σ_0 the data shows that upon turning on the Hubbard interaction the behavior is completely changed with $d\sigma/dT < 0$, characteristic of metallic behavior. This is the regime of interest for the MIT.

To ascertain that the phase produced by repulsive interactions at low T is not an insulating phase with a localization length larger than the system size but a true metallic phase one can study the conductivity response for varying lattice sizes. There is a markedly different size dependence for the $U = 0$ insulator and the $U = 4$ metal, resulting in a confirmation of the picture given above. For $U = 0$, the conductivity on a larger (12×12) system is *lower* than that on a smaller (8×8) system (see Fig. 23), consistent with insulating behavior in the thermodynamic limit, whereas for $U = 4$ the conductivity on the larger (8×8) system is *higher* than that on the smaller (4×4) system (data not shown), indicative of metallic behavior. Thus the enhancement of the conductivity by repulsive interactions becomes more pronounced with increased lattice size [75].

Concerning finite-size effects for the non-interacting system, note that at lower values of Δ, where the localization length exceeds the lattice size, σ_0 shows "metallic" behavior

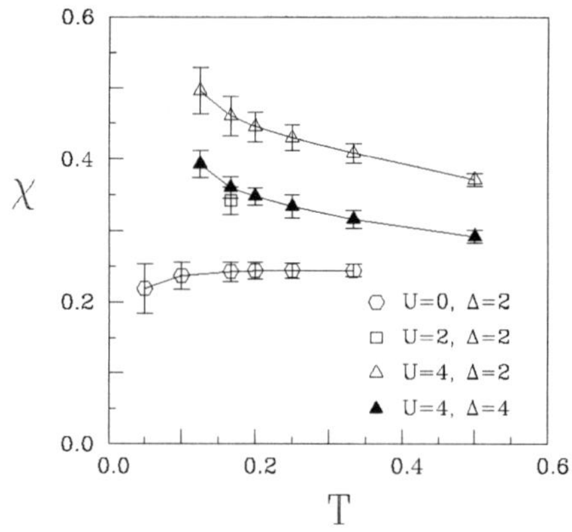

FIGURE 24. Spin susceptibility χ as a function of temperature T at $\langle n \rangle = 0.5$ comparing interaction strengths $U = 0,2,4$ and disorder strengths $\Delta = 2,4$. Calculations are performed on 8×8 square lattices; error bars are from disorder averages over up to 8 realizations.

which is diminished upon turning on the interactions [76]. Based on the analysis above, one would predict that at low enough T and large enough lattice size, the conductivity curves for the non-interacting σ_0 and interacting σ cross with $\sigma > \sigma_0$ at sufficiently low T.

To obtain information on the spin dynamics of the electrons and because it is a quantity often discussed in connection with the localization transition, one can also compute the spin susceptibility χ as a function of T (through $\chi(T) = \beta S_0(T)$ where S_0 is the magnetic structure factor at wavevector $\mathbf{q} = 0$). Fig. 24 shows two things: 1) $\chi(T)$ is enhanced by interactions with respect to the non-interacting case (at fixed disorder strength), and 2) starts to diverge when T is lowered, both on the metallic ($\Delta = 2$) and insulating ($\Delta = 4$) sides of the alleged transition. This is in agreement with experimental and theoretical work on phosphorus-doped silicon, where a (3D) MIT is known to occur and the behavior is explained by the existence of local moments[77], and also with diagrammatic work on 2D disordered, interacting systems[14].

To establish definitively the existence of a possible quantum phase transition in the disordered Hubbard model requires: (i) Extending the present data at $T = 0.1 = W/80$, where W is the non-interacting bandwidth, to lower T, which is however difficult because of the sign problem. (ii) A more detailed analysis of the scaling behavior in both linear dimension and some scaled temperature. (iii) A more accurate analytic continuation procedure to extract the conductivity. The condition for the validity of the approximate formula (Eq. 64) for $\sigma_{dc}(T)$, requires that T be less than an appropriate energy scale which is fulfilled within the two phases, but breaks down close to a quantum phase transition where the energy scale vanishes.

Given the well-known difficulty of DQMC simulations resolving the issue of d-wave superconductivity in the clean Hubbard model because of the sign problem mentioned in (i) above, it is worth discussing briefly why more definitive results are obtainable in this study of the MIT. The most important reason concerns the higher temperature scale of the phenomenon: interactions enhance the conductivity and lead to metallic behavior in a temperature range (about $1/10$ of the Fermi energy) here, similar to that of experiments. In the cuprates, although the transition temperatures are very high by the standards of other superconductors, they are of the order of $1/100$ of the Fermi energy, a factor of ten lower. Furthermore, the region of highest transition temperature is, very unfortunately, at a doping $\rho \approx 1 - \frac{1}{8}$ where the sign problem in the simulations is worst.

In summary, the temperature-dependent conductivity $\sigma(T)$ and spin susceptibility $\chi(T)$ of a model for two-dimensional electrons containing both disorder and interactions has been described. The Hubbard repulsion can enhance the conductivity and lead to a clear change in sign of $d\sigma/dT$. More significantly, a finite size scaling analysis demonstrates that repulsive interactions can drive the system from one phase to a different phase. $\sigma(T)$ has the opposite behavior as a function of system size in the two phases indicating that the transition is from a localized insulating to an extended metallic phase. The $\chi(T)$ data further suggests the formation of an unusual metal, a non-Fermi liquid with local moments. While the simplicity of the model studied prevents any quantitative connection to recent experiments on Si-MOSFETs, there is nevertheless an interesting qualitative similarity between Fig. 22(b) and the experiments. (See Fig. 1.) Varying the disorder strength Δ at fixed carrier density $\langle n \rangle$, as in the calculations, can be thought of as equivalent to varying carrier density at fixed disorder strength, as in experiments, since in a metal–insulator transition one expects no qualitative difference between tuning the mobility edge through the Fermi energy (by varying Δ) and *vice-versa* (by varying $\langle n \rangle$). This work then suggests that electron-electron interaction induced conductivity plays a key role in the 2D metal–insulator transition.

B. Effect of Zeeman Field

Journal Reference: "Interacting electrons in a two-dimensional disordered environment: Effect of a Zeeman magnetic field", P.J.H. Denteneer and R.T. Scalettar, Phys. Rev. Lett. **90**, 246401 (2003).

A hundred years after the Nobel prize was awarded in 1902 for the discovery of the Zeeman effect and the subsequent explanation by Lorentz, applying a magnetic field continues to be a powerful means to elucidate puzzling phenomena in nature. One of the most recent examples is to the topic of these notes: the interplay of interactions and disorder in electronic systems and the pioneering experiments on the metal–insulator transition (MIT) in effectively two–dimensional low-density silicon metal-oxide-semiconductor field-effect transistors (MOSFETs) discussed earlier [16, 17, 18].

Contrary to the well-known effect of a magnetic field in weak-localization theory to disturb interference phenomena and hence *undo* localization and insulating behavior, the negative magnetoresistance effect [2], in the Si MOSFETs and similar heterostructures, the magnetic field is found to suppress the metallic behavior and therefore *promote* insulating behavior [20, 78, 79]. The effect is present for all orientations of the magnetic field relative to the 2D plane of the electrons. In particular, a Zeeman magnetic field, applied parallel to the 2D plane of electrons and therefore coupling only to the spin, and not the orbital motion of the electrons, has been used extensively. This puts into focus the important role played by the spin degree of freedom of the electron, and its polarization [80, 81, 82, 83].

This section continues the DQMC study of the Anderson-Hubbard Hamiltonian, but now includes the effect of a Zeeman magnetic field, that is, an additional term in H of

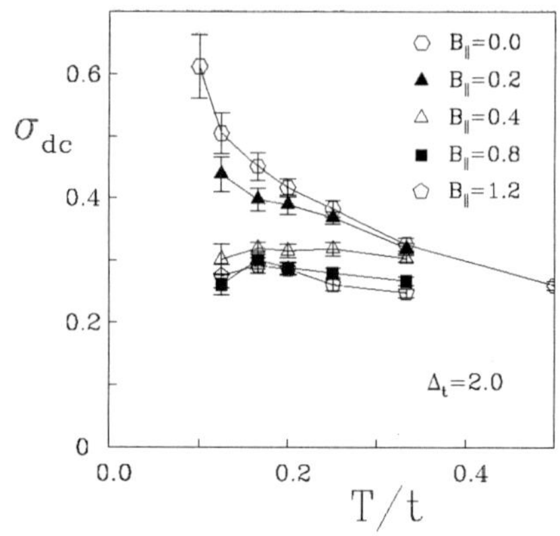

FIGURE 25. Conductivity σ_{dc} (in units of $e^2/\hbar$) as a function of temperature T for various strengths of Zeeman magnetic field $B_{\|}$. As $B_{\|}$ increases, a transition from metallic to insulating behavior is seen in σ_{dc}. Calculations are performed on 8×8 lattices for $U/t = 4$ at density $\langle n \rangle = 0.5$ with disorder strength $\Delta_t = 2.0$ (see text); error bars result from averaging over typically 16 quenched disorder realizations. $B_{\|}$ and Δ_t are given in units of t.

the form,

$$+B_{\|} \sum_{j\sigma} \sigma n_{j\sigma} \tag{65}$$

While the numerical evidence is mixed concerning the occurrence of a MIT due to interactions, [84, 85, 86, 87] there is a consensus in favor of a Zeeman magnetic field tuned transition [86, 87, 88, 89], as described in more detail below.

The main focus of this work continues to be on the conductivity, although now the $B_{\|}$–dependence is explored as well as the T–dependence. Another interesting quantity to study in conjunction with the magnetoconductivity is the degree of spin-polarization P of the electronic system:

$$P = \frac{n_{\downarrow} - n_{\uparrow}}{n_{\downarrow} + n_{\uparrow}}, \tag{66}$$

where $n_{\downarrow}, n_{\uparrow}$ are the average spin-densities of the corresponding number operators.

In order to study the effect of the Zeeman magnetic field $B_{\|}$ on the metallic behavior, one can start from the case with density $\langle n \rangle = 0.5$ and disorder strength $\Delta_t = 2.0$ for which the model exhibits clear metallic behavior: σ_{dc} rising when lowering temperature T [90]. Fig. 25 shows that turning on $B_{\|}$ reduces the conductivity and suppresses the metallic behavior; at field strength $B_{\|} = 0.4$, σ_{dc} appears T-independent (within the error bars), and at larger field strengths shows a tendency to decrease upon lowering T. σ_{dc} does not go to zero, as for a real insulator, unless very low T and very large lattices (out of reach of the computational approach) are employed. Nevertheless, Fig. 25 shows the qualitative features of a magnetic-field-driven metal–insulator transition, similar to

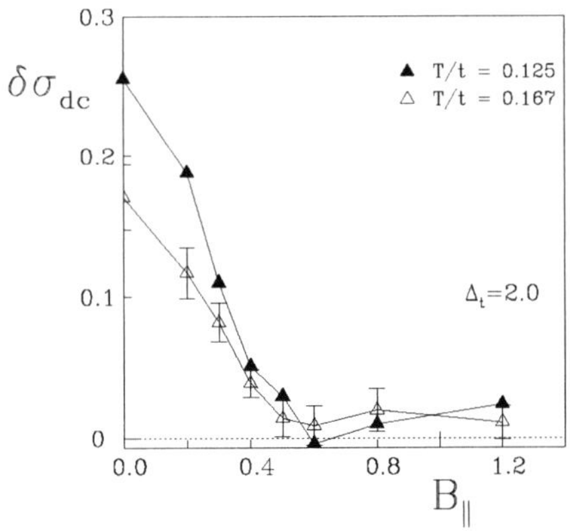

FIGURE 26. Conductivity difference $\delta\sigma_{dc} \equiv \sigma_{dc}(B_{\|},T) - \sigma_{dc}(B_{\|} = 4,T)$, with conductivity at very high B-field subtracted, as a function of $B_{\|}$ for low temperature T. A sharp onset of conductivity is seen at a Zeeman field at which the slope of $\sigma_{dc}(T)$ changes sign in Fig. 25. Computational details and units are as in Fig. 25; for clarity, only error bars on $T = t/6$-data are shown; those on $T = t/8$-data are typically slightly larger (cf. Fig. 25).

what is seen in experiment [20, 78, 79]. Previous numerical approaches using different techniques have also produced this effect [86, 87, 88].

The physical picture of this metal to insulator transition is the following: Since the effect of $B_{\|}$ is to polarize the electronic system (with the sign choice in Eq. 60), $n_{\downarrow}$ is promoted at the expense of $n_{\uparrow}$), a large enough $B_{\|}$ will result in electrons with spin down only and, because of the nature of the Hubbard interaction, in a non-interacting system [91]. Therefore, in the limit of large 2D lattices and low temperature, the hopping disorder will force the conductivity to vanish. The nonzero value of σ_{dc} at very large $B_{\|}$ is then a direct measure of the systematic errors due to finite size and non-zero T. In Fig. 27, this value is subtracted, and fields close to $B_{\|} = 0.4$ are the primary focus. The resulting $\delta\sigma_{dc}$ vs. $B_{\|}$ for the lowest temperatures ia a rather abrupt onset of $\delta\sigma_{dc}$ below $B_{\|} \approx 0.5$, which is about the field value where the curves of σ_{dc} vs. T change from insulating to metallic (Fig. 25). The data for a 2D system in Fig. 26 are consistent with a linear vanishing of $\delta\sigma_{dc}$ as the (quantum) critical point is approached. At present, the results, while presenting compelling evidence for the transition itself, are clearly not precise enough to obtain critical exponents. Interestingly, a transition from insulator to metal upon increasing magnetic field, i.e. the known negative magnetoresistance effect, occurs in an amorphous three-dimensional Gd-Si alloy (showing a MIT at zero field), also with a linear vanishing of the conductivity [92]. While it is possible that these observations are unrelated, it is also conceivable that this is an indication of a, yet hidden, connection between the two transitions.

Fig. 27 exhibits the *resistivity* ρ ($\equiv 1/\sigma_{dc}$) as a function of $B_{\|}$ for low T. The crossing point (between $B_{\|} = 0.3$ and 0.4) demarks a critical field strength B_c which separates fields for which the resistivity decreases when lowering temperature (low-

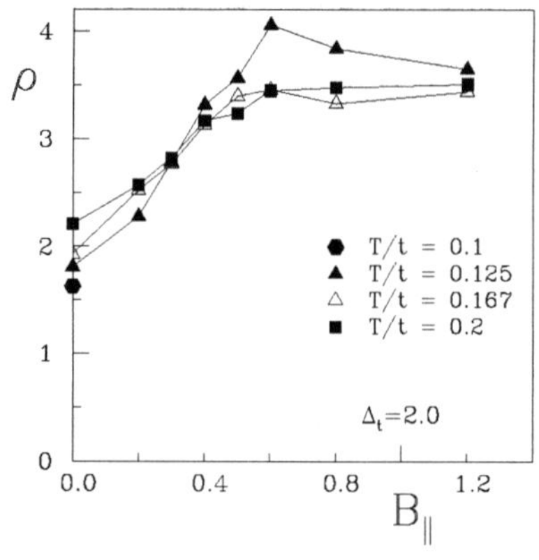

FIGURE 27. Resistivity ρ as a function of $B_{\parallel}$ for various low T. The crossing point provides another estimate for the critical field strength. Computational details and units are as in Figs. 25 and 26; for clarity the error bars have been omitted, but can be estimated from Figs. 25 and 26.

field metallic behavior) from fields for which ρ increases upon lowering T (high-field insulating behavior). It is especially noteworthy that the critical field strength (which can be roughly estimated to lie between 0.3 and 0.5 from Figs. 26 and 27) is clearly lower than the field for which full spin-polarization sets in. Indeed, Fig. 28 shows how the spin polarization P, defined in (66), behaves as a function of $B_{\parallel}$ at the lowest temperature used: there is no reflection of the critical field strength in the behavior of the polarization, and full spin-polarization only happens for $B_{\parallel} > 1.2$. This feature of the data is in agreement with recent experiments performed on 2D electron- and hole-gases in GaAs and AlAs [82, 83]. Since complete spin-polarization is equivalent to a non-interacting system, the separation of the two field strengths and the incomplete polarization at the MIT present evidence that the Zeeman field tuned MIT must be seen as a property of a fully interacting many-body system, at least in the 2D disordered Hubbard model.

Another interesting feature of Fig. 27 is what appears to be the saturation of resistivity at a field not much higher than B_c. Experiments also show this behavior, but only for AlAs, where the saturation is shown to coincide with full spin polarization [82]. The on-site nature of the interactions in the Hubbard model make the saturation happen at much reduced field strength compared to that of complete polarization: at the rather low total density the minority spin species will effectively be decoupled from the majority spin species and both spin species form non-interacting subsystems at a field where the minority spin has not disappeared completely. Increasing magnetic field further at constant total density will then not change the conducting properties anymore.

The notion of a predictable and straightforward effect of $B_{\parallel}$ is also concordant with the phenomenon that $\rho(B_{\parallel})$ behaves qualitatively the same in the metallic and insulating phases [3], and therefore the same physical mechanism seems at play in both cases. The results here suggest the reduction of the effective interaction by spin polarization as a likely candidate for this mechanism.

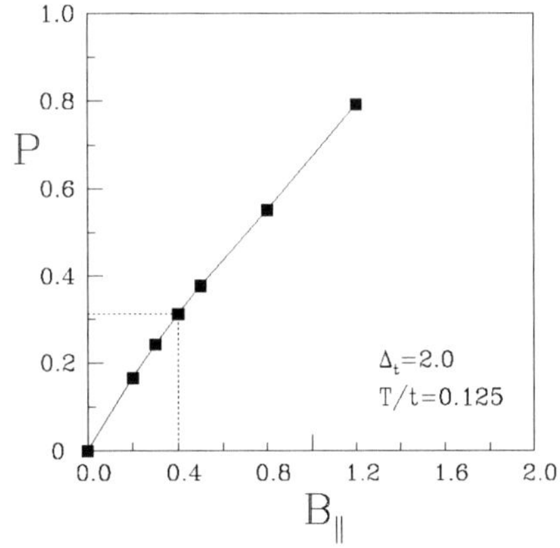

FIGURE 28. Degree of spin polarization P (defined in (66) in text) as a function of $B_{\|}$ for fixed low $T = t/8$. The polarization shows little change through the metal-insulator transition and is only around 0.3 at the critical field strength $B_{\|} = 0.4$ (dotted lines).

In summary, applying a Zeeman magnetic field in the 2D disordered Hubbard model reduces the effect of the Hubbard interaction and is able to bring about a transition from a metallic phase to an insulator at a critical field strength. This critical field is considerably less than the field required for full spin polarization, emphasizing that, for the disordered Hubbard model, the metal-insulator transition occurs in a region where a considerable degree of electronic correlation remains. This is in good qualitative agreement with experimental observations when a magnetic field is applied parallel to a 2D electron or hole gas in GaAs– and AlAs–based heterostructures [82, 83]. For Si MOSFETs, the general phenomenon of suppression of the metallic behavior is in agreement, but the issue of the critical field being smaller than a saturating field is less clear [81]. Earlier in this section, the T-dependence of $\sigma_{\rm dc}$ was studied for various Δ_t without a B-field. It was shown that the Hubbard interaction enhances $\sigma_{\rm dc}$ and leads at low T to metallic behavior that can be made insulating by sufficiently strong disorder. The present results concerning the effect of a magnetic field are consistent with that conclusion and therefore strengthen it: the rather strong interactions that caused the conducting phase at disorder strength $\Delta_t = 2.0$ (below the critical disorder strength of approximately 2.4 above which the system is insulating) without B-field are reduced by a B-field which is able to drive the system back to its insulating phase. The latter is also its natural state in the absence of interactions. This consistency indicates that the disordered Hubbard model provides a coherent, qualitative picture of the phenomena in 2D electronic, disordered systems both in the presence and absence of a Zeeman magnetic field.

C. The Role of Particle-Hole Symmetry

Journal Reference: "Particle–Hole Symmetry and the "Effect of Disorder on the Mott–Hubbard Insulator", P.J.H. Denteneer, R.T. Scalettar, and N. Trivedi, Phys. Rev. Lett. **87**, 146401 (2001).

In recent years, it has become increasingly clear that for non-interacting electrons the presence or absence of certain symmetries is crucial in determining the effect of disorder on both transport and thermodynamic properties, as well as critical properties of the localization transition.[93] Recent examples where symmetry considerations are important are given in the context of quantum wires[94] and disordered superconductors,[95, 96] where chiral, time–reversal, and spin–rotation symmetries play an important role. In the preceding sections, bond-disordered Hamiltonians which retain the particle-hole symmetry of the clean Hubbard model were studied. This section examines the effect of different types of disorder, including those which break this symmetry, on both the dynamic and equilibrium thermodynamics in the vicinity of half-filling, electron density $\langle n \rangle = 1$. The results suggest that the presence or absence of *particle–hole symmetry* determines the effect of randomness on the conductivity and the Mott gap.

In order to make the distinction between the different types of disorder clear, it is useful to write down again the 2d Anderson-Hubbard Hamiltonian,

$$\begin{aligned} H &= -\sum_{\langle \mathbf{ij} \rangle,\sigma} t_{\mathbf{ij}} c^{\dagger}_{\mathbf{i}\sigma} c_{\mathbf{j}\sigma} - \sum_{\langle\langle \mathbf{ik} \rangle\rangle,\sigma} t'_{\mathbf{ik}} c^{\dagger}_{\mathbf{i}\sigma} c_{\mathbf{k}\sigma} \\ &+ U \sum_{\mathbf{j}} (n_{\mathbf{j}\uparrow} - \frac{1}{2})(n_{\mathbf{j}\downarrow} - \frac{1}{2}) - \sum_{\mathbf{j},\sigma} \mu_{\mathbf{j}} n_{\mathbf{j}\sigma} \, . \end{aligned} \tag{67}$$

Here $t_{\mathbf{ij}}$ is a bond–dependent hopping matrix element on near-neighbor sites $\langle \mathbf{ij} \rangle$, $t'_{\mathbf{ik}}$ is a bond–dependent hopping matrix element on next-near-neighbor sites $\langle\langle \mathbf{ik} \rangle\rangle$, U is an on–site repulsion, and $\mu_{\mathbf{j}}$ is a site-dependent chemical potential. Choose $P(t_{\mathbf{ij}}) = 1/\Delta_t$ for $t_{\mathbf{ij}} \in [t - \Delta_t/2, t + \Delta_t/2]$, and zero otherwise, with $t = 1$ to set the scale of energy. Similarly, $P(t'_{\mathbf{ik}}) = 1/\Delta'_t$ for $t'_{\mathbf{ik}} \in [t' - \Delta'_t/2, t' + \Delta'_t/2]$, and $P(\mu_{\mathbf{j}}) = 1/\Delta_\mu$ for $\mu_{\mathbf{j}} \in [-\Delta_\mu/2, +\Delta_\mu/2]$, so that the various Δ measure the disorder strength. Now focus on half–filling where the effects of interactions are most prominent, as evidenced by the formation of antiferromagnetic correlations and a Mott-Hubbard charge gap at low temperatures. This Hamiltonian, Eq. 67, is particle–hole (ph) symmetric when $t'_{\mathbf{ik}} = \mu_{\mathbf{j}} = 0$. That is, under the transformation $c^{\dagger}_{\mathbf{i}\sigma} \rightarrow (-1)^{\mathbf{i}} c_{\mathbf{i}\sigma}$ the Hamiltonian is unchanged, and the system is precisely half–filled for all values of the parameters in H and also for all T. (See Appendix D.) Therefore, while near-neighbor bond and local site disorder both introduce randomness into the system, they differ fundamentally in that site disorder breaks ph–symmetry.

In addition to the conductivity and spin polarization of the preceding sections, one can also extract the *temperature-dependent* density of states at the chemical potential $N(\varepsilon = 0)$[97] from the one-electron Green function as a function of imaginary time,

$$N(0) \simeq -\beta \, G(\mathbf{r} = 0, \tau = \beta/2)/\pi \, . \tag{68}$$

This quantity allows a more clear characterization of the transport and thermodynamic properties of the system. For σ_{dc} and $N(0)$, "Trotter" errors associated with the discretization of imaginary time β are considerably less than the fluctuations associated with monte carlo and disorder averaging.

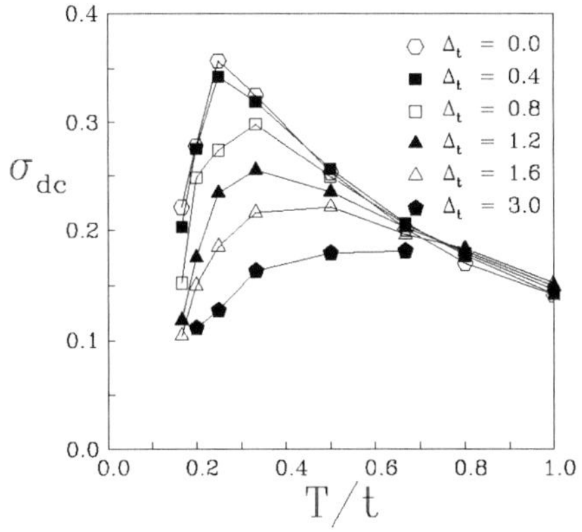

FIGURE 29. The effect of particle–hole–symmetry preserving (near-neighbor) bond disorder in the half-filled Hubbard Hamiltonian is to decrease the conductivity σ_{dc}. Data is for $U = 4t$ on a 8×8 square lattice; Δ_t measures the strength of the bond disorder.

The transport properties are given in Fig. 29, which exhibits the effect of near-neighbor hopping (bond) disorder on the conductivity. For all disorder strengths Δ_t, at temperatures greater than a characteristic temperature T_* related to the Mott gap, the system shows metallic behavior with σ_{dc} increasing upon lowering T. The conductivity turns down sharply as the temperature drops below T_* and the system shows insulating behavior with σ_{dc} decreasing upon lowering T. In the case of zero randomness, the perfect nesting of the Fermi surface in 2d leads to antiferromagnetic long range order (AFLRO) in the ground state with an associated spin density wave gap for arbitrarily small U, evolving to a Mott gap at larger U. Hopping disorder reduces AFLRO via the formation of singlets on bonds with large hopping t_{ij} and hence large coupling $J = t_{\mathrm{ij}}^2/U$ and ultimately destroys it beyond $\Delta_t \approx 1.6t$.[98] The fascinating result is that insulating behavior in the conductivity nevertheless persists to much larger Δ_t. Moreover the shift of the maximum in Fig. 29 implies that the mobility gap in fact increases with increasing Δ_t.

The situation is quite different in the case of site disorder, as shown in Fig. 30: at fixed temperature T, as site disorder Δ_μ is turned on, the conductivity increases, i.e. the Mott insulating state is weakened.[99] At weak disorder, the conductivity drops with decreasing T, reflecting again the presence of the Mott insulating phase. As the disorder strength becomes large enough to neglect U, one would expect a similar temperature dependence arising from Anderson insulating behavior. It is likely that in all cases the conductivity will ultimately turn over and go to zero at low T, but these simulations are limited to temperatures $T > W/48$ because of the fermion sign problem. Nevertheless, the data for site disorder offer a dramatic contrast to that of bond disorder (Fig. 29) where randomness decreases the conductivity.

What is the underlying reason for the distinct effects of bond and site disorder on the conductivity? There are several obvious differences in how they alter local and even longer range spin and charge correlations which could offer insight. Site disorder

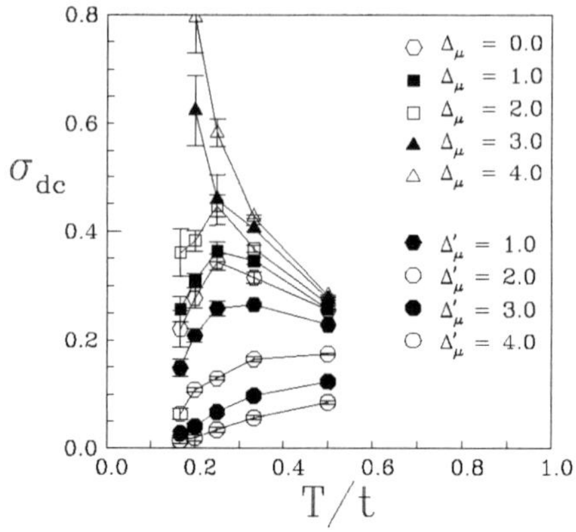

FIGURE 30. Canonical site disorder (with strength Δ_μ) enhances the conductivity. Particle–hole symmetric site disorder (with strength Δ'_μ), as with bond disorder (Fig. 29), suppresses the conductivity. Other parameters are as in Fig. 29.

enhances the amount of double occupancy on the lattice, since the energy cost U of double occupancy is compensated by differences in site energies. One explanation of why site disorder increases σ_{dc} is that the concomitant increase in empty sites leads to more mobility. This destruction of local moments ultimately also leads to the end of antiferromagnetic order. Surprisingly, the simulations suggest that bond disorder has a similar diminishing effect on local moments, so that the difference in the behavior of the conductivity arises from a different origin.

Particle–hole (ph) symmetry is the unifying criterion which underlies and determines the effect of disorder. As emphasized above, site and bond disorder have rather similar effects on the double occupancy. Moreover, the consequences of this effect for $\sigma_{\rm dc}$ are expected to become visible only above a threshold value of disorder strength, whereas $\sigma_{\rm dc}$ is affected already for weak disorder. Instead, the key distinction is in the presence or absence of ph–symmetry. In order to explore this conjecture more fully, one can study two other types of disorder: site disorder that preserves ph–symmetry and bond disorder that breaks ph–symmetry (by including next-near-neighbor hopping).

Particle–hole symmetric site disorder is introduced by adding random chemical potentials to the Hubbard model which couple with opposite sign to the density of up and down electrons, i.e. choose $\mu_{\rm j} \equiv \mu_{{\rm j}\sigma} = \sigma\mu_{\rm j}$ in Eq. 67. This type of disorder represents a random (Zeeman) magnetic field. For $U = 0$ ph–symmetric site disorder has precisely the same effect as conventional site disorder, since moving in a given random chemical potential landscape or one obtained by reversing all the site energies is entirely equivalent. However the behavior of the conductivity at finite U is dramatically different. Fig. 30 shows that ph–symmetric site disorder (with strength Δ'_μ) has the *same* effect on $\sigma_{\rm dc}$ as bond disorder, i.e. conductivity decreases with increasing Δ'_μ.

To seek final confirmation of the conjecture, one can also explore the effect of next-near-neighbor (nnn) hopping and randomness therein. Such longer ranging hybridization

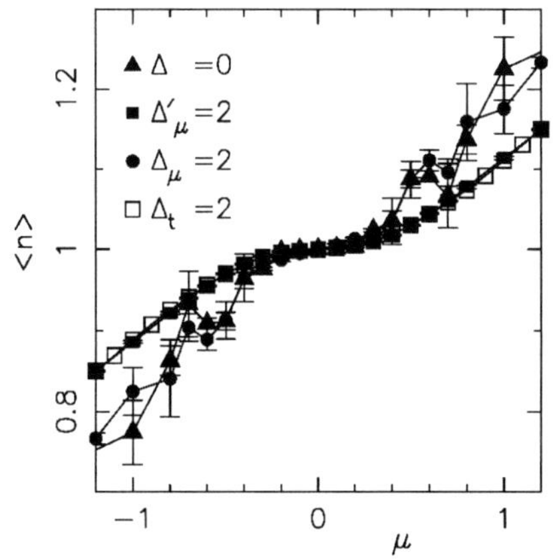

FIGURE 31. The Mott gap is made more robust by the addition of bond disorder or particle-hole symmetric site disorder (open and filled squares) of strength $\Delta = 2t = U/2$, as indicated by the response of the density to changes in the chemical potential. For canonical site disorder (filled circles) the Mott gap is practically unaffected by this strength of randomness. [19] Calculations are for $T = t/8 = W/64$ on a 8×8 lattice.

breaks ph–symmetry on a square lattice, since it connects sites on the same sublattice. Disorder has the *same* effect as conventional site randomness, i.e. it increases the conductivity at finite T. Thus in all four types of disorder, the behavior of the conductivity falls into the appropriate class based on the preservation or destruction of ph–symmetry, strengthening the evidence that it is this symmetry which is playing the crucial role in determining the effect of randomness on the transport properties.

Now consider the thermodynamics. The most direct measure of the Mott gap is from the compressibility, or from the behavior of density $\langle n \rangle$ as a function of chemical potential μ, as shown in Fig. 31. The range of μ where $\langle n \rangle$ is constant (and the system is incompressible) determines the gap in the spectrum. Hopping and ph-symmetric site disorder stabilize the plateau of the density at half–filling. On the other hand, conventional site disorder (with $\Delta_\mu = U/2$) has a compressibility which is indistinguishable (within the available computational accuracy) from the clean system.

The density of states (DOS) at the Fermi level $N(0)$ gives valuable information on the effect of disorder on the Mott gap. In the pure system, QMC studies have shown that the DOS exhibits a clear Mott gap with $N(0) \to 0$ as T is lowered to zero. The nonzero values of $N(0)$ at nonzero T reflect the small residual slopes in the plateaus in the $\langle n \rangle$ vs. μ plot (cf. Fig. 31); at lower T, $N(0)$ approaches zero just as the plateaus become perfectly flat. The behavior of $N(0)$ at a fixed low T as a function of the strength of the various types of disorder is given in Fig. 32. $N(0)$ is rather insensitive to ph–symmetric disorder (Δ_t and Δ'_μ) and is even reduced by it: the Mott gap persists. On the other hand, ph–symmetry breaking disorder (Δ_μ and Δ'_t) clearly enhances $N(0)$, i.e. fills up the Mott gap.

These results provide a clear numerical demonstration of the key role of particle–hole symmetry. The effects can also be understood qualitatively as follows: In the clean case, at $\langle n \rangle = 1$ and strong coupling, the DOS consists of an occupied lower Hubbard band (LHB) and an unoccupied upper Hubbard band (UHB), separated by a charge gap of the order of U. In the case of ph–symmetric disorder, the effect of disorder on LHB and UHB is identical. Therefore the Fermi energy remains in the middle of the gap: this enables

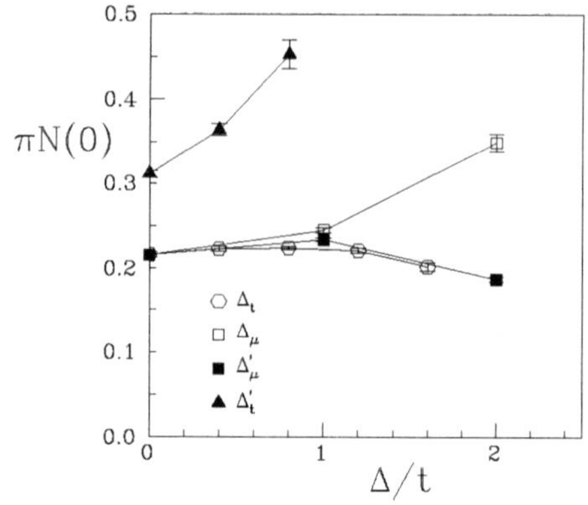

FIGURE 32. Behavior of the density of states $\pi N(0)$ at the Fermi level and at fixed low temperature as a function of disorder strength Δ/t for various types of disorder. All data are for $T = t/6$, except data for randomness in next-near-neighbor hopping (disorder strength Δ'_t) which are at temperature $T = t/5$ (the value $t' = 0$ is used). [19] Other parameters are as in Figs. 29 and 30.

the insulating behavior and Mott gap to stay intact. A stabilized charge gap for ph–symmetric *site* disorder is evident since double occupation is strongly suppressed. For nn-hopping disorder a simple argument is less obvious, but the data in Fig. 31 clearly show that these two cases fall into the same class. When ph-symmetry is broken, the LHB and the UHB will be affected differently; different numbers of states will appear at either side of the gap. As a consequence, the Fermi energy ends up in one of the tails of the DOS, resulting in an enhanced $N(0)$ (cf. Fig. 32) and increased conductivity (Fig. 30). The fact that the states introduced by disorder are localized [100] will keep the system in an insulating state (cf. Fig. 32).

A related example where symmetry plays a crucial role in the effects of disorder is the case of localization in the superconducting phase, where the quasiparticles are described by a Bogoliubov–de Gennes Hamiltonian.[96] In this case, one can classify the system according to the presence or absence of time reversal and spin rotation symmetries, and it is found in one dimension that in the absence of spin rotation symmetry, the conductance decays algebraically with system size, while in the symmetric case it decays exponentially. Therefore, in this situation as well, the extra spin rotation symmetry leads to a strengthening of insulating behavior.

The question of the behavior of the half–filled fermion Hubbard model as disorder is added is furthermore reminiscent of similar issues in the ph-symmetric boson Hubbard model.[37] At generic densities, it is believed that a new 'Bose glass' phase arises to intervene in the original ground state phase diagram between superfluid and Mott insulating phases, but the situation at the ph-symmetric tip of the Mott lobe is uniquely different. This work is a first step in the analysis of the nature of the behavior of the fermionic model.

D. Interaction driven Band Insulator to Metal Transition

Journal Reference: "Quantum Monte Carlo Study of an Interaction-Driven Band Insulator to Metal Transition", N. Paris, K. Bouadim, F. Hebert, G.G. Batrouni, and R.T. Scalettar, Phys. Rev. Lett. **98**, 046403 (2007).

Thus far these notes have emphasized the role of interaction effects in disordered, tight-binding models such as the Hubbard Hamiltonian in driving a metallic transitions from an Anderson insulating states originating in the random potential. Actually, a somewhat more simple context in which to study the possibility of interaction driven insulator to metal transitions is to begin with a band insulating state, in which the insulating behavior is caused by a periodic external potential as opposed to a random one [59, 60, 61, 62, 67]. Recently, this issue has been addressed within dynamical mean field theory (DMFT) and a number of interesting conclusions emerged[101]. However, because DMFT treats only a single site (retaining, however, all the dynamical fluctuations of the self-energy ignored in conventional, static mean field theory) it is important to undertake complementary work which is able to retain intersite fluctuations.

This section investigates such band insulator-metal transitions with DQMC. The specific Hamiltonian we study is the "ionic Hubbard model":

$$
\begin{aligned}
H &= -t \sum_{\langle lj\rangle\sigma} (c_{j\sigma}^{\dagger} c_{l\sigma} + c_{l\sigma}^{\dagger} c_{j\sigma}) + U \sum_{l} n_{l\uparrow} n_{l\downarrow} \\
&+ \sum_{l} (\Delta(-1)^{l} - \mu)(n_{l\uparrow} + n_{l\downarrow}) \,. \qquad (69)
\end{aligned}
$$

Instead of a random chemical potential, the term $\Delta(-1)^l$ provides a staggered site energy. In the noninteracting limit, $U=0$, the effect of Δ is to produces a dispersion relation, $E(k) = \pm\sqrt{\varepsilon(k)^2+\Delta^2}$ with $\varepsilon(k) = -2t[\cos k_x + \cos k_y]$, which is gapped at half-filling. (See Exercises.) A considerable amount is known concerning this model in one dimension[102], but the existence of an interaction driven metallic phase at half-filling is still unresolved even there. Metal-insulator transitions in a related system with *randomly* located site energies with a bimodal distribution have also been studied within DMFT [103, 104].

The temperature dependence of the conductivity $\sigma_{\rm dc}$ is shown for increasing values of the interaction strength for $\Delta = 0.5$ in Fig. 33. The insulating behavior at $U=0$ signalled by $d\sigma_{\rm dc}/dT > 0$ at low T is changed to metallic $d\sigma_{\rm dc}/dT < 0$ at low T when $U=1$. A further increase of the correlations to $U=2$ weakens the metallic behavior, and which is finally destroys completely in a transition to a Mott insulator at $U=4$. When the band gap is larger ($\Delta = 1$), the screening of the one-body potential is not sufficiently strong for $U=1$ to cause metallic behavior, as is shown by the corresponding data set in Fig. 33. Here, and throughout this section unless otherwise mentioned, the lattice size used in the simulations is $N = 6\times 6$ and the filling is $\rho = 1.0$ (half-filling).

In the single site ($t=0$) limit, the ionic Hubbard model is a band insulator for $U < 2\Delta$ and a Mott insulator for $U > 2\Delta$. That is, at weak coupling and half-filling, the sites with lower energy $-\Delta$ are doubly occupied and those with higher energy $+\Delta$ are empty, with a gap to further addition of particle set by $2\Delta - U$. At strong coupling, both types of sites are singly occupied, with a gap to further addition of particles set by $U - 2\Delta$. correlations close the band gap At the single special value $U = 2\Delta$ correlations close the band gap [101, 67]. Fig. 34, which presents results for $\sigma_{\rm dc}$ for $\Delta = 0.5$, shows that when t is nonzero, this single metallic point is expanded to a finite range of U values. Interestingly, however, the largest conductivity remains at $U = 2\Delta = 1$ as one might expect from the $t=0$ analysis. The band-insulator to metal transiton occurs at $U_{c1} \approx 0.4t$, where the change in the order of the three curves indicates a transition from

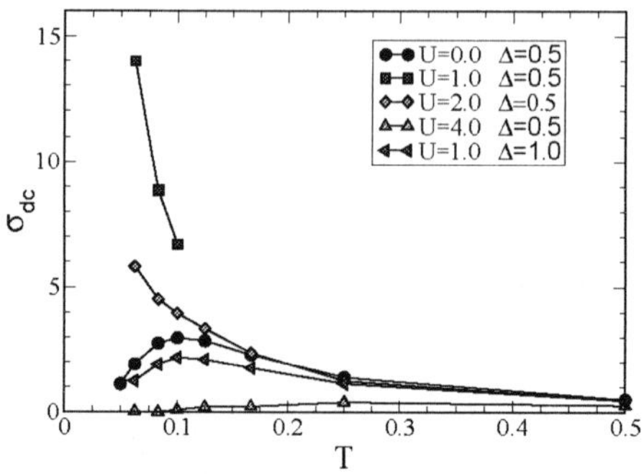

FIGURE 33. The transitions, at half-filling, from a band insulator to metal to Mott insulator with increasing U are shown for periodic potential strength $\Delta = 0.5$. At U=0 the conductivity σ_{dc} goes to zero as T is lowered. However, for at intermediate $U = 1,2$ the system is metallic. Mott insulating behavior sets in for $U = 4$. The lattice size is 6×6. When $\Delta = 1.0$, the band gap increases and $U = 1$ is no longer sufficiently large to screen the one body potential and drive the system metallic.

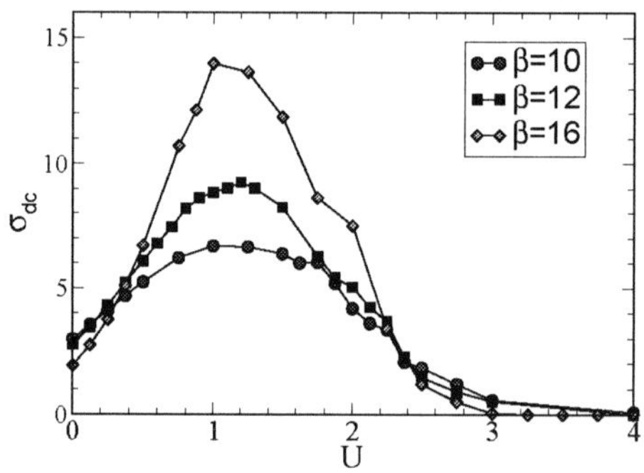

FIGURE 34. The conductivity σ_{dc} at half-filling for $\Delta = 0.5$ is shown as a function of U for three different low temperatures, $\beta = 10, 12, 16$. The band-insulator to metal transition is signaled by the crossing of the curves at $U_{c1} \approx 0.4t$. At $U_{c2} \approx 2.5t$ the three curves cross again, indicating the Mott insulator transition.

σ_{dc} decreasing as β increases to σ_{dc} increasing as β increases. The metal transiton to Mott insulator transition is at $U_{c2} \approx 2.5t$, where σ_{dc} once again decreases as β increases.

The use of DQMC to study the ionic Hubbard model allows us to examine the behavior of intersite correlations, among them the spin-spin correlations and their Fourier tranform $S(k)$. Fig. 35 shows results for the antiferromagnetic structure factor $S(\pi, \pi)$ as a function of U for $\beta = 10, 12, 16$. Comparing with Fig. 34 the band insulating and metallic phases are paramagnetic, but the transition to Mott insulating behavior is accompanied by the onset of antiferromagnetic order.

One way in which the inclusion of such intersite correlations changes the physics in a fundamental way is when the periodic potential is absent, that is, at $\Delta = 0$. In

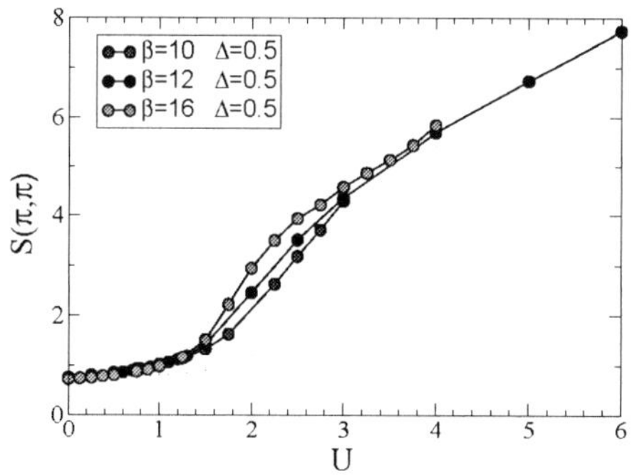

FIGURE 35. The antiferromagnetic structure factor is shown at half-filling as a function of U for two different values of the periodic potential $\Delta = 0.5$ and inverse temperature $\beta = 10, 12, 16$.

DMFT in the paramagnetic phase, the Hubbard model is a metal at weak coupling [69, 105]. However, it is known that the $d = 2$ square lattice Hubbard model being studied here is an antiferromagnetic insulator at *all* U, even weak coupling. Fig. 36 presents results for the conductivity which confirms this. At all U values shown, σ_{dc} ultimately decreases as T is lowered. Indeed, it is possible to verify that the value of T where σ_{dc} has its maximum correlates well with the temperature T_* at which antiferromagnetic correlations begin to rise rapidly. This temperature, like the Neél temperature in the $d = 3$ Hubbard model, is a non-monotonic function of U, falling to small values both at weak ($T_* \propto t\exp(a\sqrt{t/U})$ and at strong ($T_* \propto t^2/U$) coupling. This is the first time the insulating nature of the square lattice Hubbard model at weak coupling has been shown from Quamtum Monte Carlo studies of σ_{dc}. It is interesting to note that while all the $\Delta = 0$ curves share a common low temperature slope $d\sigma_{dc}/dT > 0$, a distinction between the antiferromagnetic and Mott origins of insulating behavior is clearly evident. At small U, σ_{dc} attains a much larger value before turning over as T is lowered than in the strong coupling Mott regime.

While DQMC allows us to look at intersite correlations and the associated phenomena like antiferromagnetism and insulating behavior deriving therefrom, the method employs lattices of finite size, unlike DMFT which directly probes the thermodynamic limit. Thus, it is important to verify that the observed metallic phase persists to larger lattices. Fig. 37 shows results for σ_{dc} as a function of temperature in the metallic phase for lattices up to 12×12. The rise in σ_{dc} with decreasing T is seen to occur for all the lattices studied. It is not surprising that the lattice size has a rather substantial influence on the conductivity for these parameters, since it is known that such finite size effects are larger at weak coupling.

Fig. 38 shows the conductivity as a function of U for three different temperatures at $\Delta = 1.0$. There is now a much larger band-insulating phase at weak coupling, with a critical U for the metallic transition at $U_c \approx 1.5$. It is interesting that the conductivity again appears to peak at the value $U = 2\Delta$ where the $t = 0$ analysis suggests might be most amenable to the formation of a metal. Unfortunately, as U is increased into the metallic phase, the data become extremely noisy. This is a consequence of the sign problem. Fig. 39 shows the average sign as a function of U for $\Delta = 1$. It is not possible to do DQMC simulations at low T beyond $U = 2$ or so.

The evidence for metallic behavior at $\Delta = 1.0$ in Fig. 38 is a bit lost in the noise.

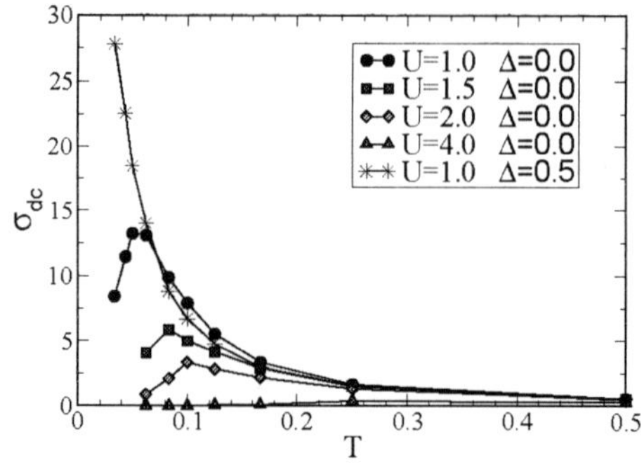

FIGURE 36. The conductivity σ_{dc} is shown as a function of temperature at half-filling. When the periodic potential, and hence the non-interacting band gap, is absent ($\Delta = 0.0$) the square lattice Hubbard model is insulating for *all* U, due to nesting of the Fermi surface. The data for $\delta = 0.5, U = 1$ from Figure 1 is redisplayed to emphasize the contrast between the metallic behavior there and the insulating behavior for all U when $\Delta = 0$.

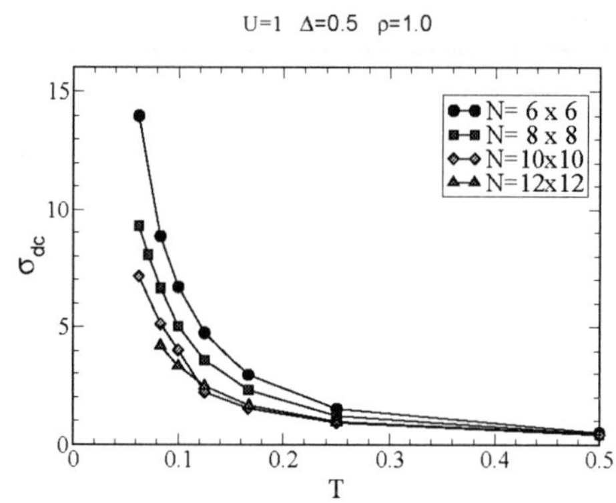

FIGURE 37. The conductivity at half-filling is shown for different lattice sizes for $U = 1$, close to the point where the system is most metallic for periodic potential $\Delta = 0.5$. (See. Fig. 2.) Although σ_{dc} decreases with increasing lattice sizes, the signature of metallic behavior ($d\sigma_{dc}/dT < 0$) is unchanged.

Fig. 40 shows the conductivity, focussing on U values just inside the metallic region where the sign problem is not yet so bad. The steady increase of σ_{dc} as T decreases shows convincingly that the system is a metal at these U values.

In this section, we have presented determinant Quantum Monte Carlo studies of the two-dimensional Hubbard Hamiltonian which demonstrate that interactions can drive a band insulator metallic. This work complements DMFT by including intersite antiferromagnetic correlations which qualitatively alter the ground state phase diagram. Most of the results have focussed on $\Delta = 0.5$. It would be interesting to attempt to construct the full phase diagram in the $\Delta/t - U/t$ plane for the $d = 2$ ionic Hubbard model, and compare with that obtained in DMFT [101]. As these notes have emphasized, the behavior along the $\Delta = 0$ axis is significantly different. At intermediate values of

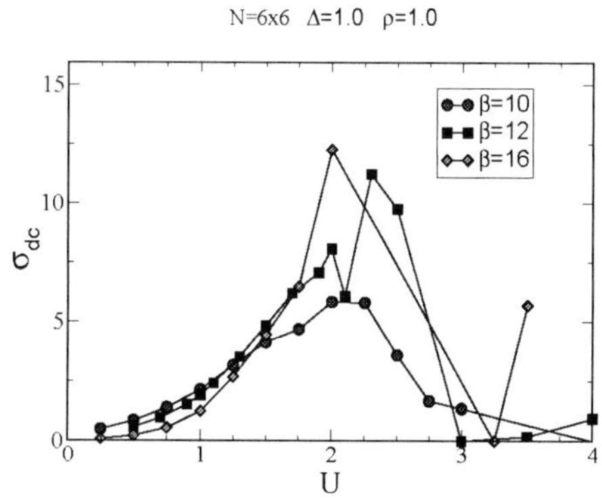

FIGURE 38. The conductivity at $\rho = 1.0$ is shown as a function of U for $\Delta = 1.0$ and $\beta = 10, 12, 16$. The system is a band insulator up to $U \approx 1.5$ whereupon metallic behavior onsets.

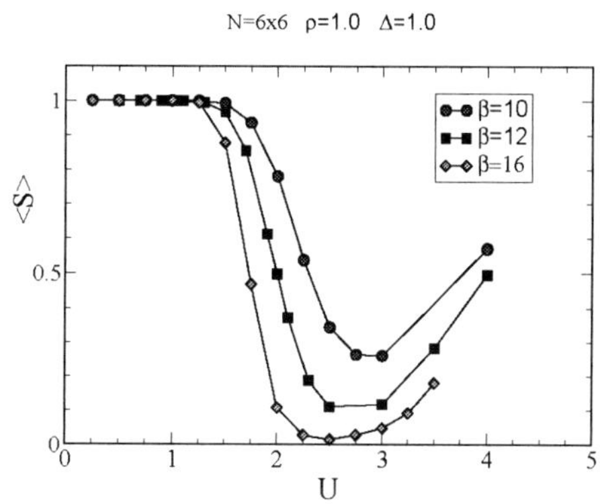

FIGURE 39. The average sign in the DQMC simulations as a function of U. $\langle S \rangle$ rapidly becomes small as the metallic region is entered. It rises again in the Mott insulator.

Δ, DMFT indicates the interesting result the metallic phase closes, with a direct band insulator to Mott insulator transition occurring beyond $\Delta/W \approx 1/8$. Can this be studied within DQMC? Unfortunately, the sign problem, which becomes serious at intermediate $\Delta \approx t$, prevents us from exploring this full phase diagram at present.

However, one limit which is accessible, that is not have an insurmountable sign problem, is the case of large Δ, where we have very widely separated bands. Related studies of the boson-Hubbard model in a "superlattice" potential, which exhibit a band-insulator to superfluid transition [59, 60, 61, 62, 67], show the appearance of insulating phases at half-integer fillings. These 'charge-transfer' insulators occur as a result of Mott splitting of the widely separated bands [103]. It would be interesting to explore this possibility in the fermion case.

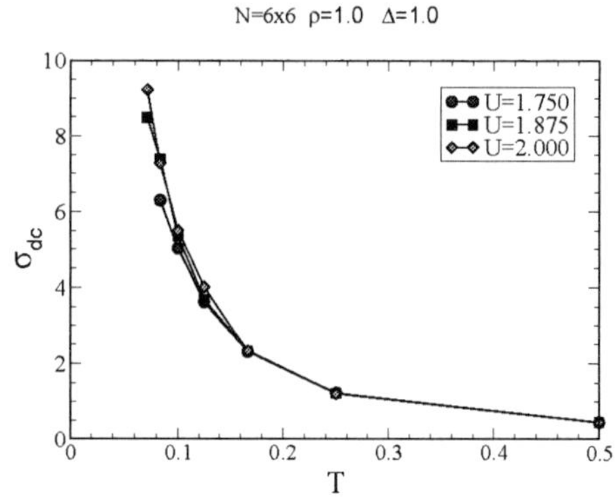

FIGURE 40. The conductivity as a function of T for $U = 1.750, 1.875$, and 2.000. The system is clearly metallic.

VI. CONCLUSIONS

These notes have attempted to provide an introduction to the Anderson-Hubbard Hamiltonian and some of the physics arising from the combination of interactions and randomness. The behavior of the clean model in the limit of no interactions, no kinetic energy, small clusters, and mean field theory was established analytically, and then the determinant quantum monte carlo method was decribed and employed for the full problem on lattices of hundreds of sites.

One significant omission concerning the physics of the Hubbard model is the idea of a 'Kondo resonance'. It turns out that as one progresses from weak to strong coupling, the spectral function of the Hubbard model does not smoothly evolve from a single blob to upper and lower Hubbard bands. Instead, in the course of changing the interaction strength a three peak structure is in evidence: nascent upper and lower Hubbard bands, but also a sharp peak at the Fermi energy. Actually, it was originally thought that such peaks only arise in variants of the Hubbard model which contain both localized and delocalized electrons. It is only relatively recently, with the introduction of dynamical mean field theory (DMFT), that it was realized this sharp peak occurs in the Hubbard model as well. This very important idea is at the heart of much of the current research into the Hubbard model and its experimental realizations.

APPENDIX A: CREATION AND DESTRUCTION OPERATORS AND THE HUBBARD HAMILTONIAN

Creation and destruction operators,

$$\begin{aligned} \hat{a} &= \sqrt{\frac{m\omega}{2\hbar}}\hat{x} + i\sqrt{\frac{1}{2m\omega\hbar}}\hat{p} \\ \hat{a}^\dagger &= \sqrt{\frac{m\omega}{2\hbar}}\hat{x} - i\sqrt{\frac{1}{2m\omega\hbar}}\hat{p} \; , \end{aligned} \tag{70}$$

are familiar from the treatment of the harmonic oscillator. They are the language in which tight-binding Hamiltonians like the Hubbard model are written, but differ in several respects. Perhaps most confusing is that the fermion operators in the Hubbard model are not introduced in terms of familiar position and momentum operators as are $\hat{a}, \hat{a}^\dagger$ above. Rather they stand on their own.

Also, instead of just one creation and one destruction operator, in the Hubbard model there is a set of such operators, which are distinguished by attaching indices $\mathbf{j}$ and σ. Thus one writes $\hat{c}^\dagger_{\mathbf{j}\sigma}$ and $\hat{c}_{\mathbf{j}\sigma}$. The index $\mathbf{j}$ labels the spatial lattice site and the index σ labels the electron spin (up or down). As a consequence, the occupation number states are no longer characterized by a single number n, as for a single harmonic oscillator, but instead by a collection of occupation numbers $n_{\mathbf{j}\sigma}$. One writes such states as $|n_{1\uparrow} n_{2\uparrow} n_{3\uparrow} \ldots . n_{1\downarrow} n_{2\downarrow} n_{3\downarrow} \ldots . \rangle$

Finally, because these operators are meant to describe fermions, they obey *anti-commutation* relations. (The anticommutator of two operators $\{\hat{A}, \hat{B}\}$ is defined to be $\hat{A}\hat{B} + \hat{B}\hat{A}$.)

$$
\begin{aligned}
\{\hat{c}_{\mathbf{j}\sigma}, \hat{c}^\dagger_{\mathbf{l}\sigma'}\} &= \delta_{\mathbf{j},\mathbf{l}}\delta_{\sigma,\sigma'} \\
\{\hat{c}^\dagger_{\mathbf{j}\sigma}, \hat{c}^\dagger_{\mathbf{l}\sigma'}\} &= 0 \\
\{\hat{c}_{\mathbf{j}\sigma}, \hat{c}_{\mathbf{l}\sigma'}\} &= 0.
\end{aligned}
\tag{71}
$$

An immediate consequence of these anticommutation relations is the Pauli principle: the maximum occupation of a particular site with a given spin is 1.

Exercise A1: Show that the Pauli exclusion principle is a consequence of the anticommutation relation amongst the fermion creation operators by considering $\mathbf{j} = \mathbf{i}$ and showing that $\hat{c}^\dagger_{\mathbf{j}\sigma}|1\rangle = 0$. Here $|1\rangle$ is the state with one electron on site $\mathbf{j}$ and with spin σ.

Note that while a fermion creation operator annihilates a state which already has a fermion in it, its action on the empty state is $\hat{c}^\dagger|0\rangle = |1\rangle$, which looks just like the bosonic operator $\hat{a}^\dagger$. Besides the Pauli principle, the anticommutation relations also ensure that the particles are fermions, that is, their wave function changes sign when two electrons with different labels $\mathbf{j}$ are exchanged. To keep track of these signs, one needs to specify a convention for the relation between a state like $|10100\ldots\rangle$ and the vacuum state $|\text{vac}\rangle = |00000\ldots\rangle$. The two possibilities, $|10100\ldots\rangle = c^\dagger_1 c^\dagger_3|\text{vac}\rangle$ and $|10100\ldots\rangle = c^\dagger_3 c^\dagger_1|\text{vac}\rangle$ differ by a sign. Either definition is fine, but in all subsequent manipulations whatever convention was chosen must be followed consistently.

Exercise A2: Suppose the indices labeling the fermion operators runs over eight possible values. Figure out the occupation number state which results from the following applications of the indicated creation and destruction operators. Choose as your convention that a given occupation number state is formed by acting on the vacuum state with the lowest indices at the right of the string of creation operators. (This corresponds to the second of the two choices discussed above.)

$$
\begin{aligned}
\hat{c}^\dagger_5 \hat{c}^\dagger_4 \hat{c}^\dagger_2 |\text{vac}\rangle &= ? \\
\hat{c}^\dagger_2 \hat{c}^\dagger_4 \hat{c}^\dagger_5 |\text{vac}\rangle &= ? \\
\hat{c}^\dagger_4 \hat{c}^\dagger_5 |11000001\rangle &= ? \\
\hat{c}^\dagger_4 \hat{c}^\dagger_5 |11001001\rangle &= ?
\end{aligned}
$$

$$\hat{c}_1^\dagger \hat{c}_2 |01001001\rangle = ?$$
$$\hat{c}_1^\dagger \hat{c}_4 |01001001\rangle = ?$$
$$\hat{c}_1^\dagger \hat{c}_5 |01001001\rangle = ?$$

Having introduced creation and annihilation operators, one can now write down the Hubbard Hamiltonian. Before doing so, think about how one might simply describe the motion and interactions of electrons in a solid. First, one needis to account for the fact that there is a regular array of nuclear positions in a solid, which for simplicity are considered to be fixed. (In other words, we will not worry about lattice vibrations.) This suggests one begin with a lattice of atoms (sites) on which the electrons move. A single atom is already a very complex structure, with many different energy levels. The most simple 'atom' would have a single energy level. Then, the Pauli principle would tell us that at most two electrons (one with spin up and one with spin down) can sit on this 'atom'. In a solid where electrons can move around, the electrons interact via a screened Coulomb interaction. The biggest interaction will be for two electrons on the same atom. For simplicity, Hubbard stops just there, so that interactions are modeled by a term which is zero if the atom is empty of electrons or has only a single electron on it, but has the value U if the atom has two electrons. There is no interaction between electrons on different sites. The kinetic energy will consist of an expression which allows electrons to move from one site to its neighbors. The energy scale t which governs this 'hopping' will be determined by the overlap of two wavefunctions on the pair of atoms. Since wavefunctions die off exponentially, one can begin by allowing hopping only between the closest atoms in the lattice.

Now let's formalize this construction. Define $c_{\mathbf{j}\sigma}^\dagger$ to be the operator which creates an electron of spin σ on lattice site $\mathbf{j}$. (Drop all the 'hats' which up to now have been used to denote operators.) Similarly, $c_{\mathbf{j}\sigma}$ is the destruction operator, and $n_{\mathbf{j}\sigma} = c_{\mathbf{j}\sigma}^\dagger c_{\mathbf{j}\sigma}$ is the number operator.

The Hubbard Hamiltonian is then,

$$H = -t \sum_{\langle \mathbf{j},\mathbf{l}\rangle\sigma} c_{\mathbf{j}\sigma}^\dagger c_{\mathbf{l}\sigma} + U \sum_{\mathbf{j}} n_{\mathbf{j}\uparrow} n_{\mathbf{j}\downarrow} - \mu \sum_{\mathbf{j}} (n_{\mathbf{j}\uparrow} + n_{\mathbf{j}\downarrow}). \tag{72}$$

The first term is the kinetic energy: It describes the destruction of an electron of spin σ on site $\mathbf{l}$ and its creation on site $\mathbf{j}$ (or *vice-versa*). The symbol $\langle \mathbf{j},\mathbf{l}\rangle$ emphasizes that hopping is allowed only between two sites which are adjacent. The second term is the interaction energy. It goes through all the sites and adds an energy U if it finds that the site is doubly occupied. The final term is a chemical potential which controls the filling. The situation where the filling is one electron per site is referred to as 'half-filling' since the lattice contains half as many electrons as the maximum number (two per site). Studies of the Hubbard model often focus on the half-filled case because it exhibits a lot of interesting phenomena (Mott insulating behavior, antiferromagnetic order, etc.)

Exercise A3: Show that the Hubbard Hamiltonian commutes with the operators $N_\uparrow = \sum_{\mathbf{j}} n_{\mathbf{j}\uparrow}$ and $N_\downarrow = \sum_{\mathbf{j}} n_{\mathbf{j}\downarrow}$. It is useful to begin by considering the commutator of the kinetic energy on a single 'link' of the lattice connecting sites $\mathbf{i}$ and $\mathbf{j}$ with the total number of electrons on those two sites. That is, begin by computing,

$$[c_{\mathbf{i}\sigma}^\dagger c_{\mathbf{j}\sigma} + c_{\mathbf{j}\sigma}^\dagger c_{\mathbf{i}\sigma}, n_{\mathbf{i}\sigma} + n_{\mathbf{j}\sigma}] \quad . \tag{73}$$

After working through the algebra, can you think of an argument that this should be the case based on the structure of H, that is, based on how the creation and destruction operators appear together?

Since the Hubbard Hamiltonian commutes with the total spin up and spin down number operators, $N_\uparrow = \sum_{\mathbf{j}} n_{\mathbf{j}\uparrow}$ and $N_\downarrow = \sum_{\mathbf{j}} n_{\mathbf{j}\downarrow}$. In finding the eigenstates, one can consider different sectors of $N_\uparrow$ and $N_\downarrow$ separately. As might be expected, in the non-interacting limit, from a solution in the single particle sector, $N_\uparrow = 1$ and $N_\downarrow = 0$ a solution of the many electron problem can be built.

Exercise A4: Suppose you have a one dimensional lattice of eight sites. Write down all the occupation number states in the sector $N_\uparrow = 1, N_\downarrow = 0$, Figure out what H does to each state. Use 'periodic boundary conditions' so that site eight is considered a neighbor of site one. Write down the matrix for H using this basis. Show that $|\,1\,1\,1\,1\,1\,1\,1\,1\rangle$ is an eigenvector of eigenvalue $-2t$. Show that $|\,1\,-1\,1\,-1\,1\,-1\,1\,-1\rangle$ is an eigenvector of eigenvalue $+2t$.

Am NxN tridiagonal matrix with "a" along the diagonal and "b" above and below the diagonal, with periodic boundary conditions has eigenvalues $\lambda_k = a + 2b\cos k$ where k takes on the discrete values $k_n = 2\pi n/N$ and $n = 1,2,3,\ldots N$.

Exercise A5: For general N, what is the eigenvector corresponding to $n = N$ so that the momentum $k = 2\pi = 0$? What is the eigenvector corresponding to $n = N/2$ which has momentum $k = 2\pi$?

Using this result for the eigenvalues of a tridiagonal matrix, one sees that the single particle, one-dimensional Hubbard Hamiltonian has energy levels $\varepsilon_k = -2t\cos k$. One can view this as a simple 'energy band' of bandwidth $W = 4t$. This establishes a rough connection between the hopping parameter t in the Hubbard Hamiltonian and physical energy scales like the bandwidth W of a real material.

The eigenvalues of the Hubbard Hamiltonian at $U = 0$ in higher particle sectors are obtained by occupying these single particle levels according to the restrictions of the Pauli principle. It is amusing explicitly to diagonalize H in the $N_\uparrow = 2, N_\downarrow = 0$ sector, computing the eigenvalues (perhaps numerically with a LAPACK routine- there are 28 states!), and verifying that they are indeed Pauli-restricted combinations of the eigenvalues obtained in the single particle case.

Let us now look at the non-interacting limit in a more formal way by transforming to momentum creation and destruction operators.

$$c^\dagger_{\mathbf{k}\sigma} = \frac{1}{\sqrt{N}} \sum_{\mathbf{l}} e^{i\mathbf{k}\cdot\mathbf{l}} c^\dagger_{\mathbf{l}\sigma}. \tag{74}$$

One can think about this process in analogy with the classical normal mode problem: Define new (momentum) creation operators as a linear combination of the old (position) ones. As shall be seen in Exercises which follow, in the Hubbard model at $U = 0$ the different momentum modes decouple from each other and behave independently, just as for classical normal modes. Notice that on a finite lattice the momentum $\mathbf{k}$ cannot be any real number but has discretized values. For a one-dimensional lattice of N sites, the periodic boundary condition $c^\dagger_{N+1} = c^\dagger_1$ yields $k_n = 2\pi n/N$. This is of course the same condition as when one considers the eigenvalues of the tri-diagonal matrix above. For a two-dimensional square or three dimensional cubic lattice, each component separately has such a discretization.

Different Fourier functions are orthogonal. The analog for these discrete site and momentum variables is given in the following Exercise:

Exercise A6: Prove the following two 'orthogonality' relations:

$$\frac{1}{N}\sum_{l} e^{i(k_n-k_m)l} = \delta_{n,m} \tag{75}$$

$$\frac{1}{N}\sum_{k_n} e^{ik_n(l-j)} = \delta_{l,j} \tag{76}$$

Exercise A7: Use the orthogonality relations invert Eq. 2, that is, to prove,

$$c^\dagger_{\mathbf{l}\sigma} = \frac{1}{\sqrt{N}}\sum_{\mathbf{k}} e^{-i\mathbf{k}\cdot\mathbf{l}} c^\dagger_{\mathbf{k}\sigma}. \tag{77}$$

Here, of course, the sum over **k** means you sum over the discrete allowed momenta $\mathbf{k}_n$. With these relations in hand, one one can show a lot of interesting things about the momentum space operators.

Exercise A8: Verify the anticommutation relations

$$\begin{aligned}
\{c_{\mathbf{k}\sigma}, c^\dagger_{\mathbf{p}\sigma'}\} &= \delta_{\mathbf{k},\mathbf{p}}\delta_{\sigma,\sigma'} \\
\{c^\dagger_{\mathbf{k}\sigma}, c^\dagger_{\mathbf{p}\sigma'}\} &= 0 \\
\{c_{\mathbf{k}\sigma}, c_{\mathbf{p}\sigma'}\} &= 0.
\end{aligned} \tag{78}$$

In other words, the anticommutation relations are "preserved" by this change in "basis" from site indices to momentum indices.

Now the $U=0$ Hubbard model can be expressed in terms of these momentum space operators, leading in a more elegant way to the same dispersion relation (energy band) obtained by considering the matrix of H in the one particle sector.

Exercise A9: Show that for $U=0$ the one dimensional Hubbard model is,

$$H = \sum_{k\sigma}(\varepsilon_k - \mu)c^\dagger_{k\sigma}c_{k\sigma} \tag{79}$$

where $\varepsilon_k = -2t\cos k$. As part of this Exercise, you will show that the sum of all the number operators over different spatial sites and spin equals the sum of all the number operators over different momenta and spin.

It is important to realize that the result that an analysis of the one-particle sector gives us full information about the model for any particle number rests only on the fact that the interactions are turned off. It is not necessary that the hopping t between different sites be the same for all pairs of sites, or that it be limited to near neighbors, or that the chemical potential be the same on all sites. All that matters is that H be a quadratic form in the fermion creation and destruction operators. To emphasize: To solve any Hamiltonian H which takes the form $H = \sum_{\mathbf{i},\mathbf{j}} c^\dagger_{\mathbf{i}} h_{\mathbf{ij}} c_{\mathbf{j}}$ with h a (symmetric) matrix of real numbers, simply diagonalize h and allow the resulting energy levels to be filled in a way which satisfies the exclusion principle. An important application of this theorem is given in the context of mean field theory, discussed in the second section of these notes.

Exercise A10: Show that for $U=0$ the two dimensional Hubbard model on a square lattice is

$$H = \sum_{\mathbf{k}\sigma}(\varepsilon_{\mathbf{k}} - \mu)c^\dagger_{\mathbf{k}\sigma}c_{\mathbf{k}\sigma} \tag{80}$$

where $\varepsilon_{\mathbf{k}} = -2t\,(\cos k_x + \cos k_y)$.

In real space, all the different (site occupation) states in the $U = 0$ model are mixed with each other. In momentum space the different states are decoupled: A fermion operator of a give **k** appears only together with operators of the same **k**. This means that the different momentum modes can be treated independently, leading to the following result.

Exercise A11: Show that the partition function of the $U = 0$ Hubbard model is given by

$$Z = \prod_{\mathbf{k}} (1 + e^{-\beta(\varepsilon_{\mathbf{k}} - \mu)}). \tag{81}$$

It is useful to remember that if a Hamiltonian is comprised of the sum of independent pieces then the partition function is the product of the associated partition functions.

Exercise A12: Compute the per site average occupation of the $U = 0$ Hubbard model and show it is given by,

$$\rho = \sum_{\mathbf{k}} (1 + e^{+\beta(\varepsilon_{\mathbf{k}} - \mu)})^{-1}. \tag{82}$$

Note this takes the form of the sum of the occupations of different pieces, and that the Fermi function, $f_{\mathbf{k}} = 1/[1 + e^{\beta(\varepsilon(k) - \mu)}]$, naturally arises. Show that when $\mu = 0$ the density $\rho = 1$ for any β if $\varepsilon_{\mathbf{k}}$ is of the form derived for the $d = 1$ or $d = 2$ square lattice Hubbard model.

The allowed k values together with the dispersion relation determine the density of states $N(E)$ which counts the number of ways in which the system can have a given energy E. Formally, $N(E)$ is defined by

$$N(E) = \frac{1}{N} \sum_{\mathbf{k}} \delta(E - \varepsilon_{\mathbf{k}}). \tag{83}$$

In the continuum limit (large number of sites) the sum over discrete momenta values is replaced by an integral according to the rule $\frac{1}{N}\sum_{\mathbf{k}} \to (2\pi)^{-d} \int d\mathbf{k}$, where d is the spatial dimension. As a simple example, consider one dimension with a relativistic dispersion relation $\varepsilon_k = ck$ for $k > 0$. One can get the density of states as follows:

$$N(E) = (2\pi)^{-1} \int dk \delta(E - ck) = (2\pi)^{-1} \int dk \frac{1}{c} \delta(E/c - k) = (2c\pi)^{-1} (1 - \theta(k)). \tag{84}$$

Here the function $1 - \theta(k)$ emphasizes that $k > 0$ is required. Likewise, for a quadratic dispersion relation $\varepsilon_k = ak^2$

$$N(E) = (2\pi)^{-1} \int dk \delta(E - ak^2) = (2\pi)^{-1} \int dk \frac{1}{2ak} \delta(\sqrt{E/a} - k) = (4a\pi)^{-1} \sqrt{a/E}. \tag{85}$$

Exercise A13: Compute (analytically) the density of states $N(E)$ of one dimensional Hubbard model. Explain why $N(E)$ diverges at $E = \pm 2t$ in terms of a picture of the dispersion relation $E(k) = -2t \cos k$. Compute $N(E)$ numerically and compare to your analytic calculation.

APPENDIX B: FORMAL FOUNDATION OF CLASSICAL MONTE CARLO

Why Monte Carlo Works: Detailed Balance, Transition Probabilities and All That

The Monte Carlo method is often referred to as a 'computer experiment'. One might think of this as a way of conveying the fact that the output of simulations is not an equation, as in conventional theory. Instead, numbers appear on the computer screen in somewhat the same way that numbers appear on a measuring device in the laboratory. Thus there is the implication that somehow simulations are a bit of a 'black box' and that the use of the computer is hiding the underlying physics. The purpose of this note is partly to emphasize some of the mathematical rigor behind Monte Carlo: It is not a happy accident that the computer is generating configurations with the desired probability distribution! Indeed, the fundamental equations underlying simulations are the same as analytic theories, and one can view simulations as a way of solving the mathematics (differential equations) when it becomes too complicated for analytic techniques.

With all that said, it is still useful to pursue the 'Monte Carlo as experiment' point of view. Consider the process of making a measurement in the laboratory. Nature prepares a 'configuration' i of the system, and the experimentalist takes that configuration and records a value for some quantity of interest. To get better statistics (or perhaps inevitably because of finite measuring time) nature actually produces many configurations, and the experimentalist averages over the values obtained. It is useful to emphasize that no matter how long the experimentalist measures, the configurations she sees are an *incredibly* small subset of those that the system is capable of exploring.

Nature uses some very complex rules for time evolving the system from configuration to configuration, for example the many particle Newton or Schroedinger equations. These rules govern the states that the experimentalist sees, and hence the data she takes.

Here's one useful way to think about computer simulations: The goal of a computer simulation is to devise a method where the *computer* plays a similar role to that of *nature* for the experimentalist. That is, the computer generates configurations upon which one makes measurements. This leaves us with the problem of devising instructions for the computer that replicate nature's way of generating configurations.

One approach to constructing a simulation would be actually coding up the microscopic equations governing the system's time evolution. Simulations of classical systems going under the name 'molecular dynamics' are actually done precisely this way. One computes the force F_n on each particle n, uses the force to compute the acceleration $a_n = F_n/m$, and then moves the velocity and position forward a small time interval dt with, $v_n \rightarrow v_n + a_n dt$; $x_n \rightarrow x_n + v_n dt$. But in the spirit of statistical mechanics, one really doesn't care about the microscopic time evolution and the paths $x_n(t)$ and $v_n(t)$ the particles take in phase space. All that is really needed is to replicate the *probability* $P(\{x_n, v_n\})$ that nature uses to generate her configurations. If one can do that, one will get get the same answers as the experimentalist!

In doing classical statistical mechanics, the probability distribution that one would be attempting to emulate would be the Boltzmann distribution $P(\{x_n, v_n\}) = Z^{-1}e^{-\beta E(\{x_n, v_n\})}$. However, let's discuss Monte Carlo within the context of a general probability distribution. This will emphasize that Monte Carlo is by no means limited to Boltzmann statistics. To make the notation less unwieldy, label the probabilities by a single index i which will be understood to represent particular values of all the degrees of freedom of the system being studied (for example i could mean a collection of positions and velocities $\{x_n, v_n\}$). In the remainder of this note the inverse temperature is denoted by $\beta = 1/T$, and Boltzmann's constant is set to unity.

As will be seen, to do Monte Carlo, it is not necessary to know p_i, but only ratios of p_j/p_i for two configurations. This is certainly an important point for statistical mechanics since $p_j/p_i = e^{-\beta(E_j - E_i)}$ is known, but the individual p_i involve the unknown partition function Z.

The goal is to figure out a rule T_{ji} to evolve the configuration i into the configuration j which will generate configurations with a desired probability. More precisely, the process will not be deterministic, but will involve random changes, so T_{ji} will be the probability to generate j from i. Because probabilities are non-negative, and sum up to unity, the rule T_{ji} satisfies,

$$\begin{aligned} T_{ji} &\geq 0 \\ \sum_j T_{ji} &= 1. \end{aligned} \tag{86}$$

A matrix obeying these two restrictions is called a 'stochastic matrix'. Its eigenvalues obey $|\lambda| \leq 1$. Also, there is an eigenvector with eigenvalue $\lambda = 1$. These facts are simple to prove.

Consider the eigenvector equation,

$$\sum_i T_{ji} v_i = \lambda v_j. \tag{87}$$

Take absolute values of both sides of the equation and use the fact that $T_{ji} \geq 0$ and the triangle inequality.

$$\sum_i T_{ji} \|v_i\| \geq \|\lambda\| \, \|v_j\|. \tag{88}$$

Now sum both sides of the equation over j and use the fact that $\sum_j T_{ji} = 1$,

$$\sum_i |v_i| \geq |\lambda| \sum_j |v_j|. \tag{89}$$

Thus $|\lambda| \leq 1$.

To show that a stochastic matrix has an eigenvalue of unity, consider the vector with all components the same, $v_j = 1$. Then, from the second line of Eq. 1,

$$\sum_j v_j T_{ji} = v_i. \tag{90}$$

So the constant vector is a left eigenvector of T of eigenvalue 1. Remember that the eigenvalues of a matrix are the same whether one considers left or right eigenvectors, but the eigenvectors themselves can be quite different unless the matrix is symmetric. This is important to keep in mind, because the right eigenvector of T is not the trivial constant vector. In fact, the right eigenvector of T is p_j.

That the eigenvalue $\lambda = 1$ is non-degenerate is a consequence of the Perron-Frobenius theorem.

Because T is the rule for generating one configuration from the next, we can view the generation of a long sequence of configurations in the Monte Carlo process[106] as the repeated application of the matrix T. When a matrix is applied repeatedly to a vector, the eigenvector of largest eigenvalue is projected out. Thus, because T is stochastic, the eigenvector of eigenvalue 1 is generated.

So now the goal is a little bit more precise: One wants to figure out T obeying Eq. 1 whose eigenvector of eigenvalue one is the vector of desired probabilities p_i. Then, repeatedly applying T will give us p and we're done.

It is easy to show that if T is constructed to obey 'detailed balance',

$$T_{ji}\, p_i = T_{ij}\, p_j \tag{91}$$

then p_i is an eigenvector of T of eigenvalue 1:

$$\sum_i T_{ji}\, p_i = \sum_i T_{ij}\, p_j = p_j \sum_i T_{ij} = p_j. \tag{92}$$

It is also easy to formulate T to obey detailed balance. The most famous prescription is due to Metropolis. The Metropolis algorithm says that one suggests a move from i to j and then accepts the move with probability one if j is more likely than i, that is, if $p_j/p_i > 1$. If j is less likely than i, accept j with probability p_j/p_i.

Exercise B1: Show that if you generate a random number r uniformly on $[0,1]$ and accept the move when $r < p_j/p_i$, then the Metropolis algorithm is correctly implemented.

The argument presented above can be summarized as follows:

```
[a] Metropolis---> Detailed Balance---> p is eigenvector of T of eigenvalue 1

[b] T is stochastic ---> max eigenvalue of T is 1

[a+b]  p is eigenvector of T of largest eigenvalue

[c]  Repeated application of T yields eigenvector of largest eigenvalue

[a+b+c]  Repeated application of T gives p, as desired.
```

(i.) Take an initial configuration i and propose a change to a new configuration j.
(ii) Accept the new configuration with probability $\min(1, p_j/p_i)$.
(iii) Measure quantities of interest.
(iv) Repeat the process many times.

There are of course many subtleties: What sort of procedure is followed to "suggest changes"? How many configurations do you need to generate and do you, as in some experiments, need to allow the system time to equilibrate? Some of these issues are addressed below.

There are a number of useful analogies between Monte Carlo and experiment. Perhaps most importantly, when first encountering Monte Carlo it is natural to wonder how the relatively small sample of configurations that a computer generates could possibly replicate the behavior of physical systems which have incredibly huge numbers of states. A partial answer is provided by noting that the exact same question could be posed to any experimentalist, where, similarly, measurements are based on a very small subset of states which are generated by the time evolution during the course of the experiment.

Second, when one collects data in a Monte Carlo, one measures observables for each configuration generated and then does a simple average over all the configurations. There are no probabilities p_i put in the averaging. The reason, of course, is that the computer is putting the p_i in for you, by generating the configurations with their correct

p_i. This is just as in an experiment. No experimentalist weights her measurements with the Boltzmann factor!

In addition to averages, error bars are generated in Monte Carlo just as in experiments. (See Sec. XI.)

One often has to wait a little while after starting a simulation to allow the system to equilibrate, much as the experimentalist will not start recording data until some time has passed following perturbing the system in some way.

If you think about it, the transition matrix T actually has two pieces in its construction. The first is the suggestion of a new state j, and the second is the decision (e.g. Metropolis) whether to accept j or not. It is the *combination* of these two factors which determines T and which must obey detailed balance. Examples of how the method of suggesting states might go awry are given in the Exercises below.

Exercise B2: The Metropolis algorithm is not the only way to go. Show that the "heat bath" prescription

$$T_{ji} = p_j/(p_i + p_j) \tag{93}$$

obeys detailed balance.

Thus the starting point 'Metropolis' of [a] in Sec. III can be replaced by 'Heat Bath'.

Exercise B3: Consider a system with two states and associated probabilities p_1, p_2. Suppose the observable A has the values A_1 and A_2 in the two states. What will you get for $\langle A \rangle$ if you measure A only when a move is accepted? What is the correct value for $\langle A \rangle$? When should you measure A to get the right answer? Argue qualitatively that, if $p_1 >> p_2$, the sequence of configurations generated gives a sensible answer when you put your measurement in the correct spot.

Exercise B4: Consider a system with three states and associated probabilities p_1, p_2, p_3. Construct the 3x3 matrix for T assuming that if you are in a state i you suggest a change to one of the other two states with probability $\frac{1}{2}$ each, and then accept or reject with Metropolis. Verify that your T obeys Eq. 1 and that the vector $(p_1\ p_2\ p_3)$ is a right eigenvector of T of eigenvalue one. Verify the other two eigenvalues are less than one in absolute value.

Exercises B5 and B6 below emphasize that the rule T for generating states is not unique. The art of Monte Carlo is in finding the T which works best, that is, which moves you around in phase space most rapidly and generates the smallest error bars.

Exercise B5: Do Exercise B4 again, but use the heat bath algorithm.

Exercise B6: Again consider the same three state system as Exercise B4. Construct the 3x3 matrix for T assuming that if you are in a state i you suggest a new state (which could be the same as the one you are in) randomly, that is, with pick the new state to be 1 or 2 or 3, with probabilty $\frac{1}{3}$, and then do Metropolis. Verify that your T obeys Eq. 1 and that the vector $(p_1\ p_2\ p_3)$ is a right eigenvector of T of eigenvalue one. Verify the other two eigenvalues are less than one in absolute value. In which case is the next smallest eigenvalues closer to $|\lambda| = 1$, here, or in Ex. 5?

Exercise B7: Consider a system with an infinite number of states $E_i = \omega i$ for $i = 0, 1, 2, 3, \ldots,$ and associated probabilities $p_i \propto e^{-\beta E_i}$. Construct the matrix for T assuming that your suggestion of a move from state i is to one of the two immediately adjacent states $j = i \pm 1$, with probability $\frac{1}{2}$. Verify $(p_1\ p_2\ p_3\ \ldots)$ is a right eigenvector of T. Be especially careful with the state $i = 0$ which has no state beneath it! Should you suggest state $j = 1$ all the time, since there is no $j = -1$? This is an example of the subtlety mentioned in Sec. VI.

Exercise B8: Write a Monte Carlo code which computes $\langle E \rangle$ for the three state system of Exercise B4, using the T described in that exercise.

Exercise B9: Write a Monte Carlo code corresponding to the T of exercise 5.

Exercise B10: Write a Monte Carlo code corresponding to the T of exercise 6.

Exercise B11: Write a Monte Carlo simulation for an energy which has a continuous degree of freedom x with $E(x) = \frac{1}{2}kx^2$. Construct T by suggesting the new state $x' = x + \Delta(r - \frac{1}{2})$ where Δ is the 'step-size', and r is a random number uniformly distributed on (0,1). Use Metropolis to accept/reject x'. What is the correct value for $\langle E \rangle$? Show your code gets it, for any value of Δ. What values of Δ are best? See also Sec. XI below.

Exercise B12: Do Exercise B11 again, except this time construct T by suggesting the new state $x' = x + \Delta(r - \frac{1}{3})$, and then use Metropolis. Show you code gets the wrong value for $\langle E \rangle$. Show that your T violates detailed balance, even though you use Metropolis after the move suggestion.

Many readers going through these notes have a particular (and presumably non-trivial) problem they wish to address with Monte Carlo. If, however, you do not, a standard Monte Carlo simulation to try is the two dimensional Ising model. Many books on simulations contain descriptions [107, 108, 109].

As with all codes, tests are essential. Here three very useful ones are mentioned.

First, Monte Carlo simulation codes can often be tested on small lattices. That is, it is often possible to enumerate all the possible states of a small system (e.g. a 4x4 Ising lattice has 2^{16} configurations) and have the computer then generate the partition function and expectation values through an explicit enumeration of all the configurations.

Second, limits of the Hamiltonian in question are often soluble. At high and low temperatures, one often can figure out the values of the energy and other observables. Likewise, by turning off one or more terms in the Hamiltonian, it may become analytically soluble. Or, if you write your code so that the interactions are different in different directions, you can sometimes turn off one or more directions and reduce the dimensionality of the system to a point where analytic solutions are known.

Finally, while you will usually write your code so that it averages observable which should be equivalent by some symmetry, looking at the unaveraged values first is often useful: Many bugs in codes are fingered by measurements which should be equal by symmetry but are not.

My experience is that if your code checks out for the entire model on small lattices, and then individual pieces on different lattice sizes, and also the high and low T limits, then it is very likely correct.

The above discussion seems to rely very heavily on the fact that the system is classical. However, it turns out to be relatively easy to generalize classical Monte Carlo to Quantum Monte Carlo. To do this, one takes the *operator* $e^{-\beta \hat{H}}$ and writes its trace as a path integral. This sum over paths involves a *classical* action, which then can be attacked with classical Monte Carlo. A nice discussion of this for the quantum harmonic oscillator is given by Creutz [110]. Another particularly interesting example is the mapping of the $d = 1$ Ising model in a transverse magnetic field onto the classical $d = 2$ Ising model.[111] Of course, the determinant QMC algorithm described in the main body is also an example of such a mapping (with an obnoxiously complicated Boltzmann weight).

Relation to Molecular Dynamics and the Langevin Equation

At the beginning of these notes, Molecular Dynamics (MD) was mentioned as one way a computer might emulate nature. Since the equations of MD keep the energy constant (to within time step errors), one way to view MD is as a Monte Carlo where the proposed move does not change the energy, and is therefore accepted with unit probability according to the Metropolis prescription. Thus MD and Monte Carlo should give the same result, under the condition that the MD energy is initialized at the same value $\langle E \rangle$ as the average energy which comes out of the Monte Carlo. This is the usual equivalence of the canonical and microcanonical ensembles in statistical mechanics.

In practice, one might want to do MD at a particular temperature T instead of at fixed E. One way to accomplish this is to evolve the positions and velocities x_n and v_n using the MD equations, but then periodically 'refresh' the velocities by replacing their current values with new ones drawn from the Gaussian distribution $e^{-m_n v_n^2/2T}$. Here m_n is the mass of the nth particle. One chooses to 'refresh' the momenta, because usually the energy takes the form $E = \frac{1}{2}\sum_n m_n v_n^2 + V(\{x_n\})$. That is, the potential energy doesn't depend on the velocities. So the probability distribution for the velocities is Gaussian, and it is known how to generate Gaussian distributed random numbers. The positions usually invovle a more complicated V than Gaussian.

This sort of finite T molecular dynamics can be understood as a Monte Carlo with two types of moves. As remarked above, the conventional MD moves keep the energy constant and hence are always accepted. The momentum refreshment moves change the energy and hence are sometimes rejected according to the Metropolis prescription.

Exercise B13: Write a Molecular Dynamics code for $E = \frac{1}{2}kx^2 + \frac{1}{2}mv^2$. (That is, use $F = -kx$ and $a = F/m$ to update v.) Verify that you travel along an ellipse in phase space at constant energy, when dt is small. Here 'dt small' really means dt is much less than the period $2\pi\sqrt{m/k}$. Compute $\langle x^2 \rangle$ and $\langle p^2 \rangle$ and show you get the correct answers (which only depend on x_0 and v_0). Note that the Molecular Dynamics integration of equation of motion introduces time step errors. In the Euler method, the accumulation of these errors will lead to an exponential increase in E. The "leap-frog" method is much more robust.

Exercise B14: Include steps which refresh the momentum. Compute $\langle x^2 \rangle$ and $\langle p^2 \rangle$ and show you get the correct answers (which now do not depend on the initial conditions as in Ex. 13 but only on temperature T.)

Monte Carlo can also be related to the Langevin Equation. Consider a system with an energy $E(x)$ which depends on the degree of freedom x. (For simplicity, consider a single degree of freedom.) Suppose a procedure is defined for updating x via

$$x' = x - \varepsilon\frac{\partial E}{\partial x} + \sqrt{4\varepsilon T}\, r \tag{94}$$

Here r is a Gaussian random number, $p(r) = e^{-r^2}/\sqrt{\pi}$, and T is the temperature. One can now show that this equation satisfies detailed balance with $p(x) \propto e^{-\beta E(x)}$.

The Langevin Equation amounts to a prescription T for getting a new state x' from an original state x. Given x' and x, the value of T is the probability of throwing the appropriate random number which will take you from x to x'.

$$T(x',x) = \frac{1}{\sqrt{\pi}} e^{-r^2} \quad (95)$$

$$r = \frac{1}{\sqrt{4\varepsilon T}} (x' - x + \varepsilon \frac{\partial E}{\partial x}) \quad (96)$$

Therefore, introducing the notation $dx = x' - x$,

$$T(x',x) = \frac{1}{\sqrt{\pi}} e^{-(dx+\varepsilon \frac{\partial E}{\partial x})^2/(4\varepsilon T)} \quad (97)$$

Likewise,

$$T(x,x') = \frac{1}{\sqrt{\pi}} e^{-(-dx+\varepsilon \frac{\partial E}{\partial x})^2/(4\varepsilon T)} \quad (98)$$

Note that actually the gradient of the energy in Eq. 99 should now be evaluated at x', but because ε is small, dx will be small and x' close to x. Since the gradient is already multiplied by ε the difference is higher order and can be dropped.

Putting this together,

$$\frac{T(x',x)}{T(x,x')} = e^{-\frac{\partial E}{\partial x} dx/T} = e^{-dE/T} = e^{-(E(x')-E(x))/T} \quad (99)$$

$$T(x,x')e^{-E(x')/T} = T(x',x)e^{-E(x)/T}, \quad (100)$$

where again terms which are higher than linear order in ε are neglected. Eq. 15 completes the demonstration that the Langevin equation obeys detailed balance. Hence by the general theorems concerning Monte Carlo, it will generate configurations of the system according to the Boltzmann distribution.

A final comment: You will notice that the prefactors of the random force and the force which depends on E in the Langevin equation are not independent. This is an example of the fluctuation-dissipation theorem.

Exercise B15: In proving detailed balance higher order terms in ε were thrown away. Consider the energy $E(x) = \frac{1}{2}kx^2$ and go through the Langevin analysis retaining terms to all orders in ε. Show that detailed balance is obeyed with a k shifted by order ε. Will $\langle x^2 \rangle$ be overestimated or underestimated?

Exercise B16: Write a computer code to implement the Langevin equation for $E(x) = \frac{1}{2}kx^2$. Verify your results from Exercise B15.

Error Analysis in Monte Carlo

This final section is intended to review error analysis in Monte Carlo, and illustrate it with the very simple example of a simulation of $E(x) = \frac{1}{2}kx^2$. Begin with the definitions,

$$\langle x \rangle = \frac{1}{N} \sum_{i=1}^{N} x_i \quad (101)$$

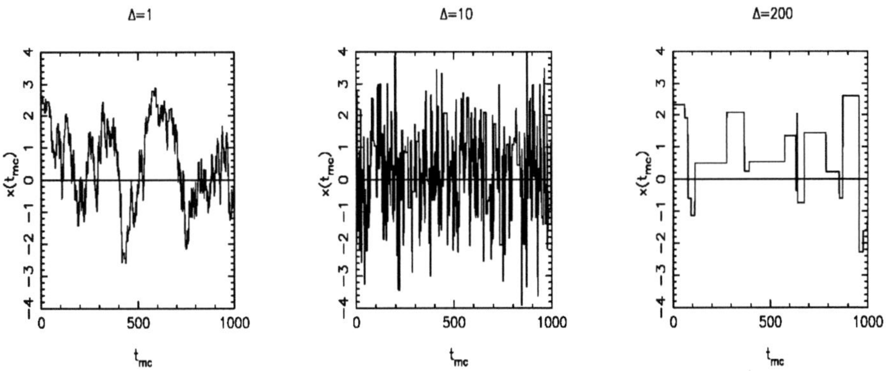

FIGURE 41. First 1000 steps in Monte Carlo time history of x for three different step sizes $\Delta = 1, 10, 200$. (Acceptance rates=0.70, 0.35, 0.02).

$$\langle x^2 \rangle = \frac{1}{N} \sum_{i=1}^{N} x_i^2 \tag{102}$$

$$\sigma = \sqrt{\frac{\langle x^2 \rangle - \langle x \rangle^2}{N-1}}. \tag{103}$$

Here x_i is the ith measured value of x, and N is the number of measurements. The definitions of $\langle x \rangle$ and $\langle x^2 \rangle$ are unambiguous. The entire content of this note is to clarify the proper denominator of the definition of σ. Specifically, the formula for σ assumes that the N measurements are all independent. Since successive values of x are generated from each other, this is never true. The x values are more and more related the less time one waits between measurements.

To quantify the correlations among successive x values, define the autocorrelation function,

$$c(l) = \frac{1}{N-l} \sum_{i=1}^{N-l} y_i y_{i+l} \tag{104}$$

Here $y(i) = x(i) - \langle x \rangle$ measures the fluctuation of x about its average value. $c(l)$ measures whether those fluctuations are related for x values l measurements apart. Saying x_i and x_{i+l} are independent means whether x_{i+l} is above or below $\langle x \rangle$ (the sign of y_{i+l}) is unrelated to the sign of y_i. If that is the case, $c(l) = 0$ (to within errors). Clearly $c(0)$ is never zero. In fact, $c(0) = \sigma^2$. It is conventional to redefine $c(0) \rightarrow c(0)/\sigma^2$ so that $c(0) = 1$.

Let's begin by looking at actual time histories of x. Choose $k = 1$ and $T = 2$ so that $\langle x^2 \rangle = 2$. The step size for suggested changes in x is Δ. Measure every Monte Carlo step and ran for $N = 400000$ sweeps. Here of course since $\langle x \rangle = 0$, $y_i = x_i$. The Monte Carlo time histories are given in Figure 1. That the data are correlated is immediately evident. If a value of x_i is positive, its successors tend to be positive and similarly if it is negative. The dependence on the step size Δ is easy to interpret.

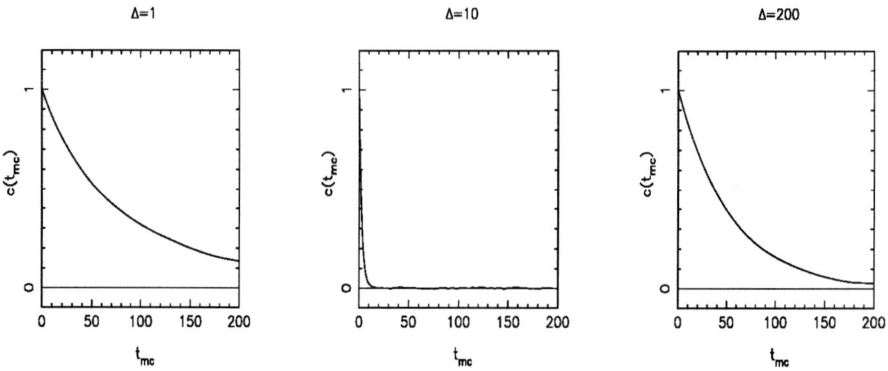

FIGURE 42. Autocorrelation functions for the complete data sets (400,000 steps) as in figure 1. (Step sizes $\Delta = 1, 10, 200$, Acceptance rates=0.70, 0.35, 0.02).

If Δ is small you do not suggest much of a change, and successive values of x are highly correlated. Likewise, if Δ is large, most suggested Monte Carlo moves take you out of the part of phase space of low energy and are rejected. (This results in the long flat regions of the evolution of x.)

The plots of $c(l)$ resulting from the same data are given in Figure 2. $c(l)$ has a characteristic decaying exponential form. Define the correlation time τ to be the point when $c(l = \tau) = e^{-1}$ and say that measurements of x are independent when l exceeds τ. (Strictly speaking, one wants c to go to zero, but $c(\tau) = e^{-1}$ is an acceptable criterion.) Notice you can almost guess the values of τ given by Figure 2 directly from the time histories of Figure 1.

As mentioned earlier, in generating the above results x was measured at every Monte Carlo step. What happens if one instead measures only every mth step? Define $c_m(l)$ to be the correlation function when measurements are only every mth Monte Carlo step. It is easy to convince yourself that $c_m(l) = c_1(ml)$, so the correlation function rescales in a trivially fashion. The point is that if one choose $m > \tau$, then the measurements all become independent.

- *So one way to ensure the value for the error bar σ is correct is to make sure measurements are separated by a waiting time $m > \tau$.*

This approach has the advantage that one does not waste time making measurements when the measurements are not independent.

An alternate (and equivalent) approach to getting the correct σ is by "rebinning" the data. Take a file containing the complete time history of a measurement, for example the data for x which is partially shown in Figure 1. Choose a "bin size" M, and average the data for x over each of the $L = N/M$ bins (remember N = total number of measurements) to create L "binned measurements" m_j.

$$m_j = \frac{1}{M} \sum_{i=M*(j-1)+1}^{M*j} x_i. \tag{105}$$

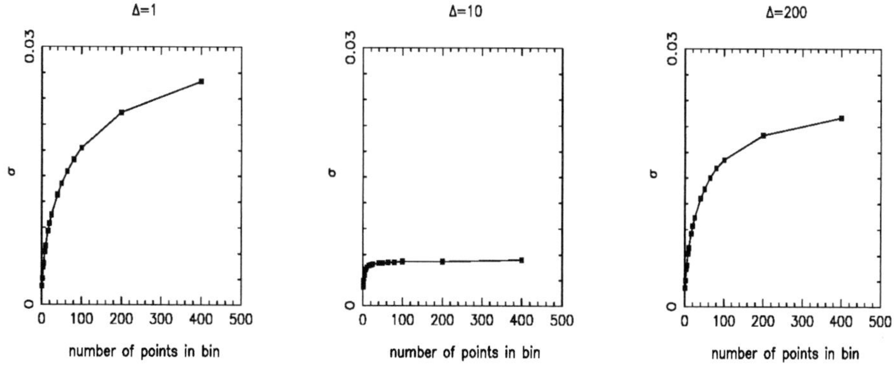

FIGURE 43. Error bars for different bin sizes M. Data is that of Figures 1,2: step sizes $\Delta = 1, 10, 200$. (Acceptance rates=0.70, 0.35, 0.02).

Treat these L values for m as your independent measurements. As seen in Equation 5, the values for m are already averages over M values of x. Define averages and error bars as in Equation 1, with L replacing N in the normalizations $1/N$ and $1/\sqrt{N-1}$. The average $\langle x \rangle$ is independent of M since all one is doing is reordering a linear sum. The average $\langle x^2 \rangle$ is however altered, and hence so is the error bar σ. Figure 3 shows values for σ as a function of the number of x values in each bin, M.

What is the interpretation of Figure 3? Consider $M = 1$. In this case only one value of x is put in each bin, so that in calculating σ it is assumed all x are independent. The error bar σ thus obtained is too small. As M becomes larger, more x values are put in each bin, the number of bins (independent measurements) decreases, and σ increases. Eventually σ goes to an asymptotic value which gives the correct error bar.

Why does σ not increase indefinitely as M increases? You might expect it to, since the denominator $\sqrt{L-1}$ is getting smaller and smaller. The answer is that as more measurements are put in each bin, the different bins fluctuate less and less. The numerator which measures these fluctuations decreases in exact compensation to the denominator. (However, to reiterate, initially for M small when you put more measurements in the bins the new values are not independent and so the numerator does *not* decrease.)

• *So a second way to ensure the value for the error bar σ is correct is to consider different binnings of the data, and use the value obtained asymptotically as each bin contains a lot of data.*

How does one see this result is consistent with the correlation function analysis? There are two checks. The first is to see that the value for M at which σ starts to flatten out should be roughly the same as the value of τ for which $c(l)$ gets small. Second, one can compare the values of σ_1 at $M = 1$, where one assumes all the x are independent, with the asymptotic value σ_∞. The claim is that these should be related by $\sigma_\infty = \sqrt{\tau}\sigma_1$. You can see this is roughly true: For $\Delta = 1$ one obtains $\sigma_1 = 0.0022$ and $\sigma_\infty = 0.030$ from Figure 3. If you assume all the measurements are independent, you underestimate σ by more than an order of magnitude. Meanwhile, from figure 2, $\tau \approx 85$, and hence $\sqrt{\tau}$ similarly reflects this order of magnitude correction factor.

The acceptance rate provides a rough guide to the choice of a good step size. If the acceptance rate is too much greater than 0.5, then one is likely in the limit of Figures (1–3)a where the correlation time is unnecessarily large due to small suggested moves. Likewise, if the acceptance rate is too much less than 0.50, then one is likely in the limit of Figures (1–3)c where the correlation time is unnecessarily large due to multiple rejections.

Careful error analysis involves the procedure outlined above, calculation of correlation times, etc. However, in practice one often has (or develops) a good intuition about these things. A very simple approach to error bars, then is simply to bin the data from your run into a small number of bins, say ten (ie. $M = N/10$) and leave it at that. This strategy assumes that you were doing a reasonably long run, so that $N/10 > \tau$. My thinking is that if this is violated, you are in more serious trouble than just getting the wrong error bars: you will have less than 10 independent measurements (and perhaps have not even properly equilibrated) so it is likely your expectation values themselves are wrong. For the particular example considered here, the reported values from the simplistic approach for $\langle x^2 \rangle$ and σ were $1.973 \pm 0.051, 1.993 \pm 0.010, 1.995 \pm 0.039$ for $\Delta = 1, 10, 200$ respectively. These error bars should be the same as the asymptotic values of σ in Fig. 3.

Does it matter which measurement you look at? Here x_i has been the focus of the analysis. Would it matter x_i^2 or some other measurement had been examined? For this simple problem, no. In more complicated simulations it is often important to calculate correlation times separately for measurements of "local" quantities (observables for degrees of freedom that are close together spatially) and "global" quantities (observables for degrees of freedom that are far apart spatially). The spatial separation of the different degrees of freedom in an observable can affect the autocorrelation time. In particular, observables containing quantities which are widely spaced generally have longer correlation times.

More generally, the size of the system you are simulating (and hence the spatial separations of the degreees of freedom) can greatly affect the correlation time, especially in the vicinity of a phase transition. Just as physical quantities like the specific heat, susceptibility etc can diverge at a critical point, the correlation time can diverge too as the system size increases near a critical point. This of course makes Monte Carlo very difficult. There is a big literature on solving this problem.

APPENDIX C: DETERMINANT QUANTUM MONTE CARLO IN DETAIL

A Useful Analogy: Multidimensional Gaussian Integration

The equations involved in determinant QMC bear many similarities with multidimensional Gaussian integrals. Reviewing these identities will help provide an intuitive feel for the formulae of determinant QMC, within a familiar context.

The generalization of the one dimensional Gaussian integral,

$$\int_{-\infty}^{+\infty} dx e^{-ax^2} = \frac{\sqrt{\pi}}{a}, \tag{106}$$

to many dimensions is,

$$Z = \int_{-\infty}^{+\infty} \int_{-\infty}^{+\infty} \cdots \int_{-\infty}^{+\infty} dx_1 dx_2 \ldots dx_N \, e^{-\vec{x} A \vec{x}^T} = \frac{\pi^{n/2}}{\sqrt{\det A}}. \tag{107}$$

Here $\vec{x}$ is an N dimensional vector of real numbers and A is a real, symmetric, N dimensional matrix. The notation Z has been used for the integral to emphasize that it would be the partition function for a set of classical variables whose action is given by the quadratic form $\vec{x}A\vec{x}^T$.

These integrals can also be done when the integrand includes factors of x_i.

$$\langle x_i x_j \rangle = Z^{-1} \int_{-\infty}^{+\infty} \int_{-\infty}^{+\infty} \cdots \int_{-\infty}^{+\infty} dx_1 dx_2 \ldots dx_N\, x_i x_j\, e^{-\vec{x}A\vec{x}^T} = \frac{1}{2}[A^{-1}]_{ij} \tag{108}$$

Again, the notation $\langle x_i x_j \rangle$ emphasizes a possible statistical mechanical interpretation of the ratio of integrals.

Further factors of x_i in the integrand generate expressions like,

$$\begin{aligned} \langle x_i x_j x_k x_l \rangle &= Z^{-1} \int_{-\infty}^{+\infty} \int_{-\infty}^{+\infty} \cdots \int_{-\infty}^{+\infty} dx_1 dx_2 \ldots dx_N x_i x_j x_k x_l e^{-\vec{x}A\vec{x}^T} \\ &= \frac{1}{4}([A^{-1}]_{ij}[A^{-1}]_{kl} + [A^{-1}]_{ik}[A^{-1}]_{jl} + [A^{-1}]_{il}[A^{-1}]_{jk}). \end{aligned} \tag{109}$$

These are similar in form to 'Wick's Theorem', which tells us that contractions of products of many fermion operators can be expressed as sums of products of contractions taken two operators at a time, in all possible permutations. While it is possible to do these integrals with arbitrary polynomials as part of the integrand, they cannot be done when a quartic term appears in the exponential. The analogies of these various statements for traces over fermion Hamiltonians will be exhibited shortly.

Basic Formalism of Determinant QMC

In solving the Hubbard model it is necessary to evaluate expressions like

$$\begin{aligned} \langle \hat{A} \rangle &= Z^{-1}\,\mathrm{Tr}[\hat{A} e^{-\beta \hat{H}}] \\ Z &= \mathrm{Tr}\, e^{-\beta \hat{H}} \end{aligned} \tag{110}$$

The "Tr" is a trace over the 4^N dimensional Hilbert space, where N is the number of sites.

In analogy with multidimensional Gaussian integration, we can do such traces if they are over *quadratic* forms of fermion operators. Suppose

$$\hat{H} = \sum_{ij} c_i^\dagger h_{ij} c_j \tag{111}$$

Here h is an NxN matrix. The identity is,

$$Z = \mathrm{Tr}\, e^{-\beta \hat{H}} = \det[I + e^{-\beta h}]. \tag{112}$$

Note that while the original "Tr" is over a quantum mechanical 4^N dimensional Hilbert space, the "det" is a usual determinant of NxN matrices. "I" is the N dimensional identity matrix and "h" is the matrix of *numbers* entering the definition of $\hat{H}$. It is worth emphasizing that because one takes the trace over the full 4^N dimensional Hilbert space, states of all occupation numbers are included. The determinant QMC method, as formulated here, works in the grand canonical ensemble. Particle density is controlled by changing the chemical potential.

It is trivial to check that Eq. 113 holds for a single fermion degree of freedom, with Hamiltonian $\hat{H} = \varepsilon c^\dagger c$. There are two states in the Hilbert space and

$$Z = \langle 0 | e^{-\beta \varepsilon c^\dagger c} | 0 \rangle + \langle 1 | e^{-\beta \varepsilon c^\dagger c} | 1 \rangle = 1 + e^{-\beta \varepsilon}. \tag{113}$$

More generally (e.g. for more than one fermion degree of freedom) Eq. 113 can be verified by going to the basis where h is diagonal. The equations can also be derived by employing the techniques of Grassman integration.

There is a more general identity. If one has a *set* of quadratic Hamiltonians $l = 1, 2, \ldots L$

$$\hat{H}(l) = \sum_{ij} c_i^\dagger h_{ij}(l) c_j \tag{114}$$

then,

$$Z = \mathrm{Tr}\,[e^{-\Delta\tau \hat{H}(1)} e^{-\Delta\tau \hat{H}(2)} \cdots e^{-\Delta\tau \hat{H}(L)}] = \det[I + e^{-\Delta\tau h(1)} e^{-\Delta\tau h(2)} \cdots e^{-\Delta\tau h(L)}]. \tag{115}$$

Here the prefactor in the exponential has been changed from β to $\Delta\tau$ for reasons which will soon be clear. It is also true that,

$$\begin{aligned} G_{ij} = \langle c_{i\sigma} c_{j\sigma}^\dagger \rangle &= Z^{-1} \mathrm{Tr}\,[c_{i\sigma} c_{j\sigma}^\dagger \, e^{-\Delta\tau H(1)} e^{-\Delta\tau H(2)} \cdots e^{-\Delta\tau H(L)}] \\ &= [I + e^{-\Delta\tau h(1)} e^{-\Delta\tau h(2)} \cdots e^{-\Delta\tau h(L)}]_{ij}^{-1}. \end{aligned} \tag{116}$$

The "fermions Greens function" is just an appropriate matrix element of the inverse of the NxN matrix whose determinant gives the partition function.

The above formulae describe how to perform traces over quadratic forms of fermion degrees of freedom. Unfortunately, the Hubbard Hamiltonian has an interaction term $U n_{i\uparrow} n_{i\downarrow} = U c_{i\uparrow}^\dagger c_{i\uparrow} c_{i\downarrow}^\dagger c_{i\downarrow}$ which is quartic in the fermion operators. To handle such terms, one employs the (discrete) Hubbard–Stratonovich transformation,

$$e^{-U\Delta\tau(n_\uparrow - \frac{1}{2})(n_\downarrow - \frac{1}{2})} = \frac{1}{2} e^{-\frac{U\Delta\tau}{4}} \sum_s e^{\lambda s (n_\uparrow - n_\downarrow)} \tag{117}$$

Here $\cosh\lambda = e^{U\Delta\tau/2}$, and s is an Ising variable which can take on the two values $S = \pm 1$. This identity can be verified by explicitly enumerating the 4 possible choices for $n_\uparrow, n_\downarrow$.

Now divide $\beta = L\Delta\tau$ and employ the Trotter decomposition. This allows us to isolate different pieces of the Hamiltonian. Write $\hat{H} = \hat{K} + \hat{V}$ where $\hat{K}$ contains all the one–body pieces and $\hat{V}$ the on–site Hubbard interaction. Then,

$$Z = \mathrm{Tr}\, e^{-\beta \hat{H}} = \mathrm{Tr}\,[e^{-\Delta\tau \hat{H}}]^L \approx \mathrm{Tr}\,[e^{-\Delta\tau \hat{K}} e^{-\Delta\tau \hat{V}} e^{-\Delta\tau \hat{K}} e^{-\Delta\tau \hat{V}} \cdots]. \tag{118}$$

The final expression is only approximate since $\hat{K}$ and $\hat{V}$ do not commute. However, the approximation becomes better and better as L increases ($\Delta\tau$ decreases). As mentioned before, the errors should be pretty small if $t\,U\,(\Delta\tau)^2 \approx 1/10$.

The $e^{-\Delta\tau\hat{K}}$ are quadratic in the fermion operators. For each factor of the L terms $e^{-\Delta\tau\hat{V}}$ above, introduce N Hubbard–Stratonovich fields, one for each of the spatial sites where there is an on–site interaction to decouple. The Hubbard–Stratonovich field $s(i,l)$ therefore has two indices, space i and imaginary–time l. Now the $e^{-\Delta\tau\hat{V}(l)}$ are also quadratic in the fermion operators. The argument l on $\hat{V}$ emphasizes that while the $\hat{K}$ are all identical, the $\hat{V}(l)$ contain different Hubbard–Stratonovich fields on the different imaginary time slices.

Applying Eqs. 22-23 allows the analytic evaluation of the trace,

$$Z = \sum_{s(il)} \det M_{\uparrow} \det M_{\downarrow}, \tag{119}$$

with,

$$M_{\sigma} = I + e^{-k} e^{-v_{\sigma}(1)} e^{-k} e^{-v_{\sigma}(2)} \cdots e^{-k} e^{-v_{\sigma}(L)}. \tag{120}$$

A determinant is obtained for each of the two spin species. The quantum partition function has now been expressed to a *classical* monte carlo problem: It is necessary to sum over the possible configurations of the real, classical, variables $s(i,l)$ with the "Boltzmann weight" which is the product of the two fermion determinants. Note that as in world-line QMC, the classical variable to be summed over has an additional index l labeling imaginary time.

Eqs. 2-4 can now be understood as coming from applying the general operator identity of Eqs. 22-23 to the specific problem of evaluating Eq. 26, with the interaction operators $\hat{V}$ made into quadratic forms by using Eq. 25.

The algorithm, as stated, scales in CPU time as N^4L. The reason is that re–evaluating the determinant of M' takes N^3 operations, and one must do that NL times to sweep through all the Hubbard–Stratonovich variables (if, as is typically done, just one at a time is changes). This scaling can be reduced to N^3L. (In what follows the spin indices will be dropped.) The idea is to write $M' = M + dM$ and the ratio of determinants as,

$$\det M'/\mathrm{detM} = \det(M^{-1}M') = \det(M^{-1}(M+dM)) = \det(I+GdM), \tag{121}$$

with the definition $G = M^{-1}$. It turns out that dM is very simple because when a Hubbard–Stratonovich field is flipped, a single diagonal entry in $v(l)$ changes. Because dM is sparse, the evaluation of $\det(I+GdM)$ takes a cpu time independent of N and L! In fact, a little bit of thought will convince you that Eq. 6 arises from Eq. 31 and the form for dM.

However, one needs $G = M^{-1}$ for this calculation, and once the Hubbard-Stratonovich field change is made, one needs to update G. This updating G does not take N^3 iterations, as one might expect of a matrix inversion, but can be done in only N^2 operations, again as a result of the simplicity of the change dM. The relevant identity which relates the new $G' = (M+dM)^{-1}$ to the old $G = M^{-1}$ is an application of the "Sherman–Morrison" formula given, for example, in Press's 'Numerical Recipes'. If you work through the Sherman-Morrison formula, as applied to the problem, you end up with Eqs. 7-8.

A final comment concerns the need for 'wrapping' Eq. 9. The use of Eq. 31 to derive Eq. 6, and the Serman-Morrison formula to derive Eqs. 7-8 require that the imaginary time slice of the Hubbard-Stratonovich variable being updated be at the end of the product in Eq. 28. The process of wrapping moves the appropriate interaction matrix to the end of the product through a cyclic permutation. That is,

$$[e^{-k}e^{-v_\sigma(L)}]\,[I+e^{-k}e^{-v_\sigma(1)}e^{-k}e^{-v_\sigma(2)}\ldots e^{-k}e^{-v_\sigma(L)}]^{-1}\,[e^{-k}e^{-v_\sigma(L)}]^{-1} \tag{122}$$

$$= [I+e^{-k}e^{-v_\sigma(L)}e^{-k}e^{-v_\sigma(1)}\ldots e^{-k}e^{-v_\sigma(L-1)}]^{-1} \tag{123}$$

Subtleties and "Tricks of the Trade"

While the above formulae allow you to write a "bare–bones" determinant QMC algorithm, there are a number of refinements which are important.
(1.) It is possible to measure correlation functions with non–zero imaginary time separation, but this requires considerably more work. Analytic continuation of such correlations is required to get the dynamical response. That is quite difficult.
(2.) The product of matrices required in constructing M and hence $G = M^{-1}$ (see Eq. 5 and Eq. 28) is numerically unstable at low temperatures and strong couplings. That is, the product has a very high ratio of largest to smallest eigenvalue. Special "stabilization" is required to do the matrix manipulations. While these add to the complexity of the code, they however have no content in the sense that all the above equations are valid, it is just a question of how best to multiply matrices on a machine of finite precision.
(3.) The determinants of the matrices can go negative. This is called the "fermion sign problem." The sign problem does not occur for certain special cases. For example, if U is negative (the "attractive" Hubbard model), the individual determinants can go negative, but the matrices are always equal and hence the determinant appears as a perfect square. This is a consequence of the fact that the appropriate Hubbard–Stratonovich transformation couples S to the charge $n_\uparrow + n_\downarrow$ as opposed to the spin as given in Eq. 25 for the repulsive model. If U is positive but the chemical potential $\mu = U/2$ ("half–filling") one is also okay. The matrices are not identical in this case, but the determinants are nevertheless related by a positive factor, that is, they again have the same sign, so their product is always positive. Some types of randomness are also acceptable. It is okay for the hoppings t and interactions U to depend on the link or site. These statements are demonstrated by various particle–hole transformations on the Hamiltonian.
(4.) Alternate Hubbard–Stratonovich transformations are possible. One can couple more symmetrically to the spin, that is not single out the z component Or, one can couple to pair creation operators. So far, all such alternatives give a worse sign problem than the transformation Eq. 25. These more complicated transformations are needed to do 'Hund's rule terms' in multi-orbital Hubbard models.
(5.) Very similar "ground state" determinant simulations exist which work at $T = 0$ and in the canonical ensemble.

What Determinant QMC Simulations Can Do and Concluding Remarks

The state of the art of determinant QMC simulations, in the absence of a sign problem, are studies of several hundred electrons down to temperatures of $\beta t = 10 - 20$. In terms of temperature and bandwidth, this means T of roughly $1/100$ of the bandwidth $W = 8t$ of the 2-d Hubbard model. This is plenty cold enough to see well developed magnetic correlations. For typical parameters, $t = 1, U = 4$ one chooses $\Delta\tau = 1/8$ so these beta values correspond to roughly $L = 100$, and the simulation involves approximately 10^4 Hubbard–Stratonovich variables.

In cases where one has a sign problem, βt is limited to 4–5. This is, unfortunately, not low enough in temperature to make conclusive statements about certain important problems, perhaps most prominently the question of the existence of long range d−wave superconducting correlations in the Hubbard model away from half-filling.

Determinant QMC is a powerful method for simulating interacting electron Hamiltonians in more than one dimension. One can easily study problems with several hundred particles, an order of magnitude greater than with exact diagonalization, and often large enough to make compelling finite size scaling analysis. The sign problem is a very significant limitation, however. For the repulsive Hubbard model, one can go to temperatures on the order of $W/30$ where W is the bandwidth. For special cases like the attractive Hubbard model or the repulsive model at half-filling, there is no sign problem, and the ground state properties can be obtained.

Algorithm development in determinant QMC currently focusses on applications to DMFT, where the Hubbard-Stratonovich field is allowed to fluctuate only in imaginary time. A number of questions are being actively explored in this field: How does one incorporate more complex (e.g. Hund's rules) interactions into simulation which include multiple orbitals? Can one re-introduce some degree of spatial fluctuations?

APPENDIX D: SUPPLEMENTARY MATERIAL

Particle-Hole Symmetry

The Hubbard Hamiltonian has a fascinating 'particle-hole' symmetry which allows us to relate its properties for different values of the parameters. Particle-Hole symmetry also plays an important role in quantum monte carlo simulations. Consider the introduction of new operators which exchange the role of creation and destruction:

$$d^{\dagger}_{\mathbf{i}\sigma} = (-1)^{\mathbf{i}} c_{\mathbf{i}\sigma} \tag{124}$$

The meaning of the $(-1)^{\mathbf{i}}$ will be explained further below.

Exercise D1: Verify that the number operator for the d particles equals one minus the number operator for c particles: $d^{\dagger}_{\mathbf{i}\sigma} d_{\mathbf{i}\sigma} = 1 - c^{\dagger}_{\mathbf{i}\sigma} c_{\mathbf{i}\sigma}$. What happens to the interaction term in the Hubbard model $U(n_{\uparrow} - \frac{1}{2})(n_{\downarrow} - \frac{1}{2})$ under the particle-hole transformation?

Before seeing what happens to the kinetic energy term under a particle-hole transformation, one can introduce the idea of a bipartite lattice. A bipartite lattice is one which can be divided into two sublattices A and B in such a way that a site in A has neighbors which are all members of B and *vice-versa*. The $(-1)^{\mathbf{i}}$ factor in the particle-hole transformation takes the value -1 on one sublattice and $+1$ on the other.

Exercise D2: Is a one dimensional chain a bipartite lattice? How about a two-dimensional square lattice?

Exercise D3: Is the triangular lattice bipartite? What about the honeycomb lattice?

The following Exercise determines what happens to the kinetic energy term when the particle-hole transformation is performed.

Exercise D4: Verify that the kinetic energy is unchanged under a particle-hole transformation. That is, it takes exactly the same form in terms of the d operators as it did in terms of the c operators. Where does the bipartite nature of the lattice come enter? What role do the $(-1)^{\mathbf{i}}$ factors play?

What has been learned? The exercises above tell us that the Hubbard model, when the interaction term is written in the particle-hole symmetric form, is invariant under particle-hole transformations when $\mu = 0$. The condition $\mu = 0$ is necessary since the number operators which μ multiplies are not invariant but go into one minus themselves. Actually, a more precise statement is that the Hubbard model with a given μ maps into the Hubbard model with the sign of the chemical potential reversed, that is, with μ replaced by $-\mu$. In fact, this implies that the whole phase diagram of the Hubbard model on a bipartite lattice is symmetric about half-filling, as the following Exercises suggest.

Exercise D5: Show that the density of the Hubbard model on a bipartite lattice obeys the relation $\rho(\mu) = 2 - \rho(-\mu)$ by starting with $\rho = \langle n_\uparrow + n_\downarrow \rangle$ and making a particle-hole transformation.

Exercise D6: Show that the local moment of the Hubbard model on a bipartite lattice obeys the relation $\langle m^2 \rangle(\mu) = \langle m^2 \rangle(-\mu)$.

Exercise D7: Look back at you pictures of the density of states $N(E)$ obtained in the Exercises and explain their behavior when reflected about $E = 0$ in terms of particle-hole symmetry.

Relation between the Attractive and Repulsive Hubbard Models

It is also interesting to consider what happens when a particle-hole transformation is performed only on one of the spin species.

Exercise D8: Show that under the transformation,

$$d_{\mathbf{i}\uparrow} = c_{\mathbf{i}\uparrow} \qquad\qquad d_{\mathbf{i}\downarrow} = (-1)^{\mathbf{i}} c^{\dagger}_{\mathbf{i}\downarrow}$$

the sign of the interaction term reverses, while the kinetic energy remains unchanges.

The Hubbard model with $-U$ is called the attractive Hubbard model because a negative value of U represents an attraction between spin up and psin down electrons on the same site. By considering various operators one can show that magnetic order in the $+U$ Hubbard model is related to superconducting and charge order in the $-U$ Hubbard model, so that an understanding of the phases of one model immediately implies considerable information about the other.

Exercise D9: Show that under a particle-hole transformation of just the down spin species, the following operator mappings occur. (Ignore constants.)

$$m_{z,\mathbf{i}} = n_{\mathbf{i}\uparrow} - n_{\mathbf{i}\downarrow} \leftrightarrow n_{\mathbf{i}} = n_{\mathbf{i}\uparrow} + n_{\mathbf{i}\downarrow}$$
$$m_{+,\mathbf{i}} = c^{\dagger}_{\mathbf{i}\uparrow} c_{\mathbf{i}\downarrow} \leftrightarrow c^{\dagger}_{\mathbf{i}\uparrow} c^{\dagger}_{\mathbf{i}\downarrow}$$
$$m_{-,\mathbf{i}} = c^{\dagger}_{\mathbf{i}\downarrow} c_{\mathbf{i}\uparrow} \leftrightarrow c_{\mathbf{i}\downarrow} c_{\mathbf{i}\uparrow}$$

The physical content of these results is that spin correlations along the z axis are interchanged with charge correlations, and spin correlations along the x and y axes (which are compinations of m_+ and m_-) are interchanged with pairing correlations.

Alternates to Determinant QMC

A discussion of the path integral formulation of quantum mechanics, starting with the harmonic oscillator and then moving to quantum spins, bosons, and fermions, with an emphasis on Quantum Monte Carlo, can be found at: http://leopard.ucdavis.edu/rts/resproj6.html by hitting the 'World-Line Quantum Monte Carlo" link (number 88). A shorter discussion which starts immediately with the Heisenberg model can be found at:

http://leopard.ucdavis.edu/rts/boulder.html by hitting the 'Lecture II (pdf file) link.

One of the reasons for the continued interest in the Hubbard model is because of a recently developed approach known as 'dynamical mean field theory' (DMFT). This new technique has allowed for very interesting solutions of the Hubbard model itself, and also, more importantly, has provided a framework for the inclusion of Hubbard-type interactions into density functional theory. A 'popular' introduction to DMFT is available in: while a much more complete technical review is in: "Strongly Correlated Materials: Insights From Dynamical Mean-Field Theory," Physics Today, March, 2004. A. Georges, G. Kotliar, W. Krauth, and M. Rozenberg, Rev. Mod. Phys. **68**, 13 (1996).

TUTORIAL ONE

On the machines in the computer laboratory, you will find three fortran codes:

hubvietri.f (determinant QMC code for $d = 2$ square lattice Hubbard model)
ueq0vietri.f (analytic solution of $U = 0$ Hubbard model, $d = 2$ square lattice)
teq0vietri.f (analytic solution of $t = 0$ Hubbard model)

I apologize that the codes are not very user-friendly and readable (eg well commented). You will also find one input file:

hubvietri.in (input for determinant QMC code)

The inputs for the other codes are simple enough just to enter while you run. The input file hubvietri.in contains comments explaining what all the parameters that you need to know about are. There are other parameters having to do with numerical stability that you should just leave alone. Two important notes: First, the name of the file which holds the output is specified as a character string in the input file (as is explained there). Second, the lattice sizes, both the linear spatial size n and the number of imaginary time slices l which determines β via $\beta = l\Delta\tau$ are set in parameter statements in the fortran code hubvietri.f. That is, they are *not* inputs. Note that when you change the settings for n and l you must change them globally. (They appear 31 times in the code!) An editor with a global replace feature is useful here. (This is one of the ways the code is not user friendly. At some point I will rewrite it so n and l are set globally at one location only.)

To compile the code

```
ifort hubvietri.f
```

(You might want to explore if ifort has any optimization swtiches.)

To run the code

```
a.out < hubvietri.in
```

Some information on the progress of the run will come to the screen. Specifically, every 10 sweeps through the lattice you will be told the accpetance rate for the Monte Carlo moves (should be around 0.5) and the "redo ratio" (should be small, like 0.0001). If you don't want this on your screen, use

```
a.out < hubvietri.in > outjunk
```

<u>TO DO:</u>

Your job is to verify the QMC code and the analytic codes agree.
For example, setting $n = 6, t = 0, U = 4, \mu = 1$ and $\Delta\tau = 0.125, l = 4$ so that $\beta = l\Delta\tau = 0.5$, I found that the QMC code run with nwarm=500 and npass=5000 gave: $\langle n_\uparrow \rangle = 0.5695 \pm 0.0009$ and $\langle n_\downarrow \rangle = 0.5659 \pm 0.0008$. The analytic answer is 0.56775. Likewise, for the QMC, $\langle n_\uparrow n_\downarrow \rangle = 0.2143 \pm 0.0002$, with an analytic answer 0.21435. Of course, since $t = 0$, the code should give the same answer (to within error bars) for all n. When $n = 6$ it was just running 36 single site simulations in parallel.
Similarly, setting $n = 6, t = 1, U = 0, \mu = -1$ and $\Delta\tau = 0.125, l = 4$ so that $\beta = l\Delta\tau = 0.5$, I found that the QMC code run with nwarm=10 and npass=100 gave: $\langle n_\uparrow \rangle = \langle n_\downarrow \rangle = 0.39906 \pm 0.00000$. The analytic answer is $\langle n_\downarrow \rangle = 0.39903$. Comments: (1) The QMC code should give the *exact* answers whenever $U = 0$ since it traces over the fermions analytically and there is no coupling to the Hubbard-Stratonovich field. That's why one could do such a short run (nwarm=10, npass=100.) The reason one does not just set npass=1, which should work in principle, is the way the code generates error bars. It assumes a certain minimal number of samples are taken. (2) The small disagreement between the codes is due to the Trotter error in the checkerboard breakup of the kinetic energy. (3) You will need to multiply the energy written by ueq0vietri.f by two, because it is does a single spin species.

Exercise 1:

What do hubvietri.f and ueq0vietri.f give for the Green's function?

Exercise 2:

Can you show that when you reduce $\Delta\tau$ (at fixed β) that the agreement between the two codes becomes perfect? The Trotter errors associated with the checkerboard breakup can thus be eliminated.

Exercise 3:

Show (numerically) that for the special case of a 4x4 lattice there are no Trotter errors with the checkerboard breakup. That is, verify that the two codes agree perfectly in this case. Can you prove the resulut analytically?

Exercise 4:

Check that the results in hubvietri.f for $G(0,1)$ are consistent with the results reported by the code for the kinetic energy. Although you do this check here at $U = 0$, the relation you will derive is perfectly general (true for any t, U, μ, β).

TUTORIAL ONE SOLUTIONS

Exercise 1:

What do hubvietri.f and ueq0vietri.f give for the Green's function?

hubvietri.f	G(0,1)=-0.10272
ueq0vietri.f	G(0,1)=-0.10246
hubvietri.f	G(0,2)=-0.00410
ueq0vietri.f	G(0,2)=-0.00411

Exercise 2:

Can you show that when you reduce $\Delta\tau$ (at fixed β) that the agreement between the two codes becomes perfect? The Trotter errors associated with the checkerboard breakup can thus be eliminated.
hubvietri.f $L = 4,\ \Delta\tau = 0.12500$ $G(0,1) = -0.10272$
hubvietri.f $L = 8,\ \Delta\tau = 0.06250$ $G(0,1) = -0.10252$
hubvietri.f $L = 16, \Delta\tau = 0.03125$ $G(0,1) = -0.10247$
teq0vietri.f $G(0,1) = -0.10246$

Exercise 3:

Show (numerically) that for the special case of a 4x4 lattice there are no Trotter errors with the checkerboard breakup. That is, verify that the two codes agree perfectly in this case. Can you prove the result analytically?
For a 4x4 lattice with $t = 1, U = 0, \mu = -1$ I find
hubvietri.f $L = 4,\ \Delta\tau = 0.12500$ $G(0,1) = -0.10136$
hubvietri.f $L = 8,\ \Delta\tau = 0.06250$ $G(0,1) = -0.10136$
hubvietri.f $L = 16, \Delta\tau = 0.03125$ $G(0,1) = -0.10136$
teq0vietri.f $G(0,1) = -0.10136$
I will leave the analytic proof of the vanishing to you. I suspect it might be related to the fact that the 4x4 lattice is topologically equivalent to the 2x2x2x2 lattice (n=2 hypercube in 4 dimensions). See also Tutorial 2 Solutions.

Exercise 4:

Check that the results in hubvietri.f for $G(0,1)$ are consistent with the results reported by the code for the kinetic energy. The code reports $G(0,1) = -0.10272$ and the kinetic energy as $KE = -0.82172$. The Green's function $\langle c_i c_j^\dagger \rangle$ contains only one of the two Hermitian conjugate terms in the kinetic energy. There are also two spatial directions (x,y) in the kinetic energy. Finally, there are two spin species. Thus one must multiply the near-neighbor Greens function by 8 to bring it into agreement with the kinetic energy. This relation is satisfied.

TUTORIAL TWO

Exercise 1:

Look for antiferromagnetism in the half-filled Hubbard model. Use $n = 4$ (a 4x4 lattice), $t = 1, U = 4, \mu = 0$, and $\Delta\tau = 0.125$. Run with $l = 4, 8, 16, 32, 48, 64$ so that so that $\beta = l\Delta\tau = 0.5, 1, 2, 4, 6, 8$. Use nwarm=500 and npass=5000. Look at the values for
zz Spin correlation function
xx Spin correlation function
The integers in the first two columns are the x and y separations between the two spins in the correlation function.
Questions:
[1] Why are only six values listed? Aren't there 16 possible separations?
[2] Is the pattern of signs in your low T data consistent with antiferromagnetism?
[3] Why is the $(0,0)$ correlation function enhanced over its $T = \infty$ value of 0.500 even at the highest temperatures, $T = 2$ $(\beta = 0.5)$ while the other separations only start to build up at much lower T?

[4] It looks like the correlation functions for (0,2) and (1,1) separations are the same to within error bars. Yet these are not the same separation in space. Or are they?!

Exercise 2:

If the computers are fast enough, try to redo Exercise 1 with $n = 6$ (a 6x6 lattice).
[1] If you compare spin correlations at the same separation and same temperature for different lattice sizes, what happens? Why?

Exercise 3:

Look for the Mott plateau and encounter the sign problem! Use $n = 4$ (a 4x4 lattice), $t = 1, U = 4$, and $\Delta\tau = 0.125$. Run with $l = 32, 48, 64$ so that so that $\beta = l\Delta\tau = 4, 6, 8$. Sweep chemical potential $\mu = 0.0, -0.2, -0.4, -0.6, -0.8, -1.0, -1.2, -1.4$ for each l. Use nwarm=500 and npass=5000.
[1] Why don't you need to get data at $\mu > 0$? Did you really need to run $\mu = 0$?
[2] Make a plot of density versus μ for each β. Do you see a Mott plateau?
[3] Make a plot of the error bars in the density as a function of the density. Also make a plot of the average sign versus density. Do these help you understand my claim that $\beta \approx 6$ is the temperature limit in these simulations?
[4] It looks like some sort of plateau might also be developing around $\rho = 0.6$. Do you have any idea where that might come from?

TUTORIAL TWO SOLUTIONS

Exercise 1:

[1] Why are only six values listed? Aren't there 16 possible separations?
On a 4x4 lattice with periodic boundary conditions, there are only six inequivalent lattice separations. The separations (0,1) and (1,0) are obviously equal by the x/y symmetry of the lattice. The separations (0,3) and (0,1) are equivalent through use of the periodic boundary conditions. One thus has to report correlation functions only for a small "wedge" of possible separations. The same is true in momentum space. The code automatically averages over the equivalent separations, just as it averages over the initial point j in any correlation function $\langle A(j+l)A(j)\rangle$.
[2] Is the pattern of signs in your low T data consistent with antiferromagnetism?
You should find the spin correlations are negative for separations (0,1), (2,1) and positive for separations (0,0), (0,2), (1,1), and (2,2). It is easy to see that in the former case the spins are on different sublattices, while in the latter case they are on the same sublattice. Thus the signs are indeed consistent with AF order.
[3] Why is the $(0,0)$ correlation function enhanced over its $T = \infty$ value of 0.500 even at the highest temperatures, $T = 2$ $(\beta = 0.5)$ while the other separations only start to build up at much lower T?
The (0,0) correlation function is the local moment, the correlation of a spin with itself. The energy scale of moment formation is the repulsion U, which suppresses double occupancy. On the other hand, the energy scale for spin order is $J = 4t^2/U$, a factor of four smaller.
[4] It looks like the correlation functions for (0,2) and (1,1) separations are the same to within error bars. Yet these are not the same separation in space. Or are they?!
This is not a coincidence. If you draw the lattice connections of a 4x4 square lattice with periodic boundary conditions, and those of a 2x2x2x2 (four dimensional hypercube of linear dimension 2 sites), you can see that all the lattice connections are equivalent.

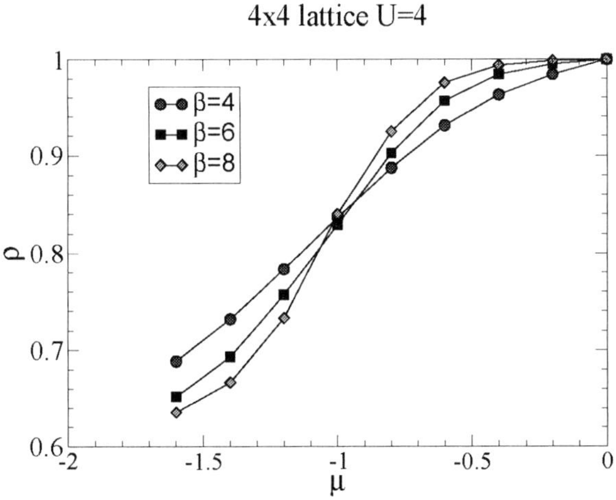

FIGURE 44. Density ρ vs chemical potential μ. As the temperature is lowered, the curve becomes increasingly flat near $\mu = 0, \rho = 1$.

On the 2x2x2x2 hypercube you can see that the (0,2) and (1,1) separations on the 4x4 lattice are identical. (This especially high levels of artificial symmetry is a danger in interpreting data on 4x4 lattices.)

__Exercise 2:__

[1] If you compare spin correlations at the same separation and same temperature for different lattice sizes, what happens? Why?
The spin correlations on the smaller lattice are larger, when all other parameters are held fixed. The reason is that on small lattices the periodic boundary conditions provide additional paths connecting sites, enhancing their correlations. I am not sure of this, but perhaps in a frustrated system things could be different. For example if you ran the code on a 3x3 lattice with periodic boundary conditions, spin correlations might be reduced over those of larger lattices.

__Exercise 3:__

[1] Why don't you need to get data at $\mu > 0$? Did you really need to run $\mu = 0$?
From particle-hole symmetry it is clear that $\langle n \rangle(\mu) = 2 - \langle n \rangle(-\mu)$. This immediately also tells us that $\langle n \rangle(\mu = 0) = 1$, so it is not necessary to run that value.
[2] Make a plot of density versus μ for each β. Do you see a Mott plateau?
See attached figure and caption.
[3] Make a plot of the error bars in the density as a function of the density. Also make a plot of the average sign versus density. Do these help you understand my claim that $\beta \approx 6$ is the temperature limit in these simulations?
See attached figure and caption.
[4] It looks like some sort of plateau might also be developing around $\rho = 0.6$. Do you have any idea where that might come from?
This is a finite size effect. $\rho = 0.625$ is a special 'shell' filling. As the $U = 0$ Fermi

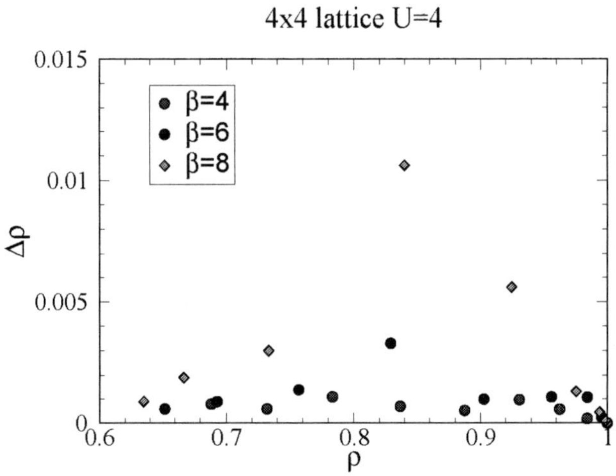

FIGURE 45. Error bars in the density as a function of the density for different temperatures. For $\beta = 8$ the error bar in the density is almost 2 percent of the density. While we can reduce the error bars by running longer (four times as long for a factor of two reduction in error bars), we also find the error bars at fixed run length gro exponentially with β. The origin is the sign problem. (See next plot.)

surface expands from the origin, it encounters four new k points at $(\pm\pi/2,0)$ and $(0,\pm\pi/2)$, and the filling jumps from $2/16$ (just the $k = (0,0)$ point occupied by the two spin species) to $10/16 = 0.625$. $U = 4$ is not sufficiently strong coupling to wipe out this vestige of the discrete k points. Obviously as the lattice size increases, such effects become less evident. However, there is an important lesson here: weak couplings are often more subject to finite size effects that stronger ones.

TUTORIAL THREE

Computational:

Exercise 1:

Tutorial 2 demonstrated evidence for AF spin correlations in the half-filled Hubbard model. Here the goal is to use simple finite size scaling, to demonstrate that there is in fact *long range* AF order. [WARNING: Trying to get this data in just a one hour training section doesn't enable us to do a very good job. The error bars on the data will be a bit too big, and the largest lattice size will be 8x8. However, if you are willing to run for a couple of days (eg after you return home), you can both beat the error bars down and also do 10x10 and even 12x12 lattices.]

Run hubvietri.f for $t = 1, U = 4, \mu = 0, \beta = 8$, using ($L = 64$ and $\Delta\tau = 0.125$. Run length should be nwarm=500 and npass=5000. Do $n = 4,6,8$ (lattices 4x4, 6x6, and 8x8). The 8x8 lattice took 45 minutes on my laptop to complete. Record the spin-spin correlations at maximal separation, ie the last value $(n/2,n/2)$ listed in

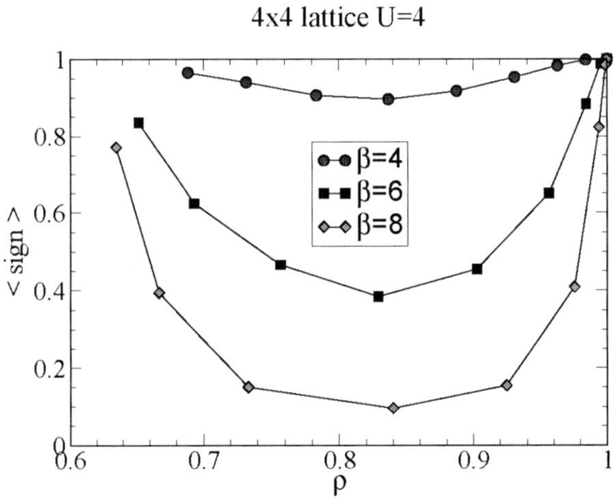

FIGURE 46. The average sign as a function of density for different inverse tempertaures β. The sign begins to become quite small between $\beta = 6$ and $\beta = 8$.

zz Spin correlation function
xx Spin correlation function

Average these two values. Also get the AF structure factors. They are called 'AF correlation function(xx)' and 'AF correlation function(zz)'. (These are not good names! One should label them the 'structure factor'!) Average them, and also divide by the volume n^2. Plot these two averaged quantities versus $1/n$.

[1] Are the extrapolations to $1/n = 0$ (that is, $n = \infty$) nonzero?

[2] Do the extrapolations agree?

[3] Again, the data will not be so great. I suggest if you are serious, try running the code with npass=20000 and also do 10x10 and 12x12 lattices. Actually, because the code can get stuck with the spin order in a particular direction, it is best to run npass=5000 with ten or so random number seeds and average the results.

Analytic:

Try some of the analytic exercises in the notes and the morning meetings. E.g.

Exercise 1:

[1] Get the dispersion relation $E(k)$ for the triangular lattice Hubbard model at $U = 0$. Does it have a van-Hove singularity? Is it symmetric about $E = 0$?

Exercise 2:

Compute $A(\omega)$ for the $t = 0$ Hubbard model $H = U(n_\uparrow - \frac{1}{2})(n_\downarrow - \frac{1}{2}) + \varepsilon(n_\uparrow + n_\downarrow)$

Exercise 3:

Get the density of states for the honeycomb lattice and show it is a semi-metal.

Exercise 4:

Solve the four site $J_1 - J_2$ Heisenberg Hamiltonian. Show you get a level crossing.

Exercise 5:

Work out the Hubbard-Stratonovich transformation for the $U < 0$ Hubbard model. Why is there no sign problem in determinant QMC simulations?
...etc...

Note on Bibliography: This list of references is not intended to acknowledge all the papers in the field. Rather it is somewhat idiosyncratic, reflecting mainly papers that connected directly to the material of section V. These papers themselves contain a more formal and complete review of the literature for those who desire it.

REFERENCES

1. "Monte Carlo calculations of coupled boson-fermion systems. I," R. Blankenbecler, D.J. Scalapino, and R.L. Sugar, Phys. Rev. **D24**, 2278 (1981).
2. "Disordered Electronic Systems," P.A. Lee and T.V. Ramakrishnan, Rev. Mod. Phys, **57**, 287 (1985); "The Anderson-Mott transition," D. Belitz and T.R. Kirkpatrick, Rev. Mod. Phys, **66**, 261 (1994).
3. "Metallic behavior and related phenomena in two dimensions," E. Abrahams, S. V. Kravchenko, and M. P. Sarachik, Rev. Mod. Phys. **73**, 251 (2001); "Metal-insulator transition in two dimensional electron systems," S.V. Kravchenko and M.P. Sarachik, Rep. Prog. Phys. **67**, 1 (2004).
4. "Lecture Notes on Electron Correlation and Magnetism," P. Fazekas, World Scientific Series in Modern Condensed Matter Physics, Vol. 5, (1999).
5. E.Y. Loh, Jr., J.E. Gubernatis, D.J. Scalapino, R.L. Sugar, S.R. White, and R.T. Scalettar, in *Interacting Electrons in Reduced Dimensions*, ed. by D. Baeriswyl and D.K. Campbell, Plenum, New York (1989).
6. "A Numerical Study of the Two-D Hubbard Model with Repulsive Coulomb Interaction," S.R. White, D.J. Scalapino, R.L. Sugar, E.Y. Loh, Jr., J.E. Gubernatis, and R.T. Scalettar, Phys. Rev. **B40**, 506 (1989).
7. E.Y. Loh, Jr. and J.E. Gubernatis, in *Electronic Phase Transitions*, ed. by W. Hanke and Yu. V. Kopaev, North-Holland, Amsterdam (1992).
8. www.ccmr.cornell.edu/ cyrus/book/muramatsu.ps.gz (This includes also a discussion of the 'world-line' method for the Hubbard Hamiltonian.)
9. "Scaling Theory of Localization– Absence of Quantum Diffusion in 2 Dimensions," E. Abrahams, P.W. Anderson, D.C. Licciardello, and T.V. Ramakrishnan, Phys. Rev. Lett. **42**, 673 (1979).
10. "Inverse participation ratio in 2+ epsilon dimensions," F. Wegner, Z. Phys. B **36**, 209 (1980).
11. "Interaction between diffusion modes in localization theory," Zh. Eksp. Teor. Fiz. **79**, 1120 (1980). K.B. Efetov, A.I. Larkin, and D.E. Khmelnitskii, Sov. Phys. JETP **52**, 568 (1980).
12. "Disordered Electronic Systems," P.A. Lee and T.V. Ramakrishnan, Rev. Mod. Phys. **57**, 287 (1985).
13. "Weak localization and Coulomb interaction in disordered systems," A.M. Finkelstein, Zeit. fur Physik **B56** 189 (1984).
14. "Interaction-driven metal-insulator transitions in disordered fermion systems," C. Castellani, C. Di Castro, P. A. Lee, and M. Ma, Phys. Rev. B **30**, 527 (1984).
15. "The Anderson-Mott transition," D. Belitz and T. R. Kirkpatrick, Rev. Mod. Phys. **66**, 261 (1994).
16. "Possible metal-insulator transition at B=0 in two dimensions," S. V. Kravchenko, G. V. Kravchenko, J. E. Furneaux, V. M. Pudalov, and M. D'Iorio, Phys. Rev. B **50**, 8039 (1994).
17. "Scaling of an anomalous metal-insulator transition in a two-dimensional system in silicon at B=0," S. V. Kravchenko, W. E. Mason, G. E. Bowker, J. E. Furneaux, V. M. Pudalov, and M. D'Iorio, Phys. Rev. B **51**, 7038 (1995).
18. "Electric Field Scaling at a B = 0 Metal-Insulator Transition in Two Dimensions," S. V. Kravchenko, D. Simonian, M. P. Sarachik, W. Mason, and J. E. Furneaux, Phys. Rev. Lett. **77**, 4938 (1996).
19. "Metal-Insulator Transition in Two Dimensions: Effects of Disorder and Magnetic Field," D. Popović, A. B. Fowler, and S. Washburn, Phys. Rev. Lett. **79**, 1543 (1997).
20. "Reflection symmetry at a B=0 metal-insulator transition in two dimensions ," D. Simonian, S. V. Kravchenko, and M. P. Sarachik, Phys. Rev. B **55**, R13421 (1997).

21. "The Metallic-like Conductivity of a Two-Dimensional Hole System," Y. Hanien, U. Meirav, D. Shahar, C. C. Li, D. C. Tsui, and Hadas Shtrikman, Phys. Rev. Lett. **80**, 1288 (1998).
22. "Metal-Insulator Transition at B = 0 in a Dilute Two Dimensional GaAs-AlGaAs Hole Gas," M. Y. Simmons, A. R. Hamilton, M. Pepper, E. H. Linfield, P. D. Rose, D. A. Ritchie, A. K. Savchenko and T. G. Griffiths, Phys. Rev. Lett. **80**, 1292 (1998).
23. "Scaling Theory of Two-Dimensional Metal-Insulator Transitions," V. Dobrosavljević, E. Abrahams, E. Miranda, and S. Chakravarty, Phys. Rev. Lett. **79**, 455 (1997).
24. "Metallic phase and metal-insulator transition in two-dimensional electronic systems," C. Castellani, C. Di Castro, and P. A. Lee, Phys. Rev. B **57**, R9381 (1998).
25. "Interactions and scaling in a disordered two-dimensional metal," S. Chakravarty, L. Yin, and E. Abrahams, Phys. Rev. B **58**, R559 (1998).
26. "Wigner glass, spin liquids and the metal-insulator transition," S. Chakravarty, S. Kivelson, C. Nayak, and K. Voelker, Phil. Mag. **B79**, 859 (1999).
27. "Superconductivity in a two-dimensional electron gas," P. Phillips, Y. Wan, I. Martin, S. Knysh, and D. Dalidovich, Nature **395**, 253 (1998).
28. "Possible triplet superconductivity in MOSFETs," D. Belitz and T. R. Kirkpatrick, Phys. Rev. B **58**, 8214 (1998).
29. "Theory of Metal-Insulator Transitions in Gated Semiconductors," B. L. Altshuler and D. Maslov, Phys. Rev. Lett. **82**, 145 (1999).
30. "A few electrons per ion scenario for the B=0 metal-insulator transition in two dimensions," T. M. Klapwijk and S. Das Sarma, Solid State Comm. **110**, 581 (1999).
31. "Coherent Propagation of Two Interacting Particles in a Random Potential," D. L. Shepelyansky, Phys. Rev. Lett. **73**, 2607 (1994).
32. "Do Interactions Increase or Reduce the Conductance of Disordered Electrons? It Depends!," T. Vojta, F. Epperlein, and M. Schreiber, Phys. Rev. Lett. **81**, 4212 (1998)]; "Delocalized Coulomb phase in two dimensions," X. Waintal, G. Benenti, and J.-L. Pichard, Europhys. Lett. **49**, 466 (2000).
33. "Consistent description of high-Tc superconductors with the three-band Hubbard model," G. Dopf, A. Muramatsu, and W. Hanke, Phys. Rev. Lett. **68**, 353 (1992).
34. "Two-dimensional Hubbard model-metal insulator transition studied by Monte Carlo calculation," N. Furukawa and M. Imada, J. Phys. Soc. Japan **61**, 3331 (1992).
35. "Numerical simulation of the 1D and 2D Hubbard models: Fermi liquid behavior and its breakdown," S. Sorella, E. Tosatti, S. Baroni, R. Car, and M. Parinello, Int. J. Mod. Phys. **B1**, 993 (1988).
36. "Two Dimensional Hubbard Model: Numerical Simulation Study," J.E. Hirsch, Phys. Rev. **B31**, 4403 (1985).
37. "Boson Localization and the Superfluid–Insulator Transition," M.P.A. Fisher, P.B. Weichman, G. Grinstein, and D.S. Fisher, Phys. Rev. **B40**, 546 (1989).
38. "Onset of superconductivity in the two-dimensional limit," D.B. Haviland, Y.Liu, and A.M. Goldman, Phys. Rev. Lett. **62** 2180, (1989).
39. "Tunneling Study of Superconductivity near the Metal-Insulator Transition," R.C. Dynes *etal*, Phys. Rev. Lett. **53**, 2437 (1984); 'Destruction of Superconductivity in Quenched Condensed Two Dimensional Films," A.E. White, R.C. Dynes, and J.P. Garno, Phys. Rev. **B33**, 3549 (1986).
40. "Magnetic Field Tuned Superconductor–Insulator Transition in Two Dimensional Films," A.F. Hebard and M.A. Paalanen, Phys. Rev. Lett. **65**, 927 (1990); and "Global phase coherence in two-dimensional granular superconductors," B.G. Orr, H.M. Jaeger, A.M. Goldman and C.G. Kuper, Phys. Rev. Lett. **56**, 378 (1986).
41. "Electron tunneling determination of the order-parameter amplitude at the superconductor-insulator transition in 2D," J.M. Valles, R.C. Dynes, and J.P. Garno, Phys. Rev. Lett. **69**, 3567 (1992).
42. "Superconducting-insulating transition in two-dimensional a-MoGe thin films," A. Yazdani and A. Kapitulnik, Phys. Rev. Lett. **74**, 3037 (1995).
43. "Pairing and Spin Gap in the Normal State of Short Coherence Length Superconductors," M. Randeria, N. Trivedi, A. Moreo, and R.T. Scalettar, Phys. Rev. Lett. **69**, 2001 (1992).
44. "Superconductor–Insulator Transition in a Disordered Electronic System," N. Trivedi, R.T. Scalettar, and M. Randeria, Phys. Rev. **B54**, 3756 (1996).
45. "Localization: Theory and Experiment," B. Kramer and A. MacKinnon, Rep. Prog. Phys. **56**, 1469 (1993).
46. J. Tinka Gammel, D.K. Campbell, and E.Y. Loh, Jr., Synthetic Metals **57**, 4437 (1993).
47. C. Lin, F.H. Zong, and D.M. Ceperley, Phys. Rev. E **64**, 016702 (2001), and references cited therein.
48. "Localization Transitions in Non-Hermitian Quantum Mechanics", N. Hatano and D.R. Nelson, Phys. Rev. Lett. **77**, 570 (1996); "Vortex pinning and non-Hermitian quantum mechanics," Phys. Rev. B **56**, 8651 (1997).
49. "Statistics of complex levels of random matrices for decaying systems, " F. Haake, F. Izrailev, N. Lehmann, D. Saher, and H.J. Sommers, Zeit. Phys. B **88**, 359 (1992).

50. "Spectrum of Large Random Asymmetric Matrices", H.J. Sommers, A. Crisanti, H. Sompolinsky, and Y. Stein, Phys. Rev. Lett. **60**, 1895 (1988).
51. "Random Matrix Model of QCD at Finite Density and the Nature of the Quenched Limit," M.A. Stephanov, Phys. Rev. Lett. **76**, 4472 (1996).
52. "Nonhermitian Random Matrix Models," R.A. Janik, M.A. Nowak, G. Papp, and I. Zahed, Nucl. Phys. B **501**, 603 (1997) and "Macroscopic Universality: Why QCD in Matter is Subtle," Phys. Rev. Lett. **77**, 4876 (1996).
53. "Two-level system with noise: Blue's function approach," E. Gudowska-Nowak, G. Papp and J. Brickmann, J. Chem. Phys **220**, 125 (1997).
54. "Almost-Hermitian Random Matrices: Eigenvalue Density in the Complex Plane," Y.V. Fyodorov, B.A. Khoruzhenko and H.J. Sommers, Phys. Lett. A **226** 46 (1997).
55. "Fermion determinants in matrix models of QCD at nonzero chemical potential," M.A. Halasz, A.D. Jackson, and J.J.M. Verbaarschot, Phys. Rev. D **56**, 5140 (1997).
56. "Non-hermitian random matrix theory: Method of hermitian reduction," J. Feinberg and A. Zee, Nucl. Phys. B **504**, 579 (1997).
57. "Non-gaussian non-hermitian random matrix theory: Phase transition and addition formalism," J. Feinberg and A. Zee, Nucl. Phys. B **501**, 643 (1997).
58. B. Zickel, unpublished.
59. "Cold Bosonic Atoms in Optical Lattices," D. Jaksch, C. Bruder, J.I. Cirac, C.W. Gardiner, and P. Zoller, Phys. Rev. Lett. **81**, 3108 (1998).
60. "Phase diagram for ultracold bosons in optical lattices and superlattices," P. Buonsante and A. Vezzani, Phys. Rev. A **70**, 033608 (2004).
61. "Fractional-filling loophole insulator domains for ultracold bosons in optical superlattices," P. Buonsante, V. Penna, and A. Vezzani, Phys. Rev. A **70**, 061603 (2004).
62. "Cell strong-coupling perturbative approach to the phase diagram of ultracold bosons in optical superlattices," P. Buonsante, A. Vezzani, Phys. Rev. A **72**, 013614 (2005).
63. J.B. Sokoloff, Phys. Rep. **126**, 189 (1985).
64. Y. Last, Proc. of XIth Int. Congress of Math. Phys. 366, 1994); S. Jitomirskaya, *ibid*, 373.
65. "Cold Atom Optical Lattices as Quantum Analog Simulators for Aperiodic One-Dimensional Localization Without Disorder," V.W. Scarola and S. Das Sarma, Phys. Rev. **A73**, 041609(R) (2006).
66. N. Trefethen and M. Embree, "Spectra and Pseudospectra - the behavior of nonnormal matrices and operators", Princeton University Press.
67. " Exact study of the one-dimensional boson Hubbard model with a superlattice potential," V.G. Rousseau, D.P. Arovas, M. Rigol, F. Hébert, G.G. Batrouni, and R.T. Scalettar, Phys. Rev. **B73**, 174516 (2006).
68. "Constrained Path Quantum Monte Carlo Method for Fermion Ground States", S. Zhang, J. Carlson, and J. E. Gubernatis Phys. Rev. Lett. 74, 3652 (1995); "Constrained path Monte Carlo method for fermion ground states", S. Zhang, J. Carlson, and J. E. Gubernatis Phys. Rev. B 55, 7464 (1997).
69. "Dynamical mean-field theory of strongly correlated fermion systems and the limit of infinite dimensions," A. Georges, G. Kotliar, W. Krauth, and M. Rozenberg, Rev. Mod. Phys. **68**, 13 (1996).
70. "Investigation of correlated electron systems using the limit of high dimensions," D. Vollhardt, in *Correlated Electron Systems*, edited by V. J. Emery (World Scientific, Singapore) 57 (1993).
71. "Quantum Cluster Theories," Th. Maier, M. Jarrell, Th. Pruschke, and M. Hettler, Rev. Mod. Phys. **77**, 1027 (2005).
72. "Monte Carlo Method for Magnetic Impurities in Metals, " J.E. Hirsch and R.M. Fye, Phys. Rev. Lett. **56**, 2521 (1986).
73. "Insulator, metal, or superconductor: The criteria," D. J. Scalapino, S. R. White, and S. C. Zhang, Phys. Rev. B **47**, 7995 (1993).
74. Consider lower temperatures and average over larger numbers of disorder realizations: for the higher temperatures generally 4 realizations are sufficient, for the lowest temperatures the fluctuations are larger and up to 20 (72) realizations are used for $U = 4$ ($U = 0$).
75. For a related enhancement of the superfluid density by interactions in a disordered Bose system, see "Superfluid-insulator transition in disordered boson systems," W. Krauth, N. Trivedi, and D. M. Ceperley, Phys. Rev. Lett. **67**, 2307 (1991).
76. For $\Delta = 1.2$ and $\beta = 5$, $\sigma_{\rm dc} = 0.88(1), 0.76(1)$ for $U = 0,4$, respectively. For $\Delta = 2.0$ and $\beta = 6$, $\sigma_{\rm dc} = 0.45(2), 0.49(3), 0.52(4)$ for $U = 0,2,4$, respectively. Numbers between brackets represent the error bar in the last digit shown.
77. "Spin Dynamics of Nearly Localized Electrons," M. A. Paalanen, S. Sachdev, R. N. Bhatt, and A. E. Ruckenstein, Phys. Rev. Lett. **57**, 2061 (1986).
78. "Spin Degree of Freedom in a Two-Dimensional Electron Liquid," T. Okamoto, K. Hosoya, S. Kawaji, and A. Yagi, Phys. Rev. Lett. **82**, 3875 (1999).
79. "Parallel Magnetic Field Induced Transition in Transport in the Dilute Two-Dimensional Hole

System in GaAs," J. Yoon, C. C. Li, D. Shahar, D. C. Tsui, and M. Shayegan, Phys. Rev. Lett. **84**, 4421 (2000).

80. "Metal-Insulator Transition in a 2D Electron Gas: Equivalence of Two Approaches for Determining the Critical Point," A. A. Shashkin, S. V. Kravchenko, and T. M. Klapwijk, Phys. Rev. Lett. **87**, 266402 (2000).
81. "Effects of a Parallel Magnetic Field on the Metal-Insulator Transition in a Dilute Two-Dimensional Electron System," K. V. Eng, X. G. Feng, D. Popović, and S. Washburn, Phys. Rev. Lett. **88**, 136402 (2002).
82. "Spin polarization and transition from metallic to insulating behavior in 2D systems," E. Tutuc, E. P. De Poortere, S. J. Papadakis, and M. Shayegan, Physica E **2-4**, 748 (2002).
83. "Magnetic Field Induced Spin Polarization of AlAs Two-dimensional Electrons," E. P. De Poortere, E. Tutuc, Y. P. Shkolnikov, K. Vakili, and M. Shayegan, Phys. Rev. **B66**, 161308 (2002).
84. "Ground-state properties of the two-dimensional disordered Hubbard model," G. Caldara, B. Srinivasan, and D. L. Shepelyansky, Phys. Rev. B **62**, 10680 (2000).
85. "Disorder and Interaction in 2D: Exact Diagonalization Study of the Anderson-Hubbard-Mott Model," R. Kotlyar and S. Das Sarma, Phys. Rev. Lett. **86**, 2388 (2001).
86. "Role of a parallel magnetic field upon a few correlated electrons trapped in 2d mesoscopic disordered clusters," F. Selva and J.-L. Pichard, Europhys. Lett. **55**, 518 (2001).
87. "Parallel magnetoconductance of interacting electrons in a two-dimensional disordered system," R. Berkovits and J. W. Kantelhardt, Phys. Rev. B **65**, 125308 (2002).
88. "The effect of parallel magnetic field on the Boltzmann conductivity and the Hall coefficient of a disordered two-dimensional Fermi liquid," I. F. Herbut, Phys. Rev. B **63**, 113102 (2001).
89. "Interaction corrections at intermediate temperatures: Magnetoresistance in a parallel field," G. Zala, B. N. Narozhny, and I. L. Aleiner, Phys. Rev. B **65**, 020201 (2002).
90. "Conducting phase in the two-dimensional disordered Hubbard model," P. J. H. Denteneer, R. T. Scalettar, and N. T. Trivedi, Phys. Rev. Lett. **83**, 4610 (1999).
91. This point of view of the effect of $B_{\|}$ in the experiments is substantiated in: "Fate of the extended states in a vanishing magnetic field: The role of spins in strongly interacting two-dimensional electron systems," M. R. Sakr, M. Rahimi, and S. V. Kravchenko, Phys. Rev. B **65**, 041303 (2002).
92. "R.C. Magnetic field induced insulator to metal transition in amorphous-Gd_xSi_{1-x}," W. Teizer, F. Hellman, and R. C. Dynes, Solid State Comm. **114**, 81 (2000).
93. "The mobility edge problem: Continuous symmetry and a conjecture," F. J. Wegner, Z. Phys. B **35**, 207 (1979).
94. "Transport Properties and Density of States of Quantum Wires with Off-diagonal Disorder," P. W. Brouwer, C. Mudry, and A. Furusaki, Physica**E9**, 333 (2001).
95. "Quasiparticle localization in superconductors with spin-orbit scattering," T. Senthil and M. P. A. Fisher, Phys. Rev. B **61**, 9690 (2000).
96. "Localization and Delocalization in Dirty Superconducting Wires," P. W. Brouwer, A. Furusaki, I. A. Gruzberg, and C. Mudry, Phys. Rev. Lett. **85**, 1064 (2000), and cited references.
97. "Deviations from Fermi-Liquid Behavior above Tc in 2D Short Coherence Length Superconductors," N. Trivedi and M. Randeria, Phys. Rev. Lett. **75**, 312 (1995). As in the case of Equation 2, this expression for the density of states is valid at temperatures below characteristic frequency scales of structures in $N(\omega)$.
98. "Magnetic correlations in the two-dimensional Anderson-Hubbard model," M. Ulmke and R. T. Scalettar, Phys. Rev. B **55**, 4149 (1997), M. Ulmke, P. J. H. Denteneer, R. T. Scalettar, and G. T. Zimanyi, Europhys. Lett. **42**, 655 (1998).
99. One can verify that beyond a certain large site disorder ($\Delta_\mu \geq 4t$) σ_{dc} decreases again.
100. "Collapse of the charge gap in random Mott insulators," Y. Otsuka, Y. Morita, and Y. Hatsugai, Phys. Rev. B **58**, 15314 (1998).
101. "Can correlations drive a band insulator metallic?," A. Garg, H.R. Krishnamurthy, and M. Randeria, Phys. Rev. Lett. **97**, 046403 (2006).
102. "Model of the Neutral-Ionic Phase Transformation," J. Hubbard and J.B. Torrance, Phys. Rev. Lett. **47**, 1750 (1981); "Lattice Effect of Strong Electron Correlation: Implication for Ferroelectricity and Superconductivity," T. Egami, S. Ishihara and M. Tachiki, Science **261**, 130 (1994); "Macroscopic polarization as a geometric quantum phase: Many-body formulation," G. Ortiz and R. Martin, Phys. Rev. B **49**, 14202 (1994); "Many-Body Effects on Polarization and Dynamical Charges in a Partly Covalent Polar Insulator," R. Resta and S. Sorella, Phys. Rev. Lett. **74**, 4738 (1995) and "Electron Localization in the Insulating State," Phys. Rev. Lett. **82**, 370 (1999); "From Band Insulator to Mott Insulator in One Dimension," M. Fabrizio, A.O. Gogolin, and A.A. Nersesyan, Phys. Rev. Lett. **83**, 2014 (1999); "Quantum Monte Carlo study of the one-dimensional ionic Hubbard model," T. Wilkens and R.M. Martin, Phys. Rev. B **63**, 235108 (2001); and "Exact Bond Ordered Ground State for the Transition between the Band and the Mott Insulator," C.D. Batista and A.A. Aligia, Phys.

Rev. Lett. **92**, 246405 (2004).
103. "Ferromagnetism and Metal-Insulator Transition in the Disordered Hubbard Model," K. Byczuk, M. Ulmke, and D. Vollhardt, Phys. Rev. Lett. **90**, 196403 (2003).
104. "Mott-Hubbard metal-insulator transition at noninteger filling," K. Byczuk, W. Hofstetter, and D. Vollhardt, Phys. Rev. B **69**, 045112 (2004).
105. "Anomalous normal-state properties of high-T_c superconductors: intrinsic properties of strongly correlated electron systems?", Th. Pruschke, M. Jarrell, and J.K. Freericks, Adv. Phys. **44**, 187 (1995).
106. There is one slightly confusing point here. From the description provided, one might think that in a Monte Carlo simulation one stores a vector p_i and applies T to it. This is completely impractical because the number of one configurations i of the system is incredibly huge and one could never allocate enough memory to hold p_i. Instead, generally only one configuration i at a time is stored, and T_{ji} gives us a probabilistic rule to get the (single) configuration j from i. The probability p_i appears through the average of all the single configurations over a long simulation.
107. "Computational Physics," R.H. Landau and M.J. Páez, Wiley, 1997.
108. "Computational Physics," N. Giordano, Prentice-Hall, 1997.
109. "Statistical Mechanics of Phase Transitions," J. Yeomans, Oxford, 1992.
110. "A statistical approach to quantum mechanics," M. Creutz and J. Freedman, Annals of Phys. **132**, 427 (1981).
111. See, http://leopard.physics.ucdavis.edu/rts/nato.html Sec. 3 of the article nato2.pdf.

Optical probes of electron correlations in solids

E. van Heumen and D. van der Marel

University of Geneva, 24 Quai E.-Ansermet, Geneva 4, Switzerland

Abstract. Classically the interaction between light and matter is given by the Maxwell relations. These are briefly reviewed and will be used as a basis to discuss several techniques that are used in optical spectroscopy. We then discuss the quantum mechanical description of the optical conductivity based on the Kubo formalism. This is used as a basis to understand how strong correlation effects can be observed using optical techniques. We will discuss the use of sum rules in the interpretation of optical experiments. Finally, we describe the effect of including interactions between electronic and collective degrees of freedom on optical spectra.

Keywords: Optical spectroscopy, correlated electron systems
PACS: 42.50.Ct, 71.36.+c, 78.20.-e

INTRODUCTION

We will discuss the physics of correlated electron systems from an experimental viewpoint, focussing on optical spectroscopy. The interaction of light and matter will be discussed first from a classical point of view, based on the Maxwell equations. This review will be the basis for a discussion of optical techniques that are most commonly used. We will then continue with a discussion of the quantum mechanical description of the interaction between light and matter, using the Kubo-formalism. We finally discuss the application of sum rules to correlated systems and what happens when interactions, like the electron-phonon interaction, become important. The first part of our review is not meant to be complete. Readers with interest for further details are referred to references [1] and [2]. In the following all fields, currents, charge densities etc. are implied to be position and time dependent if not written explicitly. Bold quantities imply vectors or matrices.

CP918, *Lectures on the Physics of Strongly Correlated Systems XI: Eleventh Training Course,* edited by A. Avella and F. Mancini

ELECTROMAGNETISM AND MATTER

Maxwell's equations

We start this review with the microscopic Maxwell equations,

$$\begin{aligned}
\nabla\cdot\mathbf{e} &= 4\pi\rho_{micro}, &(1)\\
\nabla\times\mathbf{e} &= -\frac{1}{c}\frac{\partial}{\partial t}\mathbf{b}, &(2)\\
\nabla\cdot\mathbf{b} &= 0, &(3)\\
\nabla\times\mathbf{b} &= \frac{1}{c}\frac{\partial}{\partial t}\mathbf{e}+\frac{4\pi}{c}\mathbf{j}_{micro}. &(4)
\end{aligned}$$

Here **e** and **b** are the microscopic electric and magnetic fields respectively. ρ_{micro} is the total microscopic charge distribution and j_{micro} the total microscopic current distribution (i.e. due to internal and external sources). Note that these equations are written in the C.G.S. system of units. To convert them to S.I. units, simply replace 4π by $1/\varepsilon_0$. The charge distribution for a collection of point sources with charge q_i can be written classically,

$$\rho_{micro}=\sum_i q_i\delta(\mathbf{r}-\mathbf{r}_\mathbf{i}), \tag{5}$$

or quantum mechanically as,

$$\rho_{micro}=-e\Psi^*(\mathbf{r})\Psi(\mathbf{r}). \tag{6}$$

Equations (1-4) are not very practical to work with. As a first step we rewrite them in a more familiar form. To do this we average the fields, charge and current distributions over a volume ΔV,

$$\begin{aligned}
\rho_{total}(\mathbf{r}) &= \frac{1}{\Delta V}\int_{\Delta V}\rho_{micro}(\mathbf{r}+\mathbf{r}')d^3\mathbf{r}', &(7)\\
\mathbf{J}_{total}(\mathbf{r}) &= \frac{1}{\Delta V}\int_{\Delta V}\mathbf{J}_{micro}(\mathbf{r}+\mathbf{r}')d^3\mathbf{r}', &(8)
\end{aligned}$$

and similarly for **e** and **b**. This is a sensible procedure under the condition that $a_0 \ll \Delta V \ll (2\pi c/\omega)^3$ where a_0 is the Bohr radius. Using these averaged distributions we arrive at the standard Maxwell equations in free space,

$$\begin{aligned}
\nabla\cdot\mathbf{E}_{total}(\mathbf{r},t) &= 4\pi\rho_{total}(\mathbf{r},t), &(9)\\
\nabla\times\mathbf{E}_{total}(\mathbf{r},t) &= -\frac{1}{c}\frac{\partial}{\partial t}\mathbf{B}_{total}(\mathbf{r},t), &(10)\\
\nabla\cdot\mathbf{B}_{total}(\mathbf{r},t) &= 0, &(11)\\
\nabla\times\mathbf{B}_{total}(\mathbf{r},t) &= \frac{1}{c}\frac{\partial}{\partial t}\mathbf{E}_{total}(\mathbf{r},t)+\frac{4\pi}{c}\mathbf{J}_{total}(\mathbf{r},t). &(12)
\end{aligned}$$

In order to see how matter interacts with propagating electromagnetic waves we have to distinguish between induced and external sources. We write $\mathbf{J}_{total}\equiv\mathbf{J}_{ext}+\mathbf{J}_{ind}$ and

$\rho_{total} \equiv \rho_{ext} + \rho_{ind}$. Both the induced and external charge and current distributions have to obey the continuity equations separately,

$$\nabla \cdot \mathbf{J}_{ind/ext} + \frac{\partial}{\partial t}\rho_{ind/ext} = 0. \tag{13}$$

We can distinguish three different types of macroscopic internal sources,

$$\mathbf{J}_{ind} = \mathbf{J}_{cond} + \frac{\partial \mathbf{P}}{\partial t} + c\nabla \times \mathbf{M}. \tag{14}$$

The first term on the right hand side, $\mathbf{J}_{cond}$, corresponds to the response of the free charges. The second term is the current due to changes in the polarization state of the system. Finally, we include a term representing a current due to magnetization. Note that this last term is purely transversal (the divergence of a rotation is always zero) and so is easy to distinguish from the other two terms. Since the induced free charge current due to photons is necessarily transversal, $\nabla \cdot \mathbf{J}_{cond} = 0$, we can use the continuity equations to show that the induced free charge density has to be zero and as a consequence that the total induced charge density

$$\rho_{ind} = -\nabla \cdot P. \tag{15}$$

It is convenient to introduce new fields

$$\mathbf{D}(\mathbf{r},t) \equiv \mathbf{E}_{ext}(\mathbf{r},t) \equiv \mathbf{E}(\mathbf{r},t) + 4\pi\mathbf{P}(\mathbf{r},t), \tag{16}$$

$$\mathbf{H}(\mathbf{r},t) \equiv \mathbf{B}(\mathbf{r},t) - 4\pi\mathbf{M}(\mathbf{r},t), \tag{17}$$

so that using equations (14-17) in equations (9) and (12) we find,

$$\nabla \cdot \mathbf{D}(\mathbf{r},t) = 4\pi\rho_{ext}(\mathbf{r},t), \tag{18}$$

$$\nabla \times \mathbf{H}(\mathbf{r},t) = \frac{1}{c}\frac{\partial}{\partial t}\mathbf{D}(\mathbf{r},t) + \frac{4\pi}{c}\mathbf{J}_{ext}(\mathbf{r},t) + \frac{4\pi}{c}\mathbf{J}_{cond}(\mathbf{r},t). \tag{19}$$

Linear Response Theory

In the spirit of linear response theory we assume that the response of polarization, magnetization or current are linear in the applied fields:

$$\mathbf{P} = \chi_e \mathbf{E}, \tag{20}$$

$$\mathbf{M} = \chi_m \mathbf{H}, \tag{21}$$

$$\mathbf{J} = \sigma \mathbf{E}. \tag{22}$$

The electric and magnetic susceptibilities can be expressed in terms of a dielectric function $\varepsilon' = 1 + 4\pi\chi_e$ and magnetic permittivity $\mu' = 1 + 4\pi\chi_m$. The dielectric function is a response function that connects the external field $\mathbf{E}_{ext}$ at position $\mathbf{r}$ and time t with the field $\mathbf{E}$ at all other times and positions. So in general,

$$\mathbf{E}_{ext}(\mathbf{r},t) = \int_{-\infty}^{t} \int \varepsilon'(\mathbf{r},\mathbf{r}',t,t')\mathbf{E}(\mathbf{r}',t')d^3r'dt'. \tag{23}$$

We will be mainly interested in the Fourier transform of $\varepsilon' \equiv \varepsilon(\mathbf{q}, \omega)$ however. It is an easy exercise to express the Maxwell equations in terms of $\mathbf{q}$ and ω which we leave to the reader. We can use these definitions to once again rewrite the Maxwell equations in the following form,

$$\nabla \cdot (\varepsilon' \mathbf{E}) = 4\pi\rho_{ext}, \tag{24}$$

$$\nabla \times \mathbf{E} = -\frac{1}{c}\frac{\partial}{\partial t}(\mu')\mathbf{H}, \tag{25}$$

$$\nabla \cdot \mu' \mathbf{H} = 0, \tag{26}$$

$$\nabla \times \mathbf{H} = \frac{1}{c}\frac{\partial}{\partial t}(\varepsilon' \mathbf{E}) + \frac{4\pi\sigma}{c}\mathbf{E}. \tag{27}$$

We are now in a position to study the response of a solid to an externally applied field or light wave. For simplicity we assume that our solid is homogeneous so that $\nabla\varepsilon' = 0$ and $\nabla\mu' = 0$. We can describe light waves by plane waves, i.e.

$$\mathbf{E}(\mathbf{r},t) = \mathbf{E_0} e^{i(\mathbf{q}\cdot\mathbf{r}-\omega t)}, \tag{28}$$

$$\mathbf{B}(\mathbf{r},t) = \mathbf{B_0} e^{i(\mathbf{q}\cdot\mathbf{r}-\omega t)}. \tag{29}$$

Using (29) on the right hand side of Faraday's equation (10) and rearranging we find,

$$\mathbf{B} = \frac{c}{i\omega}\nabla \times \mathbf{E}. \tag{30}$$

If we now take the curl of this equation and use the fact that we can express $\mathbf{M}$ in terms of $\mathbf{B}$ as (see equations (17) and (21)),

$$\mathbf{M} = \frac{\mu'^{-1} - 1}{4\pi}\mathbf{B}. \tag{31}$$

we find that

$$\begin{aligned} \nabla \times \mathbf{M} &= \frac{\mu'^{-1} - 1}{4\pi}\nabla \times \mathbf{B} = \frac{c}{i\omega}\frac{\mu'^{-1} - 1}{4\pi}\nabla \times \nabla \times \mathbf{E} \\ &= \frac{c}{i\omega}\frac{\mu'^{-1} - 1}{4\pi}(\nabla^2\mathbf{E} - \nabla(\nabla \cdot \mathbf{E}) = \frac{cq^2}{i\omega}\frac{\mu'^{-1} - 1}{4\pi}\mathbf{E}^T. \end{aligned} \tag{32}$$

Note that in this equation we are left with only the transversal field since the curl of a curl is transverse. We give two further identities for completeness,

$$\frac{\partial \mathbf{P}}{\partial t} = -i\omega\frac{1 - \varepsilon'}{4\pi}\mathbf{E}, \tag{33}$$

and,

$$J_{cond} = \sigma\mathbf{E}. \tag{34}$$

Finally, we note that inside the solid $\rho_{ext} = J_{ext} = 0$. With this we have all the ingredients to express equation (19) in terms of $\mathbf{E}$ and $\mathbf{J}$. We split this equation in transversal and longitudinal parts to find,

$$\frac{\mathbf{J}^T_{ind}(\mathbf{q},\omega)}{\mathbf{E}^T(\mathbf{q},\omega)} \equiv \frac{i\omega}{4\pi}\{1-\varepsilon'(\mathbf{q},\omega)-\frac{i4\pi}{\omega}\sigma(\mathbf{q},\omega)-\frac{c^2q^2}{\omega^2}(1-\frac{1}{\mu(\mathbf{q},\omega)})\}, \quad (35)$$

$$\frac{\mathbf{J}^L_{ind}(\mathbf{q},\omega)}{\mathbf{E}^L(\mathbf{q},\omega)} \equiv \frac{i\omega}{4\pi}\{1-\varepsilon'(\mathbf{q},\omega)-\frac{i4\pi}{\omega}\sigma(\mathbf{q},\omega)\}. \quad (36)$$

We can define a new dielectric function with longitudinal and transverse components and write the previous equation in a more compact form,

$$\frac{\mathbf{J}^{L,T}_{ind}(\mathbf{q},\omega)}{\mathbf{E}^{L,T}(\mathbf{q},\omega)} \equiv \frac{i\omega}{4\pi}\{1-\varepsilon^{L,T}(\mathbf{q},\omega)\}. \quad (37)$$

This new dielectric function ε is now a complex quantity: $\varepsilon \equiv \varepsilon' + i\varepsilon'' = \varepsilon' + i4\pi\sigma/\omega$. Using the last relation we can also define a complex conductivity $\hat{\sigma} \equiv \sigma' + i\sigma''$ and it is related to the dielectric function by,

$$\hat{\sigma} = \frac{i\omega}{4\pi}(1-\varepsilon). \quad (38)$$

The real part of ε is often called the reactive part and the imaginary part the dissipative part. The real and imaginary parts are also indicated with a subscript 1 and 2 instead of (') and (").

Kramers-Kronig relations

A fundamental principle in physics is the principle of causality: an effect cannot precede its cause. This principle provides us with a very useful relation between the real and imaginary parts of a response function like the optical conductivity as we now show. First we express the induced current due to an electric field in terms of a memory function,

$$j(t) = \int_{-\infty}^{t} M(t-t')E(t')dt'. \quad (39)$$

The memory function has the property that $M(\tau < 0) = 0$. This is simply a restatement of the causality principle: we switch on a driving force at time $\tau = 0$ so before that time there can be no current. We define the Fourier transform of M in equation (39) as,

$$\hat{\sigma}(\omega) = \int_{0}^{\infty} M(\tau)e^{i\omega\tau}d\tau. \quad (40)$$

To do the integral we change to the complex frequency plane, $\omega \rightarrow z = \omega_1 + i\omega_2$. The exponential in Eq. (40) now splits in a complex and real part,

$$\hat{\sigma}(\omega) = \int_{0}^{\infty} M(\tau)e^{i\omega_1\tau}e^{-\omega_2\tau}d\tau. \quad (41)$$

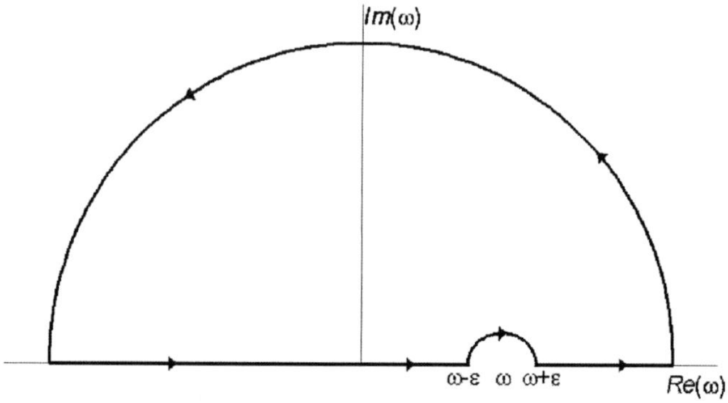

FIGURE 1. Contour used to derive the KK-relations.

The second exponent in this integral is bounded in the upper half plane for $\tau > 0$ and in the lower half plane for $\tau < 0$, so that we can evaluate the integral in the upper half plane since $M(\tau < 0) = 0$. We use the contour shown in figure 1. Since all poles occur on the real axis, the complete contour is zero,

$$\oint dz \frac{\hat{\sigma}(z)}{z-\omega} = 0. \tag{42}$$

The integral along the large semi circle is also zero. So we are left with,

$$\int_{-\infty}^{\omega-\varepsilon} dz \frac{\hat{\sigma}(z)}{z-\omega} + \int_{\omega+\varepsilon}^{\infty} dz \frac{\hat{\sigma}(z)}{z-\omega} + \int_{\pi}^{0} d(\omega + \varepsilon e^{i\phi}) \frac{\hat{\sigma}(\omega + \varepsilon e^{i\phi})}{\varepsilon e^{i\phi}} = 0. \tag{43}$$

The first two integrals give the principle value of the integral for $\varepsilon \to 0$,

$$\mathscr{P} \int_{-\infty}^{\infty} d\omega' \frac{\hat{\sigma}(\omega')}{\omega' - \omega} - \pi i \hat{\sigma}(\omega') = 0. \tag{44}$$

From which the Kramers-Kronig relations follow,

$$\sigma_1(\omega) = \frac{1}{\pi} \mathscr{P} \int_{-\infty}^{\infty} \frac{\sigma_2(\omega')}{\omega' - \omega} d\omega', \tag{45}$$

$$\sigma_2(\omega) = -\frac{1}{\pi} \mathscr{P} \int_{-\infty}^{\infty} \frac{\sigma_1(\omega')}{\omega' - \omega} d\omega'. \tag{46}$$

Using $Im(M(\tau)) = 0$ we see that $\hat{\sigma}(-\omega) = \hat{\sigma}^*(\omega)$, which implies that $\sigma_1(-\omega) = \sigma_1(\omega)$ and $\sigma_2(-\omega) = -\sigma_2(\omega)$. These relations can be used to rewrite equations (45) and (46),

$$\sigma_1(\omega) = \frac{2}{\pi} \mathscr{P} \int_{0}^{\infty} \frac{\omega' \sigma_2(\omega')}{\omega'^2 - \omega^2} d\omega', \tag{47}$$

$$\sigma_2(\omega) = -\frac{2\omega}{\pi} \mathscr{P} \int_{0}^{\infty} \frac{\sigma_1(\omega')}{\omega'^2 - \omega^2} d\omega'. \tag{48}$$

The relations (45) and (46) between the real and imaginary parts of the optical conductivity are examples of the general relations between real and imaginary parts of causal response functions and they are referred to as Kramers-Kronig (KK) relations.

Polaritons

In this section we discuss the properties of electromagnetic waves propagating through solids. Such a wave is called a polariton. A polariton is a photon dressed up with the excitations that exist inside solids. For example one can have phonon-polaritons which are photons dressed up with phonons. Although the solutions of the Maxwell equations, i.e. the fields **E** and **B**, have the same form as before (Eq. (28) and (29)) they obey different dispersion relations as we will now see. As before we assume that $\nabla\varepsilon' = \nabla\mu' = 0$ and that $\rho_{ext} = J_{ext} = 0$. Taking the curl of Eq. (25) we obtain for the left-hand side,

$$\nabla \times \nabla \times \mathbf{E} = -\nabla^2 \mathbf{E}^T, \tag{49}$$

where the T indicates that we are left with a purely transverse field. We then use Eq. (27) to work out the right-hand side of Eq. (25) and we obtain the wave equation,

$$\nabla^2 \mathbf{E}^T = \frac{\varepsilon'\mu}{c^2}\frac{\partial^2 \mathbf{E}^T}{\partial t^2} + \frac{4\pi\sigma\mu}{c^2}\frac{\partial \mathbf{E}^T}{\partial t}. \tag{50}$$

From this wave equation we easily obtain the dispersion relation for polaritons travelling through a solid by substituting Eq. (28),

$$\mu(\mathbf{q},\omega)\{\varepsilon'(\mathbf{q},\omega) + i\frac{4\pi\sigma(\mathbf{q},\omega)}{\omega}\}\omega^2 = \mu\varepsilon^L\omega^2 = \mathbf{q}^2c^2. \tag{51}$$

The dispersion relation for longitudinal waves can be found by observing that for longitudinal waves $\nabla \times \mathbf{E} = 0$ and hence the dispersion relation is simply,

$$\mu(\mathbf{q},\omega)\{\varepsilon'(\mathbf{q},\omega) + i\frac{4\pi\sigma(\mathbf{q},\omega)}{\omega}\} = 0. \tag{52}$$

The polariton solutions to Eq. (51) are of the form

$$\mathbf{E}(\mathbf{r},t) = \mathbf{E_0}e^{i(\mathbf{q}\cdot\mathbf{r}-\omega t)}, \tag{53}$$

with

$$|q| = \frac{\sqrt{\mu\varepsilon}\,\omega}{c}. \tag{54}$$

We now define the refractive index,

$$\hat{n}(\mathbf{q},\omega) = n + ik \equiv \sqrt{\mu\varepsilon}. \tag{55}$$

In all cases considered here $n > 0$ and $k > 0$. We also note that $Im(\varepsilon) \geq 0$ but it is possible to have $Re(\varepsilon) < 0$. If $k > 0$ the wave travelling through the solid gets attenuated,

$$\mathbf{E}(\mathbf{r},t) = \mathbf{E_0}e^{i\omega(nr/c-t)-r/\delta}. \tag{56}$$

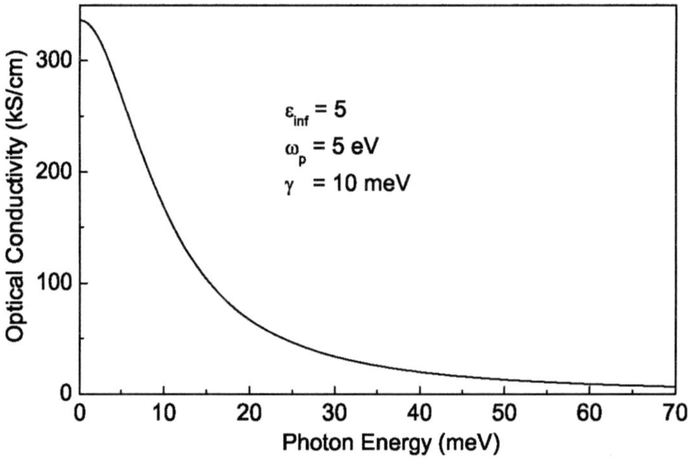

FIGURE 2. Real part of the optical conductivity for parameter values indicated in the graph. The curve is calculated using equation (58).

The extinction of the wave occurs over a characteristic length scale δ called the skin depth,

$$\delta = \frac{c}{\omega k} = \frac{c}{\omega Im\sqrt{\mu\varepsilon_1 + i4\pi\mu\sigma_1/\omega}}. \tag{57}$$

Note that we can have $k > 0$ if $Im(\varepsilon) = 0$ and $Re(\varepsilon) < 0$ so that the wave gets attenuated even though there is no absorption. In table 1 we indicate some limits of the skin depth.

TABLE 1. Some limiting cases of the general expression Eq. (57). λ in the last line is the London penetration depth.

Insulator	$\frac{4\pi\sigma_1}{\omega} \ll \varepsilon_1$	$\delta \approx \frac{c}{2\pi\sigma_1}\sqrt{\frac{\varepsilon_1}{\mu}}$
Metal	$\frac{4\pi\sigma_1}{\omega} \gg \varepsilon_1$	$\delta \approx \frac{c}{\sqrt{2\pi\mu\sigma_1\omega}}$
Superconductor	$\frac{4\pi\sigma_1}{\omega} \ll \varepsilon_1 = -\frac{c^2}{\lambda^2\omega^2}$	$\delta \approx \frac{\lambda}{\sqrt{\mu}}$

To illustrate some of the previous results we now have a look at the simplest model of a metal: the Drude-model. The optical conductivity in the Drude model is,

$$\hat{\sigma} = \frac{ne^2}{m}\frac{1}{\tau^{-1} - i\omega}. \tag{58}$$

Often $1/\tau$, the time in between scattering events, is written as a scattering rate γ. The plasma frequency is defined as $\omega_p^2 \equiv 4\pi ne^2/2m$. The dielectric function can now be

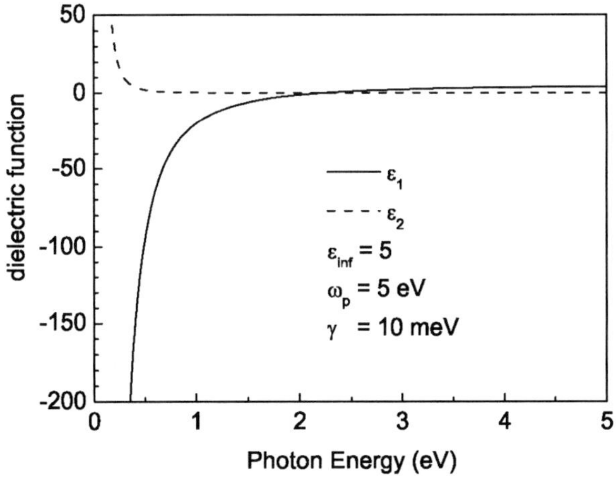

FIGURE 3. Dielectric function corresponding to equation (59) with the same parameters as in figure 2.

written as,

$$\varepsilon(\omega) = 1 + 4\pi\chi_{bound} - \frac{4\pi n e^2}{m}\frac{1}{\omega(\gamma - i\omega)} = \varepsilon_\infty - \frac{4\pi n e^2}{m}\frac{1}{\omega(\gamma - i\omega)}, \tag{59}$$

where for completeness we have included the contribution due to the bound charges, represented by a high energy contribution ε_∞. Figure 2 shows the optical conductivity given by equation (58) for parameter values typical of a metal. Using the same parameters we can calculate the dielectric function given by equation (59). The results are shown in figure 3. We note that the real part of the dielectric function is negative for $\omega < \omega_p/\sqrt{\varepsilon_\infty}$ and positive for $\omega > \omega_p/\sqrt{\varepsilon_\infty}$. The point where it crosses zero is called the screened plasma frequency ω_p^* (screened by interband transitions).

We can also easily calculate the optical constants,

$$\hat{n} = \sqrt{\varepsilon_\infty - \frac{\omega_p^2}{\omega(\omega + i\tau^{-1})}}. \tag{60}$$

The real and imaginary part are displayed in figure 4. We see that at the screened plasma frequency both n and k show a discontinuity.

The polariton dispersion follows from equation (51). Here we assume that $\mu = 1$ and frequency independent and use Eq. (59) to solve (51) for $\omega(q)$. The polariton dispersion consists of two branches the lowest one for $0 \leq \omega \leq 1/\tau$ and one for $\omega \geq \omega_p/\sqrt{\varepsilon_\infty}$.

Finally we show the skin depth in figure 6. We see that for frequencies smaller than the scattering rate, γ, light waves can enter the material. This is called the classical skin

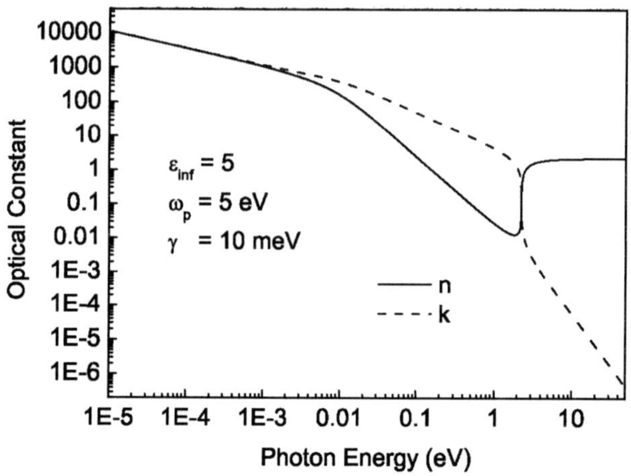

FIGURE 4. Optical constants corresponding to equation (59) with the same parameters as in figure 2.

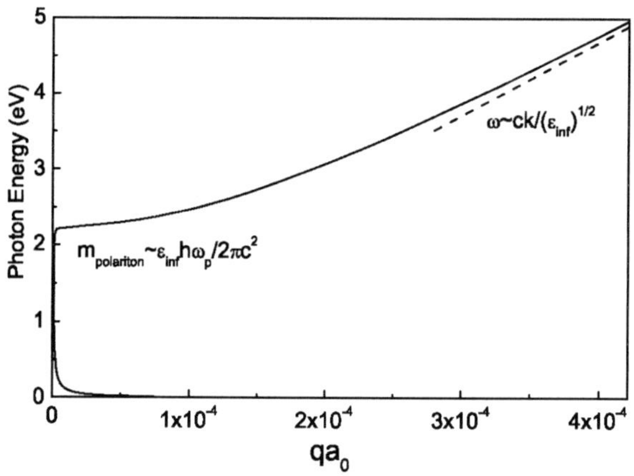

FIGURE 5. Polariton dispersion calculated with the same parameters as in figure 2.

effect. For frequencies larger than the screened plasma frequency the material becomes transparent again.

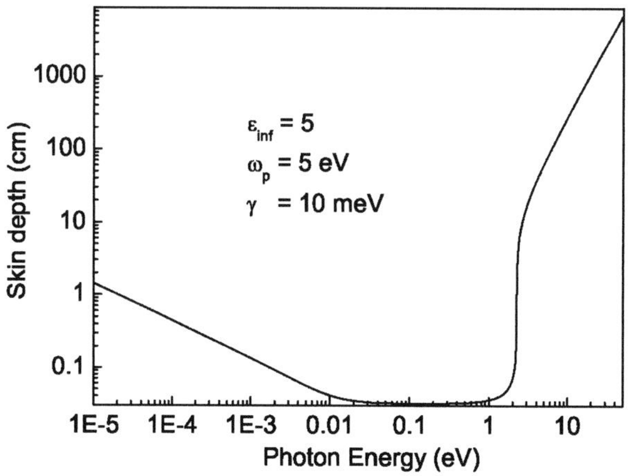

FIGURE 6. Skin depth calculated with the same parameters as in figure 2.

EXPERIMENTAL TECHNIQUES

The goal of optical spectroscopy is to determine the complex dielectric function or equivalently the complex optical conductivity. Since electromagnetic waves have small momenta compared to the typical momenta of a solid, i.e. $\mathbf{q} \ll 1/a_0$, we usually only probe the $q \to 0$ limit of the optical constants. In this limit,

$$\lim_{q\to 0}(\varepsilon^T(\mathbf{q},\omega) - \varepsilon^L(\mathbf{q},\omega)) = 0, \tag{61}$$

$$\varepsilon(\mathbf{q} \to \mathbf{0},\omega) = \varepsilon_1(\omega) + i\frac{4\pi}{\omega}\sigma_1(\omega). \tag{62}$$

In some cases we can directly obtain information on both real and imaginary components seperately, but more often we obtain information where the contributions are mixed. We then make use of some form the KK-relations to disentangle the two.

Reflection and Transmission at an interface

When we shine light on an interface between vacuum and a material, part of the light is reflected and another part is transmitted as in figure 7. At the boundary the electromagnetic waves have to obey the following boundary conditions,

$$\mathbf{E}_i + \mathbf{E}_r = \mathbf{E}_t, \tag{63}$$

$$\mathbf{E} \times \mathbf{H} \,//\, \mathbf{k}. \tag{64}$$

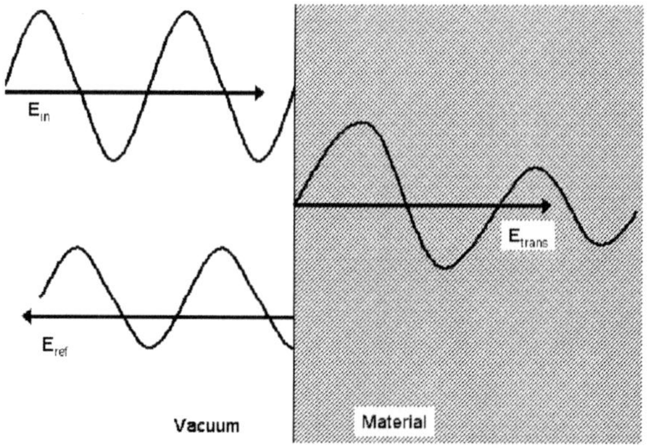

FIGURE 7. Electromagnetic waves reflecting from a material. The reflected wave has a smaller amplitude and is phase shifted with respect to the incoming wave. The transmitted wave is continuously attenuated inside the material.

From these two equations it follows that the reflected magnetic field suffers a phase shift at the boundary,

$$\mathbf{H}_i - \mathbf{H}_r = \mathbf{H}_t. \tag{65}$$

Using equation (28) in equation (25) we obtain,

$$iqc\mathbf{E}^T = i\omega\mu\mathbf{H}. \tag{66}$$

so that, using the dispersion relation (51),

$$\frac{\mathbf{H}}{\mathbf{E}^T} = \sqrt{\frac{\varepsilon}{\mu}}. \tag{67}$$

From now on we set $\mu = 1$ unless otherwise indicated. In that case $\mathbf{H}/\mathbf{E}^T = \hat{n}$. Combining this result with Eq. (65) we get,

$$\mathbf{E}_i - \mathbf{E}_r = \hat{n}. \tag{68}$$

Together with Eq. (63) we can now solve for $\mathbf{E}_r/\mathbf{E}_i$ and $\mathbf{E}_t/\mathbf{E}_i$,

$$\hat{r} \equiv \mathbf{E}_r/\mathbf{E}_i = \frac{1-\hat{n}}{1+\hat{n}}, \tag{69}$$

$$\hat{t} \equiv \mathbf{E}_t/\mathbf{E}_i = \frac{2}{1+\hat{n}}. \tag{70}$$

The two quantities $\hat{r}$ and $\hat{t}$ are the complex reflectance and transmittance.

Reflectivity experiments

The real reflection coefficient $R(\omega)$ which is measured in a reflection experiment is related to $\hat{r}$ via

$$R = |\hat{r}|^2 = |\frac{(n-1)^2 + k^2}{(n+1)^2 + k^2}|. \tag{71}$$

Note that in this experiment we obtain no information on the phase of $\hat{r}$. In these experiments the angle of incidence is as close to normal incidence as possible. To measure $R(\omega)$ one first measures the reflected intensity I_s from the sample under study. To normalize this intensity one then has to take a reference measurement. This can be done by replacing the sample with a mirror (i.e a piece of aluminum or gold) and again measure the reflected intensity, I_{ref}. The reflection coefficient is then $R(\omega) \equiv I_s(\omega)/I_{ref}(\omega)$. A better way is to evaporate a layer of gold or aluminum *in-situ* and measure the reflected intensity as a reference. This way one automatically corrects for surface imperfections and, if done properly, there are no errors due to different size and shape of the reflecting surface. To obtain the optical constants from such an experiment we have to make use of KK-relations. If we define, $\hat{r}(\omega) \equiv \sqrt{R(\omega)}e^{i\theta}$, then the logarithm of $\hat{r}(\omega)$ is

$$\ln \hat{r}(\omega) = \ln \sqrt{R(\omega)} + i\theta. \tag{72}$$

The phase θ in this expression is the unknown we want to determine. If we interpret $\hat{r}$ as a response function we can use the same arguments as in the section on KK relations and calculate $\theta(\omega)$ from,

$$\theta(\omega) = -\frac{\omega}{\pi} P \int_0^{\infty} \frac{\ln R(\omega')}{\omega'^2 - \omega^2} d\omega', \tag{73}$$

which is just the same as the KK-relation for $\hat{\varepsilon}$. The complex dielectric function is calculated from $R(\omega)$ and $\theta(\omega)$ using,

$$\hat{\varepsilon}(\omega) = \left(\frac{1 - \sqrt{R(\omega)}e^{i\theta(\omega)}}{1 + \sqrt{R(\omega)}e^{i\theta(\omega)}} \right)^2. \tag{74}$$

Although in principle exact, this technique is in practice only approximate. The reason is that we cannot measure $R(\omega)$ from zero to infinite frequency. Most experiments probe a frequency range between a few meV and a few eV. To do the integral in Eq. (74) one then has to use extrapolations in the frequency ranges where no data is available. For metals the low frequency extrapolation which is most often used is the so-called Hagen-Rubens approximation,

$$R(\omega) = 1 - \alpha\sqrt{\omega}. \tag{75}$$

For frequencies above the interband transitions one often uses an extrapolation that is proportional to ω^{-4}. As an example of a possible experimental result we show in figure 8 the reflectivity calculated from the Drude model for the same parameters as in section on polaritons.

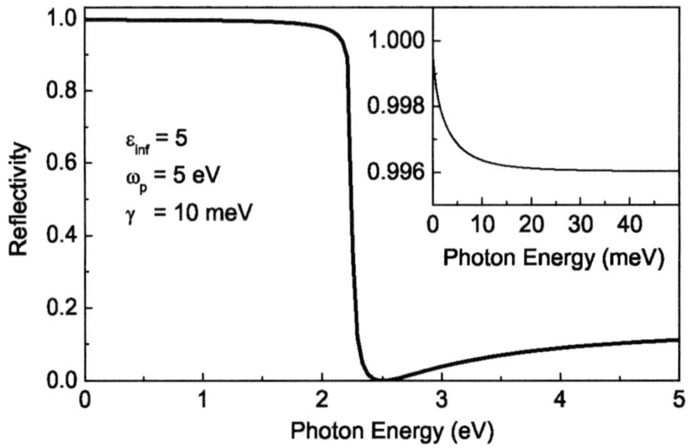

FIGURE 8. Reflectivity calculated using parameters typical for a metal. The inset shows the low energy reflectivity on an enhanced scale.

The reflectivity is close to one until just below the plasma frequency. At the zero crossing of ε_1 the reflectivity has a minimum. The inset shows a blow up of the "flat" region below 50 meV. Here one can clearly see the Hagen-Rubens behavior mentioned above. If the sample under investigation is anisotropic one has to use polarized light along one of the principle crystal axes to perform the experiment.

Grazing Incidence Experiments

A closely related technique is to measure reflectance under a grazing angle of incidence. Here one has to distinguish between experiments performed with different incoming polarizations as shown in figure 9. We distinguish between p-polarized light and s-polarized light. For p-polarization the electric field is parallel to the plane of incidence, whereas for s-polarization it is perpendicular to it (s stands for senkrecht). Since in principal the optical constants along the three crystal axes can be different, we use the labels a, b and c for the optical constants as indicated in figure 9. For p-polarized light the complex reflectance is,

$$r_{\mathrm{p}} = \frac{\hat{n}_c \hat{n}_b \cos\theta - \sqrt{\hat{n}_c^2 - \sin^2\theta}}{\hat{n}_c \hat{n}_b \cos\theta + \sqrt{\hat{n}_c^2 - \sin^2\theta}}. \tag{76}$$

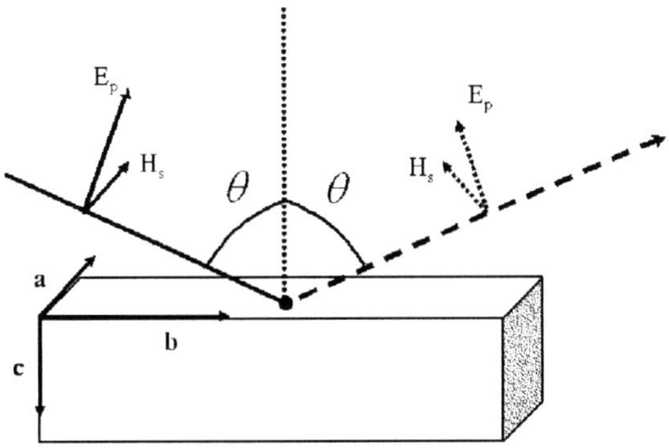

FIGURE 9. Grazing incidence experiment. The result of the experiment is extremely sensitive to the precise orientation of the crystal axes with respect to the incoming light.

The angle θ in this equation is the angle relative to the surface normal under which the experiment is performed. For s-polarized light the complex reflectance is,

$$r_s = \frac{\cos\theta - \sqrt{\hat{n}_a^2 - \sin^2\theta}}{\cos\theta + \sqrt{\hat{n}_a^2 - \sin^2\theta}}. \tag{77}$$

An example of such an experiment is shown in figure 10. In this example the samples are from the bismuth based family of cuprates [3]. They have a layered structure consisting of conducting copper-oxygen sheets, interspersed with insulating bismuth-oxygen layers. Since the bonding between layers is not very strong it is very difficult to obtain samples that are sufficiently thick along the insulating c-direction. The grazing incidence technique is used here to probe the optical constants of the c-axis without the need of a large ac-face surface area. A disadvantage in this particular experiment is that it is not possible to determine accurately the absolute value of the optical constants. It is possible however to determine the so-called loss function $Im(-1/\hat{\varepsilon}_c)$. The experiment is performed on the ab-plane of the sample using p-polarized light and we can simplify the expression for $\hat{r}_p$ by using the fact that the a and b direction are almost isotropic. The resulting expression for $\hat{r}_p$ is,

$$\hat{r}_p = \frac{\sqrt{\hat{\varepsilon}_b}\cos\theta - \sqrt{1 - \sin^2\theta/\hat{\varepsilon}_c}}{\sqrt{\hat{\varepsilon}_b}\cos\theta + \sqrt{1 - \sin^2\theta/\hat{\varepsilon}_c}}. \tag{78}$$

From this equation we can derive the following relation between the grazing incidence reflectivity and a pseudo loss-function $L(\omega)$ [4],

$$L(\omega) \equiv \frac{(1-R_p)}{(1+R_p)} \approx Im \frac{2e^{i\phi_p}}{|n_b|\cos\theta}\sqrt{1 - \frac{\sin^2\theta}{\hat{\varepsilon}_c}}. \tag{79}$$

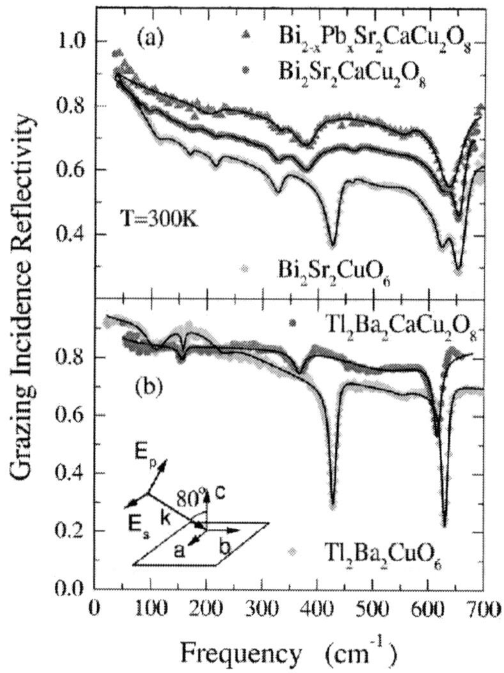

FIGURE 10. Grazing incidence reflectivity of Bi-2201, Bi-2212, Tl-2201 and Tl-2212. The inset in panel (b) indicates the measurement geometry. The figure is adapted from ref. [3].

The function $\sqrt{1-\frac{\sin^2\theta}{\hat{\varepsilon}_c}}$ has maxima at the same position as the true loss-function. In this way information was gained on the phonon structure of the c-axis of this material.

Spectroscopic Ellipsometry

The third technique we introduce here is spectroscopic ellipsometry. This relatively new technique has two major advantages over the previous techniques. Firstly, the technique is self-normalizing meaning that no reference measurement has to be done and secondly, it provides directly both the real and imaginary parts of the dielectric function.

As with the grazing incidence technique we have to distinguish between s- and p-polarized light and label the crystal axes. Instead of measuring R_p or R_s independently, we now measure directly the amplitude and phase of the ratio $\hat{r}_p/\hat{r}_s = |\hat{r}_p/\hat{r}_s|e^{i(\eta_p-\eta_s)}$. To see how this can be done we first describe the experimental setup. There are a number of different setups one can use and here we describe the simplest. This setup consists of a source followed by a polarizer. With this polarizer we can change the orientation of the polarization impinging on the sample. The light reflected from the sample passes

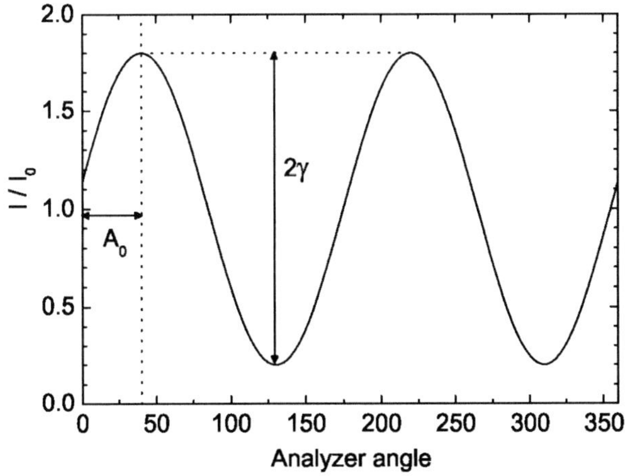

FIGURE 11. Result of an ellipsometric measurement. The phase shift A_0 and amplitude 2γ are the two quantities that we are interested in.

through another polarizer (called analyzer) and then hits the detector. Depending on the orientation of the first polarizer we can change the electric field strength of s- and p-polarized light according to,

$$E_p = |E_i| \cos(P), \tag{80}$$
$$E_s = |E_i| \sin(P). \tag{81}$$
$$\tag{82}$$

From the expressions for $\hat{r}_p$ and $\hat{r}_s$, (76) and (77), in the previous section it follows that,

$$\hat{\rho} \equiv \frac{r_\mathrm{p}}{r_\mathrm{s}} = \frac{\sqrt{\hat{n}_c^2 - \sin^2\theta} - \hat{n}_c\hat{n}_b\cos\theta}{\sqrt{\hat{n}_c^2 - \sin^2\theta} + \hat{n}_c\hat{n}_b\cos\theta} \cdot \frac{\sqrt{\hat{n}_a^2 - \sin^2\theta} + \cos\theta}{\sqrt{\hat{n}_a^2 - \sin^2\theta} - \cos\theta}. \tag{83}$$

Our task is now to invert this equation and express the optical constants in terms of measured quantities and instrument parameters. For an isotropic sample this can be done quite easily. We define the pseudodielectric function $\hat{\varepsilon}$ such that:

$$\hat{\rho} \equiv \frac{\sin\theta\tan\theta - \sqrt{\tilde{\varepsilon} - \sin^2\theta}}{\sin\theta\tan\theta + \sqrt{\tilde{\varepsilon} - \sin^2\theta}}, \tag{84}$$

where we note that $\hat{\varepsilon} = \varepsilon_a = \varepsilon_b = \varepsilon_c$ in an optically isotropic medium. This equation

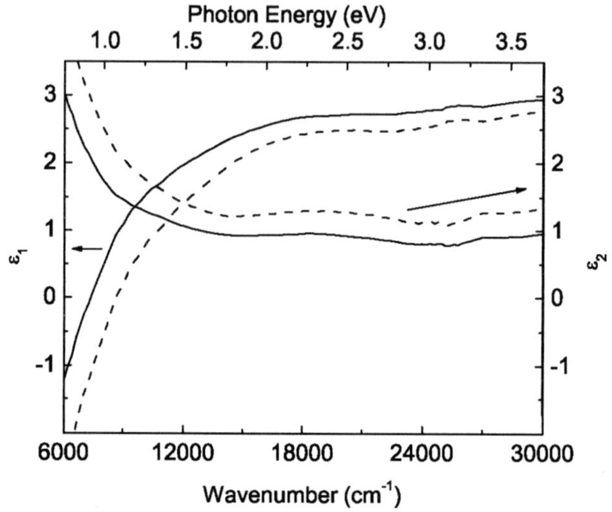

FIGURE 12. Dielectric function measured ellipsometrically on a $HgBa_2CuO_4$ sample. The true dielectric function is shown in solid lines. The pseudo dielectric function (i.e. actually measured) is shown as a dashed line. Data taken from ref. [5].

can be inverted to obtain $\tilde{\varepsilon}$,

$$\tilde{\varepsilon}(\omega) = \sin^2\theta\left[1+\tan^2\theta\left(\frac{1-\rho}{1+\rho}\right)^2\right]. \tag{85}$$

So all that is left to do is to express $\hat{\rho}$ in terms of experimental parameters. The experiment is done in the following way: we fix the polarizer at some angle $0 < P < 90$ and then we record the intensity while rotating the analyzer 360 degrees. The result is shown in figure 11. We then measure the amplitude of the resulting sine wave, γ and the phase offset with respect to zero, A_0 (we assume here that for $P = 0$ the polarizer and analyzer are aligned parallel to each other). With some goniometry and figure 11 we can show that,

$$\tan A_0 = \frac{2\tan(P)}{|\rho|^2+\tan^2(P)}\rho_1, \tag{86}$$

and

$$\sqrt{1-\gamma^2} = \frac{2\tan(P)}{|\rho|^2+\tan^2(P)}\rho_2. \tag{87}$$

Combining these two equations leads to,

$$\rho = \frac{1\pm\sqrt{\gamma^2-\tan^2 A_0}}{\tan A_0 - i\sqrt{1-\gamma^2}}\tan(P). \tag{88}$$

The combination of Eq. (88) with Eq. (85) is all we need to describe an ellipsometric experiment on an isotropic sample. For an anisotropic sample the problem is slightly more difficult. However, there exists a theorem due to Aspnes which states that the inversion of Eq. (83) results in Eq. (85) but now the dielectric function on the left-hand side is a so-called pseudo-dielectric function. This pseudo-dielectric function is mainly determined by the component parallel to the intersection of sample surface and plane of incidence (component along b in figure 9), but still contains a small contribution of the two other components. If we perform three measurements, each along a different crystal axis, we can correct the pseudo dielectric functions and obtain the true dielectric functions. If the sample is isotropic along two directions, as is the case in high temperature superconductors for example, only two measurements are required. Figure 12 shows in dashed lines the pseudo dielectric function of $HgBa_2CuO_4$. In this case the a and b axes have the same optical constants. The c-axis dielectric function was determined from reflectivity measurements and subsequently used to correct the pseudo dielectric function measured by ellipsometry on the ab-plane. The true dielectric function after this correction is shown as the solid line.

Transmission Experiments

A technique complementary to the reflection techniques is transmission spectroscopy. This technique is, obviously, most suitable for transparent samples. In principle the technique can also be applied to metallic samples but this requires very thin samples or films. The reflection experiments discussed above are usually good methods to obtain accurate estimates of the real part of the optical conductivity. In contrast the transmission experiments discussed below are more sensitive to weak absorptions or, in other words to the imaginary part of the optical conductivity. Note that the simultaneous knowledge of reflection and transmission spectra allows one to directly determine the full complex dielectric function without any further approximations. Examples of weak absorptions which are better probed in a transmission experiment are multi-phonon or magnon absorptions. The equations for transmission experiments are slightly more difficult then those for the reflection experiments. These equations simplify if we do the experiment on a wedged sample as shown in figure 13. At the boundary between vacuum and the sample, part of the light is reflected and part transmitted. The part that is transmitted is given by,

$$\hat{t}_{v,s} = \frac{2}{1+\hat{n}}. \tag{89}$$

Inside the wave propagates according to $e^{i\psi}$ where,

$$\psi \equiv \hat{n} d\omega / c. \tag{90}$$

At the next boundary again part of the beam is reflected back into the sample and part is transmitted. Now we can see the advantage of the wedged sample: the part of the light that is reflected propagates away at an angle and after another reflection the second transmitted ray is no longer parallel to and spatially separated from the first transmitted

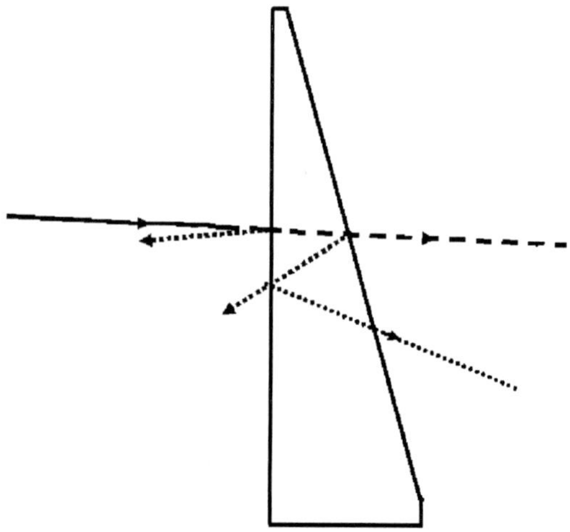

FIGURE 13. Transmission experiment on a wedged sample. After the initial ray is partially reflected back from the front surface all following rays are no longer parallel to the first transmitted ray.

ray. This means that we only have to care about the first transmitted ray. The transmission coefficient at the boundary from sample to vacuum is given by,

$$\hat{t}_{s,v} = \frac{2\hat{n}}{\hat{n}+1}, \tag{91}$$

so that the total transmission coefficient is,

$$\hat{t}_{v,s} e^{\mathrm{i}\psi} \hat{t}_{s,v}. \tag{92}$$

Putting Eq.'s (89)-(92) together and taking the absolute value to calculate the transmission T gives,

$$T = \frac{|4\hat{n}|^2}{|1+\hat{n}|^4} \exp\left\{-\frac{2d}{\delta}\right\}. \tag{93}$$

In most transmission experiments $\varepsilon_1 \gg \varepsilon_2$ and the classical skindepth δ can be approximated by,

$$\delta(\omega) \approx \frac{2\pi\sigma_1(\omega)}{c\sqrt{\varepsilon_1(\omega)}}. \tag{94}$$

Moreover in these cases ε_1 is often dispersion-less so that we can use the expression for δ to invert (93),

$$\sigma_1(\omega) = \left\{-\ln(T) + 2\ln\left(\frac{4|\hat{n}|}{|1+\hat{n}|^2}\right)\right\} \frac{c\sqrt{\varepsilon_1}}{4\pi d}. \tag{95}$$

As an example of this technique we show in figure 14 a comparison between the reflec-

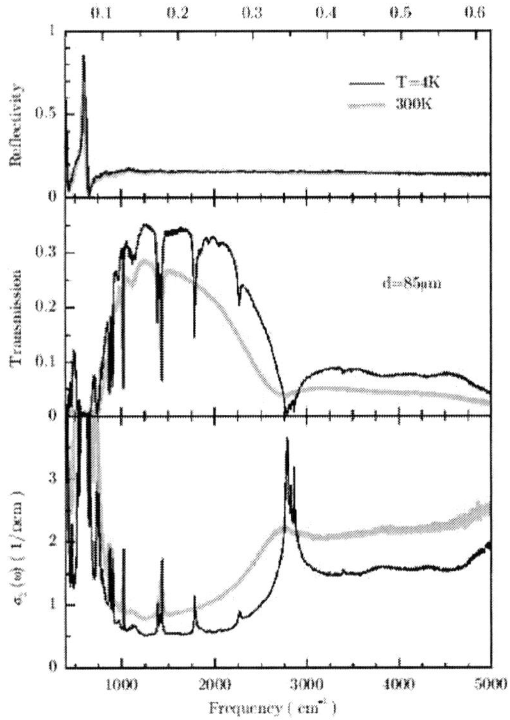

FIGURE 14. Comparison of reflectivity and transmission measured on the same sample. Note the strong absorptive features present in the transmission spectrum that are completely invisible in the reflectance spectra. Figure adapted from [6]

tivity and transmission spectra of undoped $YBa_2Cu_3O_7$ [6]. This material is an (Mott) insulator which is clearly visible from the reflectivity spectrum. The large structure at low energies is an optical phonon. At higher energy the reflectivity spectrum appears to be rather featureless. Focussing our attention on the transmission spectrum we see that it is almost zero in the phonon range but then above the phonon range a whole series of sharp dips shows up. The optical conductivity consists of a set of smaller peaks at energies between 100 meV and 300 meV which are due to multi-phonon absorptions whereas the larger peak just above 300 meV is due to a two magnon plus one phonon absorption (see also the section on spin interactions below).

We can also do the experiment on a sample with two plan-parallel sides as depicted in figure 15. We can immediately realize that for a given thickness of the sample there will be interference effects between different transmitted rays for certain frequencies. These will cause oscillations in the transmission spectra which are called Fabry-Perot resonances. We now analyze the transmission coefficients for this experiment. The coefficient for the first ray is off course the same as in Eq. (92). The coefficients for

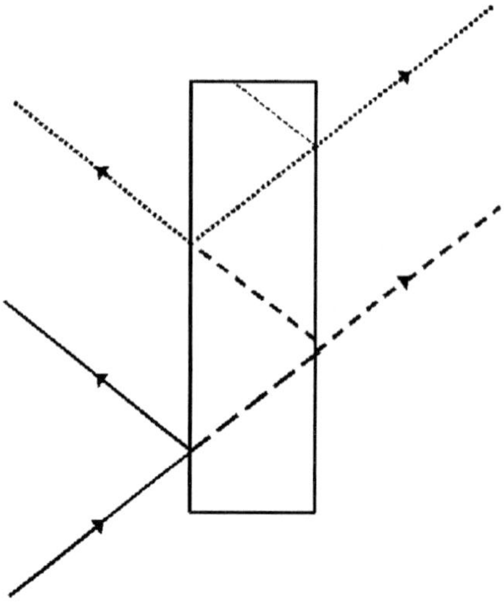

FIGURE 15. Pictorial of a transmission experiment on a plan parallel sample.

the higher order rays are formed by multiplying $\hat{t}_{v,s}e^{i\psi}$ on the right with a factor f,

$$f \equiv \hat{r}_{s,v}e^{i2\psi}\hat{r}_{s,v}, \tag{96}$$

followed by a factor $\hat{t}_{s,v}$. So the total transmission coefficient for the second transmitted ray is given by,

$$\hat{t}_{v,s}e^{i\psi}\hat{r}_{s,v}e^{i\psi}\hat{r}_{s,v}e^{i\psi}\hat{t}_{s,v} = \hat{t}_{v,s}e^{i\psi}\hat{t}_{s,v}f. \tag{97}$$

The coefficients $\hat{t}_{v,s}$ and $\hat{t}_{s,v}$ are given by Eq. (89) and (91). The coefficient for reflection on a boundary from sample to vacuum is given by,

$$\hat{r}_{s,v} = \frac{\hat{n}-1}{\hat{n}+1}. \tag{98}$$

It is easy to see that if we sum over all transmitted rays the total transmission coefficient is given by,

$$\hat{t} = \hat{t}_{v,s}e^{i\psi}\hat{t}_{s,v}\left(1+f+f^2+..\right) = \frac{2\hat{n}}{2\hat{n}\cos\psi - i(1+\hat{n}^2)\sin\psi}. \tag{99}$$

For thin films the phase factor $\psi \ll 1$ and we can simplify this equation to,

$$\hat{t} \approx \frac{1}{1+\frac{2\pi d}{c}\sigma_1 - i\frac{\omega d}{2c}(1+\varepsilon')}, \tag{100}$$

and so,

$$T(\omega) \approx \frac{1}{1+4\pi d c^{-1}\sigma_1(\omega)}. \tag{101}$$

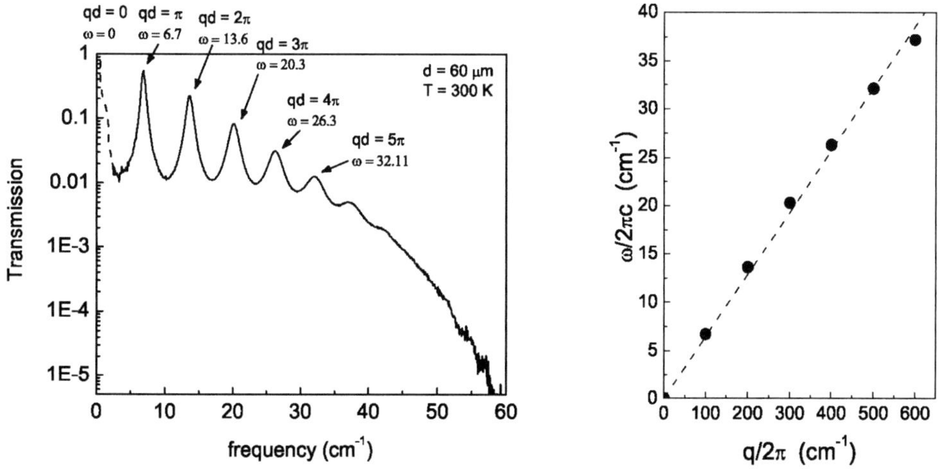

FIGURE 16. Left: Far infrared transmission spectrum for $SrTiO_3$. The positions of the peaks determine the polariton dispersion. The dashed line at low frequency is an extrapolation to zero frequency. Right: Dispersion relation of polaritons in STO as derived from transmission spectrum in the left panel.

More generally from Eq. (99) we obtain,

$$T_{LR} = \frac{4\left|\varepsilon\right|}{\left|4\varepsilon\cos^2\psi\right| + \left|(1+\varepsilon)^2\sin^2\psi\right| + 2\mathrm{Im}\left\{(1+\varepsilon)\sin 2\psi\right\}}. \tag{102}$$

In the case that the sample under investigation has only weak absorptions, i.e. $Im(\hat{n}) \approx 0$, this equation simplifies to,

$$T_{LR} \approx \frac{4n^2}{4n^2 + (1-n^2)^2\sin^2(nd\omega/c)}. \tag{103}$$

This equation gives us some insight to the occurrence of Fabry-Perot resonances: if $\omega = cm\pi/nd$ with m=0,1,2,... the sinus is equal to zero and the transmission $T = 1$. In between these maxima the transmission has minima and $T \approx 4\hat{n}^2/(1+\hat{n}^2)^2$. In reality the transmission will never reach 1 due to the fact that $Im(\hat{n}) \neq 0$ in which case our approximations are no longer valid. As an example we display in the left panel of figure 16 the transmission spectrum of $SrTiO_3$ [7]. This material is very close to being ferroelectric and as a result it has a very large dielectric constant. The non-sinusoidal shape of the peaks is due to this large dielectric constant. One can use the Fabry-Perot resonances to measure the polariton dispersion as we now show. Note that at each maximum in the transmission spectrum we know precisely the value of the argument of the sine function in Eq. (103). We can read off the value of ω from the graph and using Eq. (51) we can replace the argument in the sine function by $n\omega d/c = qd$ so the

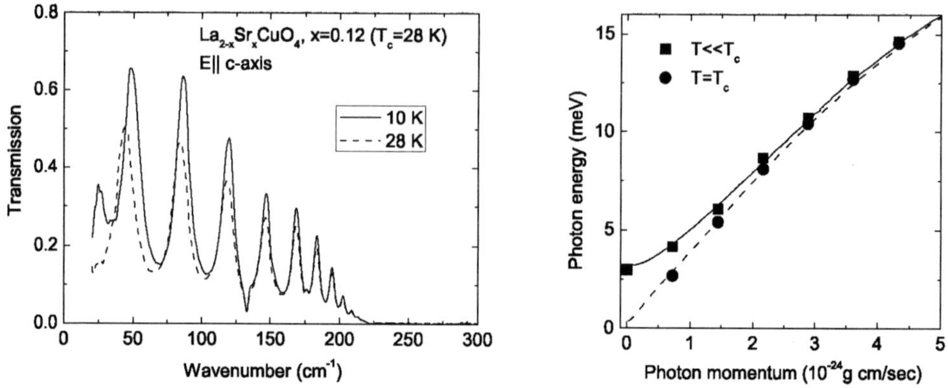

FIGURE 17. Left: Transmission spectrum of LSCO at a temperature just above T_c and one far below. Note the shift in peak positions. Figure adapted from ref. [8]. Right: Dispersion relation of polaritons in LSCO as derived from left panel. The squares are derived from the spectrum in the superconducting state whereas the circles are determined at a temperature slightly above T_c.

momentum at a given maximum is,

$$q_m = \frac{m\pi}{d} \quad m = 0, 1, 2, ... \tag{104}$$

So given the thickness of the sample we can make a plot of $\omega(q)$. The result for STO is shown in the right panel of figure 16. We see that the dispersion is linear, indicating that n is dispersion-less in this range. The slope of the curve directly gives us $n \approx 20.5$. Another interesting application of this is to superconductors. In figure 17 we show the transmission spectrum of LSCO at a temperature slightly above T_c and far below [8]. One can see that the position of the maxima has changed and this shows up in the polariton dispersion in an interesting way (see right panel figure 17). As in STO we see that in the normal state the dispersion is linear and extrapolates to zero. In the superconducting state the dispersion has acquired a q^2 dependence and no longer extrapolates to zero for $q \to 0$. This is the result of the opening of the superconducting gap and it implies that the polaritons in the superconducting state have acquired a mass. This is an example of the Anderson-Higgs mechanism[9], the same mechanism via which the Higgs-field gives a mass to the W and Z bosons in elementary particle physics. In the superconductor the order parameter plays the role of the Higgs-field and the spontaneously broken symmetry is that of the U(1) gauge symmetry.

TeraHertz time-domain spectroscopy

This relatively new technique is the last we will discuss here. This technique uses a

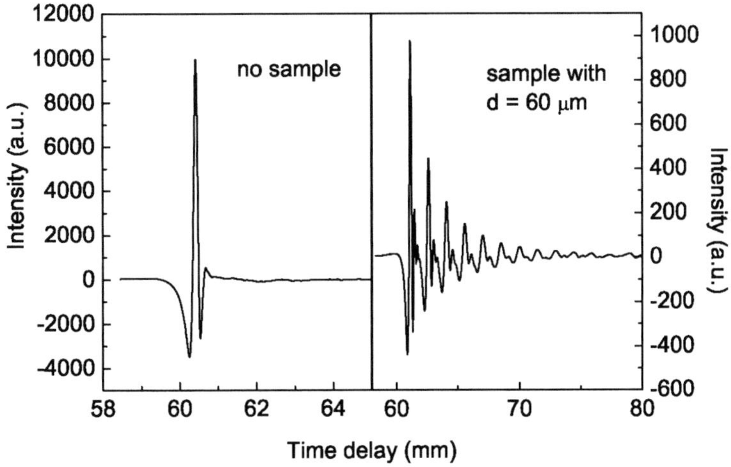

FIGURE 18. Left: recorded signal v.s. delay distance without sample. Right: recorded signal v.s. delay distance with sample. Note the extra peaks in the signal on the right due to multiple reflections in the sample.

powerful laser pulse and records the detector output as a function of time, more often expressed in an optical delay distance. The result for an experiment in vacuum is shown in figure 18 on the left. If we now insert a sample that is transparent to terahertz radiation in the path of the beam we expect that due to the different optical path length in the sample the pulse will arrive at a later time. In fact, if we use a sample with to plane parallel surfaces we expect a series of peaks due to multiple reflections in the sample (see right panel of figure 18. These peaks are just a different manifestation of the Fabry-Perot oscillations observed in transmission spectroscopy. By Fourier transforming this signal to the frequency domain and doing the same for the signal without sample we can again obtain the transmission spectrum. The frequency domain spectrum corresponding to the time domain spectra of figure 18 is shown in figure 16.

QUANTUM THEORY

We now move to the quantum theoretical description of the interaction of light and matter using the Kubo-formalism. So far we have been using a "geometrical" or macroscopic view of this interaction, but in this section we will consider the effects of the absorption of photons by electrons. Consider for simplicity a metal. The electronic states of the system are described by a set of bands, some of which are fully occupied, some partially and the rest empty, figure 19. When photons interact with these band electrons they can be absorbed and in this process the electron is excited to a higher lying band leaving

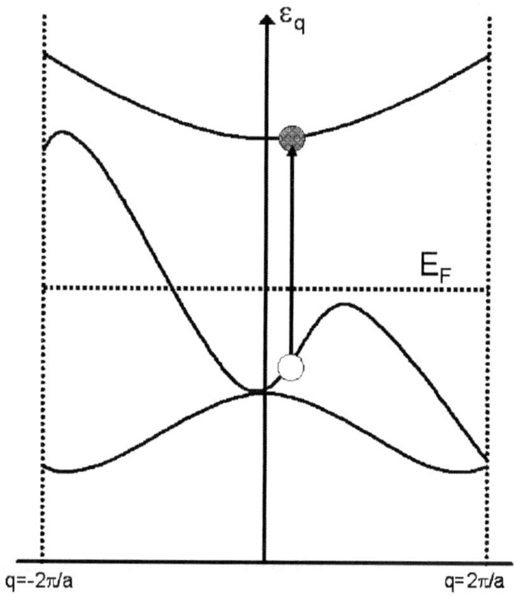

FIGURE 19. The indicated transition is an interband transition. Al states below the dashed line indicated by E_F are occupied all states above are empty.

behind a hole. In this way we create electron-hole pairs and this (dipole) transition from a state $|\Psi_\nu^N\rangle$ to a state $|\Psi_\mu^N\rangle$ is characterized by an optical matrix element,

$$M_{\mu\nu}(\vec{q}) = \left\langle \Psi_\mu^N \left| \hat{\mathbf{v}}_q \right| \Psi_\nu^N \right\rangle . \tag{105}$$

If the transition is from one band to another band we call the transition an *interband* transition and if the transition is within a band we call it a *intraband* transition.

In figure 20 we show the optical conductivity of KCl. In this compound a strong onset is seen in the optical conductivity around $\approx$8.7 eV. This onset is due to the excitation of electrons from the occupied p-band related to the Cl^- ions to the unoccupied s-band of the K^+ ions. Since this particular transition involves moving charge from the chlorine atoms to the potassium atoms this type of excitation is called a charge transfer (CT) excitation [10]. Another important feature in figure 20 are the strong peaks seen around 7.5 eV. Many theories often neglect so-called vertex corrections because these corrections cancel if the interactions between electrons are isotropic. However in real materials interactions are more often than not anisotropic and this means that these corrections have to be taken into account. The peaks seen in figure 20 are due to transitions from bound states of electron-hole pairs, called excitons, which arise due to the vertex corrections. Before we start our display of the Kubo-formalism we first introduce some notation. We introduce the field operators,

$$\psi_\sigma^\dagger(\mathbf{r}) = \sum_k e^{-i\mathbf{k}\cdot\mathbf{r}} \hat{c}_{k,\sigma}^\dagger . \tag{106}$$

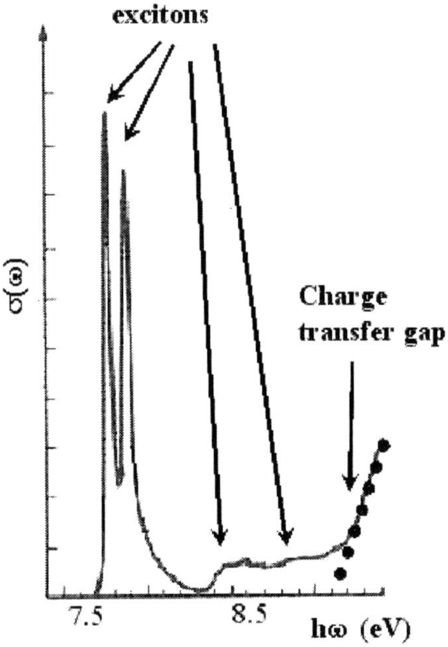

FIGURE 20. Optical conductivity of KCl. The series of strong peaks are due to excitons. The onset in absorption around 9 eV is the onset of charge transfer excitations.

The density operator is given by,

$$\hat{n}_\sigma(\mathbf{r}) = \psi_\sigma^\dagger(\mathbf{r})\psi_\sigma(\mathbf{r}). \tag{107}$$

The Fourier transform of $\hat{n}_\sigma(\mathbf{r})$ is,

$$\hat{n}_\sigma(\mathbf{r}) = \frac{1}{V}\sum_q e^{-i\mathbf{q}\cdot\mathbf{r}}\rho_q, \tag{108}$$

with

$$\rho_q = \sum_{k,\sigma} \hat{c}^\dagger_{k-q/2,\sigma}\hat{c}_{k+q/2,\sigma}. \tag{109}$$

The velocity operator is defined as,

$$\hat{\mathbf{v}}_q = \frac{\hbar}{m}\sum_{k,\sigma}\mathbf{k}\hat{c}^\dagger_{k-q/2,\sigma}\hat{c}_{k+q/2,\sigma}. \tag{110}$$

Finally, we note that the operators $\hat{n}_\sigma(\mathbf{r})$ and $\hat{\mathbf{v}}_q$ satisfy,

$$\frac{i}{\hbar}\left[\hat{n}_\sigma(\mathbf{r}),\hat{H}\right] + \nabla\cdot\hat{\mathbf{v}}_q = 0. \tag{111}$$

The Kubo-formalism

To calculate the optical conductivity from a microscopic starting point we have to add to the Hamiltonian of the system a term that describes the interaction with the electromagnetic field described by,

$$\mathbf{E}^T(\mathbf{r},t) = \frac{i\omega}{c}\sum_q \mathbf{A}_q e^{i(\mathbf{q}\cdot\mathbf{r}-\omega t)}, \tag{112}$$

Note that we have chosen the transverse gauge which we will use throughout the rest of the chapter. The interaction Hamiltonian is given by,

$$H' = -\frac{e\hbar}{c}\sum_q e^{i(\mathbf{q}\cdot\mathbf{r}-\omega t)}\mathbf{A}_q \cdot \hat{\mathbf{v}}_{-q}, \tag{113}$$

and in the presence of an electromagnetic field we use the minimal coupling,

$$\hat{\mathbf{v}}_q \rightarrow \hat{\mathbf{v}}_q - \frac{e\hbar}{mc}\mathbf{A}_q e^{i(\mathbf{q}\cdot\mathbf{r}-\omega t)}\hat{\rho}_q. \tag{114}$$

We now start by examining the current operator $\mathbf{J}(\mathbf{r},t) = \mathbf{J}^{(1)}(\mathbf{r},t) + \mathbf{J}^{(2)}(\mathbf{r},t)$. It consists of two terms the first of which is called the diamagnetic term,

$$\mathbf{J}^{(1)}(\mathbf{r},t) = -\frac{ne^2}{mc}\mathbf{A}(r,t) = \frac{ine^2}{m\omega}\mathbf{E}^T(r,t), \tag{115}$$

where in the last equality we have used Eq. (112). The second term is more difficult. It is given by,

$$\mathbf{J}^{(2)}(\mathbf{r},t) = \frac{e^2}{V}\int_{-\infty}^{t}\left\langle \mathrm{e}^{\mathrm{i}H'\tau}\mathrm{e}^{-\mathrm{i}H\tau}\hat{\mathbf{v}}(r,t)\mathrm{e}^{\mathrm{i}H\tau}\mathrm{e}^{-\mathrm{i}H'\tau}\right\rangle \mathrm{e}^{\mathrm{i}\omega(\mathrm{t}-\tau)}\mathrm{d}\tau. \tag{116}$$

We make here the approximation of using linear response theory: we expand the exponentials $e^{iH'\tau}$ to first order in $\mathbf{A}(\mathbf{r},t)$ and then stop the series expansion. After some algebra we arrive at,

$$\frac{J^{(2)}}{\mathbf{E}(\mathbf{r},t)} = \frac{ie^2}{\omega V}\sum_n \mathbf{v}_{-q}^{nm}\mathbf{v}_q^{mn}\left[\frac{1}{\omega - E_n + E_m + i0^+} - \frac{1}{\omega + E_n - E_m + i0^+}\right], \tag{117}$$

where we have defined,

$$\mathbf{v}_q^{mn} \equiv \langle \Psi_m | \hat{\mathbf{v}}_q | \Psi_n \rangle. \tag{118}$$

The result we have obtained is for zero temperature but is easily generalized to finite T if we use the grand canonical ensemble. Combining Eq. (115) and Eq. (117) we find for the optical conductivity,

$$\sigma_{\alpha,\alpha}(\mathbf{q},\omega) = \frac{iNe^2}{mV\omega} + \frac{ie^2}{V\omega}\sum_{n,m\neq n} e^{\beta(\Omega - En)}\left[\frac{v_{\alpha,q}^{nm}v_{\alpha,-q}^{mn}}{\omega - \omega_{nm} + i\eta} - \frac{v_{\alpha,-q}^{nm}v_{\alpha,q}^{mn}}{\omega + \omega_{nm} + i\eta}\right]; \tag{119}$$

where we have defined $\omega_{nm} \equiv E_n - E_m$. The optical conductivity consists now of three contributions: the diamagnetic term followed by a contribution to positive frequencies and a contribution to negative frequencies. We note that in general $\sigma_{\alpha,\alpha}(\mathbf{q},\omega)$ is a tensor as indicated by the α subscripts. We further note that the diamagnetic term does not give a real contribution to the conductivity. This term gives a δ-function contribution at zero frequency and this is exactly cancelled by two delta functions in the second part. This can be seen by using the fact that for every n we have the following relationship,

$$\sum_{n,m\neq n} \frac{v^{nm}_{\alpha,q} v^{mn}_{\alpha,-q}}{\omega_{nm}} = \frac{N}{2m}. \tag{120}$$

So we can rewrite Eq. (119) as,

$$\sigma_{\alpha,\alpha}(\mathbf{q},\omega) = \frac{ie^2}{V\omega} \sum_{n,m\neq n} e^{\beta(\Omega - En)} \left[\frac{v^{nm}_{\alpha,q} v^{mn}_{\alpha,-q}}{\omega - \omega_{nm} + i\eta} - \frac{v^{nm}_{\alpha,-q} v^{mn}_{\alpha,q}}{\omega + \omega_{nm} + i\eta} \right], \tag{121}$$

From here on we take the limit $q \to 0$ and define a generalized oscillator strength Ω_{mn} as,

$$\Omega^2_{nm} \equiv \frac{8\pi e^2 e^{\beta(\Omega - En)} |v^{nm}_{\alpha}|^2}{\omega_{nm} V}. \tag{122}$$

With this definition we are lead to the Drude-Lorentz expansion of the optical conductivity,

$$\sigma_{\alpha,\alpha}(\omega) = \frac{i\omega}{4\pi} \sum_{n,m\neq n} \frac{\Omega^2_{nm}}{\omega(\omega + i\gamma_{nm}) - \omega^2_{mn}}. \tag{123}$$

Sum Rules

Sum rules play an important role in optics. Using the equations of the previous section we derive the Thomas-Reich-Kuhn sum rule also known as the f-sum rule. The f-sum rule states that, apart from some constants, the area under σ_1 is proportional to the number of electrons and inversely proportional to their mass. This can be shown as follows: integrating Eq. (123) we have,

$$\mathrm{Re} \int_{-\infty}^{\infty} \sigma(\omega)\, d\omega = \frac{1}{4} \sum_{n,m\neq n} \Omega^2_{mn}. \tag{124}$$

Using the expression for Ω_{mn}, Eq. (122), and expression (120) we can rewrite the sum on the right hand side as,

$$\sum_{n,m\neq n} \Omega^2_{mn} = \frac{4\pi e^2 N}{mV} \sum_{n} e^{\beta(\Omega - E_n)} = \frac{4\pi e^2 N}{mV}. \tag{125}$$

So the f-sum rule states that,

$$\int_{-\infty}^{\infty} \sigma_1(\omega) d\omega = \frac{\pi e^2 N}{mV}, \tag{126}$$

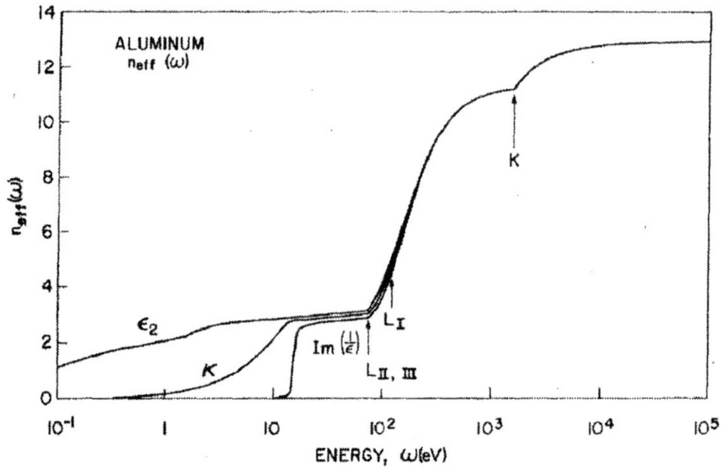

FIGURE 21. Effective number of carriers $n_{eff}(\Omega_c)$ as a function of cutoff frequency Ω_c for Al. Figure adapted from [11].

as promised. This is the full universal sum rule. It is often rewritten as an integral over positive frequencies only and using the definition of the plasma frequency ω_p,

$$\omega_p^2 \equiv \frac{4\pi e^2 N}{mV}, \tag{127}$$

as,

$$\int_0^\infty \sigma_1(\omega)d\omega = \frac{\omega_p^2}{8}. \tag{128}$$

We can also define *partial* sum rules, i.e. sum rules where we integrate up to a certain frequency cutoff Ω_c. In such a case the sum rule is not universal (this means for instance that the value of this sum rule can depend on temperature) and we can now define a plasma frequency that depends on the chosen cutoff frequency,

$$\omega_{p,valence}^2 \equiv \frac{4\pi e^2}{m} n_{eff}(\Omega_c). \tag{129}$$

A nice example of the application of the partial sum rule is shown in figure 21. Here the partial sum rule is applied to the optical conductivity of aluminum [11]. Here the effective number of carriers contributing to the sum rule is plotted as a function of Ω_c. $n_{eff}(\Omega_c)$ slowly increases to a value of roughly three around 50 eV. This means that as we increase the cutoff from zero to 50 eV we are slowly integrating over the intraband transitions and when we reach a value of 50 eV we have integrated over all transitions involving the three valence electrons. For higher energies the interband transitions start to contribute with a sharp onset near 80 eV. Finally at 10^4 eV the sum rule saturates at 13 electrons, the total number of electrons of aluminum.

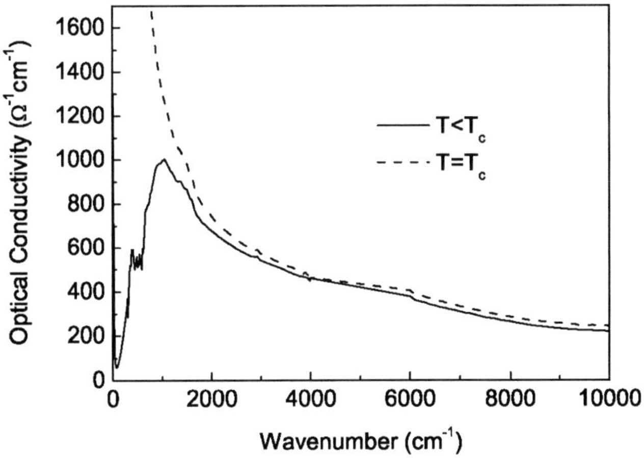

FIGURE 22. Optical conductivity of Bi-2212 at T_c and below. The difference in area between the two curves is an estimate of the superfluid density.

Another application of sum rules can be found in superconductors. In a superconductor the electrons form a superfluid condensate. This condensate shows up in the optical conductivity as a delta function at zero frequency (it contributes a diamagnetic term as in Eq. (119)). At the same time a gap opens up in low frequency part of the spectrum where the optical conductivity is (close to) zero, see figure 22. In the normal state the system is usually metallic and characterized by a Drude peak. In optical experiments we cannot measure the zero frequency response and so we cannot directly measure the spectral weight $\omega_{p,s}^2$ of the condensate. However, using sum rules we can estimate its spectral weight because the total spectral weight has to remain constant. This is summarized in the Ferrel-Glover-Tinkham (FGT) sum rule [12], which states that the difference in spectral weight between the optical conductivity in the superconducting and normal state is precisely the spectral weight of the condensate,

$$\omega_{p,s}(T)^2 = 8 \int_{0^+}^{\infty} \{\sigma(\omega, T_c) - \sigma(\omega, T)\} d\omega. \tag{130}$$

Note that we integrate here from 0^+.

There also exist sum rules for mixtures of different types of particles,

$$\int_0^{\infty} \sigma_1(\omega')d\omega' = \sum_j \frac{\pi n_j q_j^2}{2m_j}, \tag{131}$$

here the index j labels the different species. This sum rule can be applied to measure

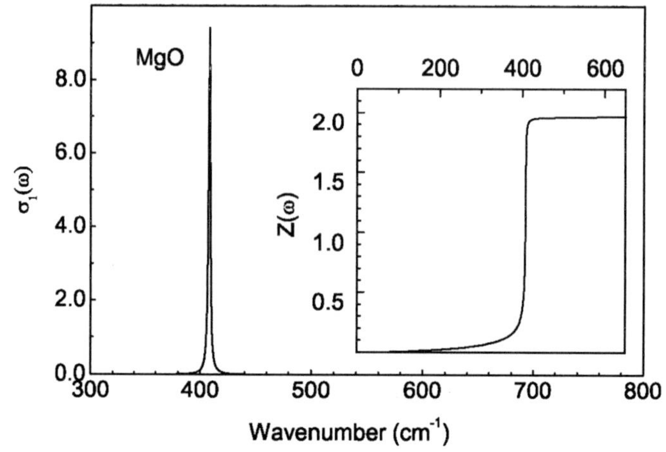

FIGURE 23. Optical conductivity due to phonon mode in MgO. The area under the peak is proportional to the effective charge of the mode. The inset shows the effective charge calculated using (132). Data from [13].

the charge of ions involved in vibrational modes. If we can separate the contribution to the optical conductivity due to the optical modes we can invert Eq. (131) to calculate the effective charge related to the mode. For example, in MgO (figure 23) both ions contribute an equal charge $q_{Mg} = -q_O$. We define the effective mass μ as $\mu^{-1} = m_{Mg}^{-1} + m_O^{-1}$ and assume that the density of the two is equal. In that case we can rewrite Eq. (131) as,

$$Z(\omega)^2 \equiv \left(\frac{q_T^*(\omega)}{e}\right)^2 \equiv \frac{2\mu}{\pi n e^2} \int_{\omega_{min}}^{\omega_{max}} \sigma_{ph}(\omega')\, d\omega', \tag{132}$$

where the integral has to be taken in a frequency range such that it includes the spectral weight of the optical phonon mode but nothing else.

We will now derive expressions for the conductivity sum rule from a more microscopic point of view. To do that we return to the Kubo expression for the optical conductivity,

$$\sigma_1(\omega) = \frac{\pi e^2}{V} Tr \langle \Psi_n | \hat{\mathbf{v}} \left\{ \frac{\delta(\omega - \hat{H} + E_n)}{\hat{H} - E_n} + \frac{\delta(\omega + \hat{H} - E_n)}{\hat{H} - E_n} \right\} \hat{\mathbf{v}} | \Psi_n \rangle . \tag{133}$$

The Hamiltonian in this expression is that of the system of interacting electrons without the interaction of light. It represents the optical conductivity for the system in an arbitrary (ground or excited) many-body state $|\Psi\rangle$. A peculiar point of this expression is that although the velocity operators create a single electron-hole pair, due to the fact that the hamiltonian in the denominator of this expression still contains the interactions between all particles in the system, the optical conductivity represents the response from the full collective system of electrons. If we integrate this expression over frequency we get,

$$\int_{-\infty}^{\infty} \sigma_1(\omega)\, d\omega = \frac{2\pi e^2}{V} Tr \langle \Psi_n | \hat{\mathbf{v}} \frac{1}{\hat{H} - E_n} \hat{\mathbf{v}} | \Psi_n \rangle . \tag{134}$$

We now take a closer look at the right-hand side of this expression. Remember that,

$$\hat{\mathbf{v}} = \frac{i}{\hbar} \left[\hat{H}, \hat{\mathbf{x}} \right] . \tag{135}$$

Using the commutator we can rewrite,

$$-2i\hbar \hat{\mathbf{v}} \frac{1}{\hat{H} - E_n} \hat{\mathbf{v}} = \left(\hat{H}\hat{\mathbf{x}} - \hat{\mathbf{x}}\hat{H} \right) \frac{1}{\hat{H} - E_n} \hat{\mathbf{v}} + \hat{\mathbf{v}} \frac{1}{\hat{H} - E_n} \left(\hat{H}\hat{\mathbf{x}} - \hat{\mathbf{x}}\hat{H} \right) . \tag{136}$$

Inserting this back into Eq. (134) we find after some rearranging

$$\int_{-\infty}^{\infty} \sigma_1(\omega)\, d\omega = \frac{i\pi e^2}{\hbar V} \langle [\hat{\mathbf{v}}, \hat{\mathbf{x}}] \rangle , \tag{137}$$

where $\langle ... \rangle$ stands for the trace over all many-body states. Here we have used that,

$$\hat{\mathbf{x}}\hat{H} \frac{1}{\hat{H} - E_n} \hat{\mathbf{v}} = \hat{\mathbf{x}} \left(\hat{H} - E_n \right) \frac{1}{\hat{H} - E_n} \hat{\mathbf{v}} + \hat{\mathbf{x}} E_n \frac{1}{\hat{H} - E_n} \hat{\mathbf{v}} = \hat{\mathbf{x}}\hat{\mathbf{v}} + \hat{\mathbf{x}} E_n \frac{1}{\hat{H} - E_n} \hat{\mathbf{v}}, \tag{138}$$

and the fact that $\hat{H}|\Psi_n\rangle = E_n|\Psi_n\rangle$. We can now obtain different expressions for the sum rule by working out the commutator on the right-hand side of Eq. (137) based on different model assumptions. In table 2 we summarize some results. The sum rule for

TABLE 2. Expressions for the commutator in Eq. (137) for three different cases. N.N. stands for Nearest Neighbors tight binding model

Free electrons	$[\hat{\mathbf{v}}, \hat{\mathbf{x}}] = \frac{\hbar}{im} \sum_{k\sigma} \hat{n}_{k\sigma}$
Band electrons	$[\hat{\mathbf{v}}, \hat{\mathbf{x}}] = \frac{\hbar}{im} \sum_{k\sigma} \hat{n}_{k\sigma} [\hat{\mathbf{v}}, \hat{\mathbf{x}}] = \frac{\hbar}{i} \sum_{k\sigma} \frac{\partial^2 \varepsilon_{k\sigma}}{\partial k^2} \hat{n}_{k\sigma}$
N.N.	$[\hat{\mathbf{v}}, \hat{\mathbf{x}}] = -\frac{\hbar a^2}{i} \sum_{k\sigma} \varepsilon_{k\sigma} \hat{n}_{k\sigma}$

band electrons is in practice the most useful. Suppose that we have a system with only a single reasonably well isolated band around the Fermi level that can be approximated by a tight binding dispersion $\varepsilon_k = -t \cos(ka)$. In that case we find an interesting relation,

$$\int_0^{\Omega_c} \sigma_1(\omega, T) d\omega = -\frac{\pi e^2 a^2}{2\hbar^2 V} \sum_{k,\sigma} \langle \hat{n}_{k\sigma} \varepsilon_k \rangle_T = -\frac{\pi e^2 a^2}{2\hbar^2 V} E_{kin}(T). \tag{139}$$

This sum rule states that by measuring the optical spectral weight we are in fact measuring the kinetic energy of the charge carriers contributing to the optical conductivity. In real systems this relation only holds approximately: usually there are other bands lying nearby and the integral on the left contains contributions from these as well. Often the bands are described by more complicated dispersion relations in which case the relation $\partial^2 \varepsilon_k / \partial k^2 = -\varepsilon_k$ does not hold. We can make some other observations from the sum rule for band electrons. Suppose again we have a single empty cosine like band (it is only necessary that the band is symmetric but it simplifies the discussion) at $T = 0$. Since the band is empty, the spectral weight is equal to zero. If we start adding electrons the spectral weight starts to increase until we reach half-filling. If we add more electrons the spectral weight will start to decrease again because the second derivative becomes negative for $k > \pi/2a$. If we completely fill the band the contributions from $k > \pi/2a$ will precisely cancel the contributions from $k < \pi/2a$ and the spectral weight is again zero. Now consider what happens if we have a half-filled band and start to increase the temperature. Due to the smearing of the Fermi-Dirac distribution higher energy states will get occupied leaving behind lower energy empty states. The result of this is that the spectral weight starts to decrease. One can show using the Summerfeld expansion that the spectral weight follows a T^2 temperature dependence. In the extreme limit of $T \longrightarrow \infty$ something remarkable happens: the Fermi-Dirac distribution is 1/2 everywhere and the electrons are equally spread out over the band. The metal has become an insulator!

Applications of sum rules to superconductors

Before we have a look at some applications of sum rules to superconductors we first summarize some results from BCS theory. We want to apply our ideas to cuprate superconductors so we use a modified version from the original theory to include the possibility of d-wave superconductivity. In other words we suppose that there is some attractive interaction between the electrons that has a momentum dependence. The energy difference between the normal and superconducting state due to interactions can be written as [14],

$$\langle \hat{H}_s^{int} \rangle - \langle \hat{H}_s^{int} \rangle = \int d^3 r g(r) V(r) = \sum_k g_k V_k, \tag{140}$$

where $g(r)$ and g_k are the pair correlation function and its fourier transform respectively. We can find an expression for g_k,

$$g_k = \sum_q \frac{\Delta_{q+k} \Delta_q^*}{4 E_{q+k} E_q}. \tag{141}$$

As usual,

$$E_k = \sqrt{(\varepsilon_k - \mu)^2 + \Delta^2}, \tag{142}$$

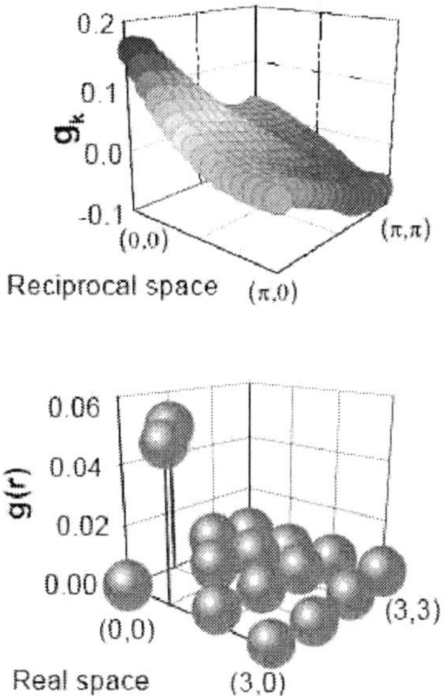

FIGURE 24. Real and momentum space picture of the correlation functions $g(r)$ and g_k. Figure adapted from [14].

and the temperature dependence of Δ_k is given by,

$$\Delta_k = \sum_q \frac{V_q \Delta_q}{2E_q} \tanh\left(\frac{E_k}{2k_b T}\right). \tag{143}$$

We now use a set of parameters extracted from ARPES measurements to do some numerical simulations. First of all we calculate g_k and fourier transform it to obtain $g(r)$. The results are shown in figure 24.

Although g_k is not so illuminating $g(r)$ is. This function is zero at the origin and strongly peaked at the nearest neighbor sites. This is a manifestation of the d-wave symmetry. We also note that the correlation function drops off very fast for sites removed further from the origin. In figure 25 we show the results for a calculation of the correlation and kinetic energy using the parameters extracted from ARPES measurements on Bi-2212. The kinetic energy is calculated from,

$$\langle \hat{H}_{kin} \rangle = \sum_k \varepsilon_k \{1 - \frac{\varepsilon_k - \mu}{E_k} \tanh\left(\frac{E_k}{2k_b T}\right)\}. \tag{144}$$

We see that the kinetic energy *increases* in the superconducting state. This can be easily understood by looking at what happens to the particle distribution function below T_c, as indicated in the left panel of figure 26: when the system enters the superconducting state the area below the Fermi energy decreases and the area above the Fermi energy increases thereby increasing the total kinetic energy of the system. Nevertheless the total internal

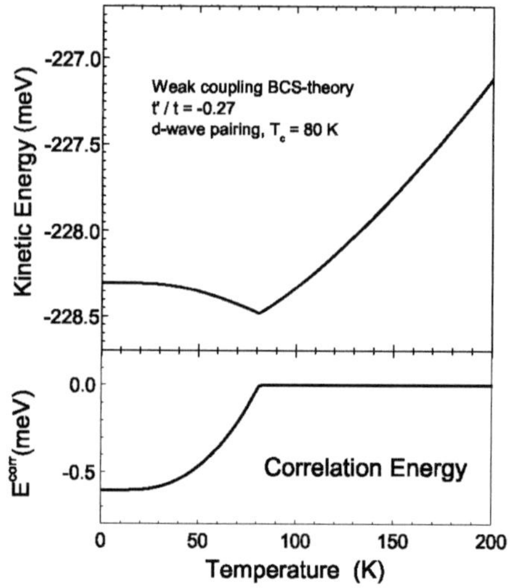

FIGURE 25. Correlation energy and kinetic energy as a function of temperature for a d-wave BCS superconductor. Figure adapted from [14].

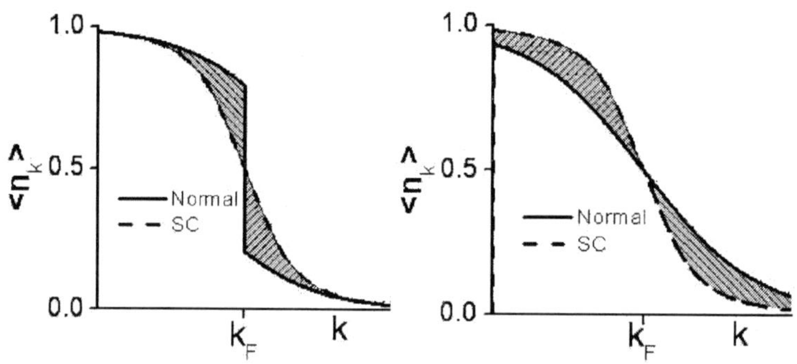

FIGURE 26. Left: Distribution function for the normal (Fermi liquid like) state and the superconducting state. Right: Distribution function for a non-Fermi Liquid like state and the superconducting state.

energy, which is the sum of the interaction energy and the kinetic energy, decreases and this is of course why the system becomes superconducting. Now let us take a look at what happens in the cuprates. In figure 27 we display the optical spectral weight $W(\Omega_c, T)$ as a function of T^2 for Bi-2223. To compare this to the BCS kinetic energy we have plotted here $-W(\Omega_c, T)$. This result is contrary to the result from our calculation: the kinetic energy decreases in the superconducting state. This experimental result,

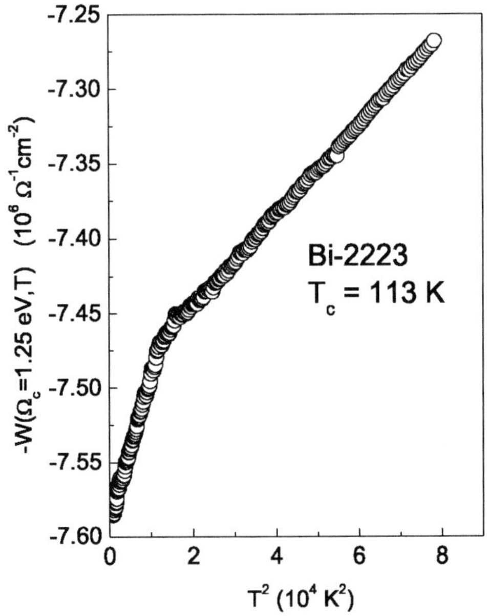

FIGURE 27. Temperature dependent spectral weight of Bi-2223. Data taken from ref. [19].

observed first by Molegraaf *et al.* [15], has sparked a lot of interest both experimentally [5, 16, 17, 18, 19] and theoretically [20, 21, 22, 23, 24, 25, 26, 27, 28]. We note that DMFT calculations with the Hubbard model as starting point have shown the same effect as observed here [27]. Roughly speaking the effect is believed to be due to the "strangeness" of the normal state (right panel figure 26). It is well known that the normal state of the cuprates shows non Fermi-liquid behavior. So if the distribution function in the normal state does not show the characteristic step of the Fermi liquid at the Fermi energy but is rather a broadened function of momentum it is very well possible that the argument we made for the increase of the kinetic energy (see above) is reversed.

Applications of sum rules: the Heitler-London model

Another interesting application of sum rules is that we can use them in some cases to extract the hopping parameters of a system. In order to see how this works we express the optical conductivity at zero temperature,

$$\sigma_1(\omega) = \frac{\pi e^2}{V} \langle \Psi_g | \hat{\mathbf{v}} \frac{\delta(\omega - \hat{H} + E_g)}{\hat{H} - E_g} \hat{\mathbf{v}} | \Psi_g \rangle, \tag{145}$$

in terms of the dipole operator. Here $|\Psi_g\rangle$ is the groundstate of the system. To do this we make use of the commutator Eq. (135) and the insertion of a complete set of states. After integrating over frequency we get

$$\int_0^\infty \sigma_1(\omega)\,d\omega = \frac{\pi e^2}{\hbar^2 V}\sum_n (E_n - E_g)\,|\langle n|\,\hat{\mathbf{x}}\,|g\rangle|^2. \tag{146}$$

We note that this can be done only for finite system sizes. Now consider the special case of a diatomic molecule with two energy levels, one on each atom and a hopping parameter t and distance d between the two atoms. We also assume that there is a splitting Δ between the two levels. The hamiltonian for such a system is,

$$H = t\sum_\sigma \left(\psi^t_{L,\sigma}\psi_{R,\sigma} + \psi^t_{R,\sigma}\psi_{L,\sigma}\right) + \frac{\Delta}{2}\left(\hat{n}_R - \hat{n}_L\right) + U\left(\hat{n}_{L\uparrow}\hat{n}_{L\downarrow} + \hat{n}_{R\uparrow}\hat{n}_{R\downarrow}\right). \tag{147}$$

The indices L and R indicate the left and right atom respectively. If we now put 1 electron in this system we have a two-level problem that is easily diagonalized. As usual we make bonding and anti-bonding states,

$$\left|\psi_{g,\sigma}\right\rangle = u\left|\psi_{l,\sigma}\right\rangle + v\left|\psi_{R,\sigma}\right\rangle, \tag{148}$$

$$\left|\psi_{e,\sigma}\right\rangle = v\left|\psi_{l,\sigma}\right\rangle - u\left|\psi_{R,\sigma}\right\rangle. \tag{149}$$

The coefficients u and v are given by,

$$u = \frac{1}{\sqrt{2}}\sqrt{1+\frac{\Delta}{E_{CT}}}; \qquad v = \frac{1}{\sqrt{2}}\sqrt{1-\frac{\Delta}{E_{CT}}}. \tag{150}$$

The bonding and anti-bonding states are split by an energy E_{CT},

$$E_{CT} = \sqrt{\Delta^2 + 4t^2}. \tag{151}$$

We are now in position to calculate the transition matrix element appearing in Eq. (146). The position operator can be represented by,

$$\hat{\mathbf{x}} = \frac{d}{2}\left(\hat{n}_R - \hat{n}_{L\cdot}\right). \tag{152}$$

So the matrix element is,

$$\begin{aligned}\left\langle \psi_{g,\sigma}\right|\hat{\mathbf{x}}\left|\psi_{e,\sigma}\right\rangle &= \left(u\langle\Psi_L| + v\langle\Psi_R|\right)\frac{d}{2}\left(\hat{n}_R - \hat{n}_L\right)\left(u|\Psi_L\rangle - v|\Psi_R\rangle\right)\\ &= -\frac{d}{2}(uv) = -\frac{t}{E_{CT}}d\end{aligned} \tag{153}$$

Using this in the sum rule Eq. (146) finally gives us the spectral weight of this model,

$$\int_0^\infty \sigma_1(\omega)\,d\omega = \frac{e^2\pi d^2}{\hbar^2 V}\frac{t^2}{\sqrt{\Delta^2+4t^2}}. \tag{154}$$

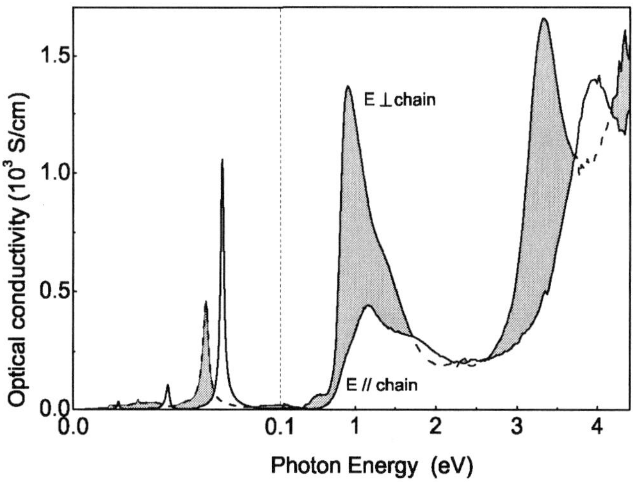

FIGURE 28. Optical conductivity of α-NaV_2O_5 for two polarizations: one with the field parallel to the chains and one perpendicular. Data taken from ref. [29].

We see that there is a very simple relation between the spectral weight of this model and the hopping parameter. This sumrule has been applied to α-NaV_2O_5 [29]. This compound is a so-called ladder compound. It consists of double chains of vanadium atoms forming ladders which are weakly coupled to each other. Each unit cell contains 4 V atoms and 2 valence electrons. The vanadiums on the rungs of the ladder are more strongly coupled than those along the legs, i.e. $t_\perp \gg t_\parallel$. The Heitler-London model we have discussed above can be applied to this system since each rung forms precisely a two level system with different levels. The only difference that we have to take into account is that this is a crystal consisting of N independent two level systems. Figure 28 shows the optical conductivity of α-NaV_2O_5.

There are two measurements shown: one with light polarized parallel to the chains and one with light polarized perpendicular to the chains. We can immediately read of $E_{CT} \approx 1$ eV. Integrating the contribution to the optical conductivity of the peaks we find that the spectral weight perpendicular to the chains is roughly 4 times larger then the spectral weight parallel to the chains, so $t_\perp \approx 4t_\parallel$. Inverting Eq. (154) we can calculate $t_\perp$ and we find $t_\perp \approx 0.3$ eV. The second strong peak at approximately 3.2 eV is a charge transfer peak from vanadium to oxygen. We will come back to this example in the last section on spin interactions.

Generalized Drude formalism

We have already encountered the Drude formula for the optical conductivity of a metal (see the section on polaritons). Even though this model is based on a classical gas of non-interacting particles it describes amazingly well the optical properties of a good metal. This is even more surprising if one realizes that in a metal electrons reside in bands and that the transitions we are making with photons are vertical due to the negligible photon momentum. So from the band picture point of view, when we consider a single band of electrons interacting with photons we should expect a single delta function at the origin. The reason that this is not what is observed is because we have neglected the other interactions in the system. Electrons in solids interact with the lattice vibrations, impurities and/or other collective modes. Due to the electron-phonon interaction for instance we can have processes where a photon creates an electron-hole pair in which the electron "shakes off" a phonon. In this process the phonon can carry away a much larger momentum then originally provided by the photon. Due to this effect we can have phonon-assisted transitions which give a width $1/\tau$ to the delta function. This width is called the scattering rate. If the interactions are inelastic, as in the interactions with impurities, this scattering rate is just a constant. Otherwise, this scattering rate can depend on frequency. However, if we define the scattering rate in Eq. (58) to be frequency dependent, $1/\tau \equiv 1/\tau(\omega)$, the KK-relations force us to introduce a frequency dependent effective mass as well. This is what is done in the generalized Drude formalism [30]. The optical conductivity is written as,

$$\sigma(\omega) = \frac{ne^2/m}{\tau^{-1}(\omega) - i\omega m*(\omega)/m}. \tag{155}$$

Having measured a conductivity spectrum we can invert these equations to calculate $1/\tau(\omega)$ or $m^*(\omega)/m$ via,

$$\tau^{-1}(\omega) \equiv \mathrm{Re}\frac{ne^2/m}{\sigma(\omega)} = \Sigma''(\omega), \tag{156}$$

and

$$\frac{m^*(\omega)}{m} \equiv \mathrm{Im}\frac{-ne^2/m}{\omega\sigma(\omega)} = 1 + \frac{\Sigma'(\omega)}{\omega}. \tag{157}$$

In the last equality of these equations we have defined an optical self-energy. Note that this quantity is *not* equivalent to the self-energy used in the context of Green's functions. We can rewrite the optical conductivity in terms of $\Sigma(\omega)$ as,

$$\sigma(\omega) = \frac{ne^2}{m}\frac{i}{\omega + \Sigma(\omega)}. \tag{158}$$

In the case of impurity scattering $\Sigma(\omega)$ is simply given by

$$\Sigma(\omega) = i/\tau_0, \tag{159}$$

so that $1/\tau(\omega) = 1/\tau_0$ and $m*(\omega)/m = 1$. We can also capture the effect of the

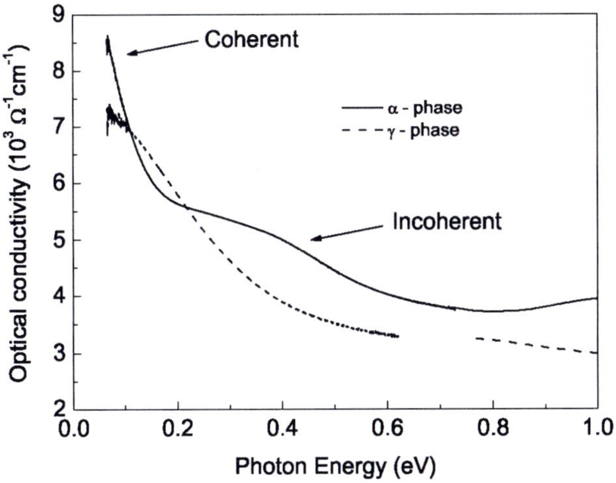

FIGURE 29. Optical conductivity of Ce in the α and γ phases. Data taken from ref. [31].

interaction of the electrons with the static lattice potential in a self-energy,

$$\Sigma(\omega) = \lambda\omega, \tag{160}$$

which gives $\tau^{-1}(\omega) = 0$ and $m^*(\omega)/m = 1 + \lambda$. This is also called static mass renormalization. Finally we consider dynamical mass renormalization where the electrons couple to a spectrum of bosons,

$$\Sigma(\omega) = \frac{\lambda\omega}{1 - i\omega/T^*}, \tag{161}$$

Here λ is a coupling constant and T^* is a characteristic temperature scale related to the bosons. In this case we find,

$$\tau^{-1}(\omega) = \lambda T^* \frac{\omega^2}{T^{*2} + \omega^2}, \tag{162}$$

and

$$\frac{m^*(\omega)}{m} = 1 + \lambda \frac{T^{*2}}{T^{*2} + \omega^2}. \tag{163}$$

As an example we will discuss the α-phase to γ-phase transition in pure Cerium. When Cerium is grown at elevated temperatures it forms in the so called γ-phase. At low temperatures a volume collapse occurs and the resulting phase is called the α-phase. This iso-structural transition is first order. The reduction in volume can be as much as 20 to 30 %. Ce has 4 valence electrons and these can be distributed between

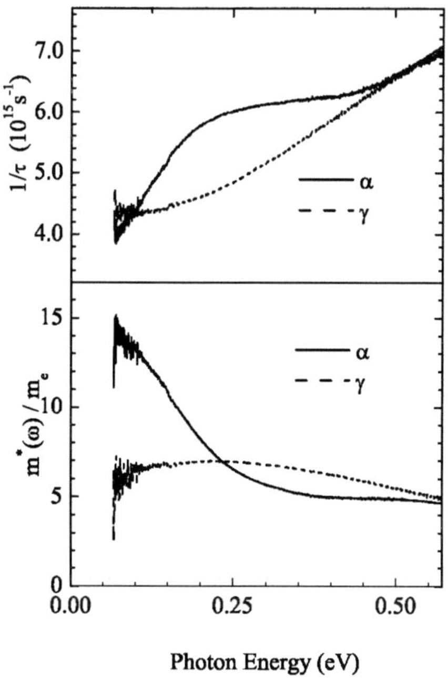

FIGURE 30. Scattering rate and effective mass of Ce in the α and γ phases. Data taken from ref. [31].

localized $4f$ states and the $5d$ states that form the conduction band. If occupied, the 4f states will act as paramagnetic impurities. In the γ-phase the Kondo temperature $T_K \approx$ 100 K whereas in the α-phase $T_K \approx 2000$ K. This difference can be understood to be simply due to the larger lattice spacing in the γ-phase: the hopping integral t is smaller and hence T_K is smaller. Figure 29 shows the optical conductivity of α- and γ-Cerium. These measurements were done by depositing Ce films on a substrate at high and low temperature to form either the α or γ phase. We see that γ-Cerium is less metallic than α-Cerium. In the γ-phase there is only a weak Kondo screening of the impurity magnetic moments and this gives rise to spin flip scattering, which is the main source of scattering. In the α-phase the moments are screened and form renormalized band electrons in a very narrow band. This suppresses the scattering. The difference in scattering rates shows up in the optical conductivity as a narrower Drude peak for the α-phase (see figure 29). Figure 30 displays the scattering rate and effective mass extracted from the optical conductivity in figure 29 using Eq. (156) and (157). In the γ-phase $1/\tau(\omega)$ extrapolates to a finite value due to local moment or spin-flip scattering. We can rewrite the real part of the optical conductivity in Eq. (158) with the self-energy of Eq. (161) as follows,

$$\frac{4\pi}{\omega_p^{*2}}\sigma(\omega) = \frac{i}{\omega}\frac{1}{1+\frac{\lambda}{1-i\omega/T^*}}. \tag{164}$$

Note that we have defined a renormalized plasma frequency, ω_p^*, since the spectral weight is not conserved when adding $\Sigma(\omega)$ to $\sigma(\omega)$. With some simple algebra this can be rewritten as,

$$\frac{4\pi}{\omega_p^{*2}}\sigma(\omega) = \frac{i}{\omega} + \lambda \frac{T^*}{\omega^2 + i\omega(1+\lambda)T^*}. \tag{165}$$

It follows that the real part of this expression is then,

$$\frac{4\pi}{\omega_p^{*2}}Re\sigma(\omega) = \frac{\pi}{2}\delta(\omega) + \lambda \frac{T^*}{\omega^2 + (1+\lambda)^2 T^{*2}}. \tag{166}$$

We see that the optical conductivity is split into two contributions: a δ-function which represents the coherent part of the charge response and an incoherent contribution. The δ-function is usually broadened due to other scattering channels present in the system. In this case the δ-function represents the contribution due to the Kondo-peak whereas the incoherent contribution is due to the side-bands. This splitting of the conductivity in a coherent and incoherent contribution is nicely observed in the α-phase of Cerium as indicated in figure 29. This splitting of the optical conductivity in coherent and incoherent contributions is much more general however and is frequently observed in correlated electron systems.

ELECTRON-PHONON COUPLING

Electron-phonon coupling is most easily described in the framework of Migdal-Eliashberg theory. The application of the theory to optics can be found in the papers by Allen [30]. In the so-called Allen approximation the self-energy in Eq. (158) is calculated using,

$$\Sigma(\omega) = -2i \int_0^\infty d\Omega \alpha_{tr}^2 F(\Omega) K(\frac{\omega}{2\pi T}, \frac{\Omega}{2\pi T}). \tag{167}$$

Here the kernel $K(\frac{\omega}{2\pi T}, \frac{\Omega}{2\pi T})$ is given by,

$$\mathrm{K(x,y)} = \frac{\mathrm{i}}{\mathrm{y}} + \frac{y-x}{x}[\Psi(1-ix+iy) - \Psi(1+iy)] + \frac{y+x}{x}[\Psi(1-ix-iy) + \Psi(1-iy)]. \tag{168}$$

where the $\Psi(x)$ are Digamma functions. The function $\alpha_{tr}^2 F(\Omega)$ appearing in Eq. (167) is the phonon spectral function. The label "tr" stands for transport indicating that the spectral function is related to a transport property. This function is different by a multiplicative factor from the true $\alpha^2 F(\Omega)$ as measured by for instance tunnelling. The electron-phonon coupling strength is easily calculated from $\alpha_{tr}^2 F(\Omega)$ by integration,

$$\lambda_{tr} = 2\int_0^\infty \frac{\alpha_{tr}^2 F(\Omega)}{\Omega} d\Omega. \tag{169}$$

This approach was first applied by Timusk and Farnworth in a comparison of tunnelling

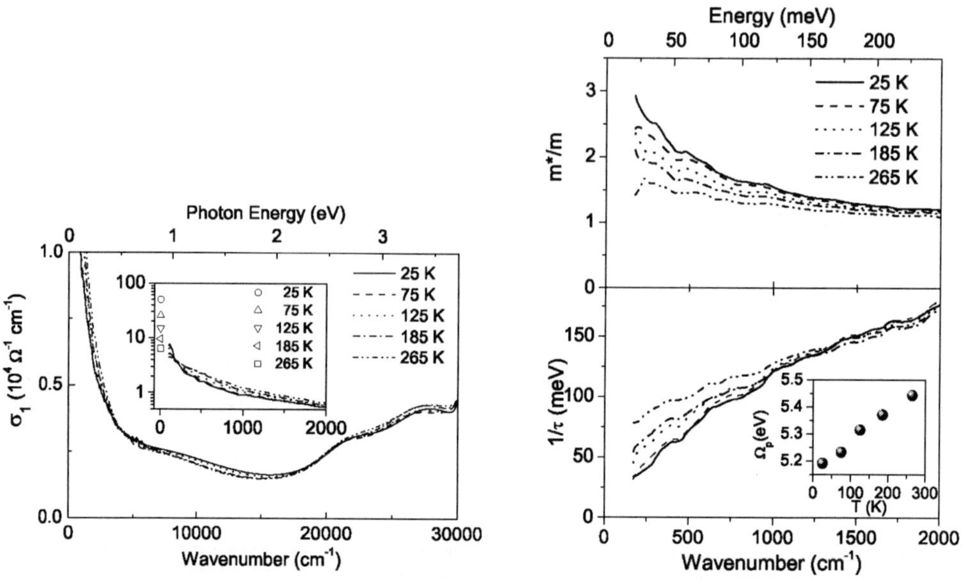

FIGURE 31. Left: Optical conductivity of ZrB_{12} at selected temperatures. Right: $1/\tau$ and m^*/m_b for several temperatures. Data taken from ref. [33].

and optical measurements on the superconducting properties of Pb [32]. As an example we discuss the application of this formalism to the optical properties of ZrB_{12} [33]. Figure 31 shows the optical conductivity of ZrB_{12}. The spectrum consists of what appears to be a Drude peak and some interband contributions. Also shown are the calculated $1/\tau(\omega)$ and $m^*(\omega)/m_b$. The temperature dependence of $1/\tau(\omega)$ is what is usually observed for a narrowing of the Drude peak with decreasing temperature whereas the strong frequency dependence is suggestive of electron-phonon interaction. Using the McMillan formula (169) the coupling strength was estimated to be $\lambda_{tr} \approx 0.7$. In figure 32 the reflectivity of ZrB_{12} together with calculations based on Eq. (158) and (167) is shown. It is clear that a simple Drude form is not capable of describing the observed reflectivity. The first fit (fit 1) is a fit where the $\alpha_{tr}^2 F(\Omega)$ that was used as input was derived from specific heat measurements [34]. Although it gives an improvement over the standard Drude fit there is still some discrepancy between the data and the fit. To make further improvements $\alpha_{tr}^2 F(\Omega)$ was modelled using a sum of δ-functions. The results of this modelling are indicated as fit 2 and fit 3. Using Eq. (169) we find coupling strengths $\lambda_{tr} \approx 1$ - 1.3. Another method to roughly estimate $\alpha_{tr}^2 F(\Omega)$ is due to Marsiglio [35, 36]. It states that a rough estimate of the shape of $\alpha_{tr}^2 F(\Omega)$ can be found by simply differentiating the optical data,

$$\alpha_{tr}^2 F(\Omega) = \frac{1}{2\pi}\frac{\Omega_p^2}{4\pi}\frac{d^2}{d\omega^2} Re(\frac{1}{\sigma(\omega)}), \tag{170}$$

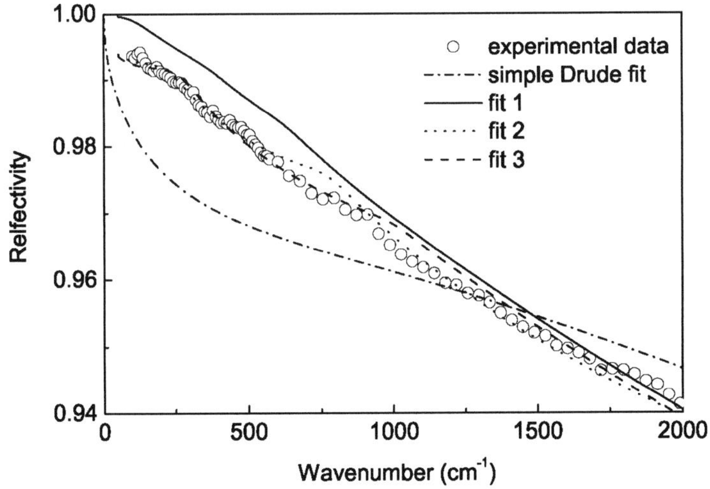

FIGURE 32. Reflectivity of ZrB_{12} around 20 K together with calculations as explained in the text. Figure adapted from ref. [33].

where Ω_p is the plasma frequency. The obvious problem with this method is that it requires the double derivative of the data. Because of the inevitable noise in the data usually some form of smoothing is required. Applied to ZrB_{12} the extracted $\alpha_{tr}^2 F(\Omega)$ shows peaks at the same positions as the ones extracted before and a coupling strength $\lambda_{tr} \approx 1.1$. These results indicate a medium to strong electron-phonon coupling for ZrB_{12}.

POLARONS

There exist many definitions of what is a polaron. Electrons coupled to a phonon have been called polaron as have free electrons moving around in an insulator. Here we will consider the Landau-Pekar approximation for a polaron [37, 38]. The idea is that when an electron moves about the crystal it polarizes the surrounding lattice and this in turn leads to an attractive potential for the electron. If the interaction between electron and lattice is sufficiently strong this potential is capable of trapping the electron and it becomes more or less localized. The new object, electron plus polarization cloud is called polaron. This self-trapping of electrons can occur in a number of different situations and different names are used. For instance, one talks about small polarons in models where only short range interactions are considered, because this typically leads to polaron formation with polarons occupying a single lattice site. From the Landau-Pekar formalism we can get some feeling of when polarons form and what their properties will be. First of all, the coupling constant α is given by,

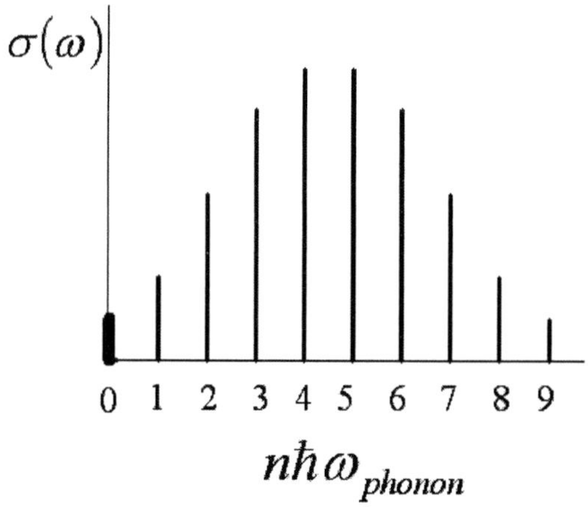

FIGURE 33. Schematic of the optical conductivity of electrons interacting with a single Einstein mode.

$$\alpha^2 = \frac{Ry}{\hbar\omega_0\tilde{\varepsilon}_\infty^2}\frac{m_b}{m_e}, \tag{171}$$

where Ry stands for the unit Rydberg (1 Rydberg = $m_e e^4/2\hbar^2$ = 13.6 eV), ω_0 is the oscillator frequency of the (Einstein) phonon mode involved m_b and m_e are the band and free electron mass respectively and $\tilde{\varepsilon}$ is given by,

$$\frac{1}{\tilde{\varepsilon}} = \frac{1}{\varepsilon_{\infty,IR}} - \frac{1}{\varepsilon(0)}. \tag{172}$$

For strong coupling (small polarons) the polaron mass is expressed in terms of the coupling constant as,

$$m_{pol} = m_b(1+0.02\alpha^4), \tag{173}$$

with the polaron binding energy given by,

$$E_{pol} = 0.1\frac{Ry}{\tilde{\varepsilon}_\infty^2}\frac{m_b}{m_e}. \tag{174}$$

The polaron binding energy typically is of the order of a few 100 meV. The polaron mass is typically of the order of 50-100 times the electron mass. The effect of polaron formation on the optical conductivity can be described by assuming a gas of non-interacting polarons (i.e. low polaron density). This results in a spectrum that can be described by a Drude peak and a so-called Holstein side-band. If we assume that the electrons interact with a single Einstein mode the spectrum will look as in figure 33. The spectrum consists of a zero-phonon, coherent part (n = 0) with a spectral weight $1/(1+0.02\alpha^4)$ followed by a series of peaks that describe the incoherent movement of polarons assisted by n=1,2,3.. phonons. In real solids the peaks are smeared out due to the fact that phonons form bands. The real part of the optical conductivity can thus be

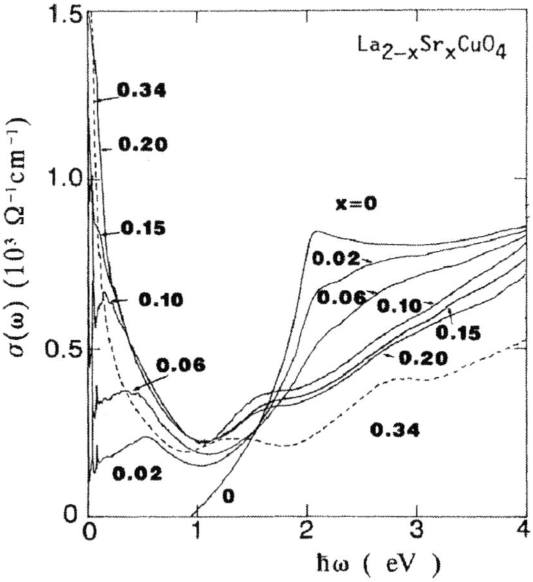

FIGURE 34. Doping dependence of the room temperature optical conductivity of $La_{2-x}Sr_xCuO_4$. Figure adapted from Uchida *et al.*, ref. [39].

described as,

$$\mathrm{Re}\frac{4\pi}{\omega_p^{*2}}\sigma(\omega) = \frac{1}{1+0.02\alpha^4}\frac{\pi\delta(\omega)}{2} + \frac{0.02\alpha^4}{1+0.02\alpha^4}E_{pol}^{-1}\exp\left\{-\left(\frac{\omega^2-E_{pol}^2}{cE_{pol}^2}\right)^2\right\}. \tag{175}$$

The first term in this expression describes the coherent part of the spectrum, which in real solids will also be smeared out to finite frequency by other forms of scattering, and an incoherent term given by the second term which is called the Holstein band. The shape of the side-band can be qualitatively understood by imagining how a polaron has to move through the lattice. In order to move from one site to another the lattice deformation around the original site has to relax and be adjusted on the new site. This relaxation process results in the multi-phonon side-bands of the Drude peak.

The observation of the Holstein side-band is somewhat complicated because it is not possible to distinguish between normal interband transitions and the effects due to polaron formation. There have been some claims that a band observed in the mid infrared region ($\approx$ 100 - 500 meV) of the spectrum of high-T_c superconductors is due to polaron formation but many other interpretations exist. Figure 34 shows the doping dependence of $La_{2-x}Sr_xCuO_4$. The peak that occurs around 0.5 eV for the 0.02 doped sample has been interpreted as the Holstein side-band. Another example where polarons could play a role is in $LaTiO_{3.41}$ [40]. In this material the resistivity (figure 35) shows a quasi one dimensional behavior with an upturn of the resistivity at lower temperatures. This could

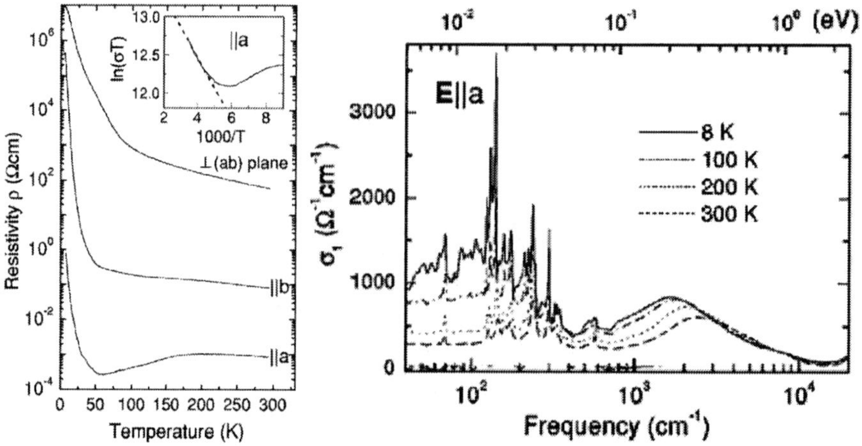

FIGURE 35. Left panel: temperature dependent resistivity of $LaTiO_{3.41}$. Right panel: Optical conductivity for selected temperatures. Figure adapted from ref. [40].

be due to polaron formation but it has also been interpreted as due to a charge density wave. The optical conductivity at low temperatures shows that a large part of the spectral weight is contained in a side-band around 300 meV (see figure 35). If this peak would be due to polarons we expect that when we warm up the system to higher temperatures its spectral weight should be diminished. This is because the increased temperature unbinds electrons from their self-trapping potential and therefore shifts spectral weight from the Holstein band to the Drude peak. This is also what is observed and at the same time explains the decrease of resistivity with increasing temperature.

The last example we will discuss is NaV_6O_{15}. The structure of this compound is build up out off octahedra and

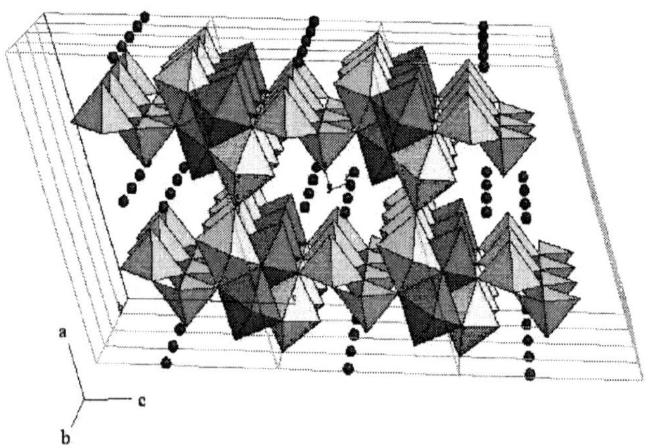

FIGURE 36. Crystal structure of NaV_6O_{15}.

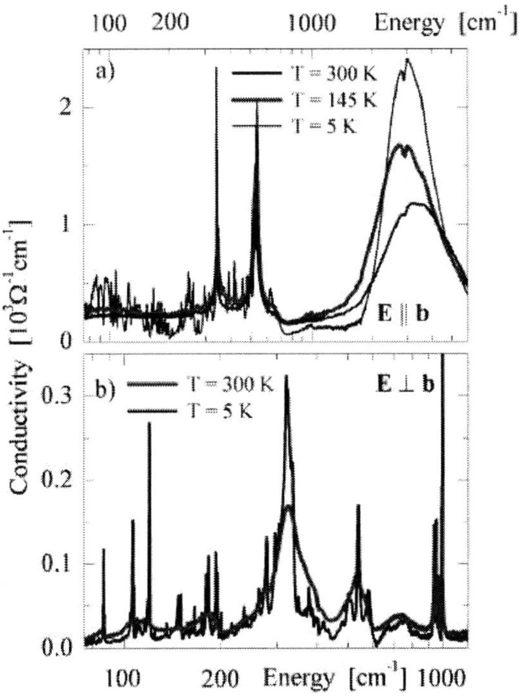

FIGURE 37. Optical conductivity of β-NaV_6O_{15} for light polarized along and perpendicular to the b-axis. Figure adapted from [41]

tetrahedra of vanadium and oxygen atoms where the tetrahedra form quasi 1-dimensional zig-zag chains (see figure 36). There are 3 different types of vanadium sites in this structure: 2 of them are ionic with a charge 5+ on the vanadium which has then a $3d^0$ configuration. The third site has half an electron more leading to a charge of 4.5+ on the vanadium atom in a $3d^{1/2}$ configuration. Because of this we expect a quarter filled band and metallic behavior. Figure 37 shows the optical conductivity of β-NaV_6O_{15}. The chains are along the direction labelled b. At energies around 3000 cm^{-1} we observe a broad peak for light polarized along the b-direction which could be due to polarons although these transitions also correspond well with the energies predicted by the Hubbard model for d-d transitions. If we compare the conductivity with polarization parallel and perpendicular to the b-axis, we see that the conductivity perpendicular to the b-axis is insulating whereas the one along the b-axis is conducting. This conducting behavior is due to the quarter filled bands.

Are polarons playing an important role in the above examples ? It is nearly impossible to answer this question experimentally due to the above mentioned difficulty in separating polaronic behavior from normal interband transitions. Moreover, in most cases where polarons are invoked, other theories are also able to reproduce the experimental results. To close this section we briefly discuss what happens if the density of polarons becomes larger. Imagine what happens if we increase the density of polarons such that we are getting close to a system with one polaron on each site. In that case the original lattice will almost be completely deformed and one can wonder wether the electrons are still capable to self-trap. It seems reasonable that in this limit the polaron picture no longer applies. Another possibility is the formation of bipolarons. Since the deformation energy of the lattice is proportional to the electron charge $E_{pol} \propto -1/2Cq^2$, the binding energy of two polarons is $\propto -Cq^2$. The binding energy of a bipolaron (two electrons trapped by the same polarization cloud) is twice as large however $E_{bipol} \propto -1/2C(2q)^2$. This binding energy is usually not enough to overcome the Coulomb repulsion between the electrons.

SPIN INTERACTIONS

As mentioned in the previous section the signatures for the presence of polarons can often be interpreted with different ideas. Most often these models are based on coupling to magnetic interactions. Consider for example the spectrum of the parent (undoped) compound $YBa_2Cu_3O_6$ which is a Mott insulator (see bottom panel of figure 14). Below 100 meV we see a series of peaks which are due to phonons. But what about the structure between 100 meV and 1 eV ? One of the difficulties in explaining this structure is that light does not directly couple to spin degrees of freedom. It is however possible to indirectly make spin flips with photons (see figure 38).

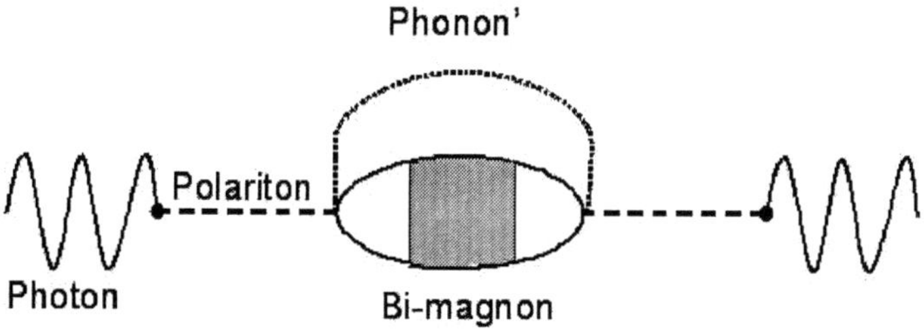

FIGURE 38. Interaction diagram for the indirect interaction of light with spin degrees of freedom.

For this process to occur we have to include phonon-magnon interaction. When a photon enters the material it gets dressed with phonons forming a polariton which is then coupled to the spin degrees by the phonon-magnon interaction. This leads to the possibility of so-called phonon assisted absorption of spin-flip excitations [42]. We see from figure 38 that the polariton creates a bi-magnon. This is because the intermediate state has to have spin S = 0. The dashed square represents all magnon-magnon interactions. The coupling constant for this process was first calculated by Lorenzana and Sawatzky and is

$$J_{ph-mag} = \frac{1}{2J} < \frac{d^2J}{du^2} >< u^2 >, \tag{176}$$

where J is the superexchange constant and u is the atomic displacement vector. In the

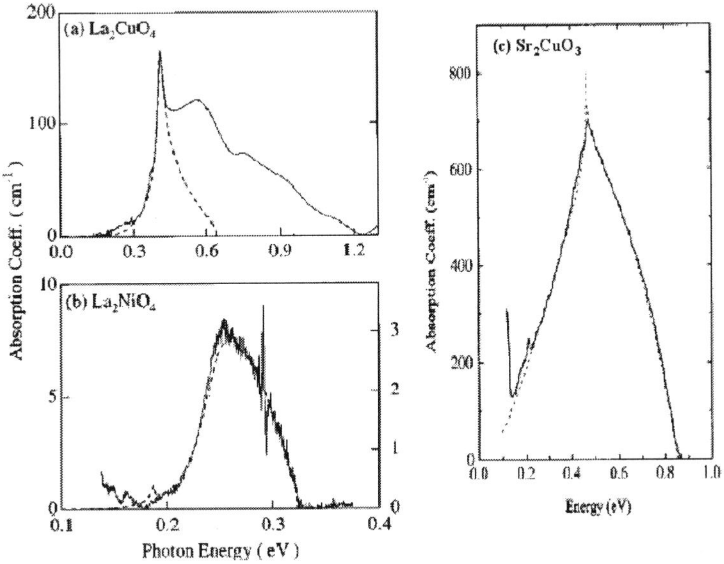

FIGURE 39. Optical conductivity of (a): La_2CuO_4, (b): La_2NiO_4 and (c): Sr_2CuO_4. Dashed lines are fits using the Lorenzana-Sawatzky model.

process momentum and energy have to be conserved and this leads to

$$k_{\text{magnon 1}} + k_{\text{magnon 2}} + k_{\text{phonon}} = k_{\text{photon}} \approx 0. \tag{177}$$

and

$$\omega_{\text{magnon 1}} + \omega_{\text{magnon 2}} + \omega_{\text{phonon}} = \omega_{\text{photon}}. \tag{178}$$

for the process in figure 38. This gives constraints on the possible absorptions. In figure 39 some examples are shown of materials in which we believe this process to play a role. One of the compounds where the predicted optical conductivity fits the spectrum very well is in the case of Sr_2CuO_3. To make the fit the magnon dispersion as measured with neutron scattering was used. The reason that this theory works so well for Sr_2CuO_3 is that the conduction is nearly one dimensional. This gives a good starting point because the magnon spectrum is completely understood. On the contrary, the theory is not completely capable of predicting the spectrum of La_2CuO_4. Most likely the peaks around 0.6 and 0.75 eV are due to 4 and 6 magnon absorption. In the case of $YBa_2Cu_3O_6$ the situation gets even more complicated due to the presence of two layers per unit cell. Because of the doubling of the unit cell, there are now acoustic and optical magnon branches just as what would happen in the case of phonons. The effect of this on the optical conductivity was first discussed by Grueninger *et al.* [6, 43].

Another example of probing of spin excitations occurs in NaV_2O_5. As already discussed in the previous section this compound has quasi one dimensional chains as shown in figure 36. These chains can be seen to form a so-called ladder structure, with the

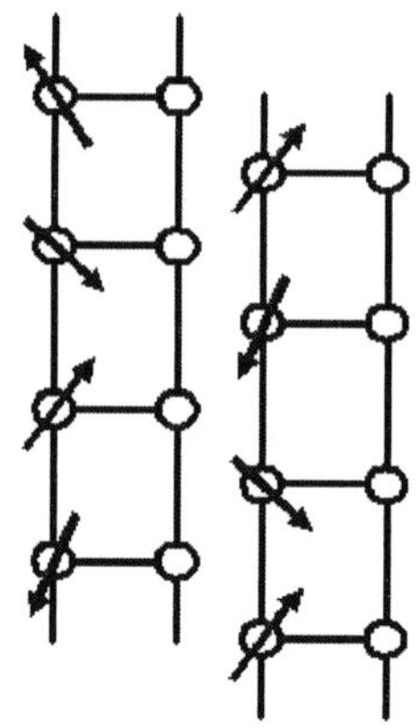

FIGURE 40. Schematic of the ladder structure of α- NaV_2O_5. Arrows indicate the position of the lectrons and their spin orientation.

ladders parallel to the *b* direction. Each adjacent ladder is shifted with respect to the previous such that the rungs of one ladder fall in between those of the next (figure 40). The vanadium atoms that form the ladders have an average charge of +4.5. It has been claimed [44] that the charge distribution is inhomogeneous with most of the charge on one side of the ladder as indicated in figure 40. The temperature dependence of the magnetic susceptibility can be modelled pretty well using a Bonner-Fischer model for a spin-1/2 Heisenberg chain [45] for temperatures higher then 34 K (see figure 41). Below 34 K, X-ray analysis shows a doubling of the a- and b- axes and a quadrupling of the c-axis. It indicates that the new unit cell consists of 64 vanadium atoms and 32 valence electrons. At the same temperature the susceptibility shows an abrupt drop. An explanation for this transition is in terms of a spin-Peierls transition. In the high temperature phase ($T > T_{SP}$) the left side of the ladder has a uniform spin distribution,

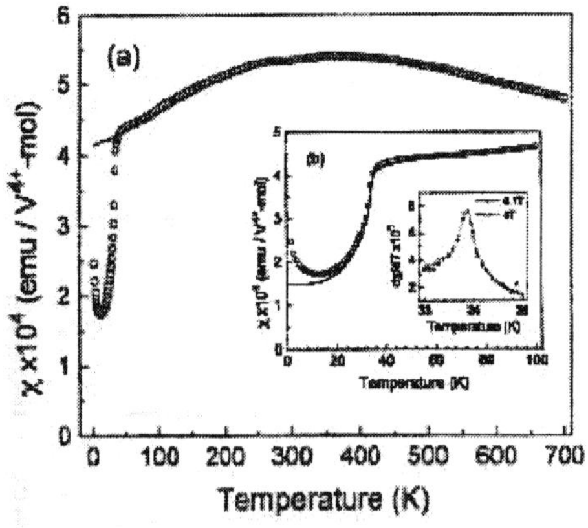

FIGURE 41. Magnetization of α- NaV_2O_5. Figure adapted from ref. [45].

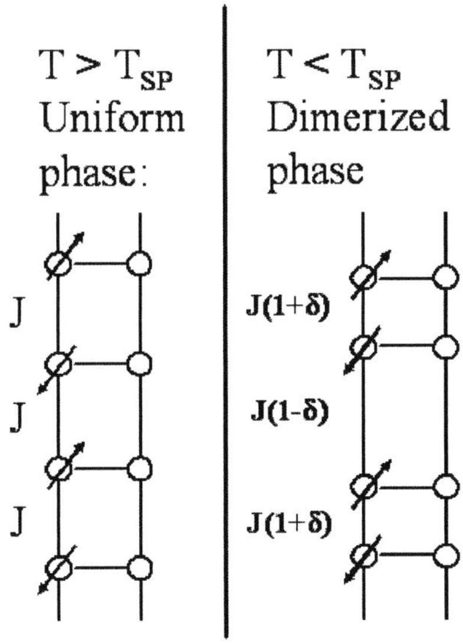

FIGURE 42. Schematic representation of the low and high temperature phase of α- NaV_2O_5.

as indicated in the left panel of figure 42, which is reasonably well described with an anti-ferromagnetic (AF) S = 1/2 Heisenberg spin chain with uniform exchange coupling J. For T < T_{SP} the system dimerises due to a deformation of the lattice, leading to an alternation of exchange couplings (see right panel figure 42). Here we focus on the high temperature phase. If the charge inhomogeneity is present it would lead to a breaking of the inversion symmetry which in turn leads to a non-zero optical matrix element for two magnon absorption [46]. The idea is similar to the Lorenzana-Sawatzky model discussed above. In the latter case the phonon effectively lowers the symmetry making the process optically allowed. The optical conductivity of α- NaV_2O_5 is shown in figure 43. We can model α- NaV_2O_5 with independent ladders where the hopping probability along a rung ($t_\perp$) is much larger than that along the ladder ($t_\parallel$). Furthermore we assume a large on-site repulsion U. We can then model a ladder by independent rungs. Assuming a quarter filled ladder (one electron per rung) we have a simple two level problem leading to bonding and anti-bonding levels (see also the discussion in the section on applications of sum rules). If we also include a potential energy difference Δ between the sites the wavefunctions become asymmetric with higher probability on the low potential site and one can show that this again leads to bonding and anti-bonding solutions which are split by an energy [46],

$$E_{CT} = \sqrt{\Delta^2 + 4t_\perp^2}. \tag{179}$$

Transitions from the bonding to anti-bonding band are optically active and involve charge transfer (CT) from the left side of the ladder to the right. The large peak seen in figure 43 around 1 eV is due to these transitions. The energy position of the peak indirectly gives evidence for the charge inhomogeneity: band structure calculations and

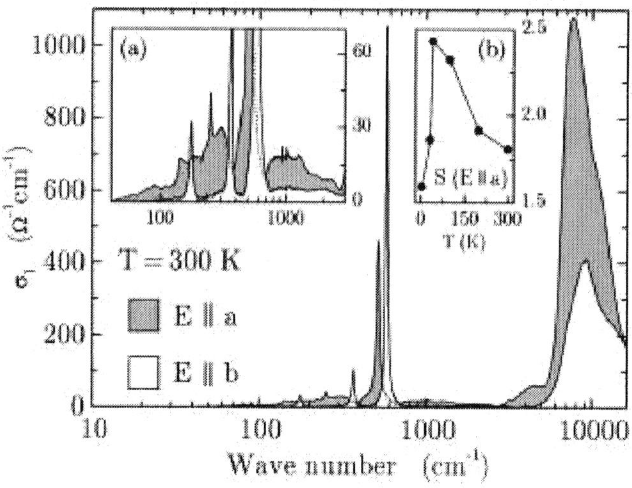

FIGURE 43. Optical conductivity of α- NaV_2O_5. Inset (a) shows the low energy continuum attributed to charged bi-magnon excitations and inset (b) shows the temperature dependence of the spectral weight of this continuum. Figure adapted from ref. [46].

exact diagonalization of finite clusters give $t_\perp \approx 0.35$ eV, which would put the charge transfer peak around 0.7 eV. The observed value of 1 eV thus indicates $\Delta \neq 0$. The spectral weight of the peak allows us to make an estimate for Δ. One can show that,

$$\int_{peak} \sigma_1(\omega) = \pi e^2 N d_\perp^2 t_\perp^2 \hbar^{-2} E_{CT}^{-1}. \tag{180}$$

Using Eq. (179) we find $t_\perp \approx 0.3$ eV and $\Delta \approx 0.8$ eV. Besides the large CT peak seen in figure 43 there is also a broad continuum in the infrared region of the spectrum for $E \parallel a$ (see inset). This part of the spectrum can be understood if we include the coupling between rungs of the ladder. For parallel spins on different rungs this coupling would have no effect since the Pauli principle would forbid hopping between the sites. For an anti-parallel spin configuration the system can gain some kinetic energy from virtual hopping of an electron from one rung to the next, putting two electrons on one rung. For very large U this electron would occupy the righthand side of the rung. Staring from an anti-parallel configuration a spin-flip transition on one rung thus leads to a net dipole displacement which leads to optical activity of this transition. We note that because of spin conservation rules we have to make two spin-flips. These excitations have been dubbed charged bi-magnon excitations.

REFERENCES

1. F. Wooten, Optical Properties of Solids (Academic Press, New York, 1972).
2. M. Dressel and G. Grunner, Electrodynamics of Solids: Optical Properties of Electrons in Matter (Cambridge University Press, 2002)

3. A.A. Tsvetkov *et al.*, Phys. Rev. B **60**, 13196 (1999).
4. D. van der Marel *et al.*, Proc. of the 10th Anniv. HTS Workshop on Physics, Materials and Applications, Houston, Texas, 1996, edited by B. Batlogg, C.W. Chu, W.K. Chu, D.U. Gubser, and K.A. Muller, World Scientific, Singapore, 1996, pp. 357.
5. E. van Heumen *et al.*, Phys. Rev. B **75**, 054522 (2007).
6. M. Grueninger, Ph.D. thesis University of groningen.
7. J.L.M. van Mechelen, unpublished.
8. A.B. Kuzmenko et al., Phys. Rev. Lett. **91**, 37004 (2003).
9. P.W. Anderson, Phys. Rev. 110, 827 (1958); P.W. Anderson, Phys. Rev. 130, 439 (1963); P.W. Higgs, Phys. Rev. Lett. 13, 509 (1964); P. W. Higgs, Phys. Rev. Lett. 145, 1156 (1967). See also D. van der Marel, Journal of Superconductivity, **17**, 559 (2004).
10. J. Zaanen, G.A. Sawatzky, J.W. Allen, Phys. Rev. Lett. **55**, 418 (1985).
11. D. Y. Smith, E. Shiles, Phys. Rev. B **17**, 4689 (1978).
12. M. Tinkham, and R. A. Ferrell, Phys. Rev. Lett. **2**, 331 (1959).
13. A. Damascelli, K. Schulte, D. van der Marel, and A.A. Menovsky, Phys. Rev. B **55**, R4863 (1997)
14. D. van der Marel, H.J.A. Molegraaf, C. Pressura, Chapter in "Concepts in electron correlation", Edited by A. Hewson and V. Zlatic, Kluwer (2003), p 7-16. ; cond-mat/0302169.
15. H.J.A. Molegraaf *et al.*, Science **295**, 2239 (2002).
16. D.N. Basov, S.I. Woods, A.S. Katz, E.J. Singley, R.C. Dynes, M.Xu, D.G. Hinks, C.C. Homes, M. Strongin, Science **283**, 49 (1999).
17. A.F. Santander-Syro, R.P.S. Lobo, N. Bontemps, Z. Konstatinovic, Z.Z. Li, H. Raffy. Europhys. Lett. **62**, 568 (2003).
18. A.B. Kuzmenko, H.J.A. Molegraaf, F. Carbone, and D. van der Marel, Phys.Rev.B **72**, 144503 (2005).
19. F. Carbone *et al.*, Phys. Rev. B **74**, 024502 (2006).
20. J. E. Hirsch, Physica C **199**, 305 (1992).
21. P.W. Anderson *et al.*, J. Phys.: Condens. Matter **16**, 755(R) (2004).
22. T. Eckl, W. Hanke and E. Arrigoni, Phys. Rev. B **68**, 014505 (2003).
23. P. Wróbel, R. Eder and P. Fulde, J. Phys.: Condens. Matter **15**, 6599 (2003).
24. K. Haule and G. Kotliar, cond-mat/0601478.
25. A. Toschi *et al.*, Phys. Rev. Lett. **95**, 097002 (2005).
26. F. Marsiglio, Phys. Rev. B **73**, 064507 (2006)
27. Th.A. Maier, M. Jarrell, A. Macridin, and C. Slezak, Phys. Rev. Lett. **92**, 27005 (2004).
28. M.R. Norman, C. Pépin, Phys. Rev. B **66**, 100506(R) (2002)
29. A. Damascelli *et al.*, Phys. Rev. Lett. **81**, 918 (1998).
30. J.W. Allen and J. C. Mikkelsen, Phys. Rev. B **15**, 2952 (1977).
31. J.E. van der Eb *et al.*, Phys. Rev. Lett. **86**, 3407 (2001).
32. B. Farnworth and T. Timusk, Phys. Rev. B **14**, 5119 (1976).
33. J. Teyssier *et al.*, accepted for Phys. Rev. B, cond-mat/0609426.
34. R. Lortz *et al.*, Phys. Rev. B 72, 024547 (2005).
35. F. Marsiglio, T. Startseva, J. Carbotte, Phys. Lett. A **245**, 172 (1998).
36. F. Marsiglio, J. of Superconductivity, **12**, 163, (1999).
37. L.D. Landau, Phys. Z. Sowjetunion, **3**, 644 (1933).
38. L.D. Landau and S. Pekar, J. Exptl. Theor. Phys. **18**, 419 (1948).
39. Uchida *et al.*, Phys. Rev. B **43**, 7942 (1991).
40. C. A. Kuntscher, D. van der Marel, M. Dressel, F. Lichtenberg and J. Mannhart, Phys. Rev. B **67**, 035105 (2003).
41. C. Presura *et al.*, Phys. Rev. Lett. **90**, 026402 (2003)
42. J. Lorenzana and G.A. Sawatzky, Phys. Rev. Lett. **74**, 1867 (1995).
43. M. Grueninger, D. van der Marel, A. Damascelli, A. Erb, T. Nunner and T. Kopp, Phys. Rev. B **62** 12422 (2000).
44. A. Carpy et al, Acta Cryst. **31**, 1481 (1975)
45. M. Isobe, and Y. Ueda, J. Phys. Soc. Japan **65**, 1178 (1996)
46. A. Damascelli *et al*, Phys. Rev. B **61**, 2535 (2000). See also A. Damascelli, PhD thesis, Rijks Universiteit groningen, http://irs.ub.rug.nl/ppn/181459132.

PART II

PARTICIPANT CONTRIBUTIONS

DMRG applied to critical systems: spin chains

J. Almeida*, M.A. Martin-Delgado* and G. Sierra†

*Departamento de Física Teórica I, Universidad Complutense, 28040 Madrid, Spain
†Instituto de Física Teórica CSIC-UAM, Cantoblanco, 28049 Madrid, Spain

Abstract.
The DMRG (Density Matrix Renormalizarion Group) is a powerful method to study the low energy properties of many body systems which are difficult to treat by other methods due to the large dimension of their Hilbert space and their strongly correlated nature. Usually these systems are not analitically solvable and numerical approaches become indispensable to shed some light upon their physical properties.
In this paper we will review the staggered Heisenberg antiferromagnetic chain with $S=1$ and $S=1/2$ and will use the DMRG algorithm to obtain some properties relative to their critical behavior, in particular concerning the critical points appearing in the phase diagram of these systems.

Keywords: S=1 staggered antiferromagnetic Heisenberg chain, S=1/2 staggered antiferromagnetic Heisenberg chain, DMRG, critical exponent
PACS: 75.10.Jm, 75.10.-b,74.20.Mn

1. INTRODUCTION

Intuitively we could say that strongly correlated systems are those in which the interactions taking place between their different constituents have such a nature that it is not possible to decompose the system in smaller relevant pieces.

Depending on the particular system we are studying we could attempt to give a definition of a strongly correlated system, but in general we can say in a practical fashion that a strongly correlated system is that one governed by a Hamiltonian of the form

$$\hat{H} = \lambda_1 \hat{H}_1 + \lambda_2 \hat{H}_2$$

Where $\lambda_1 \sim \lambda_2$, so we have to cope with it as a whole and can not apply other techniques proven to be useful in some particular limits, such as the strong or weak coupling ones in which applies $\lambda_1 \gg \lambda_2$ or vice-versa.

These systems offer very often a rich set of features and interesting physics. In condensed matter, strongly correlated systems are rather numerous and great efforts are being devoted to develop suitable tools to handle them.
Just to give some examples of actual systems of this kind, we can mention one dimensional electron systems, the insulating state of bosonic atoms in a periodic potential, fractionary quantum hall systems, high temperature superconductors

CP918, *Lectures on the Physics of Strongly Correlated Systems XI: Eleventh Training Course,* edited by A. Avella and F. Mancini

and many more.

What is most remarkable of all these systems is that we do not have a general method to cope with them all. We can sketch some techniques used to treat separately each system, such as mean field theory, non linear sigma model, conformal field theory or in the best case integrable models of which we know the exact solution. In any case these methods are uncapable to deal with a great deal of strongly correlated systems and we have to resort to computational techniques to shed light upon their properties. Among these methods we can mention exact diagonalization and lanczos algorithms for small system sizes, and more sophisticated and powerful algorithms like quantum montecarlo and renormalization group(RG) methods for larger ones.

In this paper we will study some spin-chain like systems, namely Heisenberg antiferromagnets as paradigmatic examples of strongly correlated systems with rich and even surprising physical properties. These systems have been studied for long, but nowadays the situation is more than ever in the edge thanks to the discovery of many compounds with this structure and interesting applications.

The organization of the paper is as follows: in section two and three we will survey the operation of the RG methods and the DMRG in particular. In section four we will apply the DMRG to different systems to study their properties near the critical point. These systems are the staggered $S = 1/2$ and $S = 1$ antiferromagnetic Heisenberg (HAF) chains. The results displayed in this last section is work in progress at this moment and we shall report on them elsewhere.

2. RG METHODS

Suppose that we have a system with a Hamiltonian written in a general form

$$H = \sum_{\alpha,i,j} J^{\alpha}_{i,j} \hat{O}^{\alpha}_i \hat{O}^{\alpha}_j + \sum_{\alpha,i} B^{\alpha}_i \hat{O}^{\alpha}_i \tag{1}$$

Where $J^{\alpha}_{i,j}$ and B^{α}_i are coupling constants and $\hat{O}_i$ is the set of operators acting on the i-th site.
Notice that the operator $\hat{O}_i$ may itself consist of one only operator like the usual Heisenberg spin operators or, by the contrary, consist on a product of several operators all of them acting on the same site, like the number operator. The above Hamiltonian is general in the sense that it describes any one-site and two-site interactions.

We are interested in the low temperature properties of our system, described by the low energy spectrum of its Hamiltonian. This spectrum allows us to study fundamental features like the possible existence of an excitation gap and the universality class to which our system belongs.

However the problem of diagonalizing a many-body hamiltonian is a hard one because of the exponentially growth of the Hilbert space with the number of sites. The RG methods try to acomplish this task by means of an iterative scheme in which local degrees of freedom are added at each step while truncating the Hilbert space in such a way that our truncated basis can represent faithfully the low energy physics of the whole growing system.

The key ideas in the numerical renormalizatin group methods consist then on a computationaly feasible adding scheme, and an efficient truncation prescription that truly retain the physics we are interested in as we grow our system. These two ingredients make different each member of the RG family.

Next we will discuss the Wilson RG algorithm, which was the first one to come on stage, and after that the DMRG, which overcomes the limitations of the previous methods and constitutes today the most powerful and celebrated algorithm of this family.

2.1. Wilson RG scheme

As we mentioned before, this algorithm was the first one appearing on the stage of the numerical RG algorithms. It was proposed and used by K. Wilson on the Kondo problem where this method revealed as very successful.

In this scheme the addition of degrees of freedom is such that we add one site (described by its own local degrees of freedom) at each step. Since we are interested only in the low energy physics of the system, we will truncate the Hilbert space keeping only the m lowest energy states of our system, where m is a fixed number that will determine the precision of the final result. Each complete step of the process englobes an addition action followed by a truncation of the Hilbert space, until our system contain as many sites as we want.

It is worth noticing that if we take m at each step to be the total size of the Hilbert space, the Wilson algorithm is nothing else that a way to construct the original Hamiltonian and so the procedure is exact. Otherwise the precision of the result depends strongly on the value of m, being more accurate as it takes larger values.

The scheme to implement this algorithm is the following:

1. Begin with an initial block consisting of a small number of sites, say N, and construct its Hamiltonian and the set of one-body operators acting on each of its sites:
$$\hat{H}_N,\ \{\hat{S}_i^\alpha\}_\alpha \quad i=1\ldots N$$

2. Add one site to the block, this structure(block plus one site) will be named *superblock*. Write the Hamiltonian of the superblock from the energy of the previous block, the energy of the new site and the interaction energy between them:

$$\hat{H}_{SB} = \hat{H}_b \otimes \hat{1}_s + \hat{H}_{bs} + \hat{1}_b \otimes \hat{H}_s$$

3. Diagonalize the superblock Hamiltonian in order to obtain the m lowest eigenvectors with their associated eigenvalues:

$$\hat{H}_{SB} \cdot \phi_i = E_i^{\text{low}} \cdot \phi_i, \quad i = 1 \ldots m$$

4. Construct the truncation matrix making use of the computed eigenvectors:

$$U = (\phi_0, \phi_1, \ldots, \phi_m)$$

5. Truncate both the superblock Hamiltonian and the superblock one-body operators:

$$\begin{array}{rcl} \hat{H}_{SB} & \leftarrow & U^\dagger \hat{H}_{SB} U \\ \hat{S}_i^\alpha & \leftarrow & U^\dagger \hat{S}_i^\alpha U \end{array}$$

6. Stop if the superblock already contains the desired number of sites, otherwise iterate from step 2 taking the new truncated superblock Hamiltonian and operators as the operators of the next block.

The Wilson scheme yielded very accurate results in the original problem it was devised for, but definitely failed when it was attempted to move to other physical systems showing long range order.
Historically the reason for this remarkable failure was understood resorting to quantum one-body problems and it was related to the strong constraints it imposed in the boundary conditions[1]. In many-body physics we can say that for long-range ordered systems it is no longer true that we can construct the low energy physics of our system from the low energy states of isolated parts within it.

To overcome these problems we will have to resort to another algorithm: the density matrix renormalization group.

2.2. Basics on the DMRG method.

The DMRG introduces both a new addition mechanism and an optimized truncation scheme with respect to the Wilson RG.
With the previous scheme it was clear that the fact that our algorithm did not take into account the whole lenght of our system caused a definite failure of the process. To overcome this difficulty DMRG will split our total system in two blocks, one acting as the effective representation of the whole system and a second one playing the role of an environment to which the system is coupled.

As we let our system grow we have to choose the basis that represent with the most accuracy the low energy physics (i.e. the Hamiltonian, operators and wavefunction) of the future complete system. On this purpose DMRG proposes as the best solution the eigenvectors of the partial density matrices of both blocks. In fact, it can be proven that this choice provides the most faithful representation of the whole system[2].

The fine details of the method are rather technical as regards the computer implementation and, as usually happens with the sophisticated numerical algorithms, they are susceptible of quite a large number of improvements that contribute in the end to make very costly to implement a competitive one. This point together with the fact that the DMRG power only shows up in one dimension are probably the major drawbacks of the method.

Next we give a brief explanation of the steps involved in a DMRG run. We do not intend to be exhaustive at all but only to give an idea about its behavior. We refer to the original papers of S. White[2][3] for more details.

2.2.1. Infinite size DMRG.

This version of the algorithm allows us to begin with a small size system and make it grow in a controlled and feasible fashion until we reach the wanted size. Despite the name of the algorithm of course we will never be able to go a truly infinite size, but we can reach easily sizes of hundreds or even thousand of sites where finite size efects have vanished completely.

Giving all the details of the method would be really cumbersome and scapes the scope of this paper. Just to get an idea about the method we sketch next a brief survey about its operation:

1. Begin with two blocks (left and right blocks) with the same number of sites, say L, whose Hamiltonians can be exactly diagonalized and with an exact representation of each operator acting on every site.

$$\hat{H}^{\{R,L\}}(L), \hat{O}_i^{\{R,L\}}(L)$$

2. Form a *superblock* consisting on the previous blocks and two sites between them. Compute the operators and Hamiltonian of the blocks of size $L+1$ using the Wilson scheme.

$$\hat{H}^{\{R,L\}}(L+1), \hat{O}_i^{\{R,L\}}(L+1)$$

3. Compute the Hamiltonian of the superblock and obtain numerically, using a Lanczos-like algorithm, the ground state of the system $|\psi_0\rangle$.
4. Form the reduced density matrices of each block. Diagonalize them and obtain their full spectrum.

$$\rho_{L+1}^{\{L,R\}} \cdot \phi_i^{\{L,R\}} = w_i \cdot \phi_i^{\{L,R\}}$$

5. Form the truncation matrices from the eigenvectors of each reduced density matrix, using the m eigenvectors associated with the largest eigenvalues.

$$U^{\{R,L\}} = (\phi_0^{\{R,L\}}, \phi_1^{\{R,L\}}, \ldots, \phi_m^{\{R,L\}})$$

6. Change the one-body operators and Hamiltonians of both blocks to the new truncated basis.

$$\begin{aligned} \hat{H}^{\{L,R\}} &\leftarrow U^\dagger \hat{H}^{\{L,R\}} U \\ \{\hat{S}_i^\alpha\}^{\{L,R\}} &\leftarrow U^\dagger \{\hat{S}_i^\alpha\}^{\{L,R\}} U \end{aligned}$$

7. Stop the procedure if the system has reached the desired size, otherwise iterate from 2, using as seed the recently computated Hamiltonians and operators from last point.

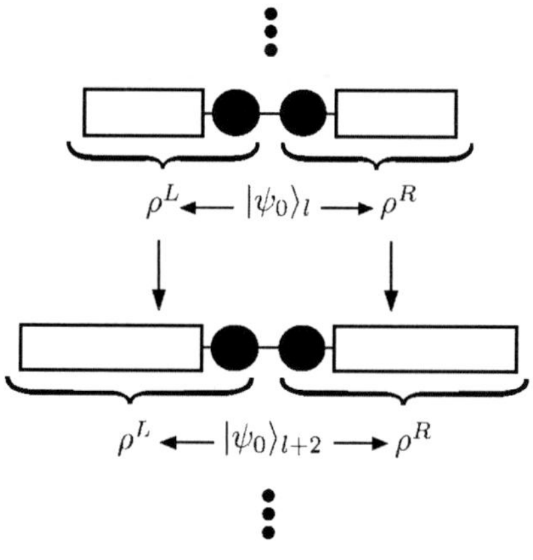

FIGURE 1. This figure ilustrate schematically the procedure of addition and truncation involved in an infinite DMRG step: from the wavefunction of the whole system we compute the reduced density matrices of both blocks which are then used to change from the previous block-plus-site to a new block basis.

2.2.2. Finite size DMRG

The infite DMRG achieve a great acuracy in the low energy spectrum of the Hamiltonian, but we can obtain even better convergence by means of the finite size algorithm.

In the previous algorith we managed to grow our system using superblocks consisting on a right and left blocks joined by two sites in the middle, where the

sizes of the left and right blocks were the same. Now in this scheme we will keep fixed the lenght of the total system, and the renormalization will be performed making only one of the blocks (the active block) grow, while decreasing the other one (the passive block), also using in this case two sites to join them.
Now the sizes of both blocks are different, but the sum of their sizes plus the two sites is always equal to the fixed size of the whole system.

Notice that we have to interchange the role of the passive and the active blocks as we reach the left and right ends of the chain.
To perform an entire sweep we begin with both blocks containing the same number of sites (i.e. $L/2-1$, where L is the number of sites of the total system), then make the active block grow up to the maximum possible size ($L-3$ in order to keep the block+site+site+block structure of the superblock), after that the passive block becomes active and begin to grow, up to the maximum size as before. To finalize the sweep, interchange again the roles of the blocks and make the active one grow to the size it had in the begining of the sweep, i.e. $L/2-1$ sites.

For more details about both finite and infinite DMRG we refer to [2][3][4]. A very complete review with updated information can be found in [5].

3. APPLICATIONS

3.1. Spin 1/2 AF staggered Heisenberg chain

The spin 1/2 antiferromagnetic Heisenberg chain has been widely studied for long now. A remarkable hit came in 1931 with the exact solution of this system found by Bethe, which predicted a ground state with gapless excitations. Many years later, in 1975, the bosonization approach to this case by Lutther and Peschel, combined with the Bethe-ansatz solution gave a suitable framework to understand some other important properties such as the potentially decaying behavior of the correlations. At this point the physics of this system was somehow settled.

An interesting and not well known phenomena ocurr however when we introduce some staggering into the system, alternating strong and weaker couplings between neighbouring sites. In this situation the original Hamiltonian transforms to a new one with the form:

$$\hat{H} = \sum_i \hat{S}_i \hat{S}_{i+1} \quad \rightarrow \quad \hat{H} = J \sum_i \{1 - \gamma(-1)^i\} \hat{S}_i \hat{S}_{i+1}$$

By simple inspection of this Hamiltonian we can differentiate two clear limits: in the case when $\gamma = 0$ we shall recover the critical phase of the non-staggered Heisenberg chain, while taking $\gamma = 1$ implies that the weaker couplings are equal to zero and so our chain is composed of indivual pairs of spins strongly coupled between them but isolated of the rest. It is not difficult to see that in the ground state these pairs of spins have to be forming individual singlets, so the first excited state appears when we break one of these singlets to give a triplet. This break requires a finite energy and so the system in this phase is gapped.

Taking into account the reasoning above we deduce that a quantum phase transition have to arise at any point between these limits. In fact, a renormalization group study of the point $\gamma = 0$ reveals that the new staggering operator is relevant and so, arbitrarily small values of staggering destabilizes the critical phase to a gapped one.
On the other hand, bosonization near the point $\gamma = 0$ predicts a power-law behavior of the gap with logarithmic corrections and the following form:

$$\Delta = C \frac{\gamma^{2/3}}{|log(\gamma)|^{1/2}}$$

Despite this formula has been known and accepted for long, the lack of suitable numerical schemes and powerful enough computers postponed numerical evidences. In fact, as far as we know, no studies have been done to test this result.

Our goal in this section was to confirm this relation and find the range of validity in which bosonization applies. On that purpose we have used the DMRG algorithm to perform finite size scaling on the triplet-singlet gap as a function of the size of the chain L, for fixed values of the dimerization parameter γ. For each values of γ the considered lenghts range from 6 to 500 sites. The number of retained states was $m = 500$ for the larger chains. With this choice of parameters we can be razonably confident that the computed values of the gap in each run will have at least four reliable decimals.
In Fig.2a we plot the computed values of the gap for $L \to \infty$ as a function of γ and the function that we have found that best fit that data, which corresponds to a simple potential law with the following form:

$$\Delta_{\text{fit}} = C \cdot \gamma^{\alpha}, \quad C = 1.988(6), \quad \alpha = 0.744(2)$$

This results contradict the expected bosonization predictions and also point towards a slight deviation from the values $C = 2$ and $\gamma = 3/4$ obtained in [6] and [7].
In our work we have studied the range from $\gamma = 0$ to $\gamma = 0.5$ and have noticed that the largest deviations from the fit ocurr in the region near $\gamma = 0$ which could be a signal that we are entering there in the bosonization regime. Our attempts however to find the range in which bosonization applies are not conclusive up to date, nonetheless the range of validity seems to be surprisingly rather small.
This unexpected deviation of the bosonization predictions for so small values of γ raises some uncertainty about not only the true limits for this regime, apparently well beyond experimental limits, but also about its range of universality.
These and other questions are work in progress at the moment.

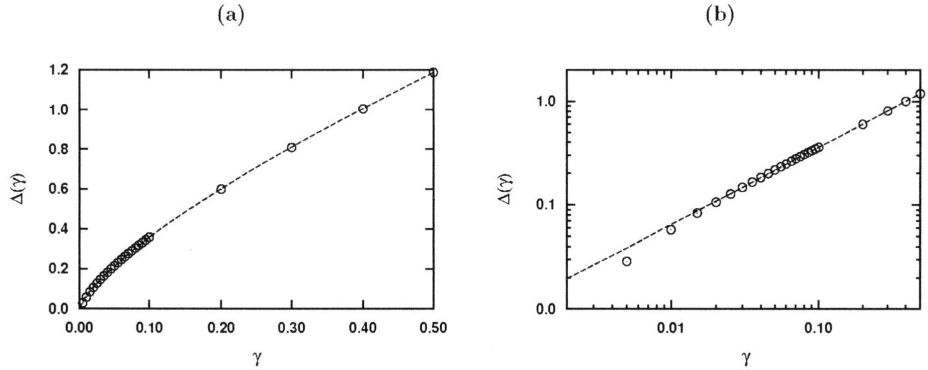

FIGURE 2. **(a)** The relation between the gap and γ is well fitted by a simple power law with exponent 0.744(2) **(b)** Using logarithmic scales, deviations in the region of low γ become more apparent.

3.2. S=1 Heisenberg AF staggered chain

The Hamiltonian of this system is the same as the last section but, as we learnt from F.D.M. Haldane, the physics of the spin chains makes a great difference between systems with half-integer and integer spin[8][9].

We know that the ground state of the homogeneus S=1 HAF is a spin liquid with short range spin correlations and a gap in the excitation spectrum.[8] If we introduce bond alternation, the arguments exposed in the case of spin $S = 1/2$ for the limit $\gamma = 1$ are still valid since we can form singlets with either spin one or one half, but the situation for the region $\gamma = 0$ is different: in the limit of weak staggering the ground state of the system is in the so called Haldane phase. This phase can be understood in terms of the valence bond picture, where we can imagine every spin one as a projection of two spins one half over the subspace with total spin equal one. This phase is gapped in contrast to the real spin one half case, and it is characterised by two effective 1/2 spins in the end of the chain and a non-vanishing string order parameter.

It becomes apparent from the discussion above that both phases of the S=1 chain with alternated bonds are gapped spin liquids. It is also known that these different phases are separated by a quantum critical point. The aim of our work at this stage is to compute the exact location of this point as well as the critical exponent which governs the behavior of the gap in its vicinity.

In order to obtain the location of the critical point we have performed a finite size scaling analysis of the gap for different values of the staggering. The values of the gap for finite sizes have been computed with DMRG, going as far as $L = 500$ and

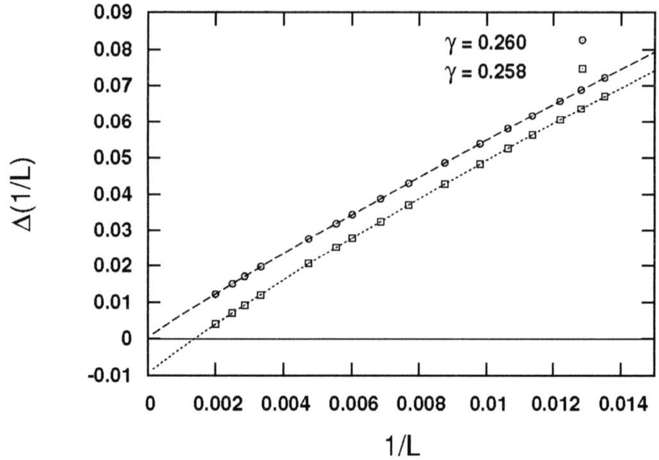

FIGURE 3. Finite size scaling for $\gamma = 0.260$ still predicts a positive value for the gap in the limit $L \to \infty$ while for $\gamma = 0.258$ the scaling is no longer valid. The value of γ_c has to lay within this interval.

retaining as many states as $m = 450$ in the most demmanding cases. Numerically it is not possible to obtain the value of γ corresponding to a exact vanishing gap so we have selected that one whose scaling yielded the smallest positive value. In Fig.3 we plot the graphs corresponding to those values of γ in which the extrapolated gap changes from positive values to negatives. Thus we estimate the critical value of γ as

$$\gamma_c = 0.259(1)$$

Which is in excelent agreement with[11].

As regards the estimation of the critical exponent, the case for spin one is numerically more demmanding than the studied in the previous section. DMRG runs are in progress at the moment and we shall report on these issues elsewhere.

4. CONCLUSSIONS

Spin chains and similar structures like ladders and other spin systems constitute good examples of strongly correlated systems with interesting physical implications. Despite they have been known and studied for long they are recovering importance these days for several reasons. First of all we have to mention the discovery of compounds implementing these quasi one dimensional estructures, like cuprate compounds, the vanadyl phyrophosphate, NTENP, NENP and NDMAP families, etc. But even more important than that is the fact that with the new improvements of computational resources and the development of powerful numerical algorithms, we are today able to put to test some of the accepted knowledge, revealing in some occasions that there are still things which are not completely understood.

Among the numerical algorithms on the stage today, the DMRG is one of the most poweful ones and this work shows how good it is to cope with the difficulty of the strongly correlated systems.

ACKNOWLEDGMENTS

J. A. acknowledges A. Avella and the rest of the organizers of the "XI Training course on Strongly Correlated Systems" for their kind invitation to lecture on the course and for their partial financial support.
This work is partially supported by the DGES under contract BFM2003-05316-C02-01.

REFERENCES

1. S. R. White, and R. M. Noack, Phys. Rev. Lett. **68**, 3487 (1992).
2. S. R. White, Phys. Rev. Lett. **69**, 2863 (1992).
3. S. R. White, Phys. Rev. B. **48**, 10345 (1993).
4. R. M. Noack and S. R. White, in *Density-Matrix Renormalization. A new Numerical Method in Physics* edited by I. Peschel, X. Wang, M. Kaulke and K. Hallberg, (Springer, Dresden workshop, Germany, 1998) pp. 27-66.
5. K. Hallberg, Adv.Phys. **55** (2006) 5-6.
6. T. Barnes, J. Riera and D. A. Tennant, Phys. Rev. B. **59**, 11384 (1999).
7. M. Kumar, S. Ramasesha, D. Sen and Z. G. Soos, arXiv:cond-mat/0608395.
8. F. D. M. Haldane,Phys. Lett. **93A**, 464 (1983).
9. F. D. M. Haldane,Phys. Rev. Lett. **50**, 1153(1983).
10. M. Hagiwara, Y. Narumi, K. Kindo, M. Kohno, H. Nakano, R. Sato, and M. Takahashi. Phys. Rev. Lett. **80**, 1312 (1998).
11. M. Kohno, M. Takahashi, Phys. Rev. B. **57**, 1046 (1998).

A diagrammatic solution of the single-impurity Anderson model in the Kondo regime

V. Janiš and P. Augustinský

Institute of Physics of the Academy of Sciences of the Czech Republic, Na Slovance 2, CZ-182 21 Praha, Czech Republic

Abstract. We study the single-impurity Anderson model with diagrammatic techniques. We employ the parquet approach to determine the electron-hole and electron-electron irreducible vertices self-consistently. We demonstrate that when the dominant contributions from the critical region of the singularity driven by multiple electron-hole scatterings are properly taken into account we make the parquet equations soluble and recover the Kondo asymptotics in the symmetric as well as in the asymmetric cases.

Keywords: Anderson impurity model, strong-coupling regime, parquet equations, quantum criticality, Kondo scale
PACS: 72.15.Qm, 75.20.Hr

INTRODUCTION

Single-impurity Anderson (SIAM) and lattice Hubbard models play an exceptional role in the description and understanding of electron correlations. Their exact solutions demonstrate that they capture most of the relevant quantum features of electron correlations. Exact solutions for electron correlations in realistic situations beyond specific limits are missing and hence we have to resort to approximations. Any approximation intended for a reliable description of strong electron correlations must at least qualitatively correctly describe the strong coupling regimes in the exactly soluble situations in the Anderson and Hubbard models.

Recently we proposed an approximate scheme aiming at interpolating between the weak and strong couplings of models with electron correlations [1]. The construction uses the parquet-type renormalization of the diagrammatic expansion for two-particle irreducible functions from which the self-energy is obtained via the Schwinger-Dyson equuation. We showed that this approximation qualitatively correctly reproduces the Kondo strong-coupling scale for the symmetric SIAM. Here we extend the solution of the parquet equations beyond the electron-hole symmetric situation.

SIMPLIFIED PARQUET EQUATIONS

The fundamental object of interest in the parquet theory is the two-particle vertex (one-particle irreducible two-particle Green function). The reason for this is that divergences in the perturbation expansion emerge as poles in the two-particle vertex. The poles in the vertex function reflect singularities in Bethe-Salpeter (BS) equations representing the full vertex via two-particle irreducible ones. The representation of the full vertex via

CP918, *Lectures on the Physics of Strongly Correlated Systems XI: Eleventh Training Course,* edited by A. Avella and F. Mancini

a Bethe-Salpeter equation is ambiguous according to which type of multiple two-particle scatterings we choose to sum explicitly. The BS equation in the electron-hole channel reads

$$\Gamma_{\uparrow\downarrow}(k,k';q) = \Lambda^{eh}_{\uparrow\downarrow}(k,k';q) - \frac{1}{\beta N}\sum_{k''}\Lambda^{eh}_{\uparrow\downarrow}(k,k'';q)G_{\uparrow}(k'')G_{\downarrow}(q+k'')\Gamma_{\uparrow\downarrow}(k'',k';q)\,. \quad (1)$$

where $k = (\mathbf{k}, \omega_n)$ and $q = (\mathbf{q}, \nu_m)$ are frequency-momentum dynamic variables with fermionic and bosonic Matsubara frequencies ω_n and ν_m, respectively. In Eq. (1) the one-particle propagators $G_\sigma(k)$ are input parameters that are at first not related to the two-particle vertex. An alternative BS equation with multiple electron-electron scatterings couples the intermediate variables in a different way [2].

The parquet approach takes advantage of the topological nonequivalence of different representations of the full vertex via BS equations. The nonequivalence means that *reducible* contributions in one two-particle scattering channel (the second part on the r.h.s. of Eq. (1)) are *irreducible* (first term on the r.h.s. of Eq. (1)) in the other channels. This fact is expressed as a basic parquet equation $\Gamma_{\uparrow\downarrow} = \Lambda^{eh}_{\uparrow\downarrow} + \Lambda^{ee}_{\uparrow\downarrow} - \mathscr{I}$ where $\mathscr{I}$ is the vertex irreducible in all considered scattering channels. The resulting parquet equations are then BS equations from different scattering channels from which the full vertex Γ was excluded by using the basic parquet equation.

Parquet equations for the irreducible vertices Λ^{eh} and Λ^{ee} are practically insoluble even for the simplest choice of the fully irreducible vertex $\mathscr{I} = U$, the bare local interaction. To be of practical use we have to introduce simplifications. To this purpose we realize that the parquet-type renormalizations become relevant with significant corrections to simple approximations in critical regions of singularities in the BS equations. Such a singularity in models with repulsive interactions is caused my multiple electron-hole scatterings. We hence assume that we are in the critical region of a pole in the BS equation in the *eh* channel and employ the standard renormalization-group reasoning to simplify the vertex functions. Near the critical point a new long-range scale emerges. We separate the long range and short-range fluctuations and eventually neglect the latter. With this assumption we can keep dynamical only those functions explicitly depending on the new diverging long-range scale. The short-range fluctuations are neglected in that the regular functions are replaced by constants. Using this simplification scheme we find that $\Lambda^{eh} = \overline{U}$ goes over to a static effective interaction, since there is no singularity in the BS equation in the *ee* channel. The irreducible vertex from the electron-electron scattering channel can be explicitly evaluated with this ansatz from a FLEX-type equation. Its singular part in SIAM reads [1]

$$\Lambda^{sing}_{\uparrow\downarrow}(i\nu_m) = -\frac{\overline{U}^2\chi_{eh}(i\nu_m)}{1+\bar{U}\chi_{eh}(i\nu_m)}\,. \quad (2a)$$

We denoted the electron-hole bubble $\chi_{eh}(i\nu_m) = \beta^{-1}\sum_n G_{\uparrow}(i\omega_n)G_{\downarrow}(i\omega_{n+m})$. The effective interaction $\overline{U}$ is then determined from the other parquet equation with $\Lambda^{ee} = \Lambda^{sing}$ averaged appropriately over the short-range time fluctuations to end up with a consistent equation. We obtain [1]

$$\bar{U} = U - \frac{\left\langle G_\uparrow G_\downarrow L^2_{\uparrow\downarrow}\right\rangle}{\chi_{ee}\left(\chi_{ee} + \left\langle G_\uparrow G_\downarrow L_{\uparrow\downarrow}\right\rangle\right)}. \tag{2b}$$

We denoted $\chi_{ee} = \beta^{-1}\sum_n G_\uparrow(i\omega_n)G_\downarrow(i\omega_{-n})$ the electron-electron bubble and $L_{\uparrow\downarrow}(i\omega_n) = \beta^{-1}\sum_{n'} G_\uparrow(i\omega_{n'})G_\downarrow(i\omega_{-n'})\Lambda^{sing}_{\uparrow\downarrow}(i\nu_{-n-n'})$, $\left\langle G_\uparrow G_\downarrow X\right\rangle = \beta^{-1}\sum_n G_\uparrow(i\omega_n)G_\downarrow(i\omega_{-n})X(i\omega_n)$.

These simplified parquet equations can be easily solved numerically for arbitrary set of input parameters including the one-electron propagators. The complexity of the numerical solution of these simplified parquet equations is comparable with FLEX-type approximations [3], however, we obtain significantly improved, more reliable, and numerically stable strong-coupling behavior.

Kondo asymptotics

Although the simplified parquet equations (2) can be formally applied in the whole region of the phases space, its derivation is based on the concept of separation of long and short-range fluctuations. The critical region of a singularity in our case is reached when the denominator on the r.h.s of Eq. (2a) approaches zero. A new small energy becomes relevant that we will call the Kondo scale. At zero temperature and in the rotationally invariant case, $G_\uparrow = G_\downarrow$, we introduce a dimensionless constant controlling the distance from the critical point $a = 1 + \bar{U}\chi_{eh}(0) = 1 - \overline{U}\int_{-\infty}^{0} d\omega \Im[G_+(\omega)^2]/\pi \to 0$, where $G_+(\omega) \equiv G(\omega + i0^+)$. The denominator on the r.h.s. of Eq. (2a) can be expanded only to the lowest (linear) order in frequency if we are interested in the leading singular behavior. We then obtain $1 + \bar{U}\chi_{eh}(\omega + i0^+) = a - i\pi\overline{U}\rho_0^2\omega$, where we denoted the density of one-particle states at the Fermi energy $\rho_0 = -\pi^{-1}\Im G_+(0)$. Putting $\overline{U} = -1/\chi_{eh}(0)$, that is choosing the critical point, we obtain an equation for the logarithm of the Kondo scale. A quite general expression was found in Ref. [1]. Here we want to be more explicit and choose the Hartree one-particle propagators in the parquet equations. They are in SIAM $G(x+iy) = 1/[x + i\text{sign}(y)(|y| + \Delta)]$. This choice of one-particle propagators allows us to evaluate explicitly the two-particle integrals and we reach at an explicit expression for the dimensionless Kondo scale for arbitrary filling

$$a \doteq \exp\left\{-4\pi^2\rho_0^2\bar{\mu}^2 \frac{\chi_{ee}(\chi_{ee} + \chi_{eh})\Delta}{3(\chi_{ee} + \chi_{eh})\Delta - 2\pi\bar{\mu}^2\chi_{eh}^2} U\right\}. \tag{3}$$

We used an abbreviation $\chi_{eh} = \chi_{eh}(0)$. The effective chemical potential $\bar{\mu} = E_d + (U/2 + \Delta x)$ with a phase $x = \arctan(\bar{\mu}/\Delta)$ measures a distance from the symmetric half-filled case $\bar{\mu} = 0$. With the Hartree propagators we also have explicit representations for the two-particle bubbles

$$\rho_0 = \frac{\Delta}{\pi\left[(\bar{\mu} + \Sigma_0)^2 + \Delta^2\right]}, \qquad \chi_{eh} = -\frac{1}{\pi}\frac{\Delta}{\bar{\mu}^2 + \Delta^2}, \qquad \chi_{ee} = \frac{x}{\bar{\mu}} \tag{4}$$

where $\Sigma_0 = \Sigma(0)$ is the self-energy (including the Hartree static term) at the Fermi energy. It is easy to verify that in the symmetric case, $\bar{\mu} = 0$, we reconstruct the result from Ref. [1].

It is clear from eq. (3) that the singularity in the electron-hole BS equation is moved to infinite interaction strength for arbitrary filling. The asymptotics with which the critical point is approached is exponential with the exponent linearly depending on the interaction strength, which corresponds to the exact solution [4]. It is also evident that the further from the symmetric case we are the stronger interaction is needed to reach the critical region of the singularity in Λ^{ee}. If we look at the extreme limits of electron or hole saturation ($|\bar{\mu}| \to \infty$, $|x| \to 1/2$) we obtain in the strong-coupling limit $U/\Delta \to \infty$

$$a \doteq \exp\left\{ -\frac{2\Delta^2}{3|\bar{\mu}|^3} U \right\} = \exp\left\{ -\frac{2U}{3\Delta} \cot^3(|x|) \right\} . \tag{5}$$

The strong-coupling regime with exponentially small Kondo scale ($a \ll 1$) is reached for small electron densities $n \to 0$ only if $Un^3 \gg \Delta$.

Self-energy

We defined the Kondo scale as a distance to the critical point in the BS equation in the electron-hole scattering channel. To demonstrate that this is also the scale determining the width of the Kondo quasiparticle peak in the density of states we now evaluate the self-energy corresponding to the solution of the parquet equations (2). To this purpose we use the Schwinger-Dyson equation. The simplified parquet equations deliver a two-particle vertex depending on only a single frequency. For SIAM we then obtain a FLEX-type representation of the self-energy

$$\Sigma_\sigma(i\omega_n) = \frac{U}{\beta} \sum_{n'} \frac{G_{-\sigma}(i\omega_{n'})}{1 + \overline{U}\chi_{eh}(i\nu_{n-n'})} . \tag{6}$$

The principal difference of Eq. (6) to the FLEX solution is that the bare interaction U in the FLEX vertex is replaced by the renormalized one, $\overline{U}$, in the parquet approach. Due to the two-particle self-consistency in the determination of the effective interaction $\overline{U}$, the one-particle propagators in the Schwinger-Dyson equation can be bare (Hartree) ones. Unlike the FLEX-type approximations, the non-self-consistent one-particle propagators in the parquet and Schwinger-Dyson equation deliver better results than the fully renormalized one-particle propagators [1]. This seemingly surprising fact is known also from other parquet-type renormalizations [5]. One-particle self-consistency spreads the original energy band to an infinite interval and thereby destroys the natural energy scale introduced by the bandwidth. Although the one-particle self-consistency does not qualitatively change the singular low-energy behavior it practically wipes out the high-energy features of the solution. As a consequence no Hubbard satellite peaks can be observed in the one-particle self-consistent theory. That is why we prefer to use the Hartree propagators in our approximation.

We plotted in Fig. 1 the density of states (DOS) in the strong-coupling regime. Formation of the Kondo resonant peak is manifested for the symmetric (half-filled) as well

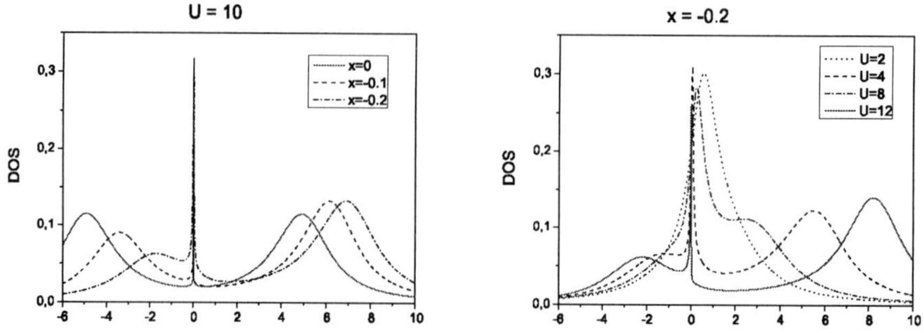

FIGURE 1. Formation of the Kondo peak at the Fermi energy in SIAM. Left pane: dependence of the Kondo resonance on the asymmetry factor x. Right pane: dependence on the interaction strength at $x = -0.2$.

as asymmetric situations. In the latter case we can observe how the central peak in DOS gets narrower and moves toward the Fermi energy with increasing interaction strength. The satellite Hubbard peaks become asymmetrically weighted in the asymmetric solution. While the critical point in the symmetric case corresponds to a metal-insulator transition, DOS at the Fermi energy remains finite even when the Kondo peak is suppressed in the asymmetric model. Kondo strong-coupling regime is hence a ubiquitous feature of SIAM.

CONCLUSIONS

We presented a solution of the SIAM derived from the parquet equations with separated short and long-range time scales. With this simplification we succeeded within the complexity of FLEX-type approximations to construct a universally applicable scheme *qualitatively correctly* interpolating between the weak and strong coupling regimes.

ACKNOWLEDGMENTS

We acknowledge financial support from the Grant Agency of the Czech Republic under Grant No. 202/04/1055.

REFERENCES

1. V. Janiš and P. Augustinský, e-print *cond-mat*/0607052.
2. V. Janiš, *Phys. Rev.* B**60**, 11345 (1999).
3. E. N. Bickers and D. J. Scalapino, *Ann. Phys. (N.Y.)* **93**, 5469 (1989).
4. A. M. Tsvelick and P. B. Wiegmann, *Adv. Phys.* **32**, 453 (1983).
5. Y. M. Vilk and A. M. S. Tremblay, *J. Phys. (Paris) Colloq.*, **7**, 1309 (1997).

Coherent spectral weights of Gutzwiller-projected superconductors

Samuel Bieri and Dmitri Ivanov

Institute of Theoretical Physics, Ecole Polytechnique Fédérale de Lausanne (EPFL), CH-1015 Lausanne, Switzerland

Abstract. We analyze the electronic Green's functions in the superconducting ground state of the t-J model using Gutzwiller-projected wave functions, and compare them to the conventional BCS form. Some of the properties of the BCS state are preserved by the projection: the total spectral weight is continuous around the quasiparticle node and approximately constant along the Fermi surface. On the other hand, the overall spectral weight is reduced by the projection with a momentum-dependent renormalization, and the projection produces electron-hole asymmetry in renormalization of the electron and hole spectral weights. The latter asymmetry leads to the bending of the effective Fermi surface which we define as the locus of equal electron and hole spectral weight.

Keywords: HTSC, lattice fermions, strongly correlated systems, unconventional superconductivity
PACS: 71.10.Li,74.72.-h,71.18.+y,71.10.Fd

INTRODUCTION

High temperature superconductivity (HTSC) is one of the most intriguing phenomena in modern solid state physics. Experimentally, HTSC is observed in layered cuprate compounds. The undoped cuprates are antiferromagnetically ordered insulators which develop the characteristic superconducting "dome" upon doping with charge carriers.

HTSC is interesting not only for promising technological applications, but also from a theoretical point of view. The relevant ingredients for HTSC are believed to be the following.

- Low dimensionality (2d).
- Strong short-range repulsion between electrons.
- Doped Mott insulator.

Taking these 3 ingredients, HTSC is modeled in the tight-binding description by large-U Hubbard models or t-J models on the square lattice:

$$H_{tJ} = -t \sum_{\langle i,j\rangle} P_G c^{\dagger}_{i\sigma} c_{j\sigma} P_G + J \sum_{\langle i,j\rangle} (\boldsymbol{S}_i \cdot \boldsymbol{S}_j - \frac{n_i n_j}{4}) \qquad (1)$$

where $n = c^{\dagger}_{\sigma} c_{\sigma}$, $\boldsymbol{S} = \frac{1}{2} c^{\dagger}_{\sigma} \boldsymbol{\sigma}_{\sigma\sigma'} c_{\sigma'}$ and $\boldsymbol{\sigma}$ are the Pauli matrices.[1] The Gutzwiller projector $P_G = \Pi_i (1 - n_{i\uparrow} n_{i\downarrow})$ prevents electrons from occupying the same lattice site.

---

[1] Repeated indices are summed over.

CP918, *Lectures on the Physics of Strongly Correlated Systems XI: Eleventh Training Course,* edited by A. Avella and F. Mancini

The non-perturbative nature of the t-J model makes it an outstanding problem to solve in dimensions larger than one. Analytical techniques (renormalized or slave-boson mean-field theories [1, 2]) are very crude and numerical techniques (e.g. exact diagonalization or cluster DMFT) are restricted to very small clusters or infinite dimensions, or they fail on the sign problem (QMC). An alternative approach was suggested by Anderson shortly after the experimental discovery of HTSC, when he proposed a Gutzwiller-projected BCS wave function as superconducting ground state for cuprates [3]. Following this conjecture, many variational studies have been performed on the basis of what is called Anderson's (long range) RVB state. This state turned out to have very low variational energy, close to exact ground state energies, as well as high overlap with the true ground states of small t-J clusters [4, 5]. On the other hand, many experimental facts about cuprate superconductors can be reproduced and are consistent with the variational results: e.g. clearly favored d-wave pairing symmetry, doping dependency of the nodal Fermi velocity and the nodal quasiparticle weight. Many of these successful efforts following Anderson's proposal are summarized in the "plain vanilla RVB theory" of HTSC, recently reviewed in [6].

With help of the relatively recent technique of angle-resolved photoemission spectroscopy (ARPES), experimentalists can probe the electronic structure of low-lying excitations inside the copper planes. The intensity measured in ARPES is proportional to the one–particle electronic spectral function: $I_{PES}(\boldsymbol{k},\omega) \propto A(\boldsymbol{k},\omega)$ [7]. It is therefore interesting to explore spectral properties within the framework of Gutzwiller-projected variational quasiparticle (QP) excitations.

In this contribution we will discuss some of our results reported in [8]. For more details, in particular for more reference to experimental studies, we invite the reader to consult that paper.

COHERENT SPECTRAL WEIGHTS

Anderson's RVB state is given by

$$|H\rangle \propto P_H P_G |d\mathrm{BCS}(\Delta,\mu)\rangle \,. \tag{2}$$

We further define projected BCS quasiparticle excitations in a similar way,

$$|H,\boldsymbol{k},\sigma\rangle \propto P_H P_G \gamma^\dagger_{\boldsymbol{k}\sigma} |d\mathrm{BCS}\rangle \,. \tag{3}$$

The unprojected states in Eqs. (2) and (3) are the usual ingredients of the BCS theory, $|d\mathrm{BCS}\rangle = \Pi_{\boldsymbol{k},\sigma}\gamma_{\boldsymbol{k}\sigma}|0\rangle$, $\gamma_{\boldsymbol{k}\sigma} = u_{\boldsymbol{k}}c_{\boldsymbol{k}\sigma} + \sigma v_{\boldsymbol{k}} c^\dagger_{-\boldsymbol{k}\bar{\sigma}}$, $u_{\boldsymbol{k}}^2 = \frac{1}{2}\left(1-\frac{\xi_{\boldsymbol{k}}}{E_{\boldsymbol{k}}}\right) = 1-v_{\boldsymbol{k}}^2$, $E_{\boldsymbol{k}} = \sqrt{\xi_{\boldsymbol{k}}^2+\Delta_{\boldsymbol{k}}^2}$, $\xi_{\boldsymbol{k}} = -2[\cos(k_x)+\cos(k_y)]-\mu$, $\Delta_{\boldsymbol{k}} = \Delta[\cos(k_x)-\cos(k_y)]$. P_G is the Gutzwiller projector and P_H projects on the subspace with H holes. $|H\rangle$ and $|H,\boldsymbol{k},\sigma\rangle$ are normalized to one. The wave functions have two variational parameters, Δ and μ, which we adjusted to minimize the energy of the t-J Hamiltonian (1) for the experimentally relevant value $J=0.3$ and every doping level. Note that in the RVB theory, Δ and μ are variational parameters without direct physical significance; physical quantities like excitation gap, superconducting order, or chemical potential must be calculated explicitly.

The spectral weights of the coherent low-lying quasiparticles (3) can be written as

$$Z_{\boldsymbol{k}}^{+} = |\langle H-1,\boldsymbol{k},\sigma|c_{\boldsymbol{k},\sigma}^{\dagger}|H\rangle|^2 \tag{4a}$$

$$Z_{\boldsymbol{k}}^{-} = |\langle H+1,\boldsymbol{k},\sigma|c_{-\boldsymbol{k},\bar{\sigma}}|H\rangle|^2 . \tag{4b}$$

These weights are measured in ARPES experiments as the residues of the spectral function $A(\boldsymbol{k},\omega)$ [7]. Note that in conventional BCS-theory, $Z_{\mathbf{k}}^{+} = u_{\boldsymbol{k}}^{2}$ and $Z_{\mathbf{k}}^{-} = v_{\boldsymbol{k}}^{2}$.

METHOD: VMC

The variational Monte Carlo technique (VMC) allows to evaluate fermionic expectation values of the form $\langle\psi|O|\psi\rangle$ for a given state $|\psi\rangle$. In order to calculate the spectral weights (4) by VMC, the following exact relations can be used.

$$Z_{\boldsymbol{k}}^{+} = \frac{1+x}{2} - \langle c_{\boldsymbol{k}\sigma}^{\dagger}c_{\boldsymbol{k}\sigma}\rangle , \tag{5a}$$

$$Z_{\boldsymbol{k}}^{+}Z_{\boldsymbol{k}}^{-} = |\langle H+1|c_{\boldsymbol{k}\uparrow}c_{-\boldsymbol{k}\downarrow}|H-1\rangle|^2 , \tag{5b}$$

where x is hole doping [9, 8].

We use VMC to calculate the superconducting order parameter $\Phi_{\boldsymbol{k}} = |\langle c_{\boldsymbol{k}\uparrow}c_{-\boldsymbol{k}\downarrow}\rangle|$, as well as diagonal matrix elements in the optimized t-J ground state (2). Using relations (5), we can then derive the spectral weights (4). The disadvantage of this procedure is large errorbars around the center of the Brillouin zone where both $Z_{\boldsymbol{k}}^{+}$ and $\Phi_{\boldsymbol{k}}$ are small.

RESULTS

In Fig. 1, we plot the nearest-neighbor superconducting orderparameter $\langle c_{i\uparrow}c_{j\downarrow}\rangle$ as a function of doping. The curve shows close quantitative agreement with the result of Ref. [10], where the authors extracted the same quantity from the long-range asymptotics of the nearest-neighbor pairing correlator, $\lim_{r\to\infty}\langle c_0c_\delta c_r^{\dagger}c_{r+\delta}^{\dagger}\rangle$. With the method employed here, we find the same qualitative and quantitative conclusions of previous authors [4, 10]: vanishing of superconductivity at half filling, $x\to 0$, and at the superconducting transition on the overdoped side, $x_c \simeq 0.3$. The optimal doping is near $x_{opt} \simeq 0.18$. In the same plot we also show the commonly used Gutzwiller approximation where the BCS orderparameter is renormalized by the factor $g_t = \frac{2x}{1+x}$ [1]. The Gutzwiller approximation underestimates the exact value by approximately 25%.

In Fig. 2, we plot the spectral weights $Z_{\boldsymbol{k}}^{+}$, $Z_{\boldsymbol{k}}^{-}$, and $Z_{\boldsymbol{k}}^{tot}$ along the contour $0\to(0,\pi)\to(\pi,\pi)\to 0$ in the Brillouin zone for different doping levels. Figure 3 shows the contour plots of $Z_{\boldsymbol{k}}^{tot}$ in the region of the Brillouin zone where our method produces small errorbars. From these data, we can make the following observations.

- In the case of an unprojected BCS wave function, the total spectral weight is constant and unity over the Brillouin zone. Introducing the projection operator,

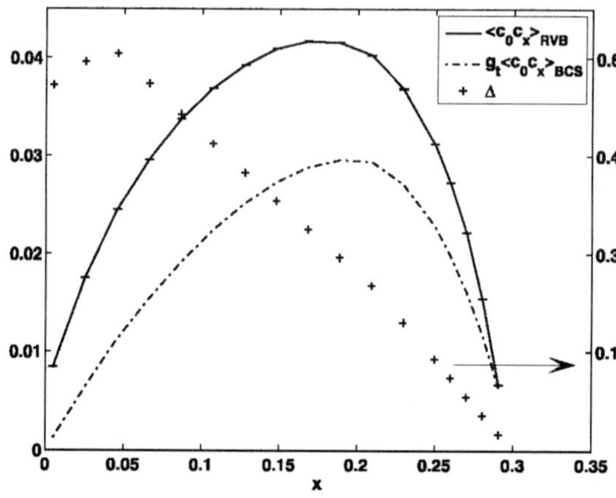

FIGURE 1. Doping dependency of the nearest-neighbor superconducting order parameter Φ_{ij} (calculated in the 14×14 system). The errorbars are smaller than the symbol size. The same quantity calculated in the Gutzwiller approximation is also shown for comparison. The variational parameter Δ is shown with the scale on the right.

we see that for low doping ($x \simeq 3\%$), the spectral weight is reduced by a factor up to 20. The renormalization is asymmetric in the sense that the electronic spectral weight $Z_{\boldsymbol{k}}^{+}$ is more reduced than the hole spectral weight $Z_{\boldsymbol{k}}^{-}$. For higher doping ($x \simeq 23\%$), the spectral weight reduction is much smaller and the electron-hole asymmetry decreases.

- Since there is no electron-hole mixing along the zone diagonal (d-wave), the spectral weights $Z_{\boldsymbol{k}}^{+}$ and $Z_{\boldsymbol{k}}^{-}$ have a discontinuity at the nodal point. Our data shows that the total spectral weight is continuous across the nodal point. Strong correlation does not affect these features of uncorrelated BCS-theory.

Effective Fermi surface

In strongly interacting Fermi systems, the notion of a Fermi surface (FS) is not clear at all. There are, however, several experimental definitions of the FS. Most commonly, $\boldsymbol{k}_F$ is determined in ARPES experiments as the maximum of $|\nabla_{\boldsymbol{k}} n_{\boldsymbol{k}}|$ or the locus of minimal gap along some cut in the $\boldsymbol{k}$-plane. The theoretically better defined locus of $n_{\boldsymbol{k}} = 1/2$ is also sometimes used. The various definitions of the FS usually agree within the experimental uncertainties [7]. The different definitions of the FS in HTSC were recently analyzed theoretically in Refs. [11, 12].

Here, we propose an alternative definition of the Fermi surface based on the ground state equal-time Green's functions. In the unprojected BCS state, the underlying FS is determined by the condition $|u_{\boldsymbol{k}}|^2 = |v_{\boldsymbol{k}}|^2$. We will refer to this as the *unprojected FS*. Since $|u_{\boldsymbol{k}}|^2$ and $|v_{\boldsymbol{k}}|^2$ are the residues of the QP poles in the BCS theory, it is natural to

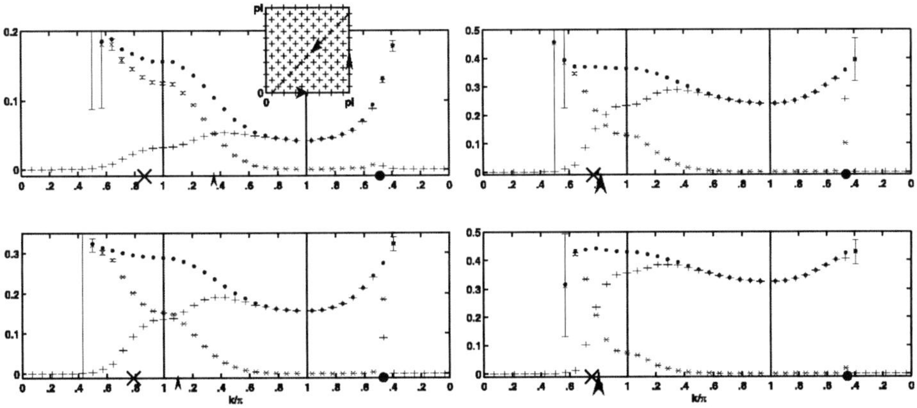

FIGURE 2. QP spectral weights for 6 holes (upper left plot, $x \simeq 3\%$), 22 holes (lower left plot, $x \simeq 11\%$), 34 holes (upper right plot, $x \simeq 17\%$), and 46 holes (lower right plot, $x \simeq 23\%$) on 196 sites. The spectral weights are plotted along the contour $0 \to (0,\pi) \to (\pi,\pi) \to 0$ (shown in inset). Plus signs (+) denote the spectral weight $Z_{\boldsymbol{k}}^{+}$, crosses ($\times$) denote $Z_{\boldsymbol{k}}^{-}$, errorbars are shown. Solid dots denote their sum, the total spectral weight $Z_{\boldsymbol{k}}^{tot}$, errorbars not shown. On the horizontal axis, the star ($*$) denotes the intersection with the unprojected Fermi surface along the $0 \to (0,\pi)$ direction; the thick dot is the nodal point. Both $Z_{\boldsymbol{k}}^{+}$ and $Z_{\boldsymbol{k}}^{-}$ jump at the nodal point, while $Z_{\boldsymbol{k}}^{tot}$ is continuous. The intersection with the effective Fermi surface (see text) is marked by an arrowhead. On the diagonal (last segment), k/π is given in units of $\sqrt{2}$.

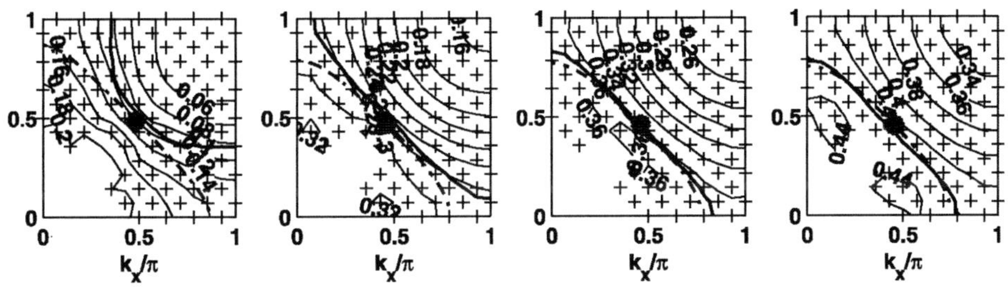

FIGURE 3. Contour plots of the total QP spectral weight $Z_{\boldsymbol{k}}^{tot}$. The effective FS (full line) and unprojected FS (dashed line) are also shown. The doping levels are $x \simeq 3\%$, 11%, 17%, and 23% (from left to right). The + signs indicate points where the values are known within small errorbars (see Section *Method: VMC*).

replace them in the interacting case by $Z_{\boldsymbol{k}}^{+}$ and $Z_{\boldsymbol{k}}^{-}$, respectively. We will therefore define the *effective FS* as the locus $Z_{\boldsymbol{k}}^{+} = Z_{\boldsymbol{k}}^{-}$.

In Fig. 3, we plot the unprojected and the effective FS which we obtained from VMC calculations. The contour plot of the total QP weight is also shown. It is interesting to note the following points.

- In the underdoped region, the effective FS is open and bent outwards (hole-like FS). In the overdoped region, the effective FS closes and embraces more and more the unprojected one as doping is increased (electron-like FS).

- Luttinger's rule [13] is clearly violated in the underdoped region, i.e. the area enclosed by the effective FS is not conserved by the interaction; it is larger than that of the unprojected FS.
- In the optimally doped and overdoped region, the total spectral weight is approximately constant along the effective FS within errorbars. In the highly underdoped region, we observe a small concentration of the spectral weight around the nodal point ($\simeq 20\%$).

Large "hole-like" FS in underdoped cuprates has also been reported in ARPES experiments by several groups [14, 15, 16].

It should be noted that a negative next-nearest hopping t' would lead to a similar FS curvature as we find in the underdoped region. We would like to emphasize that our original t-J Hamiltonian as well as the variational states do not contain any t'. Our results show that the outward curvature of the FS is due to strong repulsion, without need of t'. The next-nearest hopping terms in the microscopic description of the cuprates may not be necessary to explain the FS topology found in ARPES experiments. Remarkably, if the next-nearest hopping t' is included in the variational ansatz (and not in the original t-J Hamiltonian), a finite and negative t' is generated, as it was shown in Ref. [17]. Apparently, in this case the unprojected FS has the tendency to adjust to the effective FS. A similar bending of the FS was also reported in the recent analysis of the current carried by Gutzwiller-projected QPs [18]. A high-temperature expansion of the momentum distribution function $n_{\boldsymbol{k}}$ of the t-J model was done in Ref. [19] where the authors find a violation of Luttinger's rule and a negative curvature of the FS. Our findings provide further evidence in this direction.

A natural question is the role of superconductivity in the unconventional bending of the FS. In the limit $\Delta \to 0$, the variational states are Gutzwiller-projected excitations of the Fermi sea and the spectral weights are step-functions at the (unprojected) FS. In a recent paper [20], it was shown that $\lim_{\boldsymbol{k}\to\boldsymbol{k}_F^+} Z^+_{\boldsymbol{k}} = \lim_{\boldsymbol{k}\to\boldsymbol{k}_F^-} Z^-_{\boldsymbol{k}}$ for the projected Fermi-sea, which means that the unprojected and the effective FS coincide in that case. This suggests that the "hole-like" FS results from a non-trivial interplay between strong correlation and superconductivity. At the moment, we lack a qualitative explanation of this effect. However, it may be a consequence of the proximity of the system to the non-superconducting "staggered-flux" state [21, 22] or to antiferromagnetism near half-filling [10, 23].

ACKNOWLEDGEMENT

SB would like to thank the organizers of the *Eleventh Training Course in the Physics of Correlated Electron Systems and HTSC* in Salerno, Italy, where this work was presented. We would like to thank George Jackeli for many illuminating discussions and continuous support. We also thank Claudius Gros, Patrick Lee, and Seiji Yunoki for interesting discussions. This work was supported by the Swiss National Science Foundation.

REFERENCES

1. F. C. Zhang, C. Gros, T. M. Rice, and H. Shiba, Supercond. Sci. Technol. **1**, 36 (1988).
2. G. Kotliar and J. Liu, Phys. Rev. B **38**, 5142 (1988).
3. P. W. Anderson, Science **235**, 1196 (1987).
4. C. Gros, Phys. Rev. B **38**, 931 (1988); Ann. Phys. (N.Y.) **189**, 53 (1989).
5. Y. Hasegawa and D. Poilblanc, Phys. Rev. B **40**, 9035 (1989).
6. P. W. Anderson, P. A. Lee, M. Randeria, M. Rice, N. Trivedi, and F. C. Zhang, J. Phys.: Condens. Matter **16**, R755 (2004).
7. A. Damascelli, Z. Hussain, and Z.-X. Shen, Rev. Mod. Phys. **75**, 473 (2003); J. C. Campuzano, M. R. Norman, and M. Randeria, in *Physics of Superconductors*, edited by K. H. Bennemann and J. B. Ketterson (Springer, Berlin, 2004), Vol. II, pp. 167-273; cond-mat/0209476.
8. S. Bieri and D. Ivanov, Phys. Rev. B. **75** 035104 (2007); cond-mat/0606633.
9. S. Yunoki, Phys. Rev. B **72**, 092505 (2005); Phys. Rev. B **74**, 180504 (2006).
10. A. Paramekanti, M. Randeria, and N. Trivedi, Phys. Rev. Lett. **87**, 217002 (2001); Phys. Rev. B **70**, 054504 (2004).
11. C. Gros, B. Edegger, V. N. Muthukumar, and P. W. Anderson, Proc. Natl. Acad. Sci. U.S.A **103**, 14298 (2006); cond-mat/0606750.
12. R. Sensarma, M. Randeria, and N. Trivedi, cond-mat/0607006.
13. J. M. Luttinger, Phys. Rev. **121**, 942 (1961).
14. A. Ino, C. Kim, M. Nakamura, T. Yoshida, T. Mizokawa, A. Fujimori, Z.-X. Shen, T. Kakeshita, H. Eisaki, and S. Uchida, Phys. Rev. B **65**, 094504 (2002).
15. T. Yoshida, X. J. Zhou, T. Sasagawa, W. L. Yang, P. V. Bogdanov, A. Lanzara, Z. Hussain, T. Mizokawa, A. Fujimori, H. Eisaki, Z.-X. Shen, T. Kakeshita, and S. Uchida, Phys. Rev. Lett. **91**, 027001 (2003).
16. K. M. Shen, F. Ronning, D. H. Lu, F. Baumberger, N. J. C. Ingle, W. S. Lee, W. Meevasana, Y. Kohsaka, M. Azuma, M. Takano, and Z.-X. Shen, Science **307**, 901 (2005).
17. A. Himeda and M. Ogata, Phys. Rev. Lett. **85**, 4345 (2000).
18. C. P. Nave, D. A. Ivanov, and P. A. Lee, Phys. Rev. B **73**, 104502 (2006).
19. W. O. Putikka, M. U. Luchini, and R. R. P. Singh, Phys. Rev. Lett. **81**, 2966 (1998).
20. H. Yang, F. Yang, Y.-J. Jiang, and T. Li, cond-mat/0604488.
21. P. A. Lee and X.-G. Wen, Phys. Rev. B **63**, 224517 (2000).
22. D. A. Ivanov and P. A. Lee, Phys. Rev. B **68**, 132501 (2003).
23. D. A. Ivanov, Phys. Rev. B **74**, 24525 (2006).

Random phase approximation and the finite size errors in many body simulations

Simone Chiesa*, D. M. Ceperley†, R. M. Martin† and Markus Holzmann**

*Dept. of Physics, University of California Davis, CA 95616
†Dept. of Physics, University of Illinois Urbana-Champaign, Urbana, IL 61801
**LPTMC, UMR 7600 of CNRS, Université P. et M. Curie, Paris, France

Abstract. We address the problem of finite size errors on the energy computed in quantum Monte Carlo simulations. We introduce a scheme based on the random phase approximation that, within a single calculation, allows to retrieve the leading order correction and produces practically converged results for modest numbers of particles. Applications to the electron gas and silicon are presented.

Keywords: Long range interaction Many-body simulation, Finite size error
PACS: 71.15.Nc, 02.70.Ss

In the simulation of bulk systems via quantum Monte Carlo methods[1] the main source of systematic error is related to the relatively small number of particles that can be practically handled. The aim of this paper is to show that such a finite size error on the energy can be corrected using the information collected in a single simulation performed using a small number of particles.

In order to simulate an infinitely extended system, one defines a superlattice with basis vectors $\{\mathbf{L}_\alpha\}_{\alpha=1,2,3}$ and explicitly considers only a finite number of particles $N = \Omega/n$ where $\Omega = |\mathbf{L}_1 \cdot (\mathbf{L}_2 \times \mathbf{L}_3)|$ and n is the desired average density. The (infinitely many) remaining particles are periodic images of these N so that the potential energy is periodic and given by

$$V_{e-e} = \sum_{i<j} \sum_{\mathbf{L}} \frac{e^2}{|\mathbf{r}_i - \mathbf{r}_j + \mathbf{L}|} + V_{\text{background}} = \frac{2\pi e^2}{\Omega} \sum_{\mathbf{k} \neq 0} \frac{1}{k^2} (\rho_{\mathbf{k}} \rho_{-\mathbf{k}} - N) \qquad (1)$$

where $\rho_{\mathbf{k}} \equiv \sum_i \exp(i\mathbf{k}\mathbf{r}_i)$ and $\mathbf{k}$, the reciprocal lattice vectors, satisfy $\exp(i\mathbf{k}\mathbf{L}_\alpha) = 1$. In practice, none of the forms in Eq.1 is used in simulations: the Coulomb interaction is split in a divergent short range part treated in real space and a smooth long range part handled in reciprocal space[2]. For the purpose of this paper however, it is most convenient to focus on the reciprocal space form. Because of the periodicity of the Hamiltonian,the boundary conditions on the wave function can be chosen as $\Psi(..\mathbf{r}_i + \mathbf{L}_\alpha..) = \exp(i\theta_\alpha)\Psi(..\mathbf{r}_i..)$ where θ_α is the "twist" of the phase in the αth direction and lies in the interval $[-\pi, \pi)$[3, 4]. When using a single twist the expectation value of the potential energy is determined by the static structure factor $S_{\boldsymbol{\theta},N}(\mathbf{k}) = \langle \rho_{\mathbf{k}} \rho_{-\mathbf{k}} \rangle_{\boldsymbol{\theta},N}/N$. As the system size increases, the mesh of $\mathbf{k}$ vectors gets finer and the series eventually converges to an integral corresponding to the exact thermodynamic limit. At finite N, the only effect of averaging over $\boldsymbol{\theta}$ is to replace $S_{\boldsymbol{\theta},N}$ with $S_N(\mathbf{k}) = (2\pi)^{-3} \int d\boldsymbol{\theta} S_{\boldsymbol{\theta},N}(\mathbf{k})$. This does not change the reciprocal lattice mesh and twist-averaging is therefore expected to have a

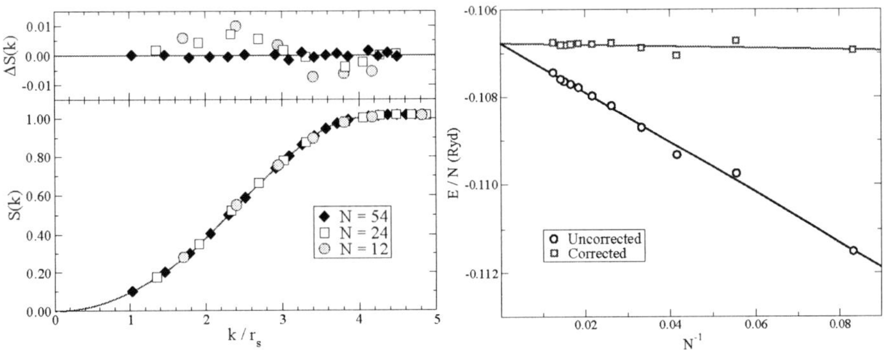

FIGURE 1. Left: Static structure factor for the electron gas at $r_s = 10$(lower panel) and $\Delta S = S_N(k) - S_{66}(k)$. The difference is computed using a spline function interpolation of S_{66}. Right: Energies per particle of the electron gas at $r_s = 10$ in Rydberg as a function of the inverse particle number. Circles are the Monte Carlo energies averaged over twist angles. Squares are the energies after the additional $\hbar\omega_p/2N$ correction (see text).

minor effect on the error in the potential energy.

The error using a simulation box with N particles is therefore given by

$$\Delta V \equiv \frac{e^2}{4\pi^2}\int \frac{S_\infty(\mathbf{k})-1}{k^2}d\mathbf{k} - \frac{2\pi e^2}{\Omega}\sum_{\mathbf{k}\neq 0}\frac{S_N(\mathbf{k})-1}{k^2}. \tag{2}$$

After the Madelung constant has been taken into account[5] the remaining part of the error is determined by i) the substitution of $S_\infty(\mathbf{k})$ by the computed $S_N(\mathbf{k})$ and ii) the discretization of the integral of $e^2S(\mathbf{k})(4\pi^2k^2)^{-1}$. In the small k region the random-phase approximation suggests $S_\infty(\mathbf{k}) \simeq S_N(\mathbf{k})$ and implies that the leading order contribution to the error comes from point ii) above: it is an integration error that originates from the omission of the $\mathbf{k} = 0$ volume element from the energy sum. Thanks to the validity of the random-phase approximation one also knows $S(\mathbf{k}) \sim k^2$ with a prefactor that can be determined either analytically or from a knowledge of the $S_N(\mathbf{k})$ computed in the simulation. Once the prefactor is known, one can accurately compute the correction.

We looked at jellium as a test case to judge to what extent the *ansatz* $S_\infty(\mathbf{k}) = S_N(\mathbf{k})$ is verified. Results for $S_N(k)$ computed in variational Monte Carlo simulations at $r_s = 10$ for 12, 24 and 54 particles are shown in Fig.1. As we increase the number of particles, the grid of k points for which S_N is defined shifts, but the values of S_N fall on a smooth curve, independent of N.

Let us now consider the kinetic energy. When using a twisted boundary condition $\boldsymbol{\theta}$ in a cubic supercell, the kinetic energy is given in terms of the momentum distribution $n_{\boldsymbol{\theta},N}(\mathbf{k}) = \langle c^\dagger_{\mathbf{k}} c_{\mathbf{k}}\rangle_{\boldsymbol{\theta},N}$ by

$$T = \frac{\hbar^2}{2m}\sum_{\mathbf{k}} n_{\boldsymbol{\theta},N}(\mathbf{k}+\boldsymbol{\theta}/L)|\mathbf{k}+\boldsymbol{\theta}/L|^2 \tag{3}$$

At variance with the potential energy case, the reciprocal lattice over which the summation is performed is shifted by $\boldsymbol{\theta}/L$ so that when the average over twist is performed one gets

$$T = \frac{\hbar^2}{2m(2\pi)^3} \int d\boldsymbol{\theta} \sum_{\mathbf{k}} n_{\boldsymbol{\theta},N}(\mathbf{k}+\boldsymbol{\theta}/L)(\mathbf{k}+\boldsymbol{\theta}/L)^2. \tag{4}$$

This is the exact thermodynamic limit if the condition $n_{\boldsymbol{\theta},N} = n_\infty$ is satisfied at those reciprocal vectors where $n_{\boldsymbol{\theta},N}$ is defined. In practice, the error in the twist angle integration can be made arbitrarily small by increasing the density of twist angles.

The assumption $n_{\boldsymbol{\theta},N} = n_\infty$ is however not satisfied in a many body system. The average over twist is important and must be performed. However, as we shall show, it does not go all the way to eliminate the finite size error on T. To understand why, we first need to remember some known features of the exact ground state wave function. In charged systems the interaction causes the wave function to have a long range charge-charge correlation factor: the Jastrow potential. Within the random phase approximation the ground state of the system is described by a collection of dressed particles interacting via short range forces and quantized coherent modes, the plasmons. Accordingly, the many-body wave function factorizes as[6]

$$\Psi = \Psi_{\text{s.r.}} \exp(J) = \Psi_{\text{s.r.}} \exp\left[-\frac{1}{2\Omega}\sum_{\mathbf{k}\neq 0} u_{\mathbf{k}} \rho_{\mathbf{k}} \rho_{\mathbf{k}}^{\dagger}\right] \tag{5}$$

where $\Psi_{\text{s.r.}}$ only contains short range correlations and $u_{\mathbf{k}}$ decays quickly to 0 as k increases and diverges as k^{-2} at small k.

We now focus on one twist and consider the Fourier transform of $n_{\boldsymbol{\theta},N}(\mathbf{k})$: the one body density matrix $\gamma(\mathbf{r}) = \frac{1}{\Omega}\sum_{\mathbf{k}} n_{\boldsymbol{\theta},N}(\mathbf{k}) \exp(i\mathbf{k}\mathbf{r})$. This quantity can be directly expressed in terms of the wave function as

$$\gamma(\mathbf{r}) = \frac{1}{N}\sum_{\alpha=1}^{N} \left\langle \frac{\Psi(\mathbf{r}_1, \mathbf{r}_2 \ldots \mathbf{r}_\alpha + \mathbf{r}, \ldots)}{\Psi(\mathbf{r}_1, \mathbf{r}_2, \ldots)} \right\rangle \tag{6}$$

and determines the kinetic energy per particle as $T = -\frac{\hbar}{2mn}\nabla^2\gamma(\mathbf{r})|_{\mathbf{r}=0}$. The correct prescription to compute $\gamma(\mathbf{r})$ consists in displacing particle α by $\mathbf{r}$ keeping all the remaining ones fixed. However, due to the periodic setup of the system, Eq.6 describes a global shift of α *and* all its images. Such an error cannot be addressed by averaging over boundary conditions since the violation of the "correct prescription" happens regardless of the value of the twist. All of this is fairly irrelevant for the short range part of the wave function since the ratio $\Psi_{\text{s.r.}}(\mathbf{r}_1, \mathbf{r}_2 \ldots \mathbf{r}_\alpha + \mathbf{r}, \ldots)/\Psi_{\text{s.r.}}(\mathbf{r}_1, \mathbf{r}_2 \ldots)$ is sensitive only to particles lying close to α (within some characteristic short range correlation length) and the images fall quickly out of this region upon increasing the supercell size. Matters are different for the long range Jastrow potential so that, as pointed out by Magro and Ceperley[7], it is necessary to retain the leading order contribution to the difference between the periodic Jastrow potential and the fixed-image one. This is given by

$$\Delta J(\mathbf{r}) \equiv \frac{1}{2\Omega}\sum_{\mathbf{k}\neq 0} u_{\mathbf{k}}(1-\exp(i\mathbf{k}\cdot\mathbf{r})) - \frac{1}{2(2\pi)^3}\int d\mathbf{k} u_{\mathbf{k}}(1-\exp(i\mathbf{k}\cdot\mathbf{r})) \tag{7}$$

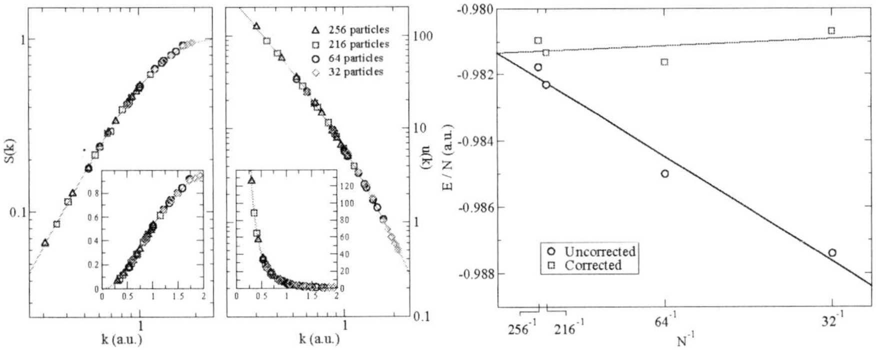

FIGURE 2. Left: Structure factor (left panel) and Jastrow potential (right panel) for diamond silicon at ambient pressure. The continuous lines are fit to the data (see text). The Jastrow potential shows a k^{-2} divergence at small k that was not explicitly imposed but obtained through energy variance minimization. Right: Diffusion Monte Carlo energies per electron in diamond Silicon at $r_s = 2.0$. Energies and the $S(k)$ and $u(k)$ used to compute the correction are all obtained in simulations with the same number of particles. The smallest cell is the conventional FCC cubic cell of diamond. The two intermediate ones are, respectively, $2\times2\times2$ and $3\times3\times3$ supercells of the primitive cell. The largest one is a $2\times2\times2$ supercell of the conventional cubic cell.

so that the corrected density matrix becomes $\gamma_{\text{corr}} = \gamma\exp(\Delta J)$. Since $\Delta J(0) = 0$, $\nabla\gamma|_{\mathbf{r}=0} = 0$ and $\gamma(0) = n$ the kinetic error is given by $\Delta T = -\frac{\hbar}{2m}\nabla^2\Delta J(\mathbf{r})\big|_{\mathbf{r}=0}$ and, by using Eq.7, can be explicitly written as

$$\Delta T = \frac{\hbar^2}{4m(2\pi)^3}\int d\mathbf{k}k^2u_k - \frac{\hbar^2}{4m\Omega}\sum_{\mathbf{k}\neq0}k^2u_k. \tag{8}$$

This is again an integration error provided u_k does not depend on the system size. As before such an error originates primarily from the omission of the $k = 0$ contribution. That u_k is independent of N is strongly suggested by the fact that a difference in u_k would necessarily imply a difference in $S(\mathbf{k})$ contrary to the random phase approximation. Within this approxiation $u_k \sim 1/k^2$ so that, once again, we need to determine a prefactor to estimate the error .

We first apply these corrections to the electron gas for which the small k limits of $S(k)$ and $u(k)$ are known from the random phase approximation as, respectively, $\hbar k^2/2m\omega_p$ and $4\pi e^2/\hbar\omega_p k^2$ where ω_p is the plasma frequency. In our tests, the wave function had a backflow-Jastrow form[8] and simulations were performed in the grand-canonical ensemble. Symmetries can be used to drastically reduce the number of needed twist angles to between $20-200$ for an unpolarized system with $N \sim 10-100$. The leading order correction due to long range correlations to kinetic and potential energy are equal and sum up to a total error $\Delta_N = \hbar\omega_p(2N)^{-1}$. Corrected and uncorrected variational energies are shown in Fig.1 for $r_s = 10$. Diffusion Monte Carlo values are uniformly shifted to lower energy by 0.6 mRyd/electron and show similar behavior. One can see that the bias due to the small size of the simulation cell is tremendously reduced, so that the $N = 12$ case is already satisfactory.

As a second example we considered the diamond structure of silicon at ambient pressure ($r_s = 2.0$). Most of the computational details can be found elsewhere[5]. Here, it is only worth stressing that the correct analytical behavior of the Jastrow potential at low k is automatically retrieved by variance minimization[9]. This is crucial for more complicated systems where there is no explicit form for the prefactors. Structure factor and Jastrow potential for different system size are presented in Fig.2.

We fit the parametric form $S(k) = 1 - \exp(-\alpha k^2)$ and $u(k) = 4\pi a[k^{-2} - (k^2 + a^{-1})^{-1}]$[10] to extract the sought prefactors. It is particularly remarkable that a 1-parameter form is capturing so well the behavior of a Jastrow potential which was optimized using more than 10 parameters. When k is expressed in atomic units, the optimal value of α and a were found to be 0.72 and 1.0 respectively, leading to corrections of $0.13/N$ and $0.092/N$ Hartree per electron for potential and kinetic energy. Results after the two corrections were applied are shown in Fig.2. Even for the smallest cell (cubic, with 8 Si atoms), the error in the energy is of the order of 1 mHartree/electron (0.1 eV/atom) when compared to the value extrapolated for the infinite size.

To summarize, we introduced a scheme to compute the correction to finite size error in the kinetic and potential energy in quantum Monte Carlo simulations. Madelung constant, Brillouin zone integration and the two $1/L$ contributions to the error addressed in this paper fall all out naturally from a reciprocal space approach as integration errors. The scheme is based on the determination of the low-k behavior of the structure factor and the Jastrow potential and greatly benefits of the random phase approximation that prescribes the power law behavior of such quantities.

This material was based upon work supported in part by the U.S Army Research Office under DAAD19-02-1-0176. Computational support was provided by the Materials Computational Center (NSF DMR-03 25939 ITR), the National Center for Supercomputing Applications at the University of Illinois at Urbana-Champaign and by CNRS-IDRIS.

REFERENCES

1. W. M. C. Foulkes, L. Mitas, R. J. Needs, and G. Rajagopal, Rev. Mod. Phys. **73**, 33 (2001).
2. V. Natoli and D. M. Ceperley, J. Comp. Phys. **117**, 171 (1995).
3. C. Lin, F. H. Zong, and D. M. Ceperley, Phys. Rev. E **64**, 16702 (2001).
4. G. Rajagopal, R. J. Needs, S. Kenny, W. M. C. Foulkes, and A. James, Phys. Rev. Lett. **73**, 1959 (1994).
5. S. Chiesa D. M. Ceperley, R. M. Martin, and M. Holzmann, Phys. Rev. Lett. **97**, 76404 (2006).
6. D. Bohm and D. Pines, Phys. Rev. **92**, 609 (1953).
7. W. R. Magro and D. M. Ceperley, Phys. Rev. Lett. **73**, 826 (1994).
8. M. Holzmann, D. M. Ceperley, C. Pierleoni, and K. Esler, Phys. Rev. E **68**, 46707 (2003).
9. N. D. Drummond, M. D. Towler, and R. J. Needs, Phys. Rev. B **70**, 235119 (2004).
10. M. S. Becker, A. A. Broyles, and T. Dunn, Phys. Rev. **175**, 224 (1968).

Jahn-Teller coupling in Ca-based layered ruthenates

Mario Cuoco*, Filomena Forte* and Canio Noce*

*Laboratorio Regionale SuperMat,CNR-INFM, Baronissi (SA), Italy
Dipartimento di Fisica "E. R. Caianiello", Università di Salerno, I-84081 Baronissi (Salerno), Italy

Abstract. The possibility of tuning the charge transfer between different orbital sectors via structural modifications is analyzed for the basic problem of two neighboring (corner sharing) octahedra with t_{2g} orbitals in $4d^4$ configuration. By means of exact diagonalization calculations, the different role played by the crystal field effect and the Jahn-Teller coupling is investigated. The analysis reveals that the former favors ferromagnetic (antiferromagnetic) correlations depending on the octahedral deformation while the latter tends to drive the system towards an antiferromagnetic phase with ferro-type orbital correlations. The relevance of these results with respect to the experimental data is also presented.

Keywords: ruthenate oxides, spin-orbital-lattice correlations, Jahn-Teller distortions
PACS: 75.50.Ee; 71.10.-w, 75.10.-b.

INTRODUCTION

The electronic, magnetic and structural properties of ruthenate oxides having perovskite-derived structures have recently attracted great attention. Indeed, by controlling both the bonding state in RuO_2 layers and the layer-stacking dimensionality, it has been shown that these compounds exhibit various kinds of ground states.[1] In this respect, it is worth mentioning the quasi-two-dimensional $Ca_{2-x}Sr_xRuO_4$ compound that provides a new type of Mott transition system that continuously connects the Mott insulator Ca_2RuO_4 with the spin triplet superconductor Sr_2RuO_4, passing through an antiferromagnetic correlated metal for $0.2< x <0.5$ and a nearly ferromagnetic metal for $x \sim 0.5$.[2, 3, 4, 5]
Moreover, recent experimental studies [6, 3, 7, 8, 9] have revealed a strong magnetoelastic coupling in the Ca rich region of the phase diagram, while it has been reported that magnetic and electronic properties are strictly correlated to the structural modifications induced by varying some specific external control parameters, such as temperature, doping and pressure.[8, 6, 10, 4] Considering the effects provided by chemical doping, the isovalent substitution of Sr for Ca primarily determines structural modifications, due to the reduced ionic size of Ca compared with Sr.[11] It has been shown that the RuO_6 rotation can enhance the ferromagnetic instability significantly, while the combination of tilting and rotation of RuO_6 is responsible for the enhancement of antiferromagnetic instability. Moreover, the flattening (elongation) of RuO_6 is the key factor to stabilize the magnetic (both the ferromagnetic and the antiferromagnetic) ground states.
Therefore, these results suggest that the basic physics governing the phase diagram of this system should be understood in terms of the strong interplay between the lattice

CP918, *Lectures on the Physics of Strongly Correlated Systems XI: Eleventh Training Course*, edited by A. Avella and F. Mancini

and the magnetic correlations, through the orbital degrees of freedom of the t_{2g} electrons.

Due to their extension, Ru $4d$ shells strongly hybridize with oxygen $2p$-derived states, and this interaction makes the structure distorted with respect to the ideal one. In this context, the c-axis distortions turn out to be considerably relevant since they control the degree of orbital fluctuations by influencing the distribution of the charge within the three orbitals in the t_{2g} sector. Two mechanisms are identified as the main responsible for the coupling between charge and orbitals. The first one is the crystal field potential, originating from the static Coulomb repulsion with the surrounding oxygens ions and by the energy shift due to the formation of antibonding molecular orbitals between Ru and O ions. The second mechanism, is the Jahn-Teller interaction, which is responsible for the dynamical coupling between the electron variables and the octahedral distortive modes, leading to a lowering of the symmetry, which further removes the d-level degeneracy, contributing possibly to the orbital ordering.
To understand the role of the t_{2g} orbital degrees of freedom, the 3-orbital Hubbard model has been commonly the starting point, and interesting orbital ordered states have been proposed as possible candidates to explain the ground state of this compounds.[12, 13, 14, 15, 16, 17] In two recent works, we have analyzed the interplay between Coulomb correlations and static octahedron distortions, described by the crystalline field potential.[18, 19] Our approach is based on exact diagonalization calculations, performed on a 3-orbital extended Hubbard model, describing the correlated dynamics of four electrons within t_{2g} orbitals. Furthermore, we have taken into account both the cases of compressed and elongated octahedra. The obtained results revealed that the crystal field is an ideal quantum control of the magnetic and orbital correlations, being able to separate ferro- and antiferro-type of magnetic and orbital patterns, depending on the character of the structural distortions.
In this frame, it may result useful to compare the role of static deformation of the octahedra with the Jahn-Teller effect. Thus, in this paper we will concentrate on the role of the electron-phonon coupling considering the coupling between electron and Jahn-Teller modes fully taking into account the correlated electron dynamics and including the effect of the phonon cloud in a variational scheme of computation. Namely, we will refer to the basic problem of two corner sharing octahedra at zero temperature. We will show that the electron-phonon coupling plays an important role in the evolution of the orbital degree of freedom, changing the energy levels of the doubly and the singly occupied orbitals. Nevertheless, unlike the crystal field potential that favors ferromagnetic (antiferromagnetic) correlations depending on the character of the octahedral deformation, the Jahn-Teller coupling tends to drive the system towards antiferromagnetism with ferro-type orbital correlations.
The outline of the paper is the following. In Sec. II the microscopic model is presented. In Sec. III we examine the interplay between Coulomb correlations and Jahn-Teller coupling, whereas the last section contains the discussion of the results and the concluding remarks.

THE MODEL

We study a three-orbital extended Hubbard model, supplemented by electron-lattice interactions

$$H = H_{kin} + H_{el-el} + H_{jt} \tag{1}$$

which effectively describes the pure dynamics of $4d$ bands, the electron variables of the oxygens being projected out. The first term in Eq. 1 is the kinetic operator that defines the connectivity between the t_{2g} orbitals through the oxygen ions,

$$H_{kin} = -\sum_{ij,\sigma} t_{ij}(d^{\dagger}_{\mathrm{i}\alpha\sigma} d_{\mathrm{j}\alpha\sigma} + h.c.) \tag{2}$$

$d^{\dagger}_{i\alpha\sigma}$ being the creation operator for an electron with spin σ at the i site in the α orbital, while t_{ij} is the hopping amplitude between i and j sites.

The second term in H stands for the local Coulomb interactions between t_{2g} electrons:

$$H_{el-el} = U\sum_{i\alpha} n_{\mathrm{i}\alpha\uparrow} n_{\mathrm{i}\alpha\downarrow} - 2J_H \sum_{i\alpha\beta} \mathbf{S}_{\mathrm{i}\alpha}\cdot\mathbf{S}_{\mathrm{i}\beta} + (U' - \frac{J_H}{2})\sum_{i\alpha\beta} n_{\mathrm{i}\alpha} n_{\mathrm{i}\beta} + J'\sum_{i\alpha\beta} d^{\dagger}_{\mathrm{i}\alpha\uparrow} d^{\dagger}_{\mathrm{i}\alpha\downarrow} d_{\mathrm{i}\beta\uparrow} d_{\mathrm{i}\beta\downarrow} \tag{3}$$

where $n_{\mathrm{i}\alpha\sigma}$, $\mathbf{S}_{\mathrm{i}\alpha}$ are the on site charge for spin σ and the spin operators for the α orbital, respectively. U (U') is the intra (inter)- orbital Coulomb repulsion, J_H is the Hund coupling, and J' the pair hopping term. Due to the invariance for rotations in the orbital space, the following relations hold: $U = U' + 2J_H$, $J' = J_H$.

The interaction between electrons and lattice is expressed through the Jahn-Teller term [20, 21] which couples the density of the electron in the different orbitals to the Q modes of the octahedron. Its expression reads as:

$$H_{jt} = g\sum_{i}(Q_{z\mathrm{i}} n_{\mathrm{i}xy} + Q_{x\mathrm{i}} n_{\mathrm{i}yz} + Q_{y\mathrm{i}} n_{\mathrm{i}xz}) + \frac{k}{2}\sum_{i}(Q^2_{2\mathrm{i}} + Q^2_{3\mathrm{i}}) \tag{4}$$

where g is the electron-lattice coupling amplitude, $Q_{x\mathrm{i}} = -\frac{1}{2}Q_{3\mathrm{i}} + \frac{\sqrt{3}}{2}Q_{2\mathrm{i}}$, $Q_{y\mathrm{i}} = -\frac{1}{2}Q_{3\mathrm{i}} - \frac{\sqrt{3}}{2}Q_{2\mathrm{i}}$, $Q_{z\mathrm{i}} = Q_{3\mathrm{i}}$, and k is the elastic constant. Here, $Q_{2\mathrm{i}} = \frac{1}{\sqrt{2}}(X_{1\mathrm{i}} - Y_{2\mathrm{i}})$ and $Q_{3\mathrm{i}} = \frac{1}{\sqrt{(6)}}(2Z_{3\mathrm{i}} - X_{1\mathrm{i}} - Y_{2\mathrm{i}})$ and X_i, Y_i, Z_i stand for the displacement of the i oxygen ion with respect to the equilibrium positions. The Jahn-Teller self-trapping energy is given by $E_{JT} = g^2/(2k)$.

By means of the above relations between Q_x ,Q_y, Q_z and the modes Q_2, Q_3, and by introducing the corresponding density operators $n_{2\mathrm{i}} = \frac{\sqrt{3}}{2}(n_{\mathrm{i}yz} - n_{\mathrm{i}xz})$, $n_{3\mathrm{i}} = n_{\mathrm{i}xy} - \frac{1}{2}(n_{\mathrm{i}xz} + n_{\mathrm{i}yz})$ it is possible to re-write H_{jt} in the following way:

$$H_{jt} = g\sum_{i}(Q_{3\mathrm{i}} n_{3\mathrm{i}} + Q_{2\mathrm{i}} n_{2\mathrm{i}}) + \frac{k}{2}\sum_{i}(Q^2_{2\mathrm{i}} + Q^2_{3\mathrm{i}}) \tag{5}$$

One can see that, depending on the orbital distribution, it is possible to activate either the Q_2 or the Q_3 mode on each octahedron, that in turn cooperates in establishing the global deformation patterns.

ROLE OF JAHN-TELLER COUPLING

Let us discuss now the role of the Jahn-Teller interaction and its interplay with the local Coulomb coupling. Our aim is to discuss how the coupling between electrons and phonons affects the magnetic character of the ground state, and to stress the differences with respect to the role played by the crystalline field potential. Due to the intrinsically complex electronic behavior in presence of Coulomb interactions, we have treated the electron-phonon coupling by assuming that the phonon modes are classical variables. The analysis performed is similar to that reported in Ref. [17], where a variational optimization has been introduced to find the minima in the octahedral configurations for the Jahn-Teller modes. Since we are limiting ourselves to two corner sharing octahedra, to allow for a non-trivial dynamics between the Jahn-Teller modes and the electronic dynamics, we have assumed that the hopping amplitude between xy orbitals is different from that of the γz ones ($t_{xy} = 2t_{\gamma z}$). This change introduces a dynamical removal of the orbital degeneracy that in turn allows for selecting the activated most stable Jahn-Teller distortions. In this way, we can easily identify a specific charge distribution associated with the possible Jahn-Teller modes, that couples to the electronic dynamics.
To analyze the effect of the Jahn-Teller coupling with respect to the Coulomb interactions, we have determined the phase diagram in the parameter plane $(J_H/t, (U' - J_H)/t)$. Hereafter, in order to have a quick reading of the phase diagrams, we introduce the following notation: A/B (m,k,l) stands for a magnetic/orbital A/B pattern with on-site charge distribution for orbital given by $\langle n_{\mathrm{i}xy} \rangle = m, \langle n_{\mathrm{i}xz} \rangle = k, \langle n_{\mathrm{i}yz} \rangle = l$; whereas (m,k) refers to a situation with equal charge occupation in the two orbitals of the γz sector. F, AF, C indicate ferromagnetic, antiferromagnetic and canted spin arrangement, respectively. FO, AFO, OD stand for ferro-orbital (double occupation on neighbors homologue orbitals), antiferro-orbital (double occupation on neighbors off-diagonal orbitals) and orbital disorder correlations (quantum superposition of diagonal and off-diagonal orbital correlations).
The magnetic character of the ground state is extracted by looking at the evolution of the average local $\langle \mathbf{S}_\mathrm{i} \cdot \mathbf{S}_\mathrm{i} \rangle$ and off-site spin correlator $\langle \mathbf{S}_\mathrm{i} \cdot \mathbf{S}_\mathrm{j} \rangle$. On the other hand, to extract the off-site distribution of double occupied configurations, we look at the equal-time doublon correlation function $\langle D_{\mathrm{i}\alpha} D_{\mathrm{i}\beta} \rangle$ evaluated on the ground, whit $D_{\mathrm{i}\alpha} = n_{\mathrm{i}\alpha\uparrow} n_{\mathrm{i}\alpha\downarrow}$. The amplitude of the doublon correlation function indicates the tendency of the two-particles state to occupy homologue and/or different orbitals on neighbor sites.

The result of the numerical calculation is reported in Fig. 1. The competition between Hund and Coulomb repulsion generates a phase diagram with three distinct regions. Below a critical value of J_H and U', the interplay between local ferromagnetic correlations induced by the Hund coupling and antiferro-type exchange due to the Coulomb repulsion, brings the system toward an incomplete ferromagnetic configuration (C) with off-diagonal orbital correlations between double occupied orbitals on different sites (AFO)(see Fig.2c).

This state evolves to a fully polarized state F/AFO or to an antiferromagnetic state AF/OD by increasing the Hund or the Coulomb interaction respectively (see Figs.2d-2a-b). Due to the difference in the kinetic energy between the electrons in the xy and γz sector, the orbital occupation is non-uniform. Having assumed a hopping amplitude for the xy orbital larger than the γz sector, it turns out to be most favorable to accommodate the double occupied configuration in the yz sector in case of dominant antiferromagnetic correlations. Indeed, in the limit of large values of the Coulomb repulsion, the gain in energy due to the superexchange is bigger in the sector with greater hopping amplitudes. For the polarized ground state, it is energetically more convenient to distribute the charge in a way that the local two-electron states show up in the xy and yz orbital sector due to the fact that a ferromagnetic exchange mechanism is favored when one double occupied configuration sits in the xy sector and the other one in the yz. Starting from this phase

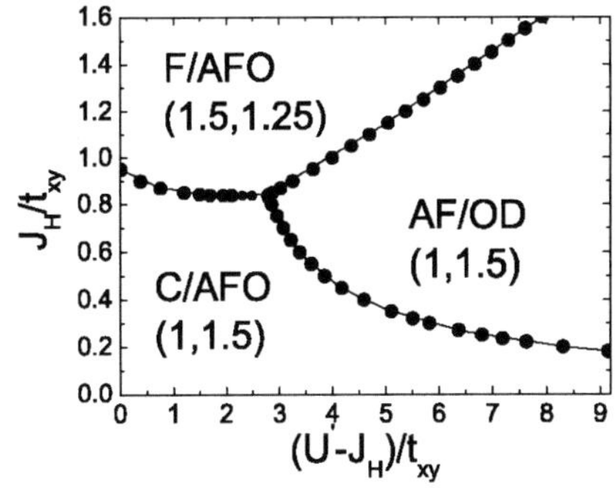

FIGURE 1. Phase diagram for the two-ion system, in the case of asymmetric hopping connection, as a function of Coulomb and Hund coupling in zero crystal field potential.

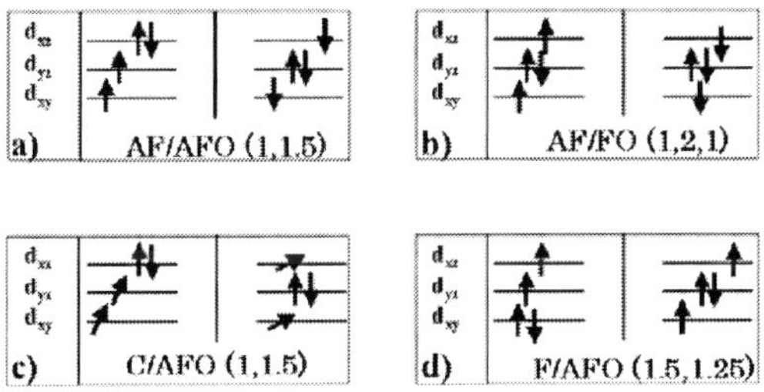

FIGURE 2. Schematic representation of possible ground states for the two corner sharing octahedra system. The AF/OD state is obtained as a linear superposition of configurations a) and b). the configurations AF/FO(1,1,2) and AF/FO(1,2,1) are degenerate in energy for the problem upon examination.

diagram, we include into the dynamics the coupling with the Jahn-Teller modes (H_{jt}). The oxygens vibrational modes are treated as classical variables and their amplitude is obtained within a variational scheme of computation. The ground state energy is calculated as a function of the Q_2 and Q_3 modes, then the saddle point solutions are extracted by an iterative procedure of self-consistency until the convergence is reached. Once the extremal solutions are determined, we perform a comparison between relative energies as given by the total Hamiltonian, in a way to get the absolute ground state of the electron correlated dynamics in presence of Jahn-Teller coupling.

As far as it concerns the minimization procedure, it is possible to show that the extremal values for Q_{i2} and Q_{i3}, obtained by minimizing the ground state energy $E_0 = \langle H^{as}_{kin} + H_{el-el} + H_{jt} \rangle$ with respect to the phonon variables, are given by $Q_{i2} = -\frac{g}{k}\langle n_{i2}\rangle$

and $\bar{Q}_{i3} = -\frac{g}{k}\langle n_{i3}\rangle$, where the average $\langle\rangle$ is performed over the correlated electronic ground state. Replacing those values in the H_{jt} part of the Hamiltonian one gets:

$$H_{jt} = \sum_i -2E_{jt}\left[\langle n_{i3}\rangle n_{i3} + \langle n_{i2}\rangle n_{i2}\right] + E_{jt}\left[\langle n_{i3}\rangle^2 + \langle n_{i2}\rangle^2\right].$$

Hence, one can see that the activation of the Jahn-Teller modes, within this approximation, manifests as an effective orbital dependent local potential, whose strength depends on the average charge distribution in the different orbital sectors. Nevertheless, as we have anticipated before, the Hamiltonian now contains terms with value of density got by averaging on its ground state, implying that the determination of the lower energy state requires a procedure of iterative convergence. Assigning a guess value to the local orbital densities, we compute the ground state and then we replace the new values of the densities in the Hamiltonian, iterating such procedure until the difference between the ground energies in the successive steps reaches the requested accuracy.

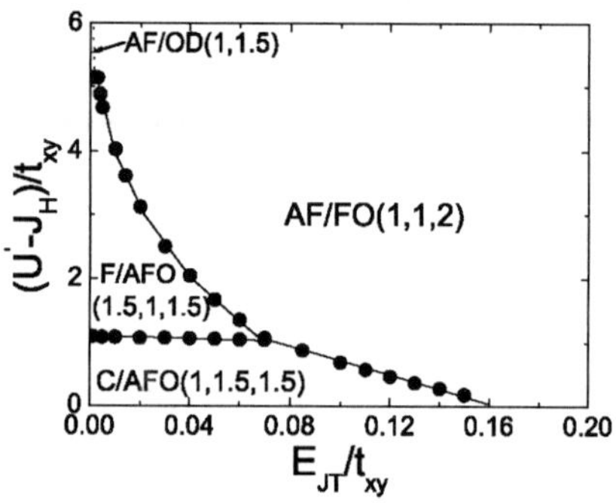

FIGURE 3. Phase diagram in presence of Jahn-Teller coupling and as a function of Coulomb interactions. We note that $(1,1,2)$ and $(1,2,1)$ configurations are degenerate in energy.

In Fig. 3 we have reported the phase diagram as a function of the Coulomb coupling and of the Jahn-Teller self-trapping energy. The strategy we follow is to keep fixed the ratio U'/J_H in a way that, by varying $(U'-J_H)/t$, it is possible to span all the regions of the phase diagram in Fig. 1. Looking at the outcome, one can see that there appears a large region where the ground state has ferro-type orbital (FO) correlations and an antiferromagnetic character as deduced from the related correlation functions.

To understand the phase diagram, it is useful to compare the contribution of the elastic energy in the different regions. By means of the expression for the effective Jahn-Teller coupling (see Eq. 6), and by taking the values of the average local densities per orbital, it is possible to obtain the following amplitudes for the elastic energy per site: $E_{0,jt}^{F/AFO} = -(1/4)E_{jt}$, $E_{0,jt}^{C/AFO} = -(1/4)E_{jt}$, $E_{0,jt}^{AF/OD} = -E_{jt}$, $E_{0,jt}^{AF/FO} = -E_{jt}$. Hence, it is clear that, from the point of view of the coupling between the electron and the Jahn-Teller modes, it is favorable to have a charge distribution with ferro-type orbital correlations (i.e. one fully occupied orbital and the remaining two with a single particle configuration). The difference between the magnetic character of the F and AF ground states is responsible for a larger elastic energy to stabilize the AF phase. On the other hand, the change from the AF/OD to the AF/FO region occurs at very small values of E_{jt} because the large gain in the local elastic energy soon induces a transition in the structural deformation and thus in the charge distribution.

Looking at the expressions of Q_{i2} and Q_{i3}, it is straightforward to extract the character of the octahedral distortions due to the coupling between the electron and the Jahn-Teller modes in the different regimes of magnetic and orbital correlations. Indeed, the Q_2 and Q_3 modes are both active and with different amplitudes in the AF/FO and F/AFO phases, while in the C/AFO and AF/OD phases only the Q_2 mode is induced. Concerning the strength of the structural deformation, we can observe that the amplitude of the Q_3 mode, per unit of Jahn-Teller self trapping energy, is larger in the AF/FO state than in the F/AFO one, while the Q_2 stays unchanged.

It is useful then to discuss on the origin of the specific magnetic and charge pattern in the AF/FO phase. As we have mentioned before, the tendency of the effective local potential, induced by the Jahn-Teller interaction, is to localize a double occupied configuration in the one orbital sector. In the present case, having chosen a lower kinetic energy for the γz orbitals, it is favorable to accommodate the local two-electron states in this orbital sector (minimum loss of kinetic energy by freezing of the orbital degree of freedom). Hence, due to the occurrence of single-particle states in the residual orbitals, there is an induced superexchange mechanism in the limit of large U' or E_{jt}, that in turn is responsible for the antiferromagnetism. We also notice that the instability towards an AF/FO configuration requires more elastic energy the lower is the Coulomb strength. This qualitative feature is a consequence of the difficulty in destabilizing the para(ferro)magnetic correlations in a regime where the antiferromagnetic ones are weak and the magnetic moment has large fluctuations due to the proliferation of intermediate double occupied local states.

It is also worth pointing out that the choice of the asymmetric kinetic term has been also taken into account to facilitate the convergence procedure. The qualitative behavior of the results does not change by modifying the kinetic term. The dominant region would have still a ferro-orbital pattern with antiferromagnetic correlations, while the only difference that occurs by changing the assumption on the kinetic term is in the charge distribution inside each region of the phase diagram.

DISCUSSION AND CONCLUSIONS

In this paper we have discussed the role of the Jahn-Teller coupling on the dynamics of correlated electrons in the framework of a three-band Hubbard model. The study has been performed within an unbiased scheme of calculation being performed in an exact two-site approximation, while the phonon oxygen modes are treated as classical variables. Though confined to a finite size problem, we have demonstrated that, when one includes the possibility of tuning the charge transfer between the xy and the γz sector via structural modifications, one can further control the occurrence of different patterns in the orbital and spin channel. The analysis has revealed that the evolution of the spin/orbital correlations is unique, and the phase diagram tends towards a ferro-orbital

pattern, where antiferromagnetic correlations are stabilized by a superexchange mechanism between singly occupied orbital sectors.
Then we have compared the role of the Jahn-Teller effect with the static octahedral distortions controlled via a tetragonal crystal field amplitude, that simulates the c-axis octahedra compression/elongation. Previous works [18, 19] show that, depending on the character of the tetragonal distortion, the crystal field response is completely asymmetric, since it can induce both ferromagnetic and antiferromagnetic correlations, which are governed by different magnetic exchange mechanisms. When the system is in the extreme flatten configuration, the double occupation is quenched in the xy sector and the spin coupling turns out be isotropically antiferromagnetic. The occurrence of this FO(2,1) configuration crucially depends on the amplitude of the Coulomb repulsion. Nevertheless, in the region of intermediate Coulomb coupling, it can be destabilized by small variations in the structural parameters and regions having weaker antiferromagnetism and enhanced antiferro-orbital correlations can be activated. The occurrence of FO pattern, and its coexistence with contributions of antiferro-orbital character, have been recently considered to explain the low temperature ground state of the AF Mott insulator Ca_2RuO_4.[13, 14, 7, 15, 22] In this context, the analysis of the Jahn-Teller effect would provide a possible scenario where the xy ferro-orbtial pattern stabilized by the compressive octahedral mode is counteracted by an inhomogeneous distribution of distorted octahedra, with the Q_2 mode active, in a way that the system separates into regions with large/small compressive octahedral distortions.

REFERENCES

1. For a review, M. Imada, A. Fujimori, and Y. Tokura, *Rev. Mod. Phys.* **70**, 1039 (1998).
2. Y. Maeno et al., *Nature (London)* **372**, 532 (1994).
3. S. Nakatsuji and Y. Maeno, *Phys. Rev. Lett.* **84**, 2666 (2000)
4. F. Nakamura et al.,*Phys. Rev. B*, **65**, 220402(R) (2002).
5. S. Perry *et al.*, *Phys. Rev. Lett.* **92**, 166602 (2004).
6. M. Braden et al., *Phys. Rev. B* **58**, 847 (1998).
7. J.S. Lee et al., *Phys. Rev. Lett.* **89**, 257402 (2002).
8. J.H. Jung et al., *Phys. Rev. Lett.* **91**, 056403 (2002).
9. M. Kubota et al., *Phys. Rev. Lett.* **95**, 026401 (2005).
10. C.S. Snow et al., *Phys. Rev. Lett.* **89**, 226401 (2002).
11. O. Friedt et al., *Phys. Rev. B* **63**,174432 (2001).
12. T. Nomura and K. Yamada, *J. Phys. Soc. Jpn.* **69**, 1856 (2000).
13. V.I. Anisimov et al., *Eur. Phys. J. B* **25**, 191 (2002).
14. Z. Fang and K. Terakura, *Phys. Rev. B* **64**, 020509(R) (2001).
15. T. Mizokawa et al., *Phys. Rev. Lett.* **87**, 077202 (2001).
16. S. Okamoto and A. Millis, *Phys. Rev. B* **70**, 195120 (2004).
17. T. Hotta and E. Dagotto, *Phys. Rev. Lett.* **88**, 017201 (2002).
18. M. Cuoco, F. Forte and C. Noce, *Phys. Rev. B* **73**, 094428 (2006).
19. M. Cuoco, F. Forte and C. Noce, *Phys. Rev. B* **74**, 195124 (2006).
20. M.D. Struge, *Solid State Phys.* **20**, 91 (1967)
21. K. Yoshida, *Theory of magnetism* (Springer, Berlin, 1998).
22. I.Zegkinoglou et al., *Phys. Rev. Lett.* **95**, 136401 (2005).

Unusual Spin-Peierls Physics in the Layered Quantum Magnet TiOCl

Sebastian Glawion[1], Michael Sing[1], Markus Hoinkis[1,2], Christine A. Kuntscher[2] and Ralph Claessen[2]

[1]*Experimentelle Physik 4 and Röntgen Research Center for Complex Materials, Universität Würzburg, D-97074 Würzburg, Germany*
[2]*Experimentalphysik II, Universität Augsburg, D-86135 Augsburg, Germany*

Abstract. We summarize the current knowledge on phases, phase transitions and the electronic structure of the S = 1/2 Mott insulator TiOCl. There are three different phases, with the low-T phase being of the spin-Peierls type. The interesting nature of these phases in conjunction with the different character of the phase transitions still leaves unanswered questions on the dimensionality of the system. Analysis of the electronic structure at room temperature by ARPES reveals pronounced differences between the orthogonal a- and b-axes within the crystal planes. In comparing with the isostructural material TiOBr we find differences in the degree of one-dimensionality, calling for inclusion of frustrated interchain interactions triggered by the underlying triangular lattice into theoretical calculations. Indications of a metallic phase at room temperature induced by pressure are reported, and possible effects of doping are discussed.

Keywords: TiOCl, ARPES, electronic structure, Mott insulator, pressure, doping
PACS: 71.20.-b, 71.27.+a, 71.30.+h, 79.60.-i

INTRODUCTION

Searching for interesting many-particle physics in strongly correlated low-dimensional systems the 3d transition metal oxide TiOCl has gained increasing interest in the past years. Beynon and Wilson rediscovered the material in the light of high-T_c superconductivity [1]. Being a Mott insulator, it has been speculated that RVB-like physics might determine the ground state, which upon doping could be eventually driven into this exotic superconducting state. Recently, quasi-one-dimensional electronic and magnetic properties have been confirmed e.g. by magnetic susceptibility measurements and LDA+U calculations [2, 3]. Additionally, a non-canonical spin-Peierls phase was identified [4], making TiOCl only the second inorganic material besides $CuGeO_3$ with this kind of ordering.

CRYSTAL STRUCTURE AND PHASE TRANSITIONS

TiOCl crystallizes in the FeOCl structure, i.e. in buckled bilayers of Ti and O separated by layers of Cl ions. Along the c-axis these structures are coupled to each other only by weak van-der-Waals forces, while chains of one atomic kind are aligned along the b-axis. In the a-b-plane, the underlying Ti mesh resembles a distorted trian-

CP918, *Lectures on the Physics of Strongly Correlated Systems XI: Eleventh Training Course,* edited by A. Avella and F. Mancini

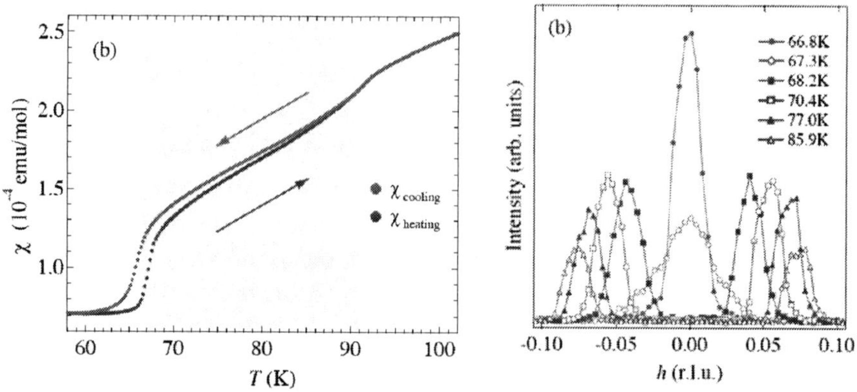

FIGURE 1. Left panel: Magnetic susceptibility of TiOCl. At T_{c1} = 91 K a kink is visible due to a second order phase transition. At T_{c2} = 67 K a first order transition can be recognized by the hysteresis between cooling and heating. Right panel: Evolution of the (δ, 1.5, 0)-Miller reflection upon crossing the low-T phase transition (from [5]). The appearance of peaks at non-integer h-values indicates the incommensurate ordering in the intermediate phase.

gular lattice. The local octahedral coordination of Ti^{3+} by four O and two Cl ions leads to a crystal field splitting into e_g- and t_{2g}-states. Distortion of these octahedra further lowers the d_{xy} orbitals by about 200 meV. It was shown to be the true ground state by LDA+U calculations and polarization-dependent PES experiments [6].

For the high-temperature phase the magnetic susceptibility can be fitted to a Bonner-Fisher curve with J ~ 660 K, thus in the respective temperature range TiOCl resembles a one-dimensional Heisenberg antiferromagnet. In Fig. 1 one sees that at T_{c1} = 91 K a second order phase transition sets in, which however does not yet lead to a (commensurate) spin-Peierls phase. Instead, another phase transition at T_{c2} = 67 K, this time of first order, finally gives this kind of ordering, as is shown by the sudden appearance of a peak at half-integer value of k in the reciprocal space by x-ray diffraction [5]. This indicates a doubled unit cell which is characteristic for the dimerization into S=0 singlets along the b-axis, as is expected in a spin-Peierls phase. The accompanying lattice distortion nevertheless is slightly shifted out of the b-direction towards the ligands (O and Cl), thus minimizing the elastic strain of the structure. By detailed analysis of the reflexes in the h-k-plane, i.e. at non-integer h- and k-values in the reciprocal plane, the intermediate phase could be identified to be incommensurate, finally locking in commensurately into the spin-Peierls order at T_{c2}. This finding is corroborated by specific heat measurements showing an insufficient release of heat at the phase transitions [7].

ELECTRONIC STRUCTURE

Single-band 1D Hubbard model calculations by DDMRG [6], DFT calculations, e.g. LDA+U [3], and combined methods like LDA+DMFT [8] have been carried out so far, shining light on the room-temperature electronic structure of TiOCl available to photoemission experiments. Early LDA+U correctly identified the energetics of the

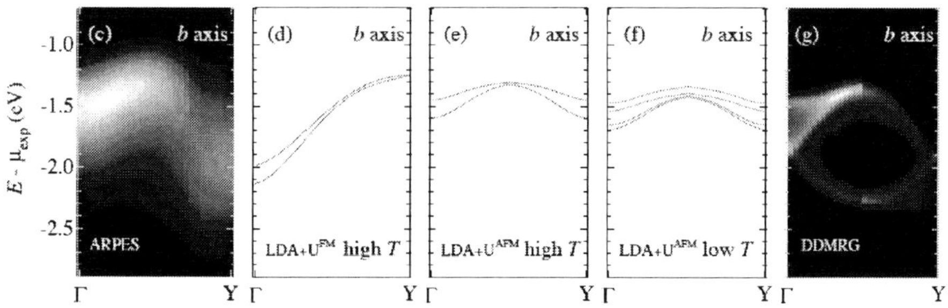

FIGURE 2. From left to right: Ti 3d-Dispersion along the b-axis from ARPES, LDA+U calculation with ferromagnetic spin alignment, LDA+U with antiferromagnetic alignment for the high-T and low-T phase, and single-band Hubbard calculation (all with dispersions along the b-axis). LDA+U^{FM} does not reproduce the downbending towards the outer zone boundary, while LDA+U^{AFM} introduces this feature artificially by means of a doubled unit cell. The shadow band in the Hubbard calculation is not reproduced in experiment (which is not due to experimental resolution). (from [6])

valence band, and also succeeded in predicting that the d_{xy}-orbital is the lowest orbital available for the 3d electron of the Ti atom. This was proven by PES studies using linearly polarized light [6]. Knowing the orientation of the orbital lobes in the crystal and making use of selection rules this procedure allowed to distinct between the respective contributions of d_{xy} and the other t_{2g} orbitals. However, the DFT methods fail to reproduce peak shapes and widths extracted from experimental determinations of the band structure, while the Hubbard model calculation, which gives reasonable agreement in this respect, shows distinct features not present in ARPES, namely the shadow band. In Fig. 2 these results are summarized. Recent cluster-DMFT calculations seem to overcome previous shortcomings and show good correspondence of Ti 3d peak shapes with experimental data [9].

We carried out ARPES measurements of the valence band structure, in particular the Ti 3d spectral weight, in the high-T phase on self-grown crystals. Results along the paths X-Γ-X (crystallographic a-axis) and Y-Γ-Y (b-axis) show a pronounced one-dimensionality in that strong dispersion is only visible in the latter case. Elevated temperatures of about 350 K had to be used in order to minimize sample charging. A comparison of a- and b-axis dispersions in TiOCl and the isostructural material TiOBr is shown in Fig. 3 [10]. Those two compounds differ only slightly in lattice spacing because of larger atomic radii of the bromide, and thus the electronic structures are equal to a large extent. Focussing on TiOCl, it can be seen that the band widths along the a-axis and the b-axis (termed w_b, as indicated in the picture) differ significantly in magnitude, in contrast to the case of TiOBr, where they are equal within experimental uncertainties. This finding has been verified over a large number of samples and is thus an intrinsic property. It seems that while qualitatively the same features appear in both compounds, TiOCl has more one-dimensional character than TiOBr. However, since the shadow band is missing (in both cases), interchain coupling is assumed to play a major role in determining the electronic structure in TiOCl as well, although the ratio of in-chain over out-of-chain hopping is roughly ~1/8. Additionally, by com-

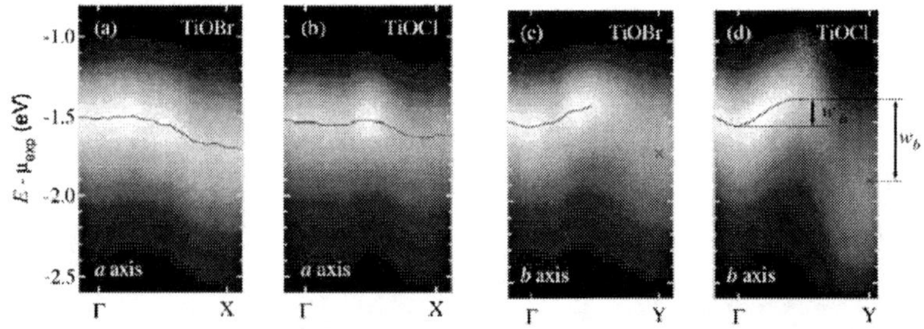

FIGURE 3. Ti 3d spectral weight dispersions (from [13]): along (a) a-axis and (c) b-axis for TiOBr, and along (b) a-axis and (d) b-axis for TiOCl. The blue lines indicate the maxima of the peaks obtained through a fitting procedure described in [10]. It can be seen that while dispersions along the a-axis are of comparable size in both compounds, the situation is significantly different along the b-axis: a much more pronounced downbending towards the zone boundary is visible in TiOCl, which in the case of TiOBr remains closer to the range of the a-axis dispersion.

TABLE 1. Comparison of TiOBr and TiOCl

	TiOBr	TiOCl	Δx/x
w'_b	0.13 eV	0.17 eV	-23%
w_b	0.26 eV	0.47 eV	
J from magnetic susceptibility	32 meV	58 meV	-45%
$t = 0.5\sqrt{(JU)}$	0.16 eV	0.22 eV	-26%
t from LDA+U	0.17 eV	0.21 eV	-19%

paring bandwidths quantitatively as is summarized in Tab. 1, we found that the they obviously scale with $t \sim \sqrt{J}$ and not with J, which can be seen from the fact that the relative differences Δx/x between TiOCl and TiOBr are almost equal when comparing w'_b with the obtained hopping integrals t, in contrast to the extracted J's. This suggests that electronic dispersions have their origin in charge excitations rather than in spin excitations.

METALLIZATION AND DOPING

In order to retain any chance of reaching a superconducting phase instead of the spin-Peierls phase at low temperatures, it is imperative to first show that it is possible to drive the Mott-insulating system into a metallic state. For TiOCl, there is strong evidence from a recent study of optical transmittance and reflectance under pressure that this can be achieved [11]. Figs. 4(a) and (b) show transmittance along the two different axes a and b for different pressures. The features around 6000 cm^{-1} and 13000 cm^{-1}, respectively, are on-site excitations into t_{2g} orbitals, which are not degenerate anymore due to the distortion of the octahedral coordination of Ti atoms. For pressures above roughly 12 GPa however, the transmittance is suppressed over almost the entire studied frequency range. In Fig. 4(c) one can see how the charge gap of initially about 2 eV closes rather abruptly at those pressures.

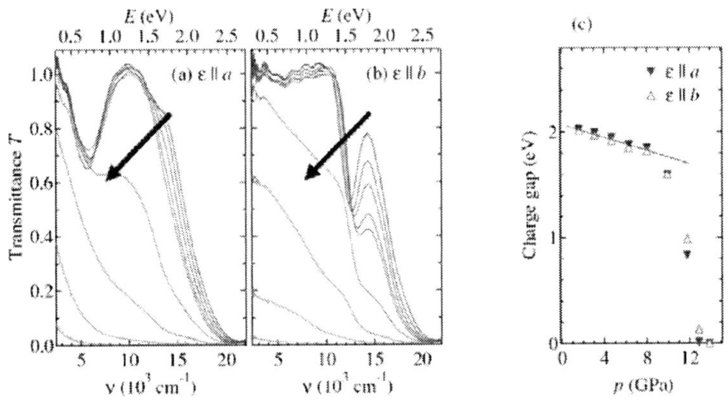

FIGURE 4. (a),(b): Transmittance spectra of TiOCl for light polarization ε along the a- and b-axis, respectively. Arrows indicate the direction of increasing pressure. The most drastic changes appear between 10 and 12 GPa. (c) Closing of the charge gap at around 12 GPa. (from [11])

In isostructural materials of the XNCl-type (X = Hf, Zr) it has been shown that a superconducting state at rather high transition temperatures (T_c = 25.5 K) can be reached by suitably doping the crystals [12]. This route seems very promising for TiOCl as well and preliminary photoemission results show the appearance of new states within the gap region, but a detailed qualitative and quantitative analysis is still under way.

REFERENCES

1. R. J. Beynon and J. A. Wilson, *Journal of Physics: Condensed Matter* **5**, 1983, 1993.
2. P. Lemmens, K. Y. Choi, R. Valenti, T. Saha-Dasgupta, E. Abel, Y. S. Lee and F. C. Chou, *New Journal of Physics* 7, 74, 2005.
3. T. Saha-Dasgupta, R. Valenti, H. Rosner and C. Gros, *Europhysics Letters* **67**(1), 63, 2004.
4. A. Seidel, C. A. Marianetti, F.C. Chou, G. Ceder and P. A. Lee, *Physical Review B* **67**, 020405(R), 2003.
5. A. Krimmel, J. Strempfer, B. Bohnenbuck, B. Keimer, M. Hoinkis, M. Klemm, S. Horn, A. Loidl, M. Sing, R. Claessen and M. v. Zimmermann, *Physical Review B* **73**, 172413, 2006.
6. M. Hoinkis, M. Sing, J. Schäfer, M. Klemm, S. Horn, H. Benthien, E. Jeckelmann, T. Saha-Dasgupta, L. Pisani, R. Valenti and R. Claessen, *Physical Review B* **72**, 125127, 2005.
7. J. Hemberger, M. Hoinkis, M. Klemm, M. Sing, R. Claessen, S. Horn and A. Loidl, *Physical Review B* **72**, 012420, 2005.
8. T. Saha-Dasgupta, A. Lichtenstein and R. Valenti, *Physical Review B* **71**, 153108, 2005.
9. T. Saha-Dasgupta, A. Lichtenstein, M. Hoinkis, S. Glawion, M. Sing, R. Claessen and R. Valenti, *(unpublished)*, 2006
10. M. Hoinkis, M. Sing, S. Glawion, L.Pisani, R. Valenti, S. van Smaalen, M. Klemm, S. Horn and R. Claessen, *cond-mat/0610106.*
11. C. A. Kuntscher, S. Frank, A. Pashkin, M. Hoinkis, M. Klemm, M. Sing, S. Horn and R. Claessen, *Physical Review B* **74**, 184402, 2006.
12. S. Yamanaka, K. Hotehama and H. Kawaji, *Nature* **392**, 580, 1998.

BEC in a star-comb graph

F. P. Mancini*, P. Sodano* and A. Trombettoni*

*Dipartimento di Fisica and Sezione I.N.F.N., Università di Perugia,
Via A. Pascoli, I-06123 Perugia, Italy

Abstract. We investigate the properties of free bosons hopping on a star-comb network, discussing the single-particle spectrum and the main thermodynamic equilibrium properties: Bose-Einstein critical temperature, fraction of condensate, and spatial boson distribution. We find an enhancement of the critical temperature with respect to other inhomogeneous networks.

Although it is well known that free bosons hopping on translationally invariant networks cannot undergo Bose-Einstein condensation at finite temperature if the space dimension d is less or equal to two, very recent studies [1, 2, 3, 4] hint to the exciting possibility that the network topology may act as a catalyst for inducing a finite temperature spatial Bose-Einstein condensation even if $d < 2$. This is indeed possible if one resorts to a suitable discrete inhomogeneous ambient space on which bosons are defined. As an example of this situation we shall investigate the properties of non-interacting bosons hopping on a star-comb shaped network, evidencing that also in this case one may have a macroscopic occupation of the ground-state at low temperatures. Ultracold bosons in optical networks represent a physical system where this setup could be experimentally implemented.

The star-comb graph is a bundled graph which can be obtained by grafting a star graph to each site of a linear chain, called backbone (see Fig. 1). The sites of the graph can be naturally labelled introducing three integer indices (x,y,z) with $x,y,z \in Z$, where $x = 1,\ldots,p$ labels the different arms on each star (excluding the links connecting the different stars), $y = 0,\ldots,L$ represents the distance from the backbone and $z = 1,\ldots,N_s$ labels the different stars. In the following we shall assume periodic boundary conditions $(x,y,1) \equiv (x,y,N_s+1)$. The Hamiltonian describing non-interacting bosons hopping on a star-comb graph can be written as:

$$H = -t \sum_{x,y,z;x',y',z'} A_{x,y,z;x',y',z'}\, \hat{a}^{\dagger}_{x,y,z} \hat{a}_{x',y',z'} \,. \tag{1}$$

In Eq. (1), t is the hopping parameter while $\hat{a}^{\dagger}_{x,y,z}$ ($\hat{a}_{x,y,z}$) is the creation (annihilation) operator for bosons; $\hat{n}_{x,y,z} = \hat{a}^{\dagger}_{x,y,z}\hat{a}_{x,y,z}$ is the number operator at site (x,y,z). The filling, i.e., the average number of particles per site, is defined as $f = N_T/N_s(pL+1)$, where N_T is the total number of bosons and $N_s(pL+1)$ is the number of sites. The adjacency matrix of a star-comb graph is given by:

$$\begin{aligned} A_{x,y,z;x',y',z'} = {} & (\delta_{y',y-1} + \delta_{y',y+1})\,(1-\delta_{y,0})\delta_{x,x'}\delta_{z,z'} + \delta_{y,0}\,\delta_{y',1}\delta_{z,z'} \\ & + \delta_{y,0}\,\delta_{y',0}\delta_{x,x'}(\delta_{z',z-1} + \delta_{z',z+1}). \end{aligned} \tag{2}$$

CP918, *Lectures on the Physics of Strongly Correlated Systems XI: Eleventh Training Course,* edited by A. Avella and F. Mancini

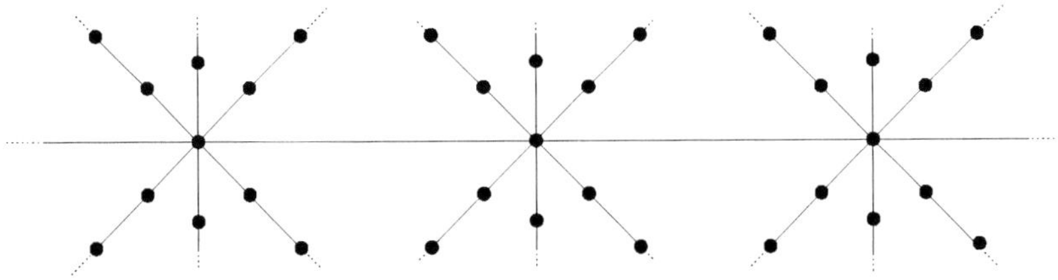

FIGURE 1. The star-comb graph.

The eigenvalue equation $-t\sum_j A_{ij}\psi(j) = E\psi(i)$, with the previous labelling of sites, reads:

$$
\begin{aligned}
-t \sum_{x',y',z'} & \left[\left(\delta_{y',y-1} + \delta_{y',y+1}\right) \left(1 - \delta_{y,0}\right) \delta_{x,x'} \delta_{z,z'} + \delta_{y,0}\, \delta_{y',1} \delta_{z,z'} \right. \\
& \left. + \delta_{y,0}\, \delta_{y',0} \delta_{x,x'} \left(\delta_{z',z-1} + \delta_{z',z+1}\right) \right] \psi(x',y',z') = E\psi(x,y,z)
\end{aligned} \tag{3}
$$

By exploiting the translation invariance in the direction of the backbone, a Fourier transform in the variable z reduces Eq. (3) to a 1-dimensional eigenvalue problem. If one defines: $\psi(x,y,k) = \sum_z e^{ikz}\psi(x,y,z)$, with $k = 2\pi n/N_s$ and $n = 1,\ldots,N_s$, the eigenvalues equation (3) becomes:

$$
\begin{aligned}
-t \sum_{x',y'} & \left[\left(\delta_{y',y-1} + \delta_{y',y+1}\right) \left(1 - \delta_{y,0}\right) \delta_{x,x'} + \delta_{y,0}\, \delta_{y',1} \right. \\
& \left. + 2\cos(k_0) \delta_{y,0}\, \delta_{y',0} \delta_{x,x'} \right] \psi(x',y') = E\psi(x,y)
\end{aligned} \tag{4}
$$

where $\psi(x,y,k) = \delta(k-k_0)\psi(x,y)$, with $k_0 = 2\pi n/N_s$, $n = 0\ldots N_s$. Equation (4) can be regarded as the equivalent problem of a quantum particle hopping on a single star and interacting with a potential in the center, $V(k_0) = -2t\cos(k_0)$. The eigenvalues and eigenvectors of Eq. (4) are determined by requiring that the free particle solutions on the arms are solutions also in $y = 0$. Indeed, for $y \neq 0$, the wave function satisfying Eq. (4) is:

$$
\psi(x,y) \quad = \quad A_x e^{ihy} + B_x e^{-ihy} \tag{5}
$$

Requiring that - on each arm - $\psi(x,y)$ is a solution of the eigenvalue equation at $y = L$ introduces restrictions on the parameters h, A_x, B_x, and on the eigenvalues E. In other words, it amounts to require that A_x, B_x and h should satisfy the p equations

$$
-t\left(A_x e^{ih(L-1)} + B_x e^{-ih(L-1)}\right) = -2t\cos(h)\left(A_x e^{ihL} + B_x e^{-ihL}\right) \quad x = 1,\cdots,p. \tag{6}
$$

The eigenstates defined in Eq. (5) should also satisfy $(p-1)$ matching conditions in the center of each star. Thus, one has

$$
A_x + B_x = A_{x+1} + B_{x+1} \qquad x = 1,\cdots,p-1. \tag{7}
$$

Using Eqs. (7), the condition in one of the centers gives one more equation

$$-t\left[\sum_{x=1}^{p}\left(A_x e^{ih}+B_x e^{-ih}\right)+2\cos(k_0)\left(A_x+B_x\right)\right]=-2t\cos(h)\left(A_{x'}+B_{x'}\right), \qquad (8)$$

with $x=1,\cdots,p$. Equations (6) and (8) may be grouped in a homogeneous linear system of $2p$ equations which allows to fix the $2p$ parameters A_x and B_x. Upon denoting with Q the $(2p\times 2p)$ matrix whose elements are the coefficients of the linear system given by Eqs. (6) and (8), and requiring that

$$\det Q=\Theta(h,L)(1-e^{2ih(L+1)})^{p-1}\cdot\{(p-2)\cos(h)-p\cot[h(L+1)]\sin(h)+2\cos(k_0)\} \qquad (9)$$

is equal to zero, the solution is unique. In Eq. (9) $|\Theta(h,L)|=1$ for any value of h.

One immediately sees that the values of h for which $h=n\pi/(L+1)$ (with $n=1,2,...,L$) provide a set of $L(p-1)$-fold degenerate eigenstates of Eq. (9) for each value of k_0 (in total N_s sets). In addition, the solutions of the transcendental equation

$$(p-2)\cos(h)-p\cot[h(L+1)]\sin(h)+2\cos(k_0)=0 \qquad (10)$$

provide the values of h associated to non-degenerate eigenstates. Equation (10) can be solved numerically and yields a set of $N_s(L-1)$ non-degenerate eigenvalues corresponding to different values of k_0 and to values of h which - in the thermodynamic limit - are equally spaced and separated by a distance $\pi/(L+1)$. As a result, each set of delocalized states is formed by $(pL-1)$ states corresponding to energies ranging between $-2t$ and $+2t$. If the potential in the center $V(k_0)$ is equal to zero, one recovers the same condition of the single star [4].

Since the total number of states should equal $N_s(pL+1)$, there are still $2N_s$ localized states in the spectrum. These states belong to the so-called hidden spectrum [1]. To find them, it is convenient to look for solutions of the eigenvalue equation (3) of the form

$$\begin{aligned}\psi_-(y)&=Ae^{-\eta y}+Be^{\eta y}\\ \psi_+(y)&=A(-1)^y e^{-\eta y}+B(-1)^y e^{\eta y},\end{aligned} \qquad (11)$$

corresponding, respectively, to the eigenvalues of the lower hidden spectrum $\sigma_-=-2t\cosh\eta$ and to the eigenvalues of the upper hidden spectrum $\sigma_+=2t\cosh\eta$. In Eqs. (11), A and B are normalization constants and $\eta\equiv 1/\xi$ is a parameter accounting for the localization of the states. By rewriting Eq. (10) for $h=i\eta$, one has to solve:

$$(p-2)\cosh(\eta)-p\coth[\eta(L+1)]\sinh(\eta)+2\cos(k_0)=0, \qquad (12)$$

together with the condition that $\sum_{x,y}|\psi_0(x,y)|^2=1$. In the thermodynamic limit, $L\to\infty$, Eq. (12) becomes: $(p-2)\cosh(\eta)-p\sinh(\eta)+2\cos(k_0)=0$, which is solved by $\eta=\log\left[p-1+2\cos k_0(p-1+\cos^2 k_0)+2\cos^2 k_0\right]$, yielding, for the lower hidden spectrum: $\sigma_-=-t\left[(p-2)\cos(k_0)+p\sqrt{p-1+\cos^2(k_0)}\right]/(p-1)$. The lowest energy level is obtained for $\cos(k_0)=1$; one has

$$E_0=-t\,\frac{(p-2)+p\sqrt{p}}{p-1}. \qquad (13)$$

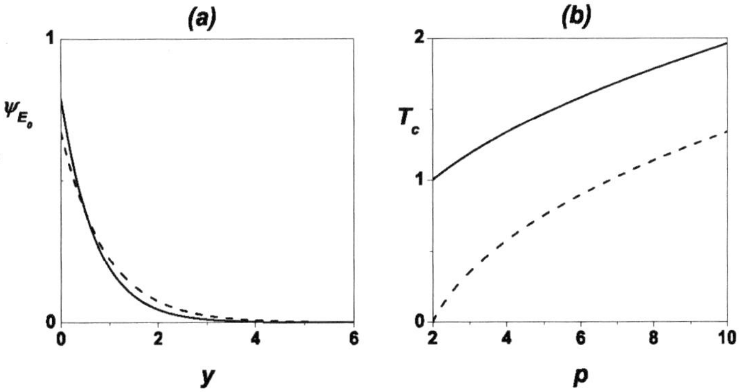

FIGURE 2. (a) The normalized single-particle ground-state wavefunction for bosons hopping on a star and star-comb graphs as a function of the distance y from the center. The number of arms is $p = 10$. (b) Critical temperature T_c (in units of k_B/E_J) as a function of the number of arms p. The solid lines are for the star-comb graph, whereas the dashed lines are for the single star.

Solving the eigenvalue equation (3) for $E = E_0$, one obtains the wavefunction of the localized ground-state:

$$\psi_{E_0}(y) = \sqrt{\frac{\sqrt{p}+2}{2\left(\sqrt{p}+1\right)}}\, e^{-y/\xi}. \tag{14}$$

$\xi = 1/\log\left(\sqrt{p}+1\right)$ provides an estimate of the ground-state localization. The ground state is localized along the backbone and it decreases exponentially along the arms. When $p = 2$, one immediately recovers all the known results of the comb graph [1, 3]. For each value of k_0 ($\cos(k_0) > 0$) there is a solution of (12) with a different energy in the interval $[E_0, -2t]$. In a finite star-comb with $N_s(pL+1)$ sites there are N_s solution of this type and for $N_s \to \infty$ these solutions fill densely the interval $[E_0, -2t]$. Analogously, when $\cos(k_0) < 0$ there are N_s solution of (12) corresponding to the spectral region at high energy ($E = [2t, |E_0|]$), where the other hidden states appear. In Fig. 2a we plot the ground-state wavefunction for $p = 10$ for a single star (dashed line) and for a star-comb (solid line) as a function of the distance from the backbone. Figure 2a evidences that adding stars enhances the localization of the wavefunction around the center of each star.

The thermodynamic properties of non-interacting bosons hopping on a star-comb graph evidence also in this case a topology induced spatial BEC even if $d < 2$ [1, 2]. To elucidate this phenomenon, it is most convenient to introduce the macrocanonical ensemble to determine the fugacity $z = e^{\beta(\mu - E_0)}$ as a function of the temperature [1]: the equation determining z is given by

$$N_T = \sum_{E\in\sigma} \frac{d(E)}{z^{-1}e^{\beta(E-E_0)} - 1}. \tag{15}$$

In Eq. (15), $d(E)$ is the degeneracy of each single-particle eigenstate of the Hamiltonian (1) and $\beta = 1/k_BT$. The sum in Eq. (15) is over the entire spectrum σ. For free bosons hopping on a star-comb graph, one has

$$N_T = N_{E_0}(L,T) + N_{\sigma_-}(L,T) + N_{\sigma_+}(L,T) + \int_{E\in\sigma_0} dE \frac{N_s(pL+1)\rho(E)}{z^{-1}e^{\beta(E-E_0)}-1}, \tag{16}$$

where $N_{E_0}(L,T)$, $N_{\sigma_-}(L,T)$ and $N_{\sigma_+}(L,T)$ denote, respectively, the number of particles at a certain temperature T in the ground-state and in the two regions σ_- and σ_+ of the hidden spectrum. $\rho(E)$, with $E \in \sigma_0$, is the energy density of states of the linear chain, i.e., $\rho = 1/(\pi\sqrt{4t^2-E^2})$, with σ_0 being the region of the spectrum corresponding to delocalized states. The last term of the right-hand side of Eq. (16) is the number of bosons in the delocalized (chain-like) states. The presence of the hidden spectrum changes the behavior of the integral evaluated in the interval $\{-2t, 2t\}$, since it reduces it to the one describing non-interacting bosons on a linear chain with an impurity in one of the sites. As a result, letting $z \to 1$, the integral converges even at finite temperatures making possible a spatial BEC in $d < 2$.

If one defines T_c as the critical temperature at which BEC occurs, then for any $T < T_c$ the ground-state is macroscopically filled. Since, at the critical temperature, $N_{E_0}(T_c) = N_{\sigma_\pm}(T_c) = 0$, the equation allowing to determine T_c as a function of the parameters f and t reads:

$$\pi f = \int_{-2t}^{2t} \frac{dE}{\sqrt{4t^2-E^2}} \frac{1}{e^{(E-E_0)/(k_B T_c)}-1}. \tag{17}$$

Equation (17) can be solved numerically for any value of f. When $f \gg 1$, one may expand the exponential in Eq. (17) to the first order in the inverse of the critical temperature T_c. Upon inserting the value of the ground-state energy (13) in Eq. (17), one easily finds that the critical temperature T_c is given by

$$T_c \approx \frac{tf}{k_B} \frac{\sqrt{p}(2+\sqrt{p})}{1+\sqrt{p}}. \tag{18}$$

Equation (18) has been checked numerically and it is in excellent agreement with the numerical solution of Eq. (17) for $f \gg 1$, the error being of order $1/f$. If one compares the critical temperature of the single star and of the star-comb graphs, one immediately sees that, once the number of arms is fixed, T_c is always enhanced in the latter realization. This is shown in Fig. 2b.

One may now use Eq. (18) to determine the condensate fraction as a function of the scaled temperature T/T_c. In the thermodynamic limit, the number of particles in the delocalized states is given by

$$N_{\sigma_0} = \lim_{L\to\infty} N_s(pL+1) \int_{E\in\sigma_0} \rho(E) \frac{dE}{e^{\beta(E-E_0)}-1} \approx N_T \frac{T}{T_c}. \tag{19}$$

In the last equation the exponential has been expanded to the first order in β: this approximation holds for $f \gg 1$ and it is in very good agreement with the numerical evaluation of the integral in Eq. (19) also in a large neighborhood below T_c. From Eqs.

(16) and (19) one gets the number of particles in the localized states $N_0 = N_{E_0} + N_{\sigma_-}$: the fraction of condensate, for $T < T_c$, is then given by

$$\frac{N_0}{N_T} \approx 1 - \frac{T}{T_c}. \tag{20}$$

For f ranging from 10^3 to 10^9, the results provided by Eq. (20) differ from those obtained by the numerical evaluation of N_0 from Eq. (16) by less than 1%. Equation (20) clearly shows that the condensate has dimension 1 just as cigar-shaped one-dimensional atomic Bose condensates [5].

Due to the topology induced spatial condensation on the backbone, one should expect an inhomogeneous distribution of the bosons along the arms of the network. The average number of bosons N_B at a site (x, y, z) depends - due to the symmetry of the graph - only on the distance y from the backbone. It is not difficult to show (see also Ref. [3]) that, away from the backbone ($y \gg 1$) and once the filling is fixed, N_B depends only on the scaled temperature T/T_c and it is given by

$$\frac{N_B(y; T/T_c)}{f} \approx \frac{T}{T_c}. \tag{21}$$

Topology induced spatial BEC in a system of non-interacting bosons hopping on a star-comb graph predicts then a rather sharp decrease of the number of bosons at sites located away from the backbone. The linear dependence is consistent with the observation that, in this system, the condensate has dimension 1.

In this paper we have investigated in detail another situation where the role played by the network's topology is crucial in determining the thermodynamic properties of the system. We focused on the properties of free bosons hopping on a star-comb network, finding an enhancement - with respect to other inhomogeneous structures previously investigated [1, 2, 3, 4] - of the critical temperature at which the particles condense.

REFERENCES

1. R. Burioni, D. Cassi, I. Meccoli, M. Rasetti, S. Regina, P. Sodano, and A. Vezzani, Europhys. Lett. **52**, 251 (2000).
2. R. Burioni, D. Cassi, M. Rasetti, P. Sodano, and A. Vezzani, J. Phys. B **34**, 4697 (2001).
3. G. Giusiano, F. P. Mancini, P. Sodano, and A. Trombettoni, Int. J. Mod. Phys. B **18**, 691 (2004).
4. I. Brunelli, G. Giusiano, F. P. Mancini, P. Sodano, and A. Trombettoni, J. Phys. **B 37**, S275 (2004).
5. W. Ketterle and N. J. van Druten, Phys. Rev. A **54**, 656 (1996); A. Görlitz *et al.*, Phys. Rev. Lett. **87**, 130402 (2001).

Nodal quasiparticles and their quantum interference effects in superconductors and magnets

Yoshifumi Morita

Faculty of Engineering, Gunma University, Kiryu, Gunma 376-8515, Japan

Abstract. We review the emergence of nodal quasiparticles in two dimensions(Dirac fermion in (2+1)-dimensions) in the context of condensed matter physics. Focus is on effects of (topological) defects and related novel criticality with continuously varying exponents. In the last, comments are given on future directions e.g. dynamical problems.

Keywords: nodal quasiparticle, Dirac fermion, graphene, unconventional super., magnet
PACS: 71.10.-w, 71.10.Fd, 73.20.Fz, 73.43.-f, 74.20.-z, 81.05.Uw

INTRODUCTION

Nodal quasiparticle attracts much attention recently especially in two dimensions in the context of e.g. superconductivity, strongly-correlated electron and also carbon-based material. It behaves as (2+1)-dim. Dirac fermion in the low-energy limit and can lead to many fascinating physics due to quantum interference effects. In this paper, we employ 'pi-flux model' as a toy model for nodal quasiparticle in two dimensions. Though it is a simple model, it tells us universal aspects of nodal quasiparticle in two dimensions. And our results should apply to universal behaviour of some unconventional superconductors, magnets and also graphene as discussed below. Before going to the main section, we shall comment on how nodal quasiparticles 'emerge' in the context of condensed matter. Graphene is an isolated single atomic layer of graphite[1]. The bandstructure can be approximated as cones located at some points in the Brillouin zone. Near the apex of the cone, two energy bands touch with each other and it can be described as Dirac fermion in (2+1)-dimensions. Not only it opens a new door of condensed matter simulation of (2+1)-dim. Dirac fermion[2,3] but also it has potential for fabricating transistor-like devices, highly-sensitive solid-state sensor and so on[4]. Also in 2D d-wave super., the superconducting quasiparticle behaves as (2+1)-dim. Dirac fermion. Combined with the character of macroscopic quantum phenomena in superconductivity, it can lead to quantum information engineering etc.. Moreover, in a class of strongly-correlated electron systems or magnets(e.g. cuprates, some organics and heavy-fermion materials), possible emergence of nodal quasiparticle is proposed experimentally(e.g. ref.[5]) and theoretically(e.g. ref.[6,7]).This paper is organized as follows. First, we define our theoretical model, 'pi-flux model' and summarize its novel properties in the presence of (topological) defects. In the last, we comment on future directions; a model of topological defect and transport properties due to the motion of topological defect.

CP918, *Lectures on the Physics of Strongly Correlated Systems XI: Eleventh Training Course*, edited by A. Avella and F. Mancini

EMERGENCE OF NODAL QUASIPARTICLE AND EFFECTS OF (TOPOLOGICAL) DEFECTS

In the very beginning, let us define our model 'pi-flux model'. We employ this model, since it is a 'minimal model' for (2+1)-dim. Dirac fermion to emerge on a lattice. Though it is minimal, it has advantages that it is easy to access both numerically and analytically. The pi-flux model is a tight-binding model on two-dimensional square lattice with a magnetic field 'pi' per plaquette. We use a notation that a fundamental unit of magnetic flux(=hc/|e|) is equal to '2pi' and it is to be noted that this model preserves the time-reversal symmetry. And defects are defined by random fluctuation in the amplitude of hopping integral, which leads to the so called chiral symmetry(and the time-reversal symmetry). This kind of model for defect has applications to the topological defect in doped graphene (so called 'ripple'), as discussed in the last section. In this paper, our focus is on statics, in particular, Density of States (DoS), which is directly accessible experimentally through the STM studies. In the last section, we also discuss on future directions e.g. dynamical(transport) properties.
See Fig.1 in the next page for the DoS of this model[8]. As seen in Fig.2, we also studied the localization length[9]. We revealed that the localization length diverges, approaching the zero energy(band center). It implies 'criticality' and, moreover, the continuously varying exponent in DoS suggests that there exists 'critical line' in a family of disordered fermionic model in two dimensions[8,9]. Actually this is consistent with a theoretical approach based on Renormalization Group(RG) [10]. Here some comments are in order. At first glance, this result seems to be 'in'consistent with the famous 'theorem' of 'absence of quantum diffusion in two dimensions'[11]. Actually, when disorder is included on site(in contrast to link), we confirmed that the result is consistent with this conventional wisdom[9]. Then what is the difference? As proposed in [8,9], it is non-trivial symmetry, which remains intact even after disorder is introduced on link. It is the chiral symmetry(and the time reversal symmetry), which leads to a reflection symmetry in the energy spectrum. This proposal on the role of symmetry is also confirmed in the RG analysis[10]. Further, this point is also crucial in recent experimental progress in the effects of (topological) defects in graphene especially in the heavily-doped case (which is also discussed in the next section).

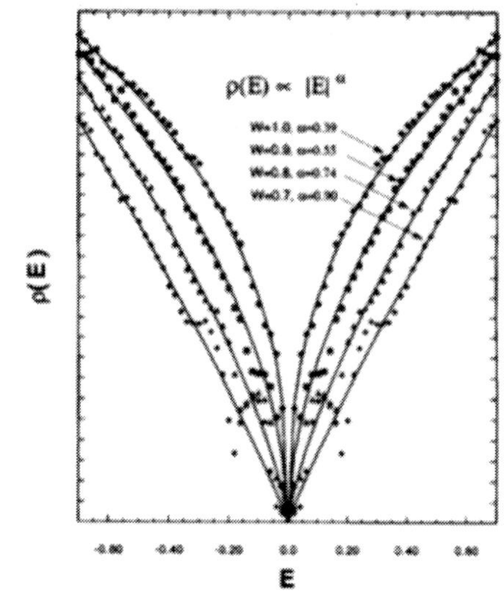

FIGURE 1. DoS for our model[8]. The exponent for DoS changes continuously as a function of the disorder strength (W). It implies the existence of ‘disordered critical line’.

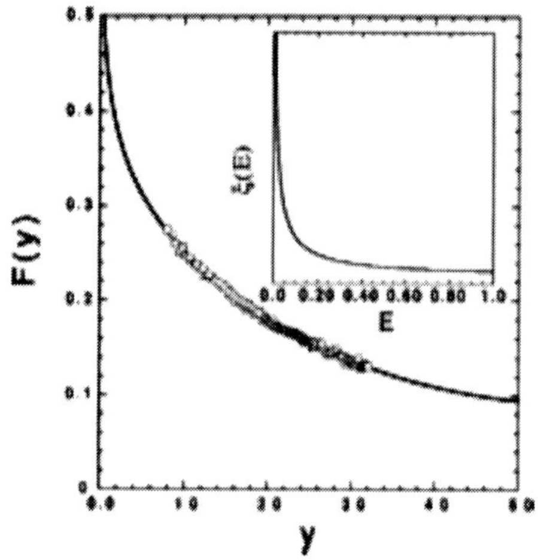

FIGURE 2. Scaling function of the Thouless number for our model as a function of the system size (divided by the localization length)[9]. The localization length is shown in the inset as a function of energy. It diverges at zero energy(band center), which implies disorder-driven criticality in our model.

DISCUSSION

In the last, we shall discuss future directions; I. a model of topological defect and II. transport properties of nodal quasiparticle in two dimensions. These two topics are closely related to each other and form a suitable point at which to end this paper.
Recent experimental progress by the Manchester's group revealed that topological defect in graphene(called 'ripple') is relevant and characterized by symmetries which is also discussed in this paper. They are the time-reversal and the chiral symmetry(plus gauge symmetry and 'locality'). Taking all these into account, we recently succeeded in modeling the topological defect on a lattice. Based on it, we can now study basic electronic structures and properties of graphene in the presence of topological defects[12]. Moreover, it can also be viewed as a model for a vortex in the 2D d-wave superconductivity.We recently calculated the vortex motion for nodal superconducting quasiparticle in the presence of electric field[13]. By analogy with the problem of 'vortex motion', it is our next project to clarify how the topological defect in graphene is controlled under an external field. We consider that it should lead to the exploration of novel carbon-based technology. Further, it would be interesting to extend our formulation for the motion of topological defects in e.g. Charge/Spin Density Wave and (bio)membrane.

ACKNOWLEDGEMENTS

The review part of this paper is based on my previous work with Y.Hatsugai. We are grateful to K.Maki for collaboration in 'Vortex Motion in Nodal Superconductors'. This work has been supported by a Grant-in Aid for Scientific Research from the Ministry of Education, Culture, Sports, Science and Technology of Japan.

REFERENCES

1.K.S.Novoselov et al., Nature **438**, 197 (2005).
2.G.W.Semenoff, *Phys.Rev.Letters* **53**, 2449 (1984).
3.F.D.M.Haldane, *Phys.Rev.Letters* **61**, 2015 (1988).
4.See, for example, F.Schedin et al., preprint(cond-mat/0610809).
5.A.Kanigel et al., Nature Physics **2**, 447 (2006).
6.S.Katayama, A.Kobayashi and Y.Suzumura, J.Phys.Soc.Jpn. **75**, 054705 (2006).
7.Y.Otsuka, Y.Morita and Y.Hatsugai, *Phys.Rev.B* **66**, 73109 (2002);Y.Morita, Y.Hatsugai and Y.Otsuka, *Phys.Rev.B* **70**, 245101 (2004).
8.Y.Morita and Y.Hatsugai, *Phys.Rev.Letters* **86**, 151 (1997).
9.Y.Morita and Y.Hatsugai, *Phys.Rev.B* **58**, 6680 (1998).
10.S.Guruswamy, A.LeClair and A.W.W.Ludwig, Nucl.Phys. **B538**, 475 (2000); See, for a review, Y.Morita et al., in 'Recent Research Developments in Physics Vol.4', pp.463-483, Transworld Research Network(2003) and references therein.
11.E.Abrahams, P.W.Anderson, D.C.Licciardello and T.V.Ramakrishnan, *Phys.Rev.Letters* **42**, 673(1979).
12.Y.Morita, work in progress.
13.Y.Morita and K.Maki, preprint in preparation.

SU(2) and SU(4) Kondo effect in double quantum dots

Jernej Mravlje*, Anton Ramšak†,* and Tomaž Rejec†,*

*Jožef Stefan Institute, Jamova 39, Ljubljana, Slovenia
†Faculty of Mathematics and Physics, University of Ljubljana, Jadranska 19, Ljubljana, Slovenia

Abstract. We investigate serial double quantum dot systems with on-site and inter-site interaction by means of Schönhammer-Gunnarsson projection-operator method. The ground state is established by the competition between extended Kondo phases and localized singlet phases in spin and charge degrees of freedom. We present and discuss different phases, as discerned by characteristic correlation functions. We discuss also how different phases would be seen in linear transport measurements.

Keywords: Kondo effect, double quantum dots, conductance
PACS: 73.23.-b, 73.63.Kv, 72.15.Qm

INTRODUCTION

In the last decade the advances in experimental techniques enabled the exploration of intriguing many-body effects occurring in solid-state systems such as the Kondo effect [1] by means of measuring the conductance of nanoscale electrical circuits. Tiny pools of electrons defined by electrodes – quantum dots (QDs) – constitute artificial atoms/molecules. Additional gates enable tuning of the orbital levels as well as the tunneling rates, which makes systematic exploration of various effects experimentally accessible. The Kondo effect is essentially the increased scattering rate (with phase shifts near $\pi/2$) at low temperatures due to magnetic impurities in host metals. In transport experiments through quantum dots it is seen in another disguise: it is discerned as the amplification of the conductance towards unitary limit. Interesting way to proceed further is to analyze the consequences of inter-impurity interaction by looking at the transport through double quantum dot (DQD) systems.

The characteristic feature of the two-impurity Kondo physics is that the two impurities either form an inter-impurity singlet, which is virtually decoupled from conduction electrons or they form a double Kondo state SU(2)×SU(2), in which each spin characterized by the SU(2) symmetry group is screened by the conduction electrons [2] depending on the scales of the energies of the inter-impurity singlet formation J and Kondo state formation T_K. When the symmetry of the Hamiltonian is larger the Kondo temperature is enhanced. For double quantum dots, which have the capacitative interaction V tuned near the value of the on-dot interaction U, the SU(4) Kondo effect occurs [3].

Here we report our results on the competition between extended Kondo and localized singlet phases in serial DQD systems with inter-dot interaction in the point of particle-hole symmetry [4] and discuss also the phases which occur outside this point. The SU(4) Kondo phase cannot be explored directly by transport experiment through a DQD as the

CP918, *Lectures on the Physics of Strongly Correlated Systems XI: Eleventh Training Course,* edited by A. Avella and F. Mancini

conductance is small irrespective of whether the system is in the SU(4) Kondo state or not. Nevertheless, the scale of the SU(4) condensation energy can be estimated by tuning the system away from the point of SU(4) symmetry until the SU(4) Kondo state collapses. The boundary is easy to discern from the conductance data as the conductance is unity whenever the crossover between the phases takes place.

MODEL AND METHOD

We model DQDs by the two-impurity Anderson Hamiltonian $H = H_{\rm d} + H_{\rm l}$, where $H_{\rm d}$ corresponds to the isolated dots

$$H_{\rm d} = \sum_{i=1,2} (\varepsilon n_i + U n_{i\uparrow} n_{i\downarrow}) + V n_1 n_2 - t \sum_{\sigma} (c^{\dagger}_{1\sigma} c_{2\sigma} + h.c.),$$

with $n_i = n_{i\uparrow} + n_{i\downarrow}$, $n_{i\sigma} = c^{\dagger}_{i\sigma} c_{i\sigma}$. The dots are coupled by a tunneling matrix element t and a capacitive V term. The on-site energies ε and the Hubbard repulsion U are taken equal for both dots. $H_{\rm l}$ describes the noninteracting left and right tight-binding leads with hopping parameter t_0 and the coupling of the leads to the DQD. We denote the characteristic tunneling rate of an isolated electron from the dot to the lead by $\Gamma = t'^2/t_0$, where t' is the parameter characterizing the dot-lead hopping.

To calculate the ground state of the system we use the Schönhammer and Gunnarsson projection-operator basis [5, 6] $|\Psi_{\lambda\lambda'}\rangle = P_{\lambda 1} P_{\lambda' 2} |\tilde{0}\rangle$, which consists of projectors $P_{\lambda i}$; $P_{0i} = \left(1 - n_{i\uparrow}\right)\left(1 - n_{i\downarrow}\right)$, $P_{1i} = \sum_{\sigma} n_{i\sigma}(1 - n_{i\bar{\sigma}})$, $P_{2i} = n_{i\uparrow} n_{i\downarrow}$ and additional operators involving the operators in leads. We used up to ~ 100 additional combinations of operators consisting of, for example, $P_{3i} = P_{0i} \widehat{v} P_{1i}$, where $\widehat{v}$ denotes the tunneling to/from dot i. These operators are applied to the state $|\tilde{0}\rangle$, which is the ground state of the auxiliary noninteracting DQD Hamiltonian of the same form as H, but with $U, V = 0$, renormalized parameters $\varepsilon, t, t' \rightarrow \tilde{\varepsilon}, \tilde{t}, \tilde{t}'$ and additional parameter $\tilde{t}''$ which corresponds to hopping from left dot to right lead and vice versa which although absent in the original Hamiltonian is present in the effective Hamiltonian in some parameter regimes.

The conductance is calculated using the sine formula [7], $G = G_0 \sin^2[(E_+ - E_-)/4t_0 L]$, where $G_0 = 2e^2/h$ and $E_{\pm}$ are the ground state energies of a large auxiliary ring consisting of L non-interacting sites and an embedded DQD, with periodic and anti-periodic boundary conditions, respectively.

GROUND STATE AND CONDUCTANCE OF DQD WITH INTER-DOT INTERACTION

Detached DQDs

The starting point towards the understanding of the ground state of DQDs are the filling properties of isolated DQDs (*i.e.* of the Heitler-London or the two-site Hubbard model). The first electron is added when $\varepsilon = t$, and the second when $\varepsilon = -t + J + [(U + V) - |U - V|]/2$, where $J = [-|U - V| + \sqrt{(U - V)^2 + 16t^2}]/2$ is the difference between singlet and triplet energies. When $n = 2$ the ground state is $[\alpha(|\uparrow\downarrow\rangle - |\downarrow\uparrow\rangle) + \beta(|20\rangle - |02\rangle)]/\sqrt{2}$, where $\alpha/\beta = 4t/(V - U + \sqrt{(U-V)^2 + 16t^2})$. The range of ε where single occupation is favorable is progressively diminished when $V \neq U$. For large t or at (and near) $V = U$ the molecular bonding and anti-bonding orbitals are formed as is seen here from $\alpha \sim \beta$.

Attached DQDs and conductance

As we attach DQDs to the leads the ground state either is or is not reminiscent of the ground state of the isolated system. Here the latter possibility is always due to some kind of the Kondo effect. In the top panels of Fig. 1 the ground state of DQDs are presented with pictograms for $V = 0, U$ on the left and right, respectively. The near vertical dividing lines correspond to values of parameters where the ground state of the isolated system is degenerate due to matching energies of states with different occupancies, for example, the rightmost line corresponds to $E(0) = 0 = E(1) = \varepsilon - t$. The horizontal U-shaped line is given by $J = 2.2T_K$, where the scale of the Kondo condensation energy is estimated by $T_K = \sqrt{U\Gamma/2}\exp(-\pi\varepsilon(\varepsilon + U)/2\Gamma)$ for $U/\Gamma = 15$.

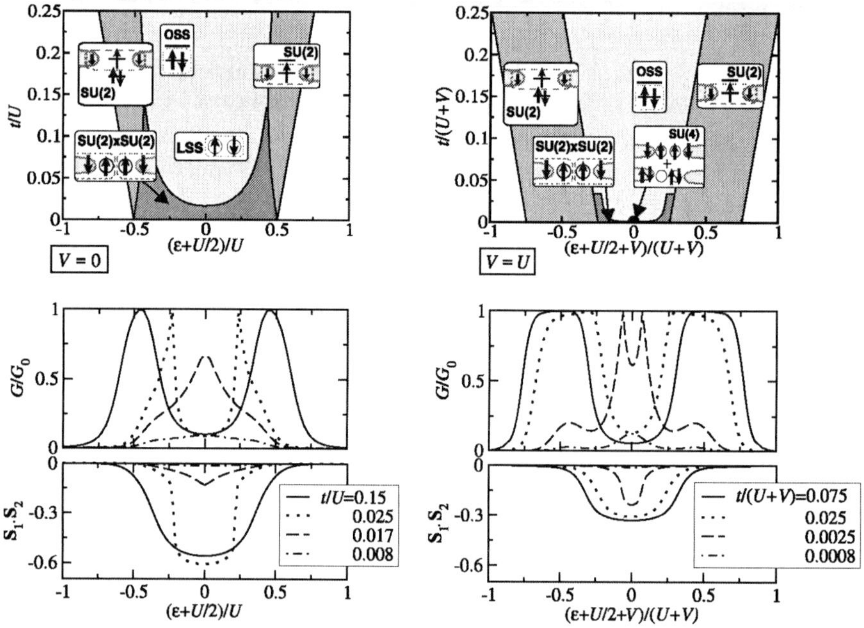

FIGURE 1. – *Top panels*: phases of serial DQDs for $V = 0$ (left) and $V = U$ (right). The occupancy of the DQD falls from left to right. Extended Kondo phases (with leads in pictograms) and localized singlet phases (without leads in pictograms) occur. – *Bottom panels*: Conductance and spin-spin correlation of the DQD for t above (full and dotted lines for $V = 0$; full, dotted and dashed lines for $V = U$) and below the localized singlet formation threshold. Note the approaching of $\mathbf{S}_1 \cdot \mathbf{S}_2$ towards $-3/8$ for large t indicating the formation of the orbital singlet.

For $n=0,4$ interaction between electrons (or holes) is not important, hence the ground state is not interesting. For $n=1,3$ the ground-state of the isolated DQD is a free spin in (anti-)bonding orbital, which is, when the leads are attached, at low-temperatures screened by conduction electrons as in 'ordinary' single impurity Anderson model. The most interesting part of the diagrams corresponds to $n \sim 2$. Here the ground state of the isolated system is a non-degenerate singlet but the tunneling to the leads breaks this singlet whenever roughly twice the Kondo condensation energy exceeds the triplet excitation energy J. For $V \sim U$ the J is enhanced hence the area corresponding to SU(2)×SU(2) Kondo is diminished. Near the symmetric point, however, another kind of the Kondo effect arises for $V \sim U$ as a consequence of larger symmetry of the $V=U$ Hamiltonian, which partially restores the occurrence of the Kondo phase.

Symmetries

The Kondo effect occurs as the consequence of the degeneracy of states of isolated impurities. If one looks at the ground state of two isolated impurities coupled by a capacitative (but not tunneling) term $V=U$, one sees that the 6 states $|\sigma_1\sigma_2\rangle$, $|20\rangle$ and $|02\rangle$ are degenerate. Indeed, by introducing the pseudospin operator [8] $\tilde{T}^i = 1/2\sum_{ll'=1,2}\sum_\sigma c^\dagger_{l\sigma}\tau^i_{ll'}c_{l'\sigma}$, where τ^i are the Pauli matrices, and the combined spin-pseudospin operators $W^{ij}=S^i\tilde{T}^j$, one sees that the Hamiltonian is SU(4) symmetric. As long as the SU(4) symmetry breaking terms are small enough $V-U, t \lesssim T_K[SU(4)]$, the ground state is an SU(4) 'spin' screened by the electrons in the leads.

Orbital representation

A complementary way is to rewrite the Hamiltonian in the basis of orbital operators $c_{b,a}=(c_1 \pm c_2)/\sqrt{2}$

$$H_{\rm d} = \sum_{\alpha=a,b}\left[\varepsilon_\alpha n_\alpha + \frac{U+V}{2}\left(n_{\alpha\uparrow}n_{\alpha\downarrow}+n_{\alpha\uparrow}n_{\bar{\alpha}\downarrow}\right)\right] + V\sum_\sigma n_{a\sigma}n_{b\sigma} + \frac{U-V}{2}\left(C_{\rm flip}-S_{\rm flip}\right),$$

where notation $\bar{a}=b$, $\bar{b}=a$ is used. The last term of $H_{\rm d}$ consists of isospin-flip $C_{\rm flip} = T^+_a T^-_b + h.c.$ and spin-flip $S_{\rm flip} = S^+_a S^-_b + h.c.$ operators, where $S^-_\lambda = c^\dagger_{\lambda\downarrow}c_{\lambda\uparrow} = (S^+_\lambda)^\dagger$ are spin and $T^-_\lambda = c_{\lambda\uparrow}c_{\lambda\downarrow} = (T^+_\lambda)^\dagger$ isospin lowering and raising operators for the orbitals $\lambda = b,a$ (or sites $\lambda = 1,2$). The full spin (isospin) algebra is closed with operators $S^z_\lambda = (n_{\lambda\uparrow}-n_{\lambda\downarrow})/2$ and $T^z_\lambda = (n_\lambda - 1)/2$, respectively.

When $V=U$, the spin- and isospin-flip terms in $H_{\rm d}$ are absent: the Hamiltonian is mapped exactly to the two-level Hamiltonian with intra- and inter-level interaction U with the bonding and anti-bonding levels coupled to even and odd transmission channels, respectively. When $V \neq U$ this mapping is no longer strictly valid: the electrons try to avoid the inter-level repulsion by occupying aligned spin-states in different orbitals, and the isospin-flip terms induce the fluctuations of charge between orbitals. Both mechanisms prohibit electrons from occupying the well-defined orbital states.

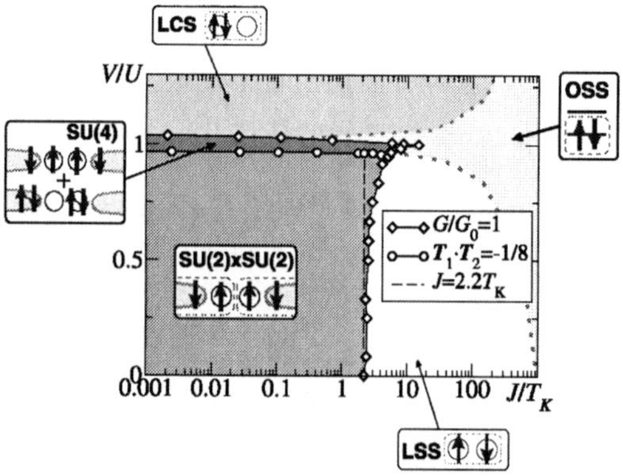

FIGURE 2. Phases of DQD in the point of particle-hole symmetry. The boundaries between Kondo and localized singlet phases are given by peaks in conductance and abrupt changes in correlation functions. The boundaries of the orbital spin singlet state are given by $\mathbf{S}_1 \cdot \mathbf{S}_2 = -3/16$ and $\Delta n_1^2 = \Delta n_b^2$ on the upper and lower side, respectively. Note the extension of the Kondo phase behind the line $J = 2.2T_K$ (dashed) at $V \sim U$.

Numerical results

In the lower panels of Fig. 1 the conductance and inter-dot spin-spin correlations are plotted. Note that the orbital picture is indeed more robust for the $V = U$ case as indicated by the broad plateaus in conductance corresponding to the SU(2) Kondo effect of a spin residing in the (anti-)bonding orbitals. Moreover, J is enhanced when compared to the $V = 0$ case: absence of singlet phase signalled by no peak with unitary conductance and minor spin-spin correlation for all ε occurs only for smaller t. Note also that conductance is small whenever the ground state is practically geometrically separable into parts. In that case the flux can be transported out of the auxiliary ring through the boundary between the parts, yielding zero conductance in our approach [7]. In Fig. 2 we indicate the phases in the $(J/T_K, V/U)$ plane. Details are given in Ref. [4].

ACKNOWLEDGMENTS

We acknowledge the support of SRA under grant PI-0044.

REFERENCES

1. A. C. Hewson, *The Kondo Problem to Heavy Fermions*, Cambridge University Press, 1993.
2. B. A. Jones, C. M. Varma, and J. W. Wilkins, *Phys. Rev. Lett.* **61**, 125–128 (1988).
3. M. R. Galpin, D. E. Logan, and H. R. Krishnamurthy, *Phys. Rev. Lett.* **94**, 186406 (2005).
4. J. Mravlje, A. Ramšak, and T. Rejec, *Phys. Rev. B* **73**, 241305(R) (2006).
5. K. Schönhammer, *Phys. Rev. B* **13**, 4336 (1976).
6. O. Gunnarsson, and K. Schönhammer, *Phys. Rev. B* **31**, 4815 (1985).
7. T. Rejec, and A. Ramšak, *Phys. Rev. B* **68**, 035342 (2003).
8. L. D. Leo, and M. Fabrizio, *Phys. Rev. B* **69**, 245114 (2004).

A non-Hermitian analysis of strongly correlated quantum systems

Yuichi Nakamura* and Naomichi Hatano†

*yuichi@iis.u-tokyo.ac.jp
†hatano@iis.u-tokyo.ac.jp

Abstract. We study a non-Hermitian generalization of quantum systems in which an imaginary vector potential is added to the momentum operator. In the tight-binding approximation, we make the hopping energy asymmetric in the Hermitian Hamiltonian. We conjecture that the non-Hermitian critical point where the energy gap vanishes is equal to the inverse correlation length of the Hermitian system. We confirm the conjecture for three exactly solved models analytically and for an unsolved model numerically.

Keywords: Non-Hermitian quantum mechanics, correlation length, dispersion relation, isotropic *XY* chain, Hubbard model, $S = 1\ 2$ antiferromagnetic *XXZ* chain, Majumdar-Ghosh model

INTRODUCTION

In the present work, we study a non-Hermitian generalization of strongly correlated quantum systems by adding an imaginary vector potential $i\mathbf{g}$ (where $\mathbf{g}$ is a real constant) to the momentum operator. In the tight-binding approximation, we multiply the right hopping energy by e^{g} and the left hopping energy by e^{-g}. Hatano and Nelson [1] applied this technique to the one-electron Anderson model. Their purpose of the non-Hermitian generalization was to obtain a *length* scale inherent in the wave function of the Hermitian model, namely the localization length, only from the non-Hermitian energy spectrum. The above study was for a *non-interacting random* electron system. We here conjecture that we can obtain in the same way the correlation length of a Hermitian *interacting non-random* model. We show for the $S = 1\ 2$ ferromagnetic isotropic *XY* chain in a magnetic field, the half-filled Hubbard model and the $S = 1\ 2$ antiferromagnetic *XXZ* chain that the non-Hermitian critical point of the ground state where the energy gap vanishes is equal to the inverse correlation length of the Hermitian system. We apply the same technique to an unsolved model, namely the $S = 1\ 2$ antiferromagnetic Heisenberg chain with nearest-neighbor and next-nearest-neighbor interactions including the Majumdar-Ghosh model [3]. We analyze numerical data of the non-Hermitian quantum system of finite size and confirm the conjecture.

CP918, *Lectures on the Physics of Strongly Correlated Systems XI: Eleventh Training Course*, edited by A. Avella and F. Mancini

NON-HERMITIAN ANALYSIS OF EXACTLY SOLVED MODELS

$S = 1\ 2$ ferromagnetic isotropic XY chain

We first consider the non-Hermitian generalization of the $S = 1\ 2$ ferromagnetic isotropic XY chain in a magnetic field. The Hermitian Hamiltonian of this model is

$$H_{XY} = -J\sum_{l=1}^{L}\ S_l^x S_{l+1}^x + S_l^y S_{l+1}^y\ \ - h\sum_{l=1}^{L} S_l^z \tag{1}$$

where we set $J > 0$. The Hamiltonian (1) is transformed into

$$H_{XY} = -\frac{J}{2}\sum_{l=1}^{L}(c_{l+1}^\dagger c_l + c_l^\dagger c_{l+1}) + \frac{hL}{2} - h\sum_{l=1}^{L} c_l^\dagger c_l \tag{2}$$

by the Jordan-Wigner transformation, where $c^\dagger$ and c are fermion creation and annihilation operators. We consider the non-Hermitian generalization of the Hamiltonian (2) in the form

$$H_{XY}(g) = -\frac{J}{2}\sum_{l=1}^{L}(e^g c_{l+1}^\dagger c_l + e^{-g} c_l^\dagger c_{l+1}) + \frac{hL}{2} - h\sum_{l=1}^{L} c_l^\dagger c_l \tag{3}$$

We can immediately diagonalize the non-Hermitian Hamiltonian (3) in the momentum space in the form

$$H_{XY}(g) = \sum_{-\pi<k<\pi} \varepsilon(k+ig)c_k^\dagger c_k + \frac{hL}{2} \tag{4}$$

with the Fourier transformation, where $\varepsilon(k) \equiv -J\cos k - h$. We calculate a non-Hermitian critical point g_c as the point where the energy gap above the ground state vanishes. The energy gap $\Delta E(g)$ for $h > J$ in the limit $L \to \infty$ is given by the elementary excitation energy in the form $\Delta E(g) = |\varepsilon(\pm\pi + ig)| = h - J\cosh g$ The energy gap vanishes at $g_c = \ln\left[h\ J + \sqrt{(h\ J)^2 - 1}\right]$ This is indeed equal to the inverse correlation length of the Hermitian system, obtained by the quantum transfer matrix method [4].

Hubbard model in the half-filled case

We next consider the non-Hermitian generalization of the Hubbard model. Our non-Hermitian Hamiltonian [2, 5] is

$$H_{\mathrm{Hubbard}}(g) = -t\sum_{l=1}^{L}\sum_{\sigma=\uparrow\downarrow}(e^g c_{l+1\,\sigma}^\dagger c_{l\,\sigma} + e^{-g} c_{l\,\sigma}^\dagger c_{l+1\,\sigma}) + U\sum_{l=1}^{L} n_{l\uparrow} n_{l\downarrow} \tag{5}$$

Fukui and Kawakami [5] solved the non-Hermitian Hubbard model (5) at the half-filled point exactly. They showed that, as we increase the non-Hermiticity g, the Hubbard gap due to the charge excitation vanishes at a point g_c given in the form

$$g_c = \operatorname{arcsinh}(U/4t) + 2i\int_{-\infty}^{\infty} \arctan\frac{\lambda + iU/4t}{U/4t}\sigma(\lambda)d\lambda \tag{6}$$

The symbol $\sigma(\lambda)$ is a distribution function of the spin rapidity given by

$$\sigma(\lambda) = \frac{1}{2\pi}\int_{0}^{\infty} \mathrm{sech}\,\frac{U}{4t}\omega\;\cos(\lambda\omega)J_0(\omega)d\omega \tag{7}$$

where $J_0(\omega)$ is the Bessel function of the first kind.

On the other hand, the inverse correlation length $1/\xi$ of the charge excitation at the Hermitian point $g = 0$ was obtained by Stafford and Millis [6] in the form

$$\frac{1}{\xi} = \operatorname{arcsinh}(U/4t) - 2\int_{0}^{\infty}\frac{J_0(\omega)\sinh((U/4t)\omega)}{\omega(1+e^{\omega U/2t})}d\omega \tag{8}$$

We can show [2] after some algebra that the expressions of the non-Hermitian critical point g_c in Eq. (6) and the inverse correlation length $1/\xi$ in Eq. (8) are equal.

$S = 1/2$ antiferromagnetic *XXZ* chain

We next consider the non-Hermitian generalization of the $S = 1/2$ antiferromagnetic *XXZ* chain. Our non-Hermitian Hamiltonian is

$$H_{XXZ}(g) = J\sum_{l=1}^{L}\frac{1}{2}\left(e^{2g}S_l^+S_{l+1}^- + e^{-2g}S_l^-S_{l+1}^+\right) + \Delta S_l^zS_{l+1}^z \tag{9}$$

The non-Hermitian Hamiltonian (9) for $\Delta = 1$ is derived by considering the second-order perturbation of the non-Hermitian Hubbard Hamiltonian

$$H = -t\sum_{l=1}^{L}\left(e^{g}c_{l+1\uparrow}^{\dagger}c_{l\uparrow} + e^{-g}c_{l\uparrow}^{\dagger}c_{l+1\uparrow} + e^{-g}c_{l+1\downarrow}^{\dagger}c_{l\downarrow} + e^{g}c_{l\downarrow}^{\dagger}c_{l+1\downarrow}\right) + U\sum_{l=1}^{L}n_{l\uparrow}n_{l\downarrow} \tag{10}$$

with respect to t. As we increase the non-Hermiticity g in Eq. (10), we can eliminate the spin gap rather than the charge gap. We then generalize the model to an arbitrary Δ. Albertini *et al.* [7] calculated for $\Delta > 1$ the non-Hermitian critical point g_c where the spin gap vanishes in the form

$$g_c = \frac{\gamma}{2} + \sum_{n=1}^{\infty}\frac{(-1)^n}{n}\tanh(n\gamma) \tag{11}$$

where $\gamma \equiv \operatorname{arccosh}\Delta$. The expression (11) is actually well known [8] as the inverse correlation length at the Hermitian point $g = 0$. We have thus confirmed our conjecture for three exactly solved models.

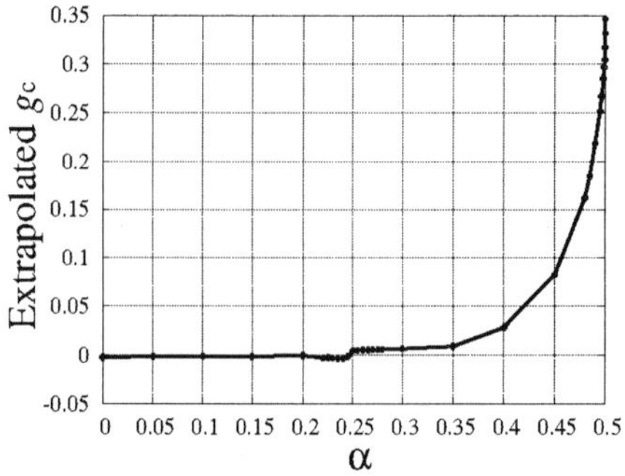

*Source: Y. N. and N. H., JPSJ **75** (2006) 104001.*

FIGURE 1. The α dependence of the estimates $g_c(\infty)$ extrapolated from second-order fitting.

NNN HEISENBERG CHAIN

We apply our non-Hermitian generalization to an unsolved model, namely the $S = 1\ 2$ antiferromagnetic Heisenberg chain with nearest-neighbor and next-nearest-neighbor interactions; we hereafter call this model the NNN Heisenberg chain. The Hermitian Hamiltonian of this model is

$$H_{\mathrm{NNN}} = J \sum_{l=1}^{L} [\mathbf{S}_l \cdot \mathbf{S}_{l+1} + \alpha \mathbf{S}_l \cdot \mathbf{S}_{l+2}] \tag{12}$$

where $J > 0$ and $\alpha \geq 0$. We require the periodic boundary conditions. At the point $\alpha = 0$, the model is the standard Heisenberg chain. The ground state is a spin liquid state and the energy gap is zero. At the point $\alpha = 1\ 2$, the model is the Majumdar-Ghosh model [3] and the energy gap is finite. Okamoto and Nomura [9] numerically showed that a massive-massless transition occurs at $\alpha_c \cong 0\ 2411$.

We here analyze the non-Hermitian Hamiltonian of the NNN Heisenberg chain

$$\begin{aligned} H_{\mathrm{NNN}}(g) = J \sum_{l=1}^{L} [\frac{1}{2}(e^{2g} S_{l+1}^{+} S_l^{-} + e^{-2g} S_l^{+} S_{l+1}^{-}) + S_l^z S_{l+1}^z] \\ + \alpha J \sum_{l=1}^{L} [\frac{1}{2}(e^{4g} S_{l+2}^{+} S_l^{-} + e^{-4g} S_l^{+} S_{l+2}^{-}) + S_l^z S_{l+2}^z] \end{aligned} \tag{13}$$

We numerically estimate the non-Hermitian "critical" point $g_c(L)$ of finite size in the subspace $S_{\mathrm{tot}}^z = 0$ where the eigenvalue corresponding to the ground state in the thermodynamic limit becomes complex. We then carry out finite-size plot of $g_c(L)$ and obtain extrapolated estimates $g_c(\infty)$.

Figure 1 shows for $0 \leq \alpha \leq 1\ 2$ the extrapolated estimates by second order fitting in the form $g_c(L) = g_c(\infty) + a\ L + b\ L^2 + \mathrm{O}(1\ L^3)$ of the finite-size data $g_c(L)$ for $L = 8\ 12$ and 16. The estimate $g_c(\infty)$ around $\alpha \cong 0$ is almost zero and is consistent with $\xi^{-1} = 0$ in the limit $L \to \infty$. The extrapolated estimates $g_c(\infty)$ at $\alpha = 1\ 2$ are around 0.35, which is consistent with the inverse correlation length $\xi^{-1} = \ln 2 \;\, 2$ ($\cong 0 \;\, 347$) obtained by the finite-size scaling of the correlation function of the ground states. The estimate $g_c(\infty)$ is almost zero in the region $0 \leq \alpha \lesssim 0 \;\, 25$ and is finite in the region $0 \;\, 25 \lesssim \alpha \leq 1 \;\, 2$. This is consistent with the massive-massless transition [9] at $\alpha \cong 0 \;\, 2411$ if we admit that $g_c(\infty)$ is equal to the inverse correlation length. It is remarkable that $g_c(\infty)$ is at least approximately equal to the inverse correlation length all through the region $0 \leq \alpha \leq 1 \;\, 2$ in spite of the fact that we do not know the exact dispersion relation of the elementary excitation of this model.

We have applied the same procedure to the Hubbard model (5) and the XXZ model (9) of finite size and confirmed that the extrapolated estimates $g_c(\infty)$ is consistent with the inverse correlation length of the Hermitian models of infinite size.

We have focused on the region $0 \leq \alpha \leq 1 \;\, 2$ so far. The non-Hermitian generalization of the form (13) is presumably not appropriate for detecting the incommensurate state in the region $\alpha > 1 \;\, 2$; we may need other types of non-Hermitian generalization.

SUMMARY

We considered the non-Hermitian generalization of strongly correlated quantum systems where the transfer energy is asymmetric. We conjectured that the non-Hermitian critical point where the energy gap vanishes is equal to the inverse correlation length. We confirmed the conjecture analytically for three strongly correlated quantum systems solved exactly and numerically for the NNN Heisenberg chain. We note that our non-Hermitian generalization is not always applicable to obtain the correlation length; the NNN Heisenberg chain in the region $\alpha > 1 \;\, 2$ and the $S = 1 \;\, 2$ transverse Ising chain are examples. We may need other types of non-Hermitian generalizations for these models.

ACKNOWLEDGMENTS

The work is supported by Grant-in-Aid for Scientific Research (No. 17340115) from the Ministry of Education, Culture, Sports, Science and Technology as well as by Core Research for Evolutional Science and Technology of Japan Science and Technology Agency.

REFERENCES

1. N. Hatano and D. R. Nelson, Phys. Rev. Lett. **77** (1996) 570; Phys. Rev. B **56** (1997) 8651.
2. Y. Nakamura and N. Hatano, Physica B **378-380** (2006) 292; J. Phys. Soc. Jpn. **75** (2006) 104001.
3. C. K. Majumdar and D. K. Ghosh, J. Math. Phys. **10** (1969) 1388; C. K. Majumdar, J. Phys. C **3** (1970) 911.
4. B. M. McCoy, Phys. Rev. **173** (1968) 531.
5. T. Fukui and N. Kawakami, Phys. Rev. B **58** (1998) 16051.
6. C. A. Stafford and A. J. Millis, Phys. Rev. B **48** (1993) 1409.
7. G. Albertini, S. R. Dahmen and B. Wehefritz, Nucl. Phys. B **493** (1997) 541.
8. R. J. Baxter, *"Exactly Solved Models in Statistical Mechanics"* (Academic Press, New York, 1982), p. 155.
9. K. Okamoto and K. Nomura, Phys. Lett. A **169** (1992) 433.

Variational cluster study of antiferromagnetism and d-wave superconductivity in layered organic superconductors

P. Sahebsara and D. Sénéchal

Département de physique and Regroupement québécois sur les matériaux de pointe, Université de Sherbrooke, Sherbrooke, Québec, Canada, J1K 2R1

Abstract. The interplay between the antiferromagnetism and d-wave superconductivity of κ-$(ET)_2X$ conductors has been investigated by the variational cluster approach (VCA), based on self-energy-functional approach. The dimer Hubbard model has been studied as a function of the on-site repulsion U and diagonal hopping amplitude t'. Our zero-temperature calculations show a transition from Néel antiferromagnetism to a spin-liquid phase at around $t' \sim 0.9$. At weak coupling we find d-wave superconductivity. and a transition between $d_{x^2-y^2}$ and d_{xy} pairing symmetry as U decreases.

Keywords: Hubbard model, superconductivity, antiferromagnetism, Mott transition, VCA, BEDT
PACS: 71.27.+a,74.70.Kn, 71.10.Fd,78.20.Bh

INTRODUCTION

Both in cuprate and organic compounds, the proximity of antiferromagnetism (AF) and unconventional superconductivity (SC) has been the focus of recent experimental and theoretical researches. κ-(BEDT-TTF)$_2$X organic superconductors are quasi-two-dimentional strongly correlated interacting electrons systems, in which holes are constrained to move mostly in the layer of BEDT-TTF molecules, and the layer of anions plays the role of barrier to the hole motion. BEDT-TTF (or simply ET) is an abbreviation for bisethylenedithio-tetrathiafulvalene, and X=$Cu(NCS)_2$, $Cu[N(CN)_2]Br$, $Cu_2(CN)_3$, I_3, etc. [1, 2]. Understanding electronic properties and the microscopic phase transition in these materials is an intriguing theoretical challenge [3, 4].

In this work we will focus on two members of namely, κ-$(ET)_2Cu[N(CN)_2]Cl$ (shortly written as κ-Cl) [5] and κ-$(ET)_2Cu_2(CN)_3$ (κ-Cu_2CN_3) [6, 7]. In these materials a competition between electron-electron Coulomb repulsion and kinetic energy results in a Mott transition without symmetry breaking. In other words the Coulomb interaction causes electrons to be localized. We used the variational cluster approximation (VCA) to probe the behavior of long range ordered AF and d-wave SC phases and the transition between them in dimer-Hubbard model, at zero temperature. This method provides both the static and dynamic information by computing the many-body one-particle Green function [8].

A schematic phase diagram for some of ET salts is drawn in the right panel of Fig. 1. Chemical pressure can be set by substitution of anion molecules and modification of donor molecules. Applied pressure is simply superimposed to this chemical pressure. Overall, pressure is assumed to be roughly proportional to the inverse screened Coulomb

CP918, *Lectures on the Physics of Strongly Correlated Systems XI: Eleventh Training Course,* edited by A. Avella and F. Mancini

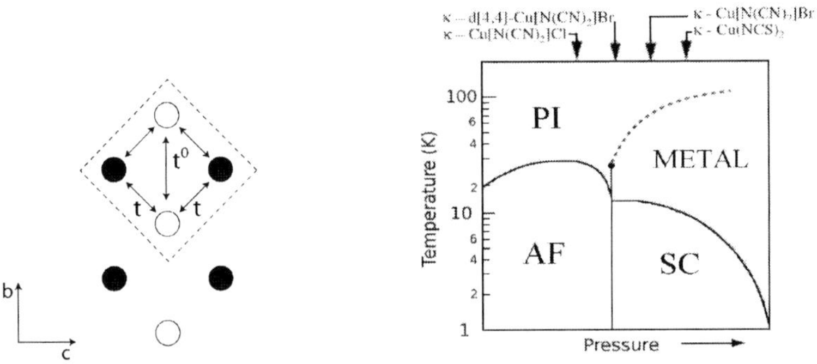

FIGURE 1. *left*: The lattice structure of the dimers of face-to-face BEDT-TTF molecules. t and t' represent the hopping integrals to nearest and next-nearest neighbor site. *right*: Generic phase diagram of κ-(ET)$_2$X organic salts. The hydrostatic pressure is proportional to the inverse Coulomb interaction U.

repulsion within a dimer, so that the higher the pressure, the lower the effects of correlations. The paramagnetic insulating (PI), antiferromagnetic (AF) insulating, unconventional superconducting (SC), and metallic (M) phases are clearly realized in the phase diagram. The positions of the various compounds are determined by their ambient-pressure ground-state properties. The left pannel of Fig. 1 demonstrates the orientation of dimers in ET lattice. Each circle represents two face-to-face ET molecule. Holes can hop between the nearest and next nearest dimers with amplitudes t and t' respectively.

THEORY

The minimum model required to study the magnetic and superconducting properties of layered organic materials is the single-band *dimer-Hubbard model* [14]. The Hamiltonian for this model can be written as

$$H = t \sum_{\langle \mathbf{r}\mathbf{r}'\rangle,\sigma} c^{\dagger}_{\mathbf{r}\sigma} c_{\mathbf{r}'\sigma} + t' \sum_{[\mathbf{r}\mathbf{r}'],\sigma} c^{\dagger}_{\mathbf{r}\sigma} c_{\mathbf{r}'\sigma} + U \sum_{\mathbf{r}} n_{\mathbf{r}\uparrow} n_{\mathbf{r}\downarrow} - \mu \sum_{\mathbf{r},\sigma} n_{\mathbf{r}\sigma} \tag{1}$$

where $c_{\mathbf{r}\sigma}$ ($c^{\dagger}_{\mathbf{r}\sigma}$) creates an electron (hole) at dimer site $\mathbf{r}$ on a square lattice with spin σ, and $n_{\mathbf{r}\sigma}=c^{\dagger}_{\mathbf{r}\sigma}c_{\mathbf{r}\sigma}$ is the hole number operator. $\langle \mathbf{r}\mathbf{r}'\rangle$ ($[\mathbf{r}\mathbf{r}']$) indicates nearest- (next-nearest)-neighbor bonds. In this model the kinetic energy term tends to delocalize the electrons (or holes) and favors a metallic state. On the other hand, the on-site interaction tends to localize the electrons to avoid doubly occupied sites and thereby favors an insulating state at half filling. In this model, the ratio t/U can be assimilated to pressure.

The dimer Hubbard model has been treated with several theoretical methods [15, 16]. Quantum cluster methods involve the exact solution of the model on a finite cluster of sites, and the embedding of this cluster into an infinite lattice [8, 9, 17, 18]. The various cluster methods differ in the way they try to connect the properties (such as the one-particle Green function) calculated exactly with one cluster, to the corresponding property of the infinite system. Cluster Perturbation Theory amounts to calculating

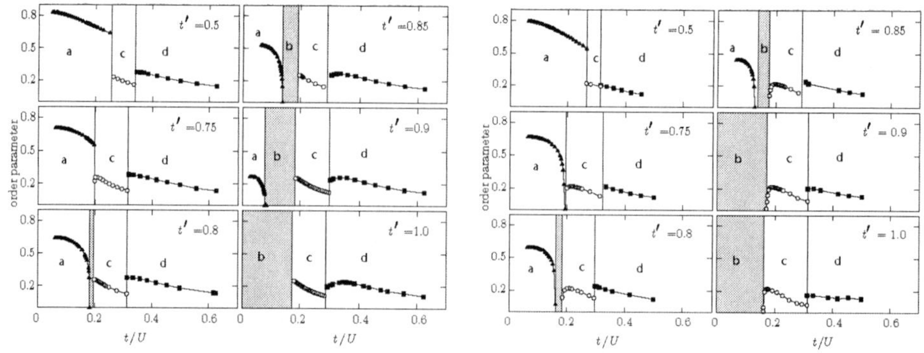

FIGURE 2. *Left*: pressure (t/U) dependence of order parameter for various values of $t_0 = t$, obtained with VCPT on 4-site clusters. **a** stands for AF (triangles), **b** for spin liquid, **c** for $d_{x^2-y^2}$ (circles, scaled by 2) and **d** for d_{xy} (squares, scaled by 5) phases, respectively. The vertical lines indicate the transition points, separating the various phases. The grey region identifies the spin liqid phase, with no long range order. *Right*: same graph, this time for 8-site clusters.

the exact self-energy of the cluster and applying this self-energy to the whole lattice, with a proper periodization of the Green function[17]. CPT is exact in both the strong and weak-coupling limits; however, it does not contain a procedure to study broken-symmetry phases (long-range order). The VCA is an extension of CPT based on the *self-energy functional approach* (SFA)[9], and enables us to overcome this problem. It provides a variational principle to use dynamical information from an exactly solvable "reference system" (here, a cluster) to describe the system on the infinite lattice.

The SFA defines a functional $\Omega_{\mathbf{t}}[\Sigma]$ of the self-energy, which is stationary at the physical self-energy: $\delta\Omega_{\mathbf{t}}[\Sigma]/\delta\Sigma = 0$. The value of the functional at the stationary point is the physical grand potential of the system. This functional can be calculated exactly, but only for a finite-dimensional space of self-energies, by defining a so-called *reference Hamiltonian* H', which has the same interaction part as the original Hamiltonian under study, but differs from it by its one-body part, in such a way that it can be solved exactly (numerically). In terms of the exactly computed Green function G' of the reference system, and the non-interacting Green function G_0 of the full Hamiltonian H, the functional $\Omega_{\mathbf{t}}$ reduces to a function of the one-body parameters of the reference Hamiltonian, i.e.:

$$\Omega_{\mathbf{t}}(\mathbf{t}') = \Omega' - \int_C \frac{d\omega}{2\pi} \sum_{\mathbf{K}} \ln\det\left(1 + (G_0^{-1} - G_0'^{-1})G'\right) \tag{2}$$

where Ω' is the exact grand potential of the cluster. The variational principle is then applied by varying the one-body part of the cluster Hamiltonian (by varying a few parameters such as symmetry breaking Weiss fields) until the function $\Omega_{\mathbf{t}}(\mathbf{t}')$ is stationary. At this stationary point, the self-energy of the cluster is taken as the self-energy of the whole lattice, like in CPT. The value of the function Ω at this point is an approximation to the physical grand potential. We refer to Refs [9, 8] for a more detailed explanation.

RESULTS AND DISCUSSION

We now present the zero temperature VCA results for the dimer Hubbard model (1) on four-site and eight-site clusters [11]. We have added symmetry-breaking terms (Weiss fields) to the cluster Hamiltonian (reference system), which have the form:

$$H'_{AF} = M\sum_{\mathbf{r}}(-1)^{\sigma}e^{i\mathbf{Q}\cdot\mathbf{r}}n_{\mathbf{r}\sigma} \quad (3)$$

$$H'_{SC} = \sum_{\mathbf{r},\mathbf{r}'}(\Delta_{\mathbf{r}\mathbf{r}'}c_{\mathbf{r}\uparrow}c_{\mathbf{r}'\downarrow} + \text{H.c.}) \quad (4)$$

where $\mathbf{Q} = (\pi,\pi)$ is the Néel wave vector. A nonzero value of M would induce Néel antiferromagnetism. The $d_{x^2-y^2}$ state is probed by letting $\Delta_{\mathbf{r}\mathbf{r}'} = \Delta$ if $\mathbf{r}' = \mathbf{r}\pm\mathbf{x}$ and $\Delta_{\mathbf{r}\mathbf{r}'} = -\Delta$ if $\mathbf{r}' = \mathbf{r}\pm\mathbf{y}$. On the other hand, a d_{xy} state is probed by setting

$$\Delta_{\mathbf{r},\mathbf{r}+\mathbf{x}+\mathbf{y}} = \Delta_{\mathbf{r},\mathbf{r}-\mathbf{x}-\mathbf{y}} = \Delta_1 \quad (5)$$

$$\Delta_{\mathbf{r},\mathbf{r}+\mathbf{x}-\mathbf{y}} = \Delta_{\mathbf{r},\mathbf{r}-\mathbf{x}+\mathbf{y}} = -\Delta_2 \quad (6)$$

These Weiss fields are not part of the lattice Hamiltonian and they are determined by looking for a stationary point of the self-energy functional.

The nature of the superconducting pairing in κ-(ET)$_2$X is still the object of a controversy among experimental and theoretical physicists [13]. The terminology $d_{x^2-y^2}$ and d_{xy} are not rigorous in the context of the dimer model. We should instead consider the point group of the model, which is C_{2v}, and look at its irreducible representations [11, 19]. This group admits two possible gap symmetries for singlet superconductivity: the so-called A_2 representation, which is odd under reflections accross the $\mathbf{x}+\mathbf{y}$ and $\mathbf{x}-\mathbf{y}$ lines, and the A_1 representation, which is even. The $d_{x^2-y^2}$ state defined above corresponds to the A_2 representation, and the d_{xy} state corresponds to the A_1 representation. The C_{2v} symmetry does not require that the two lobes of the d_{xy} gap have the same size. Thus, the expression "extended s-wave" would be more appropriate than d_{xy} [19].

Besides the four variational (broken symmetry) parameters M, Δ, Δ_1 and Δ_2 , the chemical potential μ' on the cluster is also taken as a variational parameter to guarantee thermodynamic consistency. The lattice chemical potential is adjusted so that the electron density is $n = 1$ (half-filling). The order parameters as a function of pressure (t/U) are shown for $t'/t = 0.5$ to 1.0, on the left pannel of Fig. 2 for a 4-site cluster, and right pannel of Fig. 2 for an 8-site cluster. In the case where two solutions, for instance AF and dSC, were found, the one with the lowest energy, or value of $\Omega + \mu n$, was chosen.

We find that the system is in a Néel state when U is large enough and t' smaller than a critical value roughly equal to $0.9t$. For t' smaller than ~ 0.8, the following phases are met upon decreasing U: first an AF phase, then a SC phase with $d_{x^2-y^2}$ symmetry and finally, at low coupling, a SC phase with d_{xy} symmetry. The transitions between these phases are all of the first order, which opens the door to the possibility of macroscopic coexistence. This is indeed the case in the κ-Cl compound. For t' lying between ~ 0.8 and ~ 0.9, a paramagnetic insulating phase is intercalated between the AF and $d_{x^2-y^2}$ phases. We interpret this phase as the spin liquid phase proposed recently [7, 12]. For larger values of t', the AF phase disappears, leaving the spin liquid phase dominant at strong coupling. However, we have not studied the possibility of a 120° magnetic order, and such an order might exist around $t' = t$ for large-enough U. The small differences between the two cluster shapes indicate that the results are not yet converged with respect to cluster size. For instance, the AF order parameter is slightly smaller on the larger cluster, as it should. There are also minor differences in the values of critical parameters. But the overall picture is the same.

We have shown that the essential physics of the layered organic superconductors can be obtained form the dimer-Hubbard model. A rigorous variational approach enables to recover the dSC and AF ground states at half filling for different values of the on-site repulsive potential. The phase transition between the phases appears to be of first order. This is consistent with the experimental results of these materials. Decreasing the potential U within the dSC region results in a transition from $d_{x^2-y^2}$ to d_{xy} pairing symmetry. The so-called spin liquid phase, without long range order, is found for higher values of t', because of the higher level of magnetic frustration. Our calculations indicate that a pseudogap phenomenon also occurs in these systems, due to short-range AF fluctuations, even within SC phases where no AF long range order is observed.

ACKNOWLEDGMENTS

We thank A-M. Tremblay, and C. Bourbonnais for useful discussion. This work was supported by NSERC (Canada). Computations were performed on RQCHP clusters.

REFERENCES

1. T. Ichiguro, K. Yamaji, and G. Saito: *Organic Superconductors* (Springer, Heidelberg, 1998).
2. J. Singleton, Rep. Prog. Phys. **63**, 1111 (2000).
3. H. Kino *et. al.*, J. Phys. Soc. Jpn. **64** 2726 (1995); R. H. McKenzie, Science **278**, 820 (1997); Comments Cond. Matt. **18**, 309 (1998); B. J. Powell *et. al.*, cond-mat/0607078; cond-mat/0607079.
4. K. Kanoda, Physica C **282**, 299 (1997).
5. J. M. Williams, Science **252** 1510 (1991); S. Lefebvre *et. al.*, Phys. Rev. Lett. **85**, 5420 (2000).
6. T. Komatsu, N. Matsukawa, T. Inoue, and G. Saito, J. Phys. Soc. Jpn. **65**, 1340 (1996).
7. Y. Shimizu *et. al.*. Phys. Rev. Lett. **91**, 107001 (2003).
8. C. Dahnken *et. al.*, Phys. Rev. B **70**, 245110 (2004).
9. M. Potthoff *et. al.*, Phys. Rev. Lett. 91, 206402 (2003); Eur. Phys. J. B **32**, 429 (2003).
10. D. Sénéchal *et. al.*. Phys. Rev. Lett. **94**, 156404 (2005).
11. P. Sahebsara and D. Sénéchal, Phys. Rev. Lett. **97**, 257004 (2006).
12. Y. Kurosaki *et. al.*. Phys. Rev. Lett. **95**, 177001 (2005).
13. H. Mayaffre *et. al.*. Phys. Rev. Lett. **75**, 4122 (1995); H. Elsinger *et. al.*. Phys. Rev. Lett. **84**, 6098 (2000); J. Müller *et. al.*. Phys. Rev. B **65**, 140509 (2002); T. Arai *et. al.*. Phys. Rev. B **63**, 104518 (2001); K. Ichimura and K. Nomura, J. Phys. Soc. Jpn **75**, 051012 (2006); K. Izawa *et. al.*. Phys. Rev. Lett.**88**, 27002 (2002); M. Pinteric, J. Phys. IV France **114**, 245 (2005).
14. J. Hubbard, Proc. R. Soc. London A **276**, 238 (1963); H. Kino and H. Fukuyama, J. Phys. Soc. Jpn. **65**, 2158 (1996); R. McKenzie, Comm. Cond. Mat. Phys. **18** 309 (1998).
15. E. Dagotto, Rev. Mod. Phys. **66**, 763 (1994).
16. M. Jarrell. Phys. Rev. Lett. **69**, 168 (1992); A. Georges *et. al.*. Phys. Rev. Lett. **72**, 1545 (1996).
17. D. Sénéchal *et. al.*. Phys. Rev. Lett. **84**, 522 (2000); Phys. Rev. B **66**, 075129 (2002).
18. M. H. Hettler *et. al.*. Phys. Rev. B **58**, R7475 (1998); T. Maier *et. al.*. Rev. Mod. Phys. **77**, 1027 (2005); G. Kotliar *et. al.*. Phys. Rev. Lett. **87**, 186401 (2001).
19. B. J. Powell, J. Phys.: Condens. Matter **18**, 575 (2006).

QED_3 as an effective theory of the pseudogap phase in underdoped cuprates

Iorwerth Owain Thomas

Department of Physics, Loughborough University,Loughborough, Leicestershire LE11 3TU, U.K.

Abstract. A version of non-compact QED in 2+1 dimensions (QED_3), with anisotropies introduced by differing Fermi and gap velocities in the spatial directions has been proposed [1, 2] as an effective theory of the SDW-pseudogap phase transition in underdoped cuprates at T=0. Preliminary simulations using lattice gauge theory techniques suggest that matters are a little more complicated than previously thought; in particular, measurements of the fermion propagator in Landau gauge indicate that one should be cautious in inferring properties of this form of QED_3 from those of isotropic QED_3 .

Keywords: lattice gauge theory, cuprate, phase diagram, pseudogap
PACS: 11.10.Kk, 11.15.Ha, 71.27.+a, 74.25.Dw

INTRODUCTION

QED_3 , quantum electrodynamics in $2+1$ dimensions, has recently been a focus of interest in the condensed matter community, as it has been found that various versions of it are candidate models of high temperature superconductivity.

In what follows, we examine the model presented in [1, 2], which is intended to be a depiction of the passage between the antiferromagnetic SDW (spin-density wave) and superconducting dSC phases at low temperature T as the doping fraction x is increased.

In the condensed-matter context, the chiral symmetry breaking (χSB) order parameter – corresponding to a dynamically generated mass – of QED_3 can be mapped directly into the SDW one. If χSB does not occur (i.e. $N_f > N_{fc}$), then the resulting theory of light fermion degrees of freedom interacting with massless gauge degrees of freedom is a candidate theory of the so-called "pseudogap" region of the cuprate phase diagram. The main prediction of the QED_3 approach is thus that if $N_{fc} > 2$ the dSC and SDW phases are connected in the $T \to 0$ limit [2], whereas if $N_{fc} < 2$ they are separated by a region of pseudogap phase [1].

This reasoning assumes that results from continuum QED_3 in the isotropic limit can be applied directly to the condensed-matter system, whose Lagrangian density has kinetic terms describing a single flavour of fermion with differing strengths or "velocities" in x- and y-directions as an artifact of the transformation to the relativistic spinor basis. Evidence in favour of applying predictions of the isotropic theory comes from a renormalisation-group analysis, which studied small anisotropy perturbations to the isotropic system and concluded that weak anisotropy is an irrelevant perturbation [3, 4]. This result has then been used in conjunction with an assumption that N_c is a universal constant (and so independent of irrelevant parameters) as an argument that the many and varied estimates of N_c for isotropic QED_3 in the literature are sufficient to the needs of

CP918, *Lectures on the Physics of Strongly Correlated Systems XI: Eleventh Training Course,* edited by A. Avella and F. Mancini

the cuprate problem [3].

In what follows, we summarise some results of lattice gauge simulations pertinent to the problem, performed in collaboration with Simon Hands of the University of Wales Swansea, which were originally presented in [5, 6].

THE MODEL

As described in [1, 2], the Bogluibov-deGennes model of d-wave quasiparticles may be rewritten, if $T = 0$ and we linearize the gap and quasi-particle energies near the nodes of the Fermi surface, as a QED_3 action with two flavours of relativistic nodal fermion and doping-dependent velocity anisotropies in the spatial directions.

For N flavours of staggered lattice fermion, the following is a QED_3 action with an explicit spatial anisotropy:

$$S = \sum_{i=1}^{N} \sum_{x,x'} a^3 \bar{\chi}_i(x) M_{x,x'} \chi_i(x') + \frac{\beta}{2} \sum_{x,\mu<\nu} a^3 \Theta^2_{\mu\nu}(x). \tag{1}$$

We define the fermion matrix $M_{x,x'}$ as follows:

$$M_{x,x'} = \frac{1}{2a} \sum_{\mu=1}^{3} \xi_\mu(x) [\delta_{x',x+\hat{\mu}} U_{x\mu} - \delta_{x',x-\hat{\mu}} U^\dagger_{x'\mu}] + m\delta_{\mu\nu} \tag{2}$$

where ξ_μ is $\xi_\mu(x) = \lambda_\mu \eta_\mu(x)$ and $\eta_\mu(x) = (-1)^{x_1+\ldots+x_{\mu-1}}$, where $x_1 = x$, $x_2 = y$ and $x_3 = \tau$, is the Kawomoto-Smit phase of the staggered fermion field. The physical lattice spacing is denoted by a. The λ_μ are the anisotropy factors $\lambda_x = \kappa^{-\frac{1}{2}}$, $\lambda_y = \kappa^{\frac{1}{2}}$, $\lambda_t = 1$, where $\kappa = v_F/v_\Delta$ (v_F and v_Δ are the *Fermi* and *Gap* velocities derived from the linearisation of $\xi_{\vec{k}}$ and $\Delta(\vec{k})$ respectively about the nodes). The η factors ensure that the action describes relativistic covariant fermions in the isotropic limit $\kappa = 1$.

The photon-like fields $\theta_\mu(x)$ exist on the links connecting site x to site $x + \hat{\mu}$, and so $U_\mu(x) \equiv \exp(ia\theta_\mu(x))$ in (2) defines the gauge interaction with the fermions. We utilise a non-compact gauge action, where $\Theta_{\mu\nu}(x) = \frac{1}{a^2}[\Delta^+_\mu \theta_\nu(x) - \Delta^+_\nu \theta_\mu(x)]$. The dimensionless parameter β is given in terms of the QED coupling constant via $\beta \equiv 1/g^2 a$. It is convenient to work wherever possible in 'lattice units' such that $a = 1$.

Our simulation parameters are as follows: we used a Hybrid Monte Carlo algorithm, running for around 1000 trajectories of mean length 1.0 on L^3 lattices with $L = 16$, and the gauge coupling set to $\beta = 0.2$. Even-odd partitioning was used; this allowed us to set $N = 1$, giving us $N_f = 2$ in the continuum limit. Typical acceptances were $60-70\%$ for $m = 0.01$, and $70-80\%$ for other bare mass values. Our 20^3 and 24^3 lattices typically contain 700 and 600-700 trajectories per point with acceptances of $82-94\%$ and $75-82\%$ respectively. The gauge-fixed configurations needed for the measurement of the fermion propagators were generated on a 16^3 lattice and consisted of ~30,000 trajectories per point with an acceptance rate of $79-87\%$.

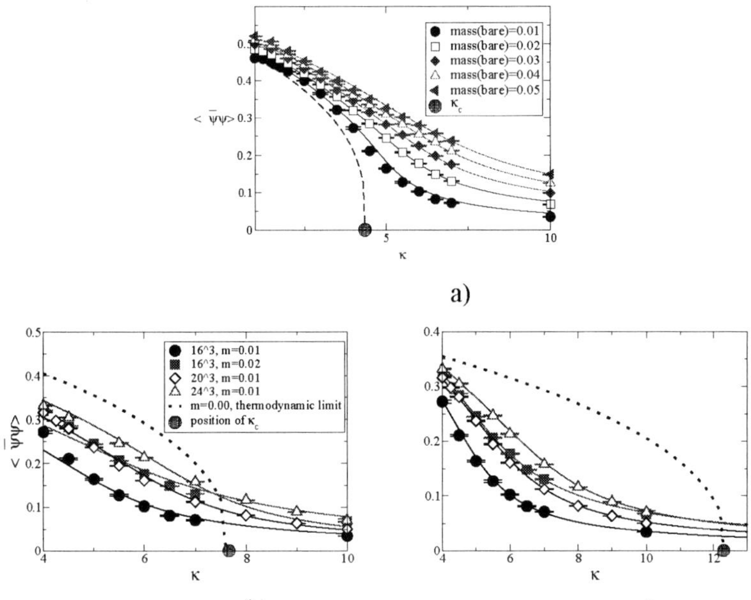

FIGURE 1. Finite size scaling for the chiral condensate. a) is for 16^3 lattices only, b) shows isotropic scaling, c) shows anisotropic scaling. Data for a) is from [5], the data for b), c) and the graphs themselves are from [6].

CHIRAL SYMMETRY BREAKING

Figure 1 shows the chiral condensate (the order parameter of the chirally-broken phase), defined as

$$\langle\bar{\chi}\chi\rangle = -\frac{1}{V}\frac{\partial \ln Z}{\partial m} = \frac{1}{V}\langle \mathrm{tr} M^{-1}\rangle, \tag{3}$$

for a variety of lattice sizes and bare fermion masses. We also include fits to the equation of state

$$m = B\langle\bar{\chi}\chi\rangle^{\delta} + A(k + CL^{-1/\nu})\langle\bar{\chi}\chi\rangle^{\rho} \tag{4}$$

where $k = (\kappa - \kappa_c)$, $\rho = \delta - \beta^{-1}$ and $C = 0$ when treating the 16^3 data only (using values where $1 \leq \kappa \leq 7$), C being allowed to vary when treating the finite size scaling as isotropic and the replacement $\frac{C}{L^{1/\nu}} \to R(\kappa; C, D, E)L^{-1/\nu}$ being made when treating it as weakly anisotropic. As can be seen from the results (Table 1), the chiral symmetry is restored at large κ; however, the anisotropic fit (which is also the best) seems to place the critical value of κ outside of the measured data set. This is perhaps an indication that there is no phase transition here, only a crossover, and that this is confusing the fitting algorithm.

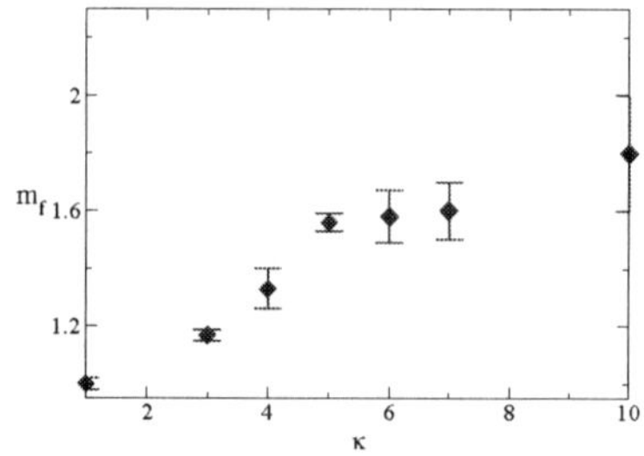

a)

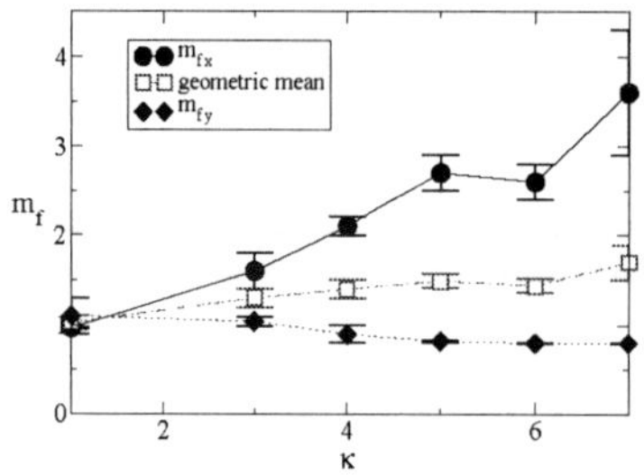

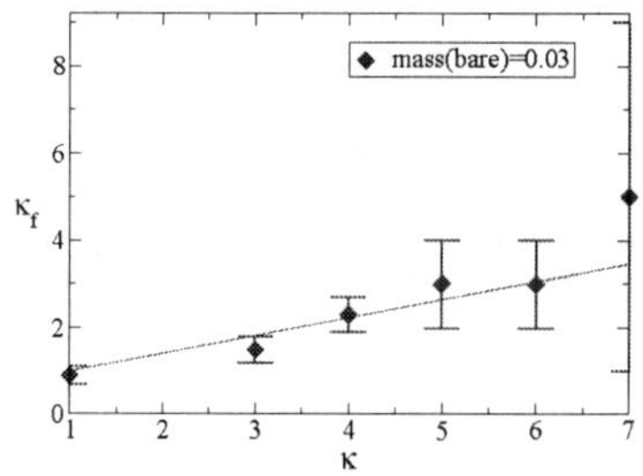

b) c)

FIGURE 2. Fermion masses extracted from the propagator. a) is that in the temporal direction, b) shows isotropic scaling, c) shows anisotropic scaling, with a line of best fi t. (Adapted from graphs in [6]).

TABLE 1. Equation of state fi t results, allowing for fi nite size scaling. (Adapted from data presented in [5, 6]).

Quantity	16^3 only	Isotropic FSS	Weakly anisotropic FSS
A	0.103(1)	.0393(8)	.0111(7)
B	2.75(6)	1.28(8)	1.02(6)
C	–	368(16)	-755(338)
D	–	–	527(78)
E	–	–	-716(447)
κ_c	4.35(2)	7.66(5)	12.3(6)
δ	3.99(3)	3.40(6)	3.33(6)
ρ	1.297(9)	.991(7)	1.01(1)
$\frac{\chi^2}{d.o.f.}$	51	162	6

Figure 2 shows measurements of the fermion mass (extracted from the fermion time/space-slice propagators – note that for technical reasons only odd numbered slices could be fitted for the τ and y directions in the chirally restored phase [6]) in modified iterative Landau gauge [7, 8]. Firstly, we see that the fermion retains its dynamically generated mass even in the chirally symmetric phase (an unusual effect first observed by Witten in the Gross-Neveau model [9]). This is unexpected; given the requirements given by Witten for the occurrence of this phenomenon, it suggests that the precise nature of the phase transition should be quite interesting.

Secondly, we note that the renormalised anisotropy (defined by the ratio of the effective masses in the x and y directions) is irrelevant, that is, $R_f = (\kappa_{Rf} - 1)/(\kappa - 1) < 1$, (fitting the data, we obtain $R_f = .41(4)$, with a $\chi^2/d.o.f.$ of 1.67) confirming the results of [3, 4]. Coupled with the apparent chiral restoration and the results in [5] demonstrating that the anisotropy is *relevant* with respect to the pions (the Goldstone bosons of the broken phase), this suggests that N_c is non-universal.

While much remains to be done regarding the study of this model, such as taking the continuum limit and examining the fermion propagator as the bare mass is taken to zero, these numerical simulations suggest that doing so could uncover some intriguing results both in relativistic field theory and condensed matter.

REFERENCES

1. M. Franz, Z. Tesanovic, and O. Vafek, *Phys. Rev.* **B66**, 054535 (2002), `cond-mat/0203333`.
2. I. F. Herbut, *Phys. Rev.* **B66**, 094504 (2002), `cond-mat/0202491`.
3. D. J. Lee, and I. F. Herbut, *Phys. Rev.* **B66**, 094512 (2002), `cond-mat/0201088`.
4. O. Vafek, Z. Tesanovic, and M. Franz, *Phys. Rev. Lett.* **89**, 157003 (2002), `cond-mat/0203047`.
5. S. Hands, and I. O. Thomas, *Phys. Rev.* **B72**, 054526 (2005), `hep-lat/0412009`.
6. I. O. Thomas, and S. Hands (2006), `hep-lat/0609062`.
7. M. Gockeler, et al., *Phys. Lett.* **B251**, 567–574 (1990).
8. S. Durr, and P. de Forcrand, *Phys. Rev.* **D66**, 094504 (2002), `hep-lat/0206022`.
9. E. Witten, *Nucl. Phys.* **B145**, 110 (1978).

Dephasing caused by the Thin Spectrum in a BCS Superconductor

Jasper van Wezel and Jeroen van den Brink

Institute-Lorentz for Theoretical Physics, Universiteit Leiden, P.O. Box 9506, 2300 RA Leiden, The Netherlands

Abstract. Recently we have shown that the thin spectrum associated with the process of spontaneous symmetry breaking will lead to a maximum coherence time for many-particle solid state qubits [1, 2, 3]. In this paper we will discuss how the general mechanism of dephasing caused by a thin spectrum can be applied to the case of BCS superconductors. We will find that the BCS superconducting state is a state of broken global U(1) symmetry, and that the thin spectrum associated with the breaking of this phase symmetry induces dephasing and decoherence within the fundamental timescale $\tau = 2\pi N\hbar/k_BT$.

Keywords: Superconductivity, Spontaneous Symmetry Breaking, Decoherence
PACS: 03.65.Yz, 11.30.Qc, 74.20.Fg

INTRODUCTION

Recently we have shown that spontaneous symmetry breaking imposes a fundamental limit for the time that a many-particle solid-state qubit can stay quantum coherent. This universal timescale is $\tau = 2\pi N\hbar/k_BT$, where N is the number of constituent particles involved in the superposed state [1]. We have shown this timescale to be applicable to qubit systems based on crystals, antiferromagnets and even superconductors [2, 3]. In the case of superconducting qubits we have considered a local pairing, negative-U Hubbard model to describe the superconducting state [3]. The goal of this paper is to show that the result on decoherence in a superconductor does not depend on the specific model used. To do so, we will consider the standard BCS model for superconductivity and show that it too has a thin spectrum which causes the dephasing of the computational states of a superconducting qubit and thus gives rise to the universal maximum coherence time τ.

The many-particle qubits that motivate us to study decoherence due to spontaneous symmetry breaking in superconductors are realized in a number of mesoscopic solid state systems. For instance, by engineering aluminum on a sub-micron length scale, superconducting Cooper pair boxes can be manufactured. A Cooper pair box is a superconducting island, containing $N \sim 10^8$ electrons, which can be brought into a superposition of two states with a different number of Cooper pairs [4, 5].

THE BCS SUPERCONDUCTOR

Previously we have shown that the presence of a thin spectrum leads to dephasing and decoherence in a local pairing model for superconductivity [3]. It could be argued that such a local pairing model is somewhat pathological, and not really representative of

CP918, *Lectures on the Physics of Strongly Correlated Systems XI: Eleventh Training Course,* edited by A. Avella and F. Mancini

real-life superconductors, even though from the point of view of symmetry the model is equivalent to a weak coupling model (because there is no phase transition which separates the two [6]). In this paper we will therefore study the symmetry breaking and dephasing in a BCS description, and show that although the picture changes slightly, the underlying physics is exactly equivalent, and in fact gives rise to the exact same conclusions regarding the thin spectrum and the timescale on which decoherence will set in. The draw-back of doing the calculation in the BCS description is that we cannot do it in a manifestly gauge invariant way, so that the role of the vector potential is obscured.

After creating Cooper pairs, we arrive in the standard BCS theory at the effective Hamiltonian [7]

$$H_{BCS} = \sum_k \varepsilon_k \left(c_k^\dagger c_k + c_{-k}^\dagger c_{-k} \right) - \sum_{k \neq k'} U_k \, c_k^\dagger c_{-k}^\dagger c_{-k'} c_{k'}.$$

Here we have adopted the convention to write $(k, \uparrow)$ as k and $(-k, \downarrow)$ as $-k$. The dispersion of the bare Fermi-sea is characterized by ε_k while U_k is the effective pairing interaction due to phonon exchange. U_k is non-zero and attractive only in a shell around the Fermi energy with a width of about the Debije energy. It is easy to see that extensivity of the model in fact requires U_k to be inversely proportional to the total number of electrons in the system. We will therefore redefine the pairing potential as $U_k = V_k/N$, where N denotes the total number of electrons in the k-space shell in which U_k is non-zero.

By writing down the Hamiltonian (1) we have assumed that there is no external magnetic field and we have fixed the gauge to ensure that the electromagnetic vector potential vanishes everywhere (to include the full vector potential, we would have to let the phase of the hopping parameter depend on the gauge field, and change the Hamiltonian accordingly). Anderson showed that the BCS Hamiltonian in this form can be rewritten as a spin problem by introducing the pseudospins [7]

$$S_k^+ = c_{-k} c_k \quad ; \quad S_k^z = \frac{1}{2} \left[1 - c_k^\dagger c_k - c_{-k}^\dagger c_{-k} \right]. \tag{1}$$

In the subspace without any quasiparticles (i.e. $n_k = n_{-k} \; \forall k$), the Hamiltonian up to an overall constant becomes

$$H_{BCS} = -2 \sum_k \varepsilon_k S_k^z - \sum_{k \neq k'} \frac{V_k}{N} \left(S_k^x S_{k'}^x + S_k^y S_{k'}^y \right). \tag{2}$$

Interpreted at face value, this Hamiltonian describes pseudo spins on a lattice which has position-label k. On this lattice, three different and independent regions can be identified. In the region $k < k_F - k_D$ (where k_F is the Fermi wavenumber and k_D the Debije wavenumber) we know that the pairing potential vanishes and ε_k is negative, so that all pseudospins in that region will point down, which corresponds to completely filled electronic states. In the region $k > k_F + k_D$ the pairing potential is zero as well, but here ε_k will be positive, causing all spins to point up, and all electronic states to be empty. In the shell of width k_D around k_F a more interesting situation occurs. There V_k is nonzero (and approximately constant), while ε_k switches sign right at k_F. The pseudo-spin structure that one would classically expect in that region is that of a magnetic domain wall: the pseudospins point up at one end of the region, then continuously fall over until they reach the xy plane exactly at k_F, and then they continue on until they point down at the other end [7]. Electronically that structure corresponds to the BCS wavefunction $\prod_k \left(u_k + v_k c_k^\dagger c_{-k}^\dagger\right) \mid \mathrm{vac}\rangle$ [8].

The Hamiltonian H_{BCS} however is invariant under rotations around the z-axis, and thus the groundstate will also obey this symmetry and have a completely delocalized projection of the pseudospins on the xy plane. To form a true domain wall, and thus the classical superconducting state, this U(1) symmetry will have to be spontaneously broken.

FIGURE 1. A schematic representation of the region of width k_D around k_F. The vectors represent the pseudospins **S**. Spontaneous symmetry breaking causes the projections of the pseudospins in the horizontal plane perpendicular to the paper to align.

Because the symmetry breaking will only have an effect in the region around k_F and because this region is fully decoupled from the other two regions of k-space, let's focus solely on that shell from now on, and define all sums over k to run from $k_F - k_D$ to $k_F + k_D$. In this region the pairing potential is approximately constant and thus we set $V_k \equiv V$ there. The collective dynamics of the system which govern the process of spontaneous symmetry breaking will in fact only be described by the singular points of the Bogoliubov transformation which diagonalizes the Hamiltonian [2]. Because of the ferromagnetic sign, the collective model in this case consists of only the $k = 0$ part of equation (2):

$$H_{coll} \simeq -\frac{V}{N}\left[\vec{S}_{tot}\cdot\vec{S}_{tot} - S^z_{tot}S^z_{tot}\right], \tag{3}$$

where $S_{tot} \equiv \sum_k S_k$ and where we have neglected terms of order $1/N^2$. We have also made a strong coupling approximation by setting $\varepsilon_{tot} = 0$. Other approximations of ε_{tot} are possible, but it turns out that after some tedious mathematics these will give the exact same form for the thin spectrum and the maximum coherence time as the simple approximation $\varepsilon_{tot} = 0$. The eigenstates of the collective Hamiltonian are trivially found to be labeled by the total spin quantum number S and its z-projection M, while the corresponding energies are given by $E_{coll}(S,M) = -V/N\left(S(S+1) - M^2\right)$. The thin spectrum in this case is labeled by M, and describes states with different total electron densities. The total spin excitations labeled by S on the other hand, are gapped with an energy $\sim V$. To break the xy-symmetry of H_{coll} we can add a symmetry breaking field $-BS^x_{tot}$ along for example the x-axis. After evaluating its matrix elements [2] and taking the continuum limit, Schrödinger's equation can be written as a harmonic oscillator equation

$$-\frac{1}{2}\frac{\partial^2}{\partial M^2}\Psi(M,x) + \frac{1}{2}\omega^2 M^2 \Psi(M,x) = \nu\Psi(M,x), \tag{4}$$

with $\omega^2 = (2V)/(BNS)$ and $\nu = 1 + (E(S,x) - E_{coll}(S,0))/(BS)$. The symmetry broken wavefunctions $|S,x\rangle \equiv \sum_M \Psi(M,x)\,|S,M\rangle$ thus have energies

$$E(S,x) \;=\; -\frac{V}{N}S(S+1) - BS + \left(x + \frac{1}{2}\right)\sqrt{VB}\sqrt{\frac{2S}{N}}. \tag{5}$$

In the ground state S will be maximal (i.e. $N/2$), and then the term $\propto NB$ in the energy signals spontaneous symmetry breaking: in the thermodynamic limit the system can gain an infinite amount of energy by aligning with an infinitesimally small symmetry breaking field [9].

DECOHERENCE

The collective excitations that make up the (dual) thin spectrum on top of the symmetry broken ground state are labeled by x. Their energies are slightly influenced by the remaining collective quantum number S. If we were to use this superconductor as a qubit then we would have to use the states labelled by S as the computational states of the qubit system. After all, the thin spectrum states labelled by x are so sparse that they cannot be seen in experiment [2]. The total spin excitations in this model correspond to changing the total number of cooper pairs in the electronic Hamiltonian (1). The superposition state that we are describing should therefore be closely related to the superposition state that is experimentally realized in Cooper-pair box qubits [4, 5]. If we make a superposition of total spin states and trace away the unobservable thin spectrum, then the small shift in the thin spectrum's energy levels will cause a dephasing of the computational states and subsequently the decoherence of the visible, reduced density matrix, in a manner completely analogous to the introduction of decoherence in a crystal, antiferromagnet or local pairing superconductor [1, 2, 3]:

$$\begin{aligned}
\rho_{t=0} &= \frac{1}{2Z}\sum_x e^{-\beta E(x,0)}\left[|x,0\rangle + |x,1\rangle\right]\left[\langle x,0| + \langle x,1|\right] \\
\rho_{t>0} &= \frac{1}{2Z}\sum_x e^{-\beta E(x,0)}\Big[|x,0\rangle\langle x,0| + |x,1\rangle\langle x,1| \\
&\quad + e^{-\frac{i}{\hbar}(E(x,0)-E(x,1))t}\,|x,0\rangle\langle x,1| + \text{h.c.}\Big], \\
\rho^{OD}_{t>t_0} &= \frac{1}{Z}\sum_x e^{-\beta E(x,0)} e^{-i(E(x,0)-E(x,1))(t-t_0)/\hbar}.
\end{aligned} \tag{6}$$

Here Z is the partition function at time $t=0$, ρ_t is the density matrix at time t, and ρ_t^{OD} is the off diagonal element of the reduced density matrix which is obtained by tracing the full density matrix over the thin spectrum states. The resulting maximum coherence time of the superposition state within the BCS superconductor is given by the half-time of ρ_t^{OD}, which turns out to be

$$\tau = \frac{2\pi\hbar}{k_B T}N, \tag{7}$$

where N counts the number of states in the k space volume of k_D around k_F, which is proportional to the number of Cooper pairs in the superconducting condensate. Notice that we find the same universal form for the expression of the coherence time set by spontaneous symmetry breaking as in the case of the crystal, the antiferromagnet and the local pairing model for superconductivity [1, 2, 3].

If we substitute values for the temperature and number of Cooper pairs as they are used in modern day Cooper-pair box qubits [4, 5], then the resulting decoherence time τ will be of the order of 500 microseconds. Other sources of decoherence typically destroy superposition states in these qubits long before the limit set by the thin spectrum states is reached. However, with the current rapid advance of technology this new source of decoherence may become important within the near future.

CONCLUSIONS.

In this paper we have shown that the superconducting state of the BCS model is characterized by a broken global U(1) phase symmetry. We have described the process of spontaneous symmetry breaking and the associated thin spectrum in the BCS model for superconductivity. The presence of the thin spectrum states associated with spontaneous symmetry breaking in qubits based on superconducting material, will lead to dephasing and decoherence of these qubits within the time $\tau = 2\pi N\hbar/k_B T$, where N counts the number of Cooper pairs involved in the superposition. This timescale is universal in the sense that it does not depend in the underlying modelparameters. Its form coincides precisely with that of the decoherence time induced by thin spectrum dynamics in antiferromagnets and quantum crystals.

The maximum coherence time that we found here for superconducting devices should apply directly to experimental realizations of the so called Cooper pair box qubits, and thus give a maximum coherence time of the order of half a millisecond. The decoherence caused by the thin spectrum at the moment is much weaker than that caused by other sources, but it may well come within experimental reach within the near future.

REFERENCES

1. J. van Wezel, J. van den Brink, and J. Zaanen, *Phys. Rev. Lett.* **94**, 230401 (2005).
2. J. van Wezel, J. Zaanen, and J. van den Brink, *Phys. Rev. B* **74**, 094430 (2006).
3. J. van Wezel, and J. van den Brink, *To Be Published* (2007).
4. Y. Nakamura, Y. A. Pashkin, and J. S. Tsai, *Nature* **398**, 786 (1999).
5. D. Vion, A. Aassime, A. Cottet, P. Joyez, H. Pothier, C. Urbina, D. Esteve, and M. H. Devoret, *Science* **296**, 886 (2002).
6. P. Nozières, and S. Schmitt-Rink, *J. Low Temp. Phys.* **59**, 195 (1985).
7. P. Anderson, *Phys. Rev.* **112**, 1900 (1958).
8. M. Tinkham, *Introduction to Superconductivity*, McGraw-Hill inc., New York, 1996.
9. J. van Wezel, J. Zaanen, and J. van den Brink, *ArXiv: Physics* p. 0609177 (2006), submitted to AM. J. Phys.

Doped Spin Ladder: Zhang-Rice Singlets or Rung-centred Holes?

Krzysztof Wohlfeld

Marian Smoluchowski Institute of Physics, Jagellonian University, Reymonta 4, PL-30059 Kraków, Poland

Abstract. We formulate charge transfer model for a single doped spin ladder, relevant for the intrinsically doped plane of coupled spin ladders in $Sr_3Ca_{11}Cu_{24}O_{41}$. Assuming the presence of the experimentally observed charge order in the system we solve the model using Hartree-Fock approximation. Our results show the profound stability of the Zhang-Rice singlets with respect to other configurations of doped holes, in agreement with recent x-ray absorption measurements.

Keywords: Zhang-Rice singlets, spin ladder, high temperature superconductors, charge order
PACS: 74.72.-h; 71.10.Fd; 75.10.Lp; 71.45.Lr

INTRODUCTION

It is a widely accepted feature of two-dimensional high temperature superconducting cuprates (HTSC) [1] that a hole doped into CuO_2 plane enters into a square of O ($2p$) orbitals and strongly binds with the central Cu ($3d$) hole forming a local, so-called Zhang-Rice [2] singlet (ZR; cf. Fig. 1). On the other hand, a hole doped into the HTSC Cu_2O_3 coupled spin ladders can *a priori* behave differently. This is due to the fact that: (*i*) spin ladders do not have the D_{4h} symmetry (the four O surrounding Cu are not identical), and (*ii*) for *coupled* spin ladders the coordination number of the leg orbitals (= 3) is different than that of the bridge orbital (= 2) resulting in lower on-site energy for the hole in the bridge orbital than in one of the leg orbitals, cf. local-density approximation studies in [3]. Hence, one could expect that e.g. a rung-centred state (rung) could be stabilized with a doped hole residing on the O ($2p$) bridge orbital and bound to the two neighbouring Cu holes via kinetic exchange interactions (cf. Fig. 1).

The aim of the present paper is to study the possible stability of *these* two states in the doped spin ladder by means of model calculations. It is an interesting theoretical task by itself, still backed by the recent x-ray absorption measurements which suggest that in the spin ladder plane of $Sr_3Ca_{11}Cu_{24}O_{41}$ (SCCO) a hole crystal with period $\lambda = 3$ has been discovered with the doped holes evenly distributed among the O $2p$ orbitals involved in the ZR singlet [4]. This stays in contrast with the other experiments which suggest that the holes are not isotropically distributed. E.g. Nücker et al. [5] also in the x-ray absorption measurements find that the doped holes go mainly into those leg orbitals which are responsible for transport properties of the single ladder (crystalographic c direction). Such result, however, not only contradicts those obtained in Ref. [4] but also a naive picture of rather localized holes in this *insulating* compound. Thus, the issue of the distribution of holes doped into spin ladders indeed needs some further studies.

CP918, *Lectures on the Physics of Strongly Correlated Systems XI: Eleventh Training Course*, edited by A. Avella and F. Mancini

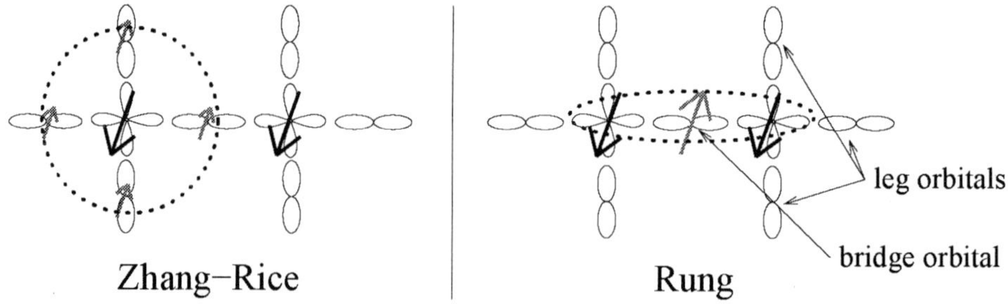

FIGURE 1. Schematic picture of the Zhang-Rice (ZR) singlet (left panel) and rung-centred (rung) hole (right panel) in a Cu_2O_7 cluster. Large (small) arrows depict the hole spins for +1.0 (+0.25) charge. The grey arrows stand for spins of doped holes.

MODEL HAMILTONIAN AND HARTREE-FOCK APPROXIMATION

Our starting point is the three band charge transfer Hamiltonian relevant for a single doped spin ladder [6]:

$$H = \Big(- \sum_{m,j \in R,L,\sigma} t_{mj} d^{\dagger}_{m\sigma} p_{j\sigma} - \sum_{m \in R,L; i\sigma} t_{mi} d^{\dagger}_{m\sigma} b_{i\sigma} + \text{h.c.} \Big) + \Delta \sum_{j \in R,L} n_{pj} + \varepsilon \sum_{i} n_{bi}$$
$$+ U \sum_{m \in R,L} n_{m\uparrow} n_{m\downarrow} + U_p \Big(\sum_{i} n_{bi\uparrow} n_{bi\downarrow} + \sum_{j \in R,L} n_{pj\uparrow} n_{pj\downarrow} \Big), \tag{1}$$

where: Δ is the charge transfer energy for the O ($2p_\sigma \equiv p$) *leg* orbital, ε is the charge transfer for the O ($2p_\sigma \equiv b$) *bridge* orbital (cf. Fig. 1), t_{mj} is the hopping integral between the nearest neighbour pairs of the Cu ($3d_{x^2-y^2} \equiv d$) orbital and p or b orbitals. We also include Coulomb on-site repulsion U (U_p) between holes in the d (p or b) orbitals, respectively. Besides, the d and p orbitals belong to two sets of sites for the Left ($d_L, p_L \in$ L) and Right ($d_R, p_R \in$ R) leg of the ladder, and n_m, n_{pj}, n_{bi} are particle number operators for d, p, b orbitals, respectively.

In the following study we assume that the charge order is present in the ladder, as one of the few features of the ladder subsystem of SCCO compound on which the experimentalists agree is the formation of the hole crystal [4]. Therefore we expect that for commensurate values of doping a doped hole forms a ZR or rung bound states which then form a hole crystal of relevant period.

The presence of the charge order enables us to safely use Hartree-Fock approximation which works well for ordered states. This, however, means that we must adopt a classical picture for ZR and rung states: a hole with the opposite S^z component of the spin with respect to the spin of the relevant Cu holes involved in kinetic exchange interactions (with no energy gain due to quantum fluctuations or phase coherence in ZR state).

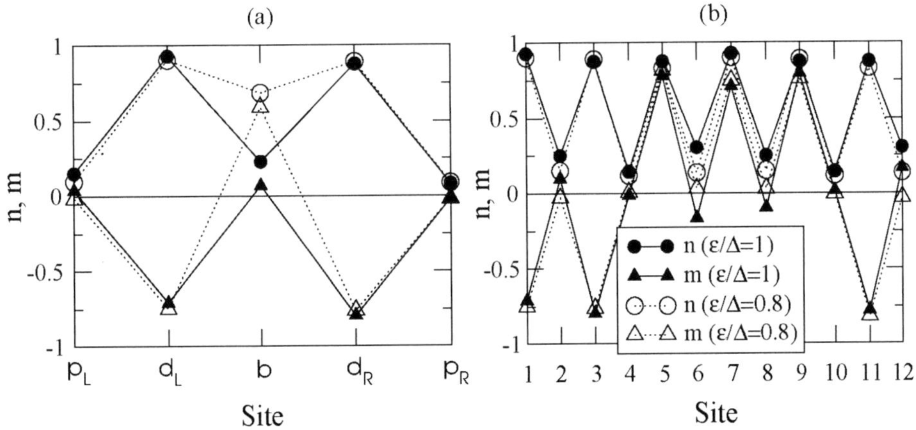

FIGURE 2. Hole density (n) and magnetization (m) for: (a) sites along the rung with enhanced hole density; (b) 12 consecutive sites along one of the legs of the ladder (starting from the d orbital, ie. for $d-p-d-\ldots$ chain). The legend applies to both of the panels.

RESULTS AND DISCUSSION

We performed self-consistent numerical calculations using 60×7 (the unit cell of the spin ladder consists of 7 orbitals [6]) clusters for the following model parameteres: $U = 9$ eV, $U_p = 4.2$ eV, $\Delta = 3.5$ eV [7], whereas $t = 1$ eV yields the correct superexchange constant $J/t \sim 0.1$, cf. Ref. [8]. We assume that the total number of holes per Cu $n_h = 1.33$ in order to make our results comparable with the recent x-ray absorption measuremernts in SCCO [4]. Though, let us stress that the stability of ZR or rung states does not depend on the value of the commensurate doping since similar results as shown below were obtained for other values of hole doping [6].

The results for two different values of the bridge orbital energy ε are shown in Fig. 2. For $\varepsilon/\Delta = 1$ the hole distribution and magnetization resembles the classical ZR state: the doped hole is distributed rather isotropically among four O sites (e.g. sites b and p_L on Fig. 2(a) and sites 2 and 12 on Fig. 2(b)) surrounding the central Cu site occupied by roughly one hole (e.g. site 1 on Fig. 2(b) or d_L on Fig. 2(a)). Also the spin of the doped hole in the O ($2p$) orbitals compensates roughly the spin of the hole in the Cu site. On the other hand, for $\varepsilon/\Delta = 0.8$ the doped hole enters mainly into the b orbital and there are much less holes in the p orbitals of the leg of the ladder suggesting a rung character of the doped hole. Let us also note that, in agreement with the assumption of charge order, we find a hole crystal with less charge per every third rung of the ladder for the solution with ZR or rung character.

In order to investigate the role of the specific spin ladder geometry on the stability of ZR and rung states we calculate the densities and magnetization of holes involved in forming: (i) ZR state: $n_{\rm ZR} = n_{p_L} + n_2 + n_{12}$ and $|m_{ZR}| = |n_{ZR\uparrow} - n_{ZR\downarrow}|$ (we exclude the b orbital from the sum to be able to distinguish between rung and ZR states), and (ii) rung state: $n_{\rm rung} = n_b$ and $|m_{\rm rung}| = |n_{rung\uparrow} - n_{rung\downarrow}|$. The results are shown in Fig.

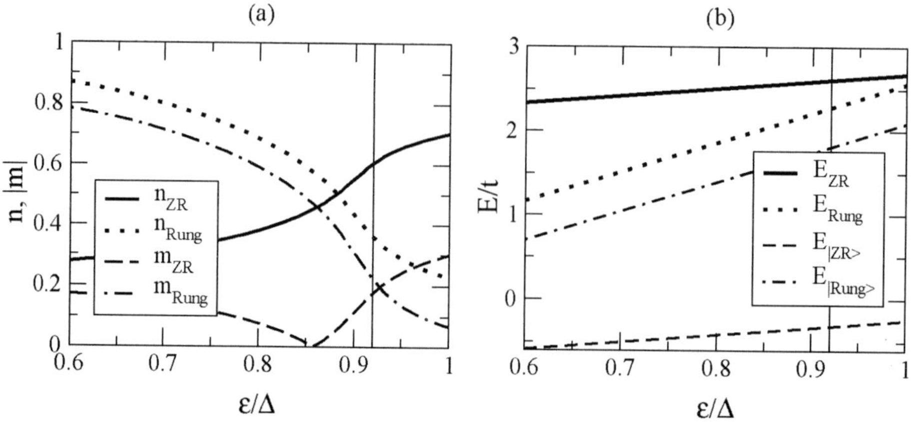

FIGURE 3. (a) Hole density (n) and absolute value of magnetization ($|m|$) for holes involved in forming ZR ($n_{\rm ZR}, |m_{\rm ZR}|$) and rung states ($n_{\rm rung}, |m_{\rm rung}|$) as a function of the of the bridge orbital energy ε. (b) Binding energies of a single hole in ZR and rung states in the classical case (E_{ZR} and E_{rung}, respectively) and quantum-mechanical case ($E_{|ZR\rangle}$ and $E_{|rung\rangle}$, respectively) as a function of the the bridge orbital energy ε. The vertical line on both panels depicts the value of $\varepsilon/\Delta = 0.92$, cf. Ref. [3].

3(a) as a function of the on-site energy of the bridge orbital ε. We find that with the decreasing value of ε doped holes tend to occupy the b orbital, and the spins of the holes in the b orbital become polarized. Besides, the spins of the holes involved in forming ZR state do not only compensate the spin of the central Cu hole but for $\varepsilon/\Delta < 0.85$ even weakly align ferromagnetically with the Cu spin. Hence, we suggest that for $\varepsilon/\Delta < 0.85$ the doped holes show a rung character while for $\varepsilon/\Delta > 0.9$ they show a distinctive ZR character separated by a crossover regime. It means that ZR state is stable for the value of $\varepsilon/\Delta = 0.92$, calculated in Ref. [3], though we are very close to the crossover regime.

Let us now pose the question to what extent our results are relevant for the stability of the real quantum-mechanical Zhang-Rice singlets or rung states. Therefore, using *second* order perturbation theory in U and $U - \Delta$ [9] we calculate the binding energy of a single hole doped into ZR and rung states: in the classical case (E_{ZR} and E_{rung}, respectively), and in the quantum-mechanical case ($E_{|ZR\rangle}$ and $E_{|rung\rangle}$, respectively). In the classical case, which resembles the states obtained in the Hartree-Fock approximation, we have:

$$E_{\rm ZR} = \frac{3}{4}\Delta + \frac{1}{4}\varepsilon + J\Big\langle \sum_{i\in ZR} \mathbf{S}_i \cdot \mathbf{S}_{\rm O} - \frac{1}{4}\Big\rangle_{\rm ZR}, \quad E_{\rm rung} = \varepsilon + J\Big\langle \sum_{i\in rung} \mathbf{S}_i \cdot \mathbf{S}_{\rm O} - \frac{1}{4}\Big\rangle_{\rm rung} \qquad (2)$$

where: the kinetic exchange $J = 2t^2(1/U + 1/(U-\Delta))$, $\mathbf{S}_{\rm O}$ is the spin of the doped O ($2p$) hole, $\mathbf{S}_i$ is the spin of the Cu ($3d$) hole, and the sum goes over those Cu sites which are involved in forming a bound state with the O ($2p$) hole in rung or ZR state. The expressions for the energies in the quantum-mechanical case look similar except for the averages of the spin operators which, unlike in the classical case, include also spin fluctuations. In addition, for the real ZR singlet we include the phase coherence of holes doped into the O ($2p$) orbitals [2]. The results as a function of the energy of the bridge orbital ε are shown in Fig. 3(b).

We find for the classical state that for $\varepsilon/\Delta < 0.97$ the energy difference ($E_{ZR} - E_{rung}$) is larger than the effective hopping energy of the O ($2p$) hole ($= t^2/U$ or $t^2/(U-\Delta)$). Hence for finite bandwidth the rung state could only be stabilized up to the above value of the bridge orbital energy, qualitatively in agreement with the previous Hartree-Fock results. However, in the quantum mechanical case the rung state could never be stabilized, and due to the large energy difference ($E_{|ZR\rangle} - E_{|rung\rangle}$) the true Zhang-Rice singlet should not be destabilized by finite bandwidth.

CONCLUDING REMARKS

In summary, we have shown the profound stability of the Zhang-Rice singlets in the hole doped spin ladders. First, using model Hartree-Fock calculations we obtain the isotropic distribution of doped holes among the O ($2p$) orbitals surrounding the central Cu ($3d$) hole to be stable for the relevant range of the model parameters. Second, quantum-mechanical calculations of the binding energy of holes forming Zhang-Rice singlets and rung states suggest the Zhang-Rice singlets to be even more stable. Hence, our results provide an interpretation of recent experimental data obtained in the x-ray absorption measumerents in the spin ladder plane of SCCO [4].

ACKNOWLEDGMENTS

I would like to thank the organising committee of the course for their financial support. I am particularly grateful to Andrzej M. Oleś for his invaluable help, ideas, and comments and to George A. Sawatzky for very stimulating discussions. This work was supported by the Polish Ministry of Science and Education under Project No. 1 P03B 068 26.

REFERENCES

1. D. J. Scalapino, arXiv:cond-mat/0610710 (unpublished). [to appear in: "Numerical Studies of the 2D Hubbard Model", in *Handbook of High Temperature Superconductivity*, edited by J. R. Schrieffer, Springer, 2006]
2. F.C. Zhang and T.M. Rice, *Phys. Rev. B* **37**, R3759 (1988).
3. T. F. A. Müller et al., *Phys. Rev. B* **57**, R12655 (1998).
4. A. Rusydi et al., *Phys. Rev. Lett.* **97**, 016403 (2006); A. Rusydi et al., arXiv:cond-mat/0604101 (unpublished).
5. N. Nücker et al., *Phys. Rev. B* **62**, 14384 (2000).
6. K. Wohlfeld, A. M. Oleś, and G. A. Sawatzky, arXiv:cond-mat/0612669 (unpublished). [to appear in Physica C: M2S-HTSC VIII Dresden 2006 Conference Proceedings, 2007]
7. J.B. Grant and A.K. McMahan, *Phys. Rev. B* **46**, 8440 (1992).
8. R.S. Eccleston et al., *Phys. Rev. Lett.* **81**, 1702 (1998).
9. J. Zaanen and A. M. Oleś, *Phys. Rev. B* **38**, 9423 (1988).

Non-Fermi-liquid properties of three-impurity Anderson models

Rok Žitko* and Janez Bonča†,*

*J. Stefan Institute, Jamova 39, SI-1000 Ljubljana, Slovenia
†FMF, University of Ljubljana, Jadranska 19, SI-1000 Ljubljana, Slovenia

Abstract. We discuss the possibility of observing non-Fermi-liquid behavior in a system of three coupled quantum dots embedded between two conduction channels. The regime where the system approaches the two-channel Kondo model fixed point in a wide temperature interval occurs when the magnetic ordering competes with Kondo screening. We study the robustness of the system with respect to various perturbations.

Keywords: triple quantum dots, two-channel Kondo model, non-Fermi-liquid behavior
PACS: 75.30.Hx, 71.10.Hf. 72.10.Fk, 72.15.Qm

INTRODUCTION

Advances in nanotechnology offer the possibility to probe systems of increasingly small sizes. Nowadays one can, for example, measure transport properties of semiconductor quantum dots, single molecules and even individual atoms adsorbed on a metal surface. In very small electronic devices the electron-electron interactions are strong and they induce interesting many-particle effects. Most notable is the Kondo screening of local moments which appears to be a relatively generic feature of nanodevices at low temperatures [1–4]. The most interesting nanodevices are probably the quantum dot (QD) systems that serve as tunable realizations of quantum impurity models in which on-site energy and hybridization strength can be easily swept in-situ. In systems of multiple coupled QDs more exotic Kondo states can be realized. In this work, we focus on the triple QD system and study a possible realization of the two-channel Kondo (2CK) model [5]; the setup consists of three QDs coupled in series and embedded between two metal electrodes, see the inset in Fig. (1a) [6].

TWO-CHANNEL KONDO MODEL

The two-channel Kondo model describes a spin coupled by exchange interaction to two independent conduction channels [5, 7]:

$$H = \sum_{k\mu\alpha} \varepsilon_k c^\dagger_{k\mu\alpha} c_{k\mu\alpha} + \vec{S} \cdot (J_1 \vec{s}_1 + J_2 \vec{s}_2), \tag{1}$$

where $\mu = \uparrow, \downarrow$ is the spin index, $\alpha = 1,2$ the channel index, $\vec{S}$ the impurity spin-1/2 operator, and $\vec{s}_\alpha$ spin density of the conduction channel α at the position of the impurity: $\vec{s}_\alpha = \sum_{kp\mu\nu} c^\dagger_{k\mu\alpha} 1/2 \vec{\sigma}_{\mu\nu} c_{p\nu\alpha}$. We combine the exchange constants as $J_{\text{avg}} = 1/2(J_1 + J_2)$

CP918, *Lectures on the Physics of Strongly Correlated Systems XI: Eleventh Training Course,* edited by A. Avella and F. Mancini

and $\Delta J = J_1 - J_2$. If ΔJ is small (to be specified below), the system undergoes a Kondo cross-over from the local-moment fixed point to a non-Fermi-liquid (NFL) strong-coupling fixed point at the Kondo temperature

$$T_K = D_{\text{eff}} \rho J_{\text{avg}} \exp\left(-\frac{1}{\rho J_{\text{avg}}}\right), \tag{2}$$

where $D_{\text{eff}} \approx 3D$ with D the half-bandwidth of each conduction band and $\rho = 1/(2D)$ is the density of states in the band. Below this temperature the local moment is screened, however the system has a $\ln 2/2$ residual entropy and exhibits unusual NFL behavior of the magnetic susceptibility, resistance, and other thermodynamic and dynamic properties [5]. If ΔJ is not strictly zero, another cross-over occurs at still lower temperature T_Δ [8]

$$T_\Delta \propto T_K \times \frac{(\Delta J)^2}{\rho^2 J_{\text{avg}}^4}, \tag{3}$$

and the system ends up in the stable Fermi-liquid fixed point of the conventional single-channel Kondo model. The criterion for observability of the NFL regime is clearly $T_\Delta \ll T_K$, or $\Delta J \ll \rho J_{\text{avg}}^2$. Since a single QD embedded between two non-interacting channels couples only to the symmetric combination of the electrons from both channels [9], such a system is governed by the single-channel Kondo model. To observe 2CK behavior, more elaborate setups with several QDs need to be used [6, 8, 10]; we study the case of three dots coupled in series.

MODEL

We model the three impurities as a three-impurity Anderson model, i.e. a Hubbard chain hybridizing with two conduction channels [see the inset in Fig. (1a)] [11, 12]:

$$\begin{aligned} H = &\sum_{k\mu\alpha} \varepsilon_k c^\dagger_{k\mu\alpha} c_{k\mu\alpha} + \sum_{k\mu} \left(V_k d^\dagger_{1\mu} c_{k\mu L} + V_k d^\dagger_{3\mu} c_{k\mu R} + \text{H.c.} \right) \\ &+ \sum_{i=1}^{3} \frac{U}{2} (n_1 - 1)^2 + \sum_{i=1}^{2} \sum_{\mu} t \left(d^\dagger_{i\mu} d_{i+1,\mu} + \text{H.c.} \right), \end{aligned} \tag{4}$$

where $\alpha = L,R$ now denotes the two conduction channels, V_k is the hybridization, $n_i = \sum_\mu d^\dagger_{i\mu} d_{i\mu}$ is the site occupancy, U is the electron-electron repulsion and t is the inter-dot charge hopping parameter. The four relevant energy scales in the three-impurity Anderson model are the repulsion U, the hybridization strength $\Gamma = \pi\rho |V_{k_F}|^2$, the superexchange $J = 4t^2/U$ and kinetic energy $\sim t$. For fixed U and Γ, the system has a number of interesting regimes as a function of t. In particular, it is found that there is a range of t where the system approaches the 2CK NFL fixed point at finite temperatures [12]. This fixed point is unstable, since electron hopping enables charge transfer between the two channels [13] which amounts to channel-symmetry breaking ($\Delta J \neq 0$). The question, therefore, is to find the regime where T_Δ is much smaller than T_K.

THE CROSSOVER REGIME

For intermediate t, the antiferromagnetic exchange $J = 4t^2/U$ is strong enough to bind the three spins into a rigid AFM spin chain for temperatures $T < J$. The spin of the chain is then screened as a whole due to the Kondo effect [11]. For very small t, however, the spins do not bind into a spin chain; instead, the two spins on left and right dot are screened by the electrons of the nearest electrode at some higher Kondo temperature $T_K^{(1)}$, while the spin on the central dot is screened by the quasiparticles of the Kondo correlated states on the side dots at some exponentially lower Kondo temperature $T_K^{(2)}$. This is the "two-stage Kondo effect" [14–16]. The smooth cross-over between the spin chain regime and the two-stage Kondo regime occurs when $T_K^{(1)} \sim J$. In this cross-over region, magnetic ordering competes with Kondo screening of the side dots. Interestingly, we find that in this region the 2CK Kondo temperature has a maximum (as a function of t) and the NFL-FL cross-over temperature T_Δ is very low [12]. This range of hopping t is therefore the most likely candidate for experimental observation of NFL physics in QD realizations of Hubbard chains.

Robustness

We studied the effect of various perturbation terms on the stability of the $\ln 2/2$ NFL plateau in the temperature dependence of the impurity contribution to the entropy. We focused on the cross-over regime, which occurs for $U/D = 1$ and $\Gamma/U = 0.045$ around $t/D = 0.005$. We find a high degree of robustness with respect to the particle-hole symmetry breaking up to $V_{123}/U \approx 0.2$, Fig. (1a), left-right symmetry (parity) breaking up to $V_{1-3}/U \approx 0.2$, Fig. (1b), and unequal e-e repulsion parameters, Fig. (1c). In this last case we notice that when $U_1 = U_3$ is decreased, the system is pushed towards the AFM regime [12], since the fluctuations on sites 1 and 3 increase and $T_K^{(1)}$ decreases below J. If, however, $U_1 = U_3$ is increased, the system goes into the two-stage Kondo regime since $T_K^{(1)}$ is higher than J. In this latter case, there will still be a $\ln 2/2$ NFL plateau, however the relevant temperature interval is shifted to considerably lower temperatures and becomes narrower [12].

The most dangerous perturbation is the channel-symmetry breaking, Fig. (1d), which rapidly wipes out the NFL plateau. It should be noted, however, that those asymmetries of the device that break the channel-symmetry can be corrected using gate voltages in experimental realizations of the three-impurity model where the on-site energies and inter-dot tunneling parameters can be controlled independently.

It is interesting to follow the behavior of the system as a function of the on-site energy, Fig. (2). This corresponds to sweeping the gate voltage in experimental setups. We observe that the occupancy of side (left and right) dots decreases and the charge fluctuations increase as we move towards the valence fluctuation regime of the side dots; the occupancy of the central dot is hardly affected due to its weak effective coupling. An important consequence is the reduction of the spin correlations: the rigid spin chain breaks near $\delta = 0.3U$ and the system enters the two-stage Kondo regime.

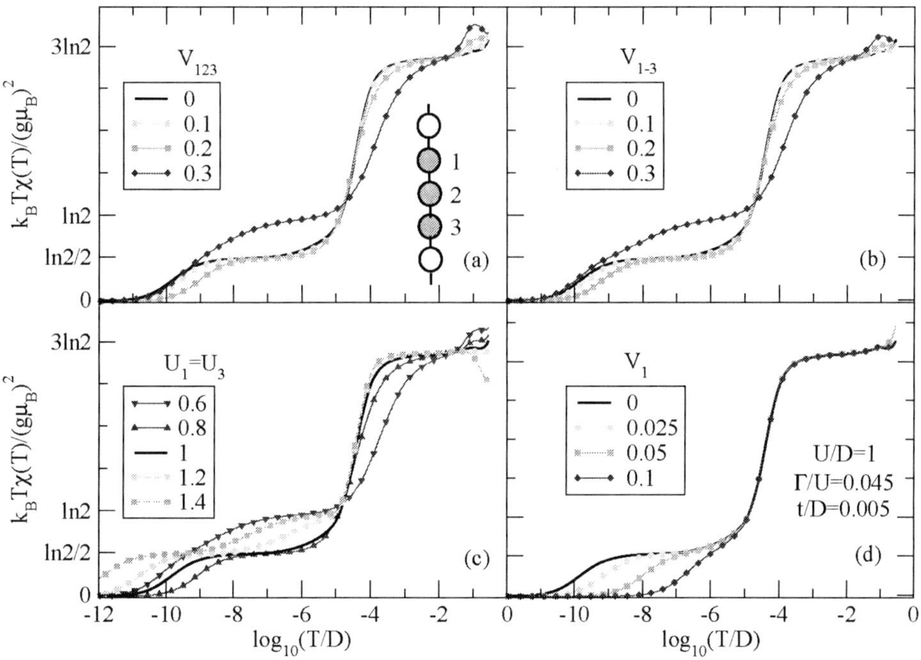

Figure 1. Effect of various perturbation terms on the impurity contribution to the entropy. (a) Particle-hole symmetry breaking $H' = V_{123} \sum_i n_i$. (b) Parity breaking $H' = V_{1-3}(n_1 - n_3)$ (note that the channel asymmetry is not affected by this form of parity breaking). (c) Unequal e-e repulsion $U_1 = U_3 \neq U_2$. (d) "Dangerous" perturbation that breaks channel symmetry $H' = V_1 n_1$.

Transport properties

The NFL region of the TQD can be detected by measuring the conductance either through the three dots or through a single side dot [12]. If conductance through a side dot increases to $1/2G_0$ with $G_0 = 2e^2/h$, while the triple QD "molecule" is non-conducting, the system is a candidate for a NFL state. Further confirmation should then come from measuring non-equilibrium current and conductance in magnetic field and comparing with predicted NFL exponents [8, 10].

CONCLUSION

The triple quantum dot in the cross-over regime between magnetic ordering and the two-stage Kondo effect is a good candidate for observation of the non-Fermi liquid physics in mesoscopic systems. Albeit the two-channel Kondo fixed point is, strictly speaking, unstable, it is nevertheless robust in the sense that a wide NFL temperature interval exists even for relatively large perturbation terms. Recently, evidence of the two-channel Kondo behavior has been found in a system of a quantum dot coupled to

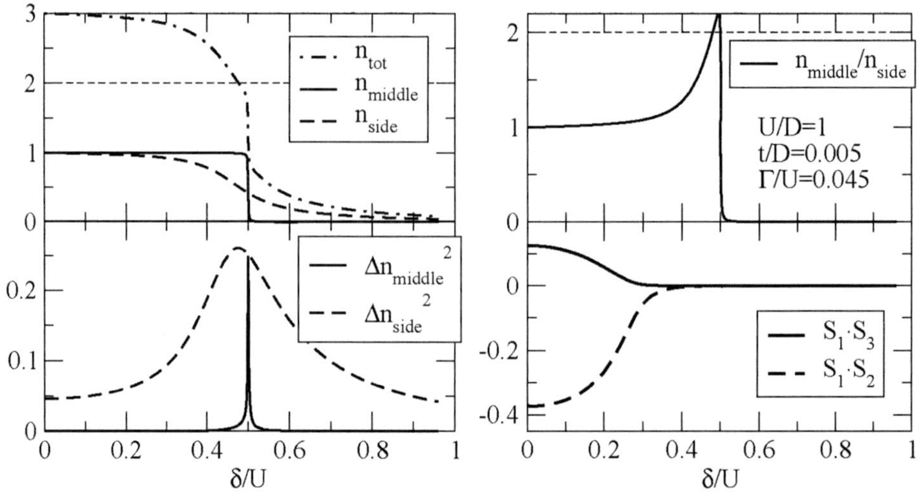

Figure 2. Site occupancy, charge fluctuations and spin correlations as a function of the on-site energy $\delta = \varepsilon_d + U/2$ for a perturbation $H' = \sum_i \delta(n_i - 1)$. Note that for $\delta = 0$ the model is particle-hole symmetric. $S_1 \cdot S_3$ is the spin correlation between left and right dot, $S_1 \cdot S_2$ is the spin correlation between two neighboring dots.

two conventional large electrodes and an additional finite electron reservoir (a quantum box) [17]. Studies of non-Fermi liquid behavior in quantum dot systems are thus within experimental reach and have the potential to become an important area of research.

REFERENCES

1. D. Goldhaber-Gordon, H. Shtrikman, D. Mahalu, D. Abusch-Magder, U. Meirav, and M. A. Kastner, *Nature* **391**, 156 (1998).
2. J. Park, A. N. Pasupathy, J. I. Goldsmith, C. Chang, Y. Yaish, J. R. Petta, M. Rinkoski, J. P. Sethna, H. D. Abruna, P. L. McEuen, and D. C. Ralph, *Nature* **417**, 722 (2002).
3. W. Liang, M. P. Shores, M. Bockrath, J. R. Long, and K. Park, *Nature* **417**, 725 (2002).
4. L. H. Yu, and D. Natelson, *Nanoletters* **4**, 79 (2004).
5. D. L. Cox, and A. Zawadowski, *Adv. Phys.* **47**, 599 (1998).
6. T. Kuzmenko, K. Kikoin, and Y. Avishai, *Europhys. Lett.* **64**, 218 (2003).
7. I. Affleck, A. W. W. Ludwig, and B. A. Jones, *Phys. Rev. B* **52**, 9528 (1995).
8. M. Pustilnik, L. Borda, L. I. Glazman, and J. von Delft, *Phys. Rev. B* **69**, 115316 (2004).
9. L. I. Glazman, and M. E. Raikh, *JETP Lett.* **47**, 452 (1988).
10. Y. Oreg, and D. Goldhaber-Gordon, *Phys. Rev. Lett.* **90**, 136602 (2003).
11. R. Žitko, J. Bonča, A. Ramšak, and T. Rejec, *Phys. Rev. B* **73**, 153307 (2006).
12. R. Žitko, and J. Bonča, Fermi-liquid versus non-fermi-liquid behavior in triple quantum dots, cond-mat/0606287 (2006).
13. G. Zarand, C.-H. Chung, P. Simon, and M. Vojta, *Phys. Rev. Lett.* **97**, 166802 (2006).
14. M. Vojta, R. Bulla, and W. Hofstetter, *Phys. Rev. B* **65**, 140405(R) (2002).
15. P. S. Cornaglia, and D. R. Grempel, *Phys. Rev. B* **71**, 075305 (2005).
16. R. Žitko, and J. Bonča, *Phys. Rev. B* **73**, 035332 (2006).
17. R. M. Potok, I. G. Rau, H. Shtrikman, Y. Oreg, and D. Goldhaber-Gordon, Observation of the two-channel kondo effect, cond-mat/0610721 (2006).

AUTHOR INDEX

V

W

Z